A text book on

# Analog Circuits

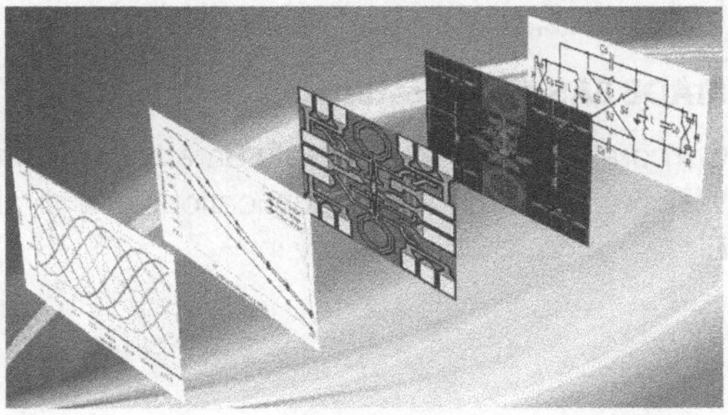

*Useful for IAS, IES, GATE, PSUs and other competitive examinations*

**A. RAJKUMAR**

*M. Tech. (JNTU Kakinada)*
*(Ph.D) JNTU Hyderabad*
*Expert Faculty, MADE EASY*



MADE EASY Publications

**Corporate Office:** 44-A/4, Kalu Sarai, New Delhi-16
**Website:** www.madeeasypublications.org | **Phone:** 011-45124660, 08860378007

**MADE EASY Publications**
Corporate Office: 44-A/4, Kalu Sarai, New Delhi-110016
Website: www.madeeasypublications.org | Phone: 011-45124660, 08860378007
E-mail: infomep@madeeasy.in

**A text book on Analog Circuits**



First Edition: 2012
Reprint: 2013
Reprint: 2014
Reprint: 2015
Reprint: 2016
**Reprint: 2019**

Typeset at: MADE EASY Publications, New Delhi-110016

Dedicated to

My Wife

***A. Smitha***

who believes, in

*The best teacher teaches from the heart,*

*.... not from the book....*

# PREFACE

I have great pleasure in writing the book of core subject of Electronics Engineering viz. **Analog Circuits.** A thorough understanding of the concepts developed in this book will prepare the reader for more advanced course on the subject. The entire syllabus of Analog circuits is presented in a simple and lucid style to make it comprehensible to an average student. This text book has been written to meet the requirements for the students of B.E./B.Tech., ECE, EEE, EIE.

In this book, I have tried to present the approach for competitive examinations like GATE, IES and IAS. While teaching various categories of students, I understood that, it becomes very easy for the students when things are explained by going through the fundamentals. So in the present book, I tried to explain most of the topics through the basics. The ***Questions with Solutions*** which already appeared in competitive examinations like GATE, IES are incorporated in each chapter.

First, I would like to thank **Mr. B. Singh** (Chief Managing Director, MADE EASY Group) for giving me the opportunity for writing the text book. His constructive suggestions and support helped me a lot.

I express my heartfelt regard and gratitude to my teacher Dr. Srinivas Rao from whom I have learnt the subject matter and which gave me inspiration to write this book.

I would also like to thank my H.O.D. Dr. Koteswara Rao and Principal Dr. Chinna Keshava Rao (CBIT), Mr. Sudarshan Reddy (Associate Prof. in CBIT) who always inspire me in discipline and hard work.

A special thanks to Mr. Sai Prasad who always, encourages me and gave me the first opportunity to teach for competitive exams.

I express my sincere gratitude to my favorite Sir, Mr. Prem R. Chadha, friends M.V. Kiran Kumar, Ramana Reddy, Krishna Kumar, Jagan, Krishna who always direct me in a right path.

Last but not the least, I also thank MADE EASY staff especially Vinod Kumar and Md. Asim who put their sincere efforts to develop this book in time.

Any comments and suggestions for the improvement of this book will be thankfully acknowledged and incorporated in the next edition.

**A. Rajkumar**

# CONTENTS

## 1 BJT Biasing and Thermal Stabilization


1.1 Operating Point and DC Load Line **2**
1.2 Temperature Dependence on Transistor Parameters **5**
1.3 Stability Factor **6**
1.4 Biasing Techniques **7**
1.5 Fixed Bias Circuit **7**
1.6 Collector to Base Bias **8**
1.7 Voltage Divider Bias or Self Bias **10**
1.8 Bias Compensation by Diode **12**
1.9 Bias Compensation by Thermistor **12**
1.10 Bias Compensation by Sensistor **12**
1.11 Thermal Run away **13**


## 2 Small Signal Analysis of BJT

### Low Frequency Analysis


2.1 Introduction **32**
2.2 Two Port Network **32**
2.3 Hybrid Model of BJT **34**
2.4 Hybrid Model in Different Configurations **34**
2.5 Typical Values of H-Parameters **35**
2.6 Conversion Formulas of H Parameters in Different Configuration **35**
2.7 Transistor Amplifier Analysis Using Exact Model **35**
2.8 Transistor Amplifier Analysis Using Approximate Model **38**


### High Frequency Analysis


2.9 Introduction **41**
2.10 Hybrid-$\pi$ Model **41**
2.11 Typical Values of Hybrid-$\pi$ Parameters **42**
2.12 Expressions for Hybrid-$\pi$ Parameters **43**
2.13 Variation of Hybrid $\pi$ Parameters With $I_c$, $V_{CE}$ and Temperature **48**
2.14 Common Emitter Short Circuit Current Gain **48**
2.15 High Frequency Current Gain with Resistive Load **50**


### Frequency Response of an Amplifier


2.16 Low Frequency Range **52**
2.17 High Frequency Range **52**
2.18 Midband Range **52**
2.19 Application of Bodeplot in Frequency Response of Amplifier **53**

# 3 Feedback Amplifiers


3.1 Introduction **66**

3.2 Limitations of Basic Amplifier **66**

3.3 Feedback Principle **66**

3.4 Classification of Amplifiers **67**

3.5 General Theory of Feedback **69**

3.6 Concept of Positive Feedback and Negative Feedback **71**

3.7 Advantages of Negative Feedback Amplifiers **72**

3.8 Topologies of Feedback **76**

3.9 Voltage Series Feedback **79**

3.10 Voltage Shunt Feedback **79**

3.11 Current Series Feedback **81**

3.12 Current Shunt Feedback **82**

3.13 Method of Analysis of a Feedback Amplifier **83**


# 4 Oscillators


4.1 Introduction **96**

4.2 Types of Oscillators **96**

4.3 Essentials of Transistor Oscillator **97**

4.4 Barkhausen Criterion **98**

4.5 RC Phase Shift Oscillator **99**

4.6 Wien Bridge Oscillator **103**

4.7 Comparison of RC Oscillators **105**

4.8 LC Oscillators **106**

4.9 Hartley Oscillator **107**

4.10 Colpitts Oscillator **109**

4.11 Clapp Oscillator **111**

4.12 Crystal Oscillator **112**


# 5 Power Amplifiers


5.1 Introduction **124**

5.2 Difference Between Voltage Amplifier and Power Amplifier **124**

5.3 Power Amplifiers **126**

5.4 Power Amplifier Classes **126**

5.5 Comparison of Amplifier Classes **128**

5.6 Class-A Power Amplifier **128**

5.7 Distortion in Amplifiers **133**

5.8 Class-B Power Amplifier **134**

5.9 Class-B Push Pull Amplifier **136**

5.10 Complementary Symmetry Class-B Amplifier **140**

5.11 Comparison Between Push Pull and Complementary Symmetry **140**
5.12 Cross Over Distortion **141**
5.13 Class-AB Amplifier **141**
5.14 Safe Operating Area (SOA) For a Transistor **142**
5.15 Thermal Analogy of Power Transistors **143**
5.16 Class-C Power Amplifier **144**
5.17 Class-D Amplifiers **145**
5.18 Class-S Power Amplifier **146**

6 **Multistage Amplifiers**

6.1 Introduction **154**
6.2 Characteristics of Cascaded Amplifier **154**
6.3 Gain in Decibels **155**
6.4 Selection of an Amplifier Configuration For Cascade Connection **156**
6.5 Methods of Coupling in Multistage Amplifiers **156**
6.6 RC Coupled Amplifiers **157**
6.7 Transformer Coupled Amplifier **158**
6.8 Direct Coupled Amplifiers **160**
6.9 Comparison of Different Coupling Techniques **161**
6.10 Rise Time-bandwidth Relation **161**
6.11 Cascode Amplifier (CE-CB) **163**
6.12 Darlington Pair (CC-CC) **163**
6.13 Bootstrap Emitter Follower **164**
6.14 Millers Theorem **165**
6.15 Effect of Cascading on Bandwidth (identical stages) **166**
6.16 Effect of Cascading on Bandwidth (non identical stages) **168**

7 **Linear and Non Linear Wave Shaping**

**Linear Wave Shaping**

7.1 Introduction **176**
7.2 High Pass RC Circuit **176**
7.3 RC Differentiator **177**
7.4 Sinusoidal Input **178**
7.5 Response of Step, Pulse, Square, Ramp, Exp. for High Pass ckt **179**
7.6 Low Pass RC Circuit **180**
7.7 Low Pass RC Circuit as Integrator **181**
7.8 Sinusoidal Input **181**
7.9 Response of Step, Pulse, Square, Ramp and Exp. for Low Pass ckt **183**

**Non Linear Wave Shaping**

7.10 Clippers **184**
7.11 Shunt Clipper Models **184**
7.12 Series Clipper Models **189**
7.13 Clipping at Two Independent Levels **195**
7.14 Clamping Circuits **199**

# 8 Rectifiers, Filters and Voltage Regulators

### Rectifiers

8.1 Introduction **216**
8.2 Rectifier Parameters **216**
8.3 Half Wave Rectifier **217**
8.4 Full Wave Rectifier **222**
8.5 Bridge Rectifier **226**

### Filters

8.6 Introduction **227**
8.7 Capacitor Filter **228**
8.8 Inductor Filter **231**
8.9 L Section Filter **234**
8.10 $\pi$ Section Filter **235**
8.11 Bleeder Resistor **236**

### Voltage Regulators

8.12 Introduction **236**
8.13 Block Diagram of Power Supply **236**
8.14 Load Regulation **237**
8.15 Line Regulation **237**
8.16 Series Voltage Regulator **237**
8.17 Shunt Voltage Regulator **238**
8.18 Limitations of Series and Shunt Voltage Regulators **238**

# 9 Operational Amplifiers

### Differential Amplifiers

9.1 Introduction **248**
9.2 Differential Amplifier **248**
9.3 Circuit Configurations **249**
9.4 Circuit Diagrams **250**
9.5 Operating Point (Q) for a Differential Amplifier **252**
9.6 Practical Design Problems in Differential Amplifiers **253**
9.7 Swamping Resistor Technique **254**
9.8 Constant Current Source Circuits **255**
9.9 Current Mirror Circuits **257**
9.10 Common-Mode Voltage Gain **260**
9.11 Differential Amplifier With Active Load **262**

### Operational Amplifiers

9.12 Block Diagram of Operational Amplifier **263**
9.13 Symbol Representation of Op-Amp **263**

9.14 Equivalent Circuit of an Op-Amp **264**
9.15 Ideal Voltage Transfer Characteristics **264**
9.16 Pin Diagram of 741 IC **265**
9.17 Op-amp Characteristics **265**
9.18 Importance of Negative Feedback in Op-Amp **270**
9.19 Virtual Ground **271**

**Applications of Operational Amplifier**

9.20 Inverting Amplifier **271**
9.21 Non-Inverting Amplifier **272**
9.22 Phase Shifter **273**
9.23 Voltage Follower **273**
9.24 Differential Amplifier **274**
9.25 Subtracter **275**
9.26 Inverting Adder **275**
9.27 Non Inverting Adder **276**
9.28 Current to Voltage Converter **277**
9.29 Voltage to Current Converter **278**
9.30 Voltage to Current Converter (with load) **278**
9.31 Op-Amp Integrator **279**
9.32 Op-Amp Differentiator **280**
9.33 Precision Rectifiers **281**
9.34 Current Amplifiers **285**
9.35 Instrumentation Amplifier **286**
9.36 Voltage Limiters **287**
9.37 Log Amplifiers **289**
9.38 Anti Log Amplifiers **289**
9.39 Analog Multiplier **290**
9.40 Analog Divider **291**
9.41 Peak Detector **292**
9.42 Sample and Hold Circuit **292**
9.43 Op-amp Comparator **293**
9.44 Schmitt Trigger **295**
9.45 Astable Multivibrator (free-running oscillator) **297**
9.46 Triangular-Wave Generator **298**
9.47 Monostable Multivibrator (pulse generator) **299**

**Active Filters**

9.48 Introduction **301**
9.49 Classification of Active Filters **301**
9.50 Butterworth Filter **303**
9.51 First Order Low Pass Butterworth Filter **304**
9.52 First Order High Pass Butterworth Filter **306**
9.53 Band Pass Filters **309**

9.54 Band Stop Filter **313**
9.55 All Pass Filter **315**

**555 Timer**

9.56 Introduction **318**
9.57 555 Timer Pin Configuration **318**
9.58 Block Diagram of 555 Timer **319**
9.59 555 Timer as Monostable Multivibrator **320**
9.60 555 Timer as Astable Multivibrator **323**

# 10 FET Biasing and Small Signal Analysis

**FET Biasing**

10.1 Introduction **370**
10.2 DC-Load Line and Bias Point **370**
10.3 Fixed Bias Circuit **372**
10.4 Self Bias **373**
10.5 Voltage Divider Bias **374**

**Small Signal Analysis**

10.6 Introduction **376**
10.7 Common Source Bypass JFET Amplifier **377**
10.8 Common Source Unbypass JFET Amplifier **379**
10.9 Common Drain JFET Amplifier **381**
10.10 Common Gate JFET Amplifier **382**
10.11 Comp. Between CS Bypass, CS Unbypass, CD, CG Amplifiers **385**

# 11 Multivibrators and Time Base Generators

**Multivibrators**

11.1 Introduction **394**
11.2 Bistable Multivibrator **395**
11.3 Schmitt Trigger **397**
11.4 Mono Stable Multivibrator **399**
11.5 Astable Multivibrator **401**

**Time Base Generators**

11.6 Introduction **404**
11.7 Sweep Voltage **404**
11.8 Methods of Generating a Time Base Waveform **405**
11.9 Exponential Sweep Circuit **406**
11.10 UJT Sweep Generator **406**
11.11 Bootstrap Sweep Generator **408**
11.12 Miller Sweep Circuit **410**

# CHAPTER 1

# BJT BIASING & THERMAL STABILIZATION

## About This Chapter

- Operating point and DC load line.
- AC load line.
- Stability factors S, S′ and S″.
- Various biasing circuits like fixed bias, collector to base bias and self bias.
- Compensation circuits.
- Thermal run-away and conditions to maintain thermal stability.

# 1 BJT Biasing & Thermal Stabilization

## 1.1 Operating Point and DC Load Line

- It is clear that the transistor functions most linearly when it is constrained to operate in its active region.
- To establish an operating point in this region it is necessary to provide appropriate direct potentials and currents, using external sources.
- Once an operating point Q is established as shown below:

### ANALYSIS

Assume common emitter amplifier circuit

In DC analysis,

(i) ac should be grounded

(ii) $X_c \propto \frac{1}{f}$, as $f \to 0$, $X_c \to \infty$

$\therefore$ Capacitor is open.

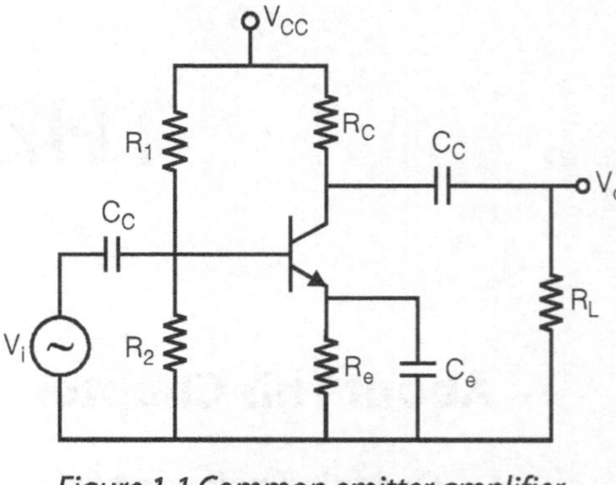


*Figure 1.1 Common emitter amplifier*

### DC Equivalent Model

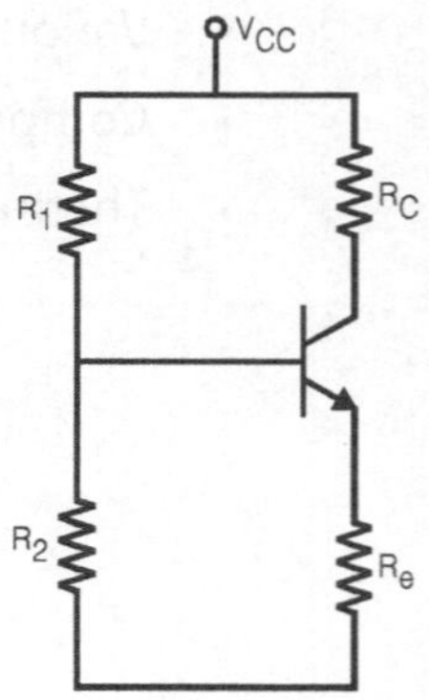


The output loop equation is given by

$$V_{CC} = I_C(R_C + R_e) + V_{CE}$$

$$I_C = \frac{V_{CC}}{R_C + R_e} - \frac{V_{CE}}{R_C + R_e}$$

Above equation represents like a straight line y = mx + C

When $I_C = 0$, $(V_{CE})_{max} = V_{CC}$

$V_{CE} = 0$, $(I_C)_{max} = \frac{V_{CC}}{R_C + R_e}$

$$\text{Slope of DC load line} = -\frac{1}{R_C + R_e}$$

## Output Characteristics of Common Emitter

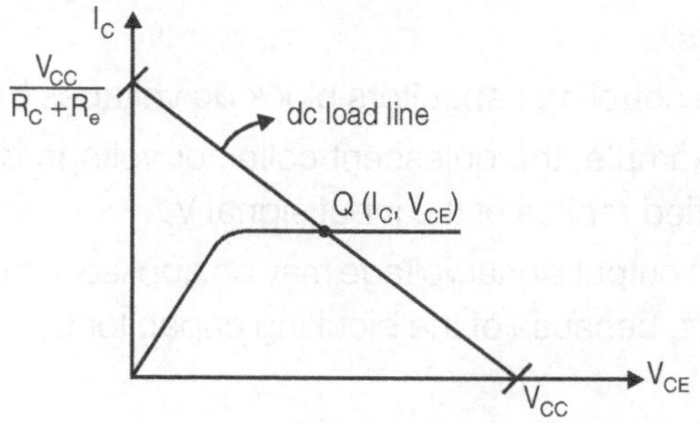


*Fig. 1.2 DC load line and Q point*

## Q Point

"The intersection point between the dc load line and sample graph of $I_C$ is called as operating point (or) Quiescent point".

## DC Load Line

"The locus of all the Q points are concentrated on a line called as DC load line".

- When time varying excursions of the input signal (base current, for example) is applied to common emitter amplifier, should cause an output signal (collector voltage or collector current) of the same wave form.
- If the output signal is not a faithful reproduction of the input signal, for example, if it is clipped on one side. The operating point is unsatisfactory and should be relocated on the collector characteristics.
- The question now naturally arises as, how to choose the operating point.
- Note that even if we are free to choose $R_c$, $R_L$, $R_b$ and $V_{CC}$, we may not operate the transistor every where in the active region, because the various transistor ratings limit the range of useful operation.
- These ratings are
  (i) $(V_{CE})_{max}$ (ii) $(I_C)_{max}$ (iii) $P_{c\,max}$

## Capacitive Coupling

- The capacitor $C_{b1}$ is to couple the input signal to the transistor, as indicated in the Figure (1.3).
- In this diagram, one end of $V_i$ is at ground potential, and the collector supply $V_{CC}$ also provides the biasing base current $I_B$.
- Under quiescent conditions (no input signal), $C_{b1}$ (called a blocking capacitor) acts as an open circuit because the reactance of a capacitor is infinite at zero frequency (dc).

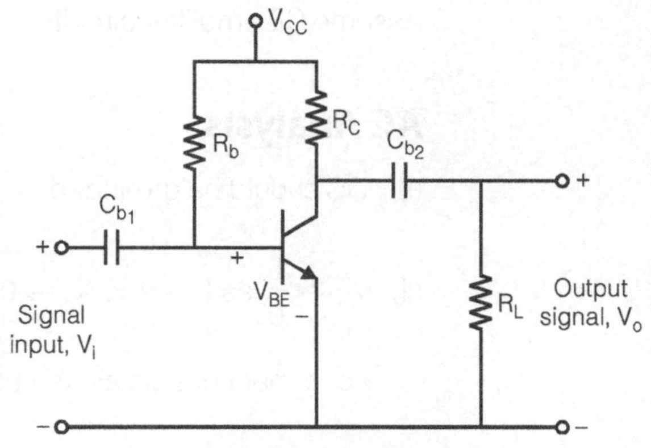


*Figure 1.3 Fixed bias circuit*

- The capacitances $C_{b1}$ and $C_{b2}$ are chosen large enough so that, at the lowest frequency of excitation their reactances are small enough so that they can be considered to be short circuits.
- These coupling capacitors block dc voltages but freely pass signal voltages.
  For example, the quiescent collector voltage does not appear at the output, but $V_o$ is an amplified replica of the input signal $V_i$.
- The ac output signal voltage may be applied to the input of another amplifier with out affecting its bias, because of the blocking capacitor $C_{b2}$.

## DC and AC Load Lines

- We noted that under dc conditions $C_{b2}$ acts as an open circuit. Hence the quiescent collector current and voltage are obtained by drawing a static (dc) load line.
- If $R_L = \infty$ and if the input signal (base current) is large and symmetrical, we must locate the operating point $Q_1$ at the centre of the DC load line. In this way the collector voltage and current may vary approximately symmetrically around the quiescent values $V_C$ and $I_C$ respectively.

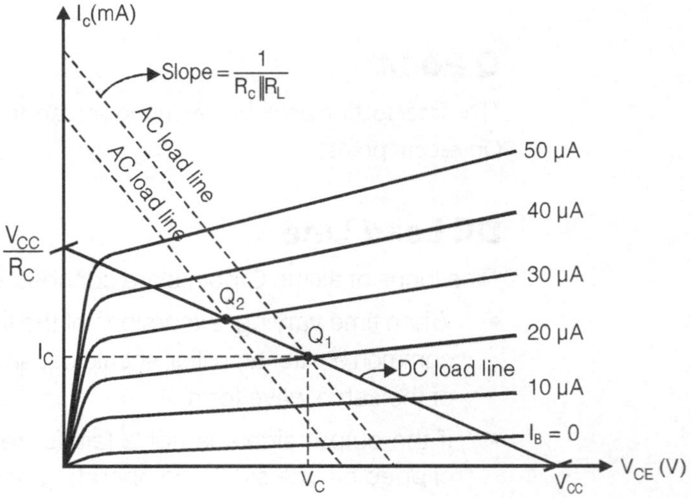


*Figure 1.4 DC and AC load line*

- If $R_L \neq \infty$, however, a dynamic (ac) load line must be drawn.
- Since, we have assumed that, at the signal frequency, $C_{b2}$ acts as a short circuit, the effective load $R_L'$ at the collector is $R_C$ in parallel with $R_L$.

## AC Load Line Analysis

Assume CE amplifier circuit:

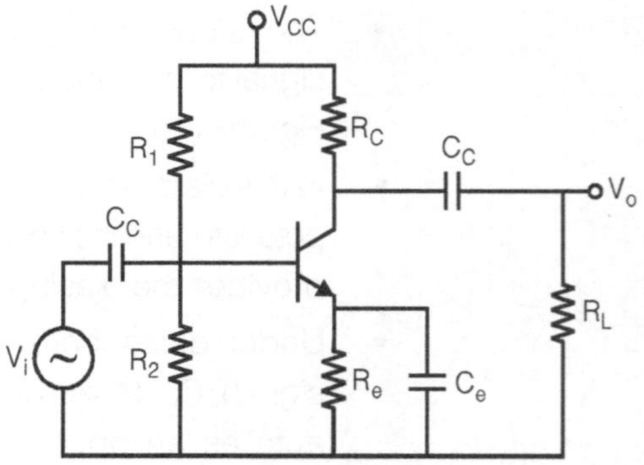


## AC Analysis

(i) DC should be grounded.

(ii) $X_C \propto \frac{1}{f}$, as $f \to \infty \therefore X_c \to 0$

i.e. capacitor acts as short circuit.

**AC Equivalent Model**

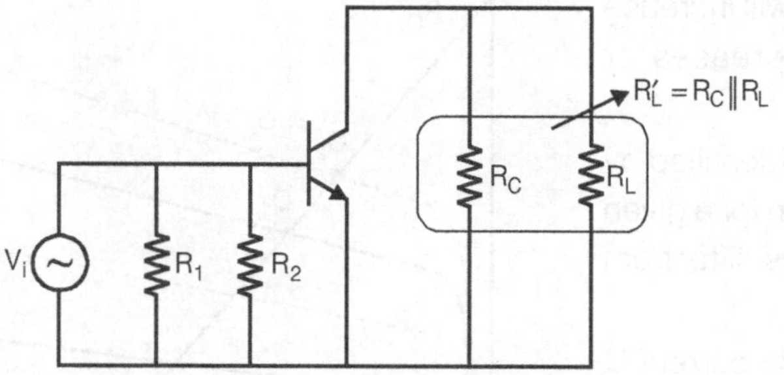


**AC Load Line**

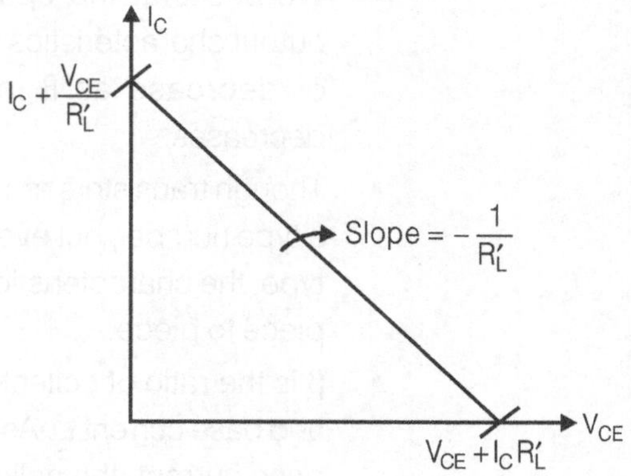


$$\text{Slope of ac load line is } -\frac{1}{R'_L} = -\frac{1}{R_C \| R_L}$$

- We observe that the input signal may swing a maximum of approximately 20 μA around $Q_1$ because, if the base current decreases by more than 20 μA, the transistor is driven off.
- If a larger input swing is available, then in order to avoid cut-off during a part of the cycle. The quiescent point must be located at a higher current.
- Suppose we locate $Q_2$ on the dc load line such that a line with a slope corresponding to the ac resistance $R'_L$ and drawn through $Q_2$ gives as large as output as possible with out too much distortion.
- The choice of $Q_2$ allows an input peak current swing of about 30 μA.

## 1.2 Temperature Dependence on Transistor Parameters

The sources of instability of $I_C$ are essentially three:

(i) Reverse saturation current $I_{CO}$

(ii) Base-emitter voltage $V_{BE}$

(iii) Current gain β

- **$I_{CO}$ Versus Temperature**

  "$I_{CO}$ increases by 7%/°C rise in temperature".

  (or)

  "$I_{CO}$ doubles for every 10°C rise in temperature".

  i.e. as temperature increases, $I_{CO}$ also increases.

- **$V_{BE}$ Versus Temperature**

  "The base to emitter voltage $V_{BE}$, which decreases at the rate of 2.5 mV/°C for both Ge and Si transistors."

  $$\frac{dV_{BE}}{dT} = -2.5 \text{ mV/°C}$$

  i.e. as temperature increases, $V_{BE}$ decreases.

- **β Versus Temperature**

  **(a)** β increases with temperature

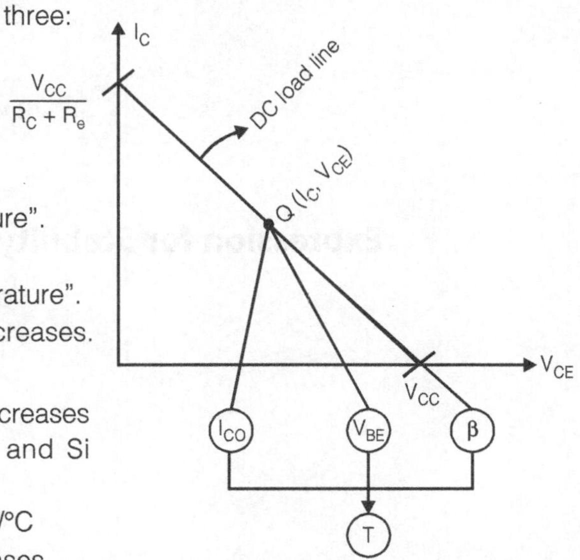


*Figure 1.5 DC load line and Q point*

(b) Replacement of Transistor

- We see that the spacing of the output characteristics will increase or decrease as $\beta$ increases or decreases.
- Though transistors are identified by a type number, but even for a given type, the characteristics differ from piece to piece.
- $\beta$ is the ratio of collector current $I_C$ and base current $I_b$. As for the same base current, the collector current of the transistor will vary from replacement of the device.
- Generally $\beta$ is greater for the replacement of transistor.

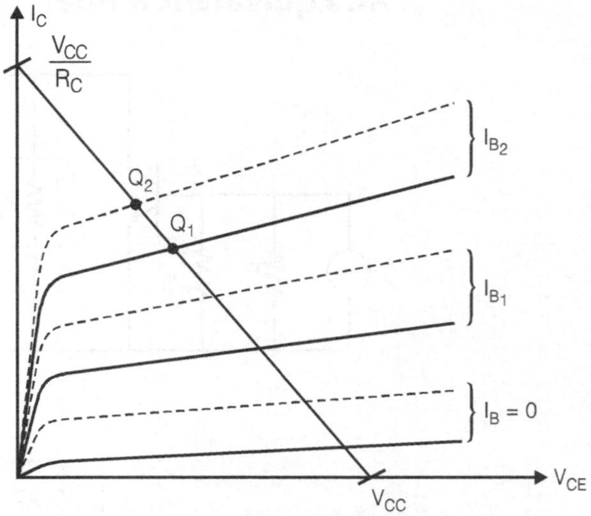

*Figure 1.6 Common emitter characteristics*

## 1.3 Stability Factor

"It is a measure of variation in the operating point with respect to the temperature is called as stability factor".

$I_C$ is a function of $I_{CO}$, $V_{BE}$ and $\beta$. It is convenient to introduce the three partial derivatives of $I_C$ with respect to these variables. These derivatives are called the stability factors S, S′ and S″.

$$S = \left.\frac{\partial I_C}{\partial I_{CO}}\right|_{V_{BE},\,\beta \text{ constant}}$$

$$S' = \left.\frac{\partial I_C}{\partial V_{BE}}\right|_{I_{CO},\,\beta \text{ constant}}$$

$$S'' = \left.\frac{\partial I_C}{\partial \beta}\right|_{I_{CO},\,V_{BE} \text{ constant}}$$

### Expression for Stability Factor

$$I_C = \beta I_b + (1+\beta) I_{CO} \quad \text{d.w.r.t } I_C$$

$$1 = \beta \frac{\partial I_b}{\partial I_c} + (1+\beta)\frac{\partial I_{CO}}{\partial I_C}$$

$$S = \frac{\partial I_C}{\partial I_{CO}} = \frac{1+\beta}{1-\beta \dfrac{\partial I_b}{\partial I_c}}$$

**Conclusion:**

- Ideally the stability factor should be zero.
- Practically stability factor should be very very less.

## 1.4 Biasing Techniques

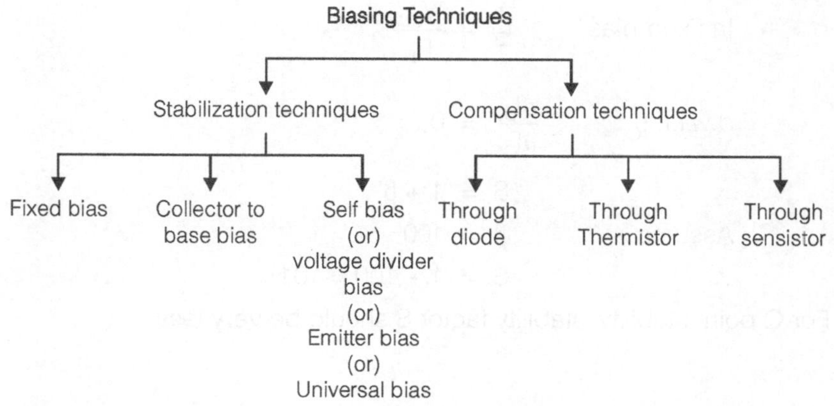


## 1.5 Fixed Bias Circuit

- From fixed bias circuit Figure (1.7) there is one resistance $R_b$ connected between base and $V_{cc}$, through which the base current $I_B$ flows.
- To make the Q point stable, base current $I_B$ should be constant.

The input loop is

$$V_{CC} = I_B R_b + V_{BE}$$

$$I_B = \frac{V_{CC} - V_{BE}}{R_b}$$

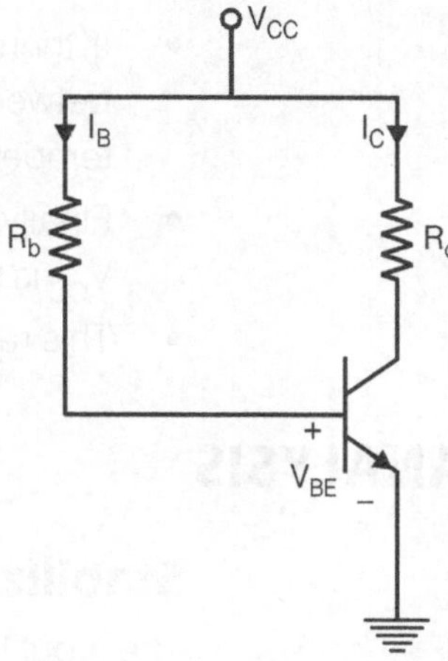


Figure 1.7 Fixed bias circuit

The voltage $V_{BE}$ across the forward biased emitter junction is approximately 0.2 V for a germanium transistor and 0.7 V for a silicon transistor in the active region.

Since $V_{CC}$ is usually much larger than $V_{BE}$, we have

$$I_B \approx \frac{V_{CC}}{R_b}$$

The current $I_B$ is constant, and the network of Fig. (1.7) is called the fixed bias circuit.

### Disadvantages of fixed bias circuit

(i)

$$I_C = \beta \; I_B$$

$I_B$ → Constant

$\beta$ → Varies with temperature and transistor replacement

∴ If $I_B$ is constant also, β may vary with respect to temperature.

(ii) Stability factor $$S = \frac{1+\beta}{1-\beta\frac{\partial I_B}{\partial I_C}}$$

In fixed bias $$I_B \approx \frac{V_{CC}}{R_b}$$

d.w.r.t. $I_C$ $$\frac{\partial I_B}{\partial I_C} = 0$$

$\therefore$ $$S = 1+\beta$$

Assume $$\beta = 100$$

$\therefore$ $$S = 1 + 100 = 101$$

For Q point stability, stability factor S should be very less.

## 1.6 Collector to Base Bias

- collector base bias circuit, shown in Figure (1.8) is an improvement over the fixed bias method.
- We know that the biasing resistance $R_b$ is connected between the collector and the base of the transistor.
- Let us assume the base current as $I_B$ and the collector current as $I_C$. So $I_B$ flows through $R_b$ and $(I_C + I_B)$ flows through collector resistance $R_C$.
- If there is a change in β (due to characteristic variation between transistors), or if there is any increase of ambient temperature, collector current $I_C$ tends to increase.
- Finally, voltage drop $I_C R_C$ increases. Since supply voltage $V_{CC}$ is constant, therefore $V_{CE}$ decreases.
- This reduced $V_{CE}$ results in reduced current $I_B$ through $R_b$.

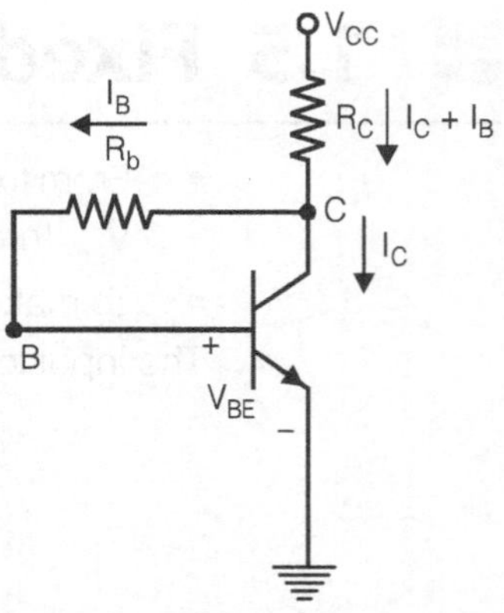

*Fig. 1.8 Collector to base bias*

## ANALYSIS

### Stabilization Concept

The input loop is given as

$$V_{CC} = (I_B + I_C) R_C + I_B R_b + V_{BE}$$

$$I_B = \frac{V_{CC} - I_C R_C - V_{BE}}{R_b + R_C} \quad \text{...(i)}$$

The output loop is given as

$$V_{CC} = I_C R_C + V_{CE} + I_b R_C$$

$$V_{CC} - I_C R_C = V_{CE} + I_b R_C \quad \text{...(ii)}$$

From equation (i) and (ii)

$$I_B \text{ (Input parameter)} = \frac{V_{CE} + I_B R_C - V_{BE} \text{ (Output parameter)}}{R_b + R_C}$$

- Since now the base current $I_B$ is reduced, the collector current is reduced. Thus, we find that the collector-to-base biasing helps to partly compensate the changes which occurred originally due to change in temperature or change in β.

## Stability Factor S for Collector to Base Bias

We can determine the stability factors S for the circuit, by applying KVL to the input loop

$$V_{CC} = (I_B + I_C) R_C + I_B R_b + V_{BE}$$

d.w.r.t $I_C$

$$0 = R_C + \frac{\partial I_B}{\partial I_C}(R_b + R_c) + 0$$

$$\frac{\partial I_B}{\partial I_C} = \frac{-R_C}{R_b + R_c}$$

$$S = \frac{1+\beta}{1-\beta\frac{\partial I_B}{\partial I_c}}$$

$$\boxed{S = \frac{1+\beta}{1+\beta\frac{R_c}{R_b + R_c}}}$$

It is an advantage compared to fixed bias circuit.

∴ This value is smaller than (1 + β), which was the stability factor S for fixed bias circuit. Thus there is an improvement of stability in the collector-to base bias circuit.

## Stabilization Against Change in β

$$I_C = (1 + \beta) I_{CO} + \beta I_B \quad \text{as} \quad \beta >> 1$$

$$I_C = \beta I_{CO} + \beta I_B \quad ; \quad I_B = \frac{I_C}{\beta} - I_{CO}$$

$$V_{CC} = (I_B + I_C)R_c + I_B R_b + V_{BE} \quad ; I_C = \beta\frac{(V_{CC} - V_{BE} + (R_b + R_C)I_{CO})}{\beta R_C + R_b}$$

$\beta R_c >> R_b$

$$I_C = \frac{V_{cc} - V_{BE} + (R_b + R_c)I_{co}}{R_c}$$

The collector current has become independent of β and hence stabilized against change in β.

## Drawback

Always, it is not possible to maintain the above condition (i.e. $\beta R_C >> R_b$). For example, if the load resistance $R_C$ is very small (as in a transformer coupled load), $\beta R_C$ is in fact less than $R_b$. Thus for low values of $R_C$, the collector to base bias offers no improvement in stabilization as compared to fixed bias circuit.

# 1.7 Voltage Divider Bias or Self Bias

- In voltage divider bias circuit, the biasing is provided by three resistors ($R_1$, $R_2$ and $R_e$).
- The resistors $R_1$ and $R_2$ act as a potential divider giving a fixed voltage to the base.
- If the collector current $I_C$ tends to increase (due to reasons like change in temperature or change in β). The emitter current $I_E$ also increases and the voltage drop across $R_e$ increases. The voltage difference between base and emitter will reduce and hence base current $I_B$ will reduce. In turn, $I_C$ will reduce, trying to partly compensate for the original change. This type of biasing improves stability.
- In order to prevent the instability of gain due to $R_e$, it is advantages to bypass $R_e$ by a large value capacitor $C_e$. This gives the advantage of self bias, while maintaining the gain equal to that of a C.E. amplifier without $R_e$.

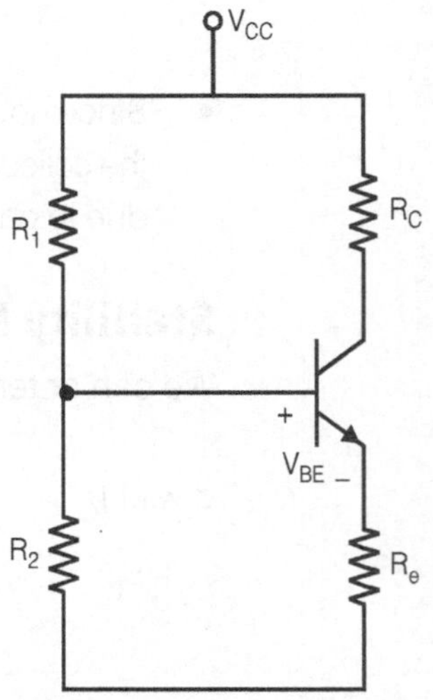

*Figure 1.9 Voltage divider bias*

## Stability Factor S

To find the stability factor S, we first replace the voltage divider network $R_1$, $R_2$ by its Thevenins equivalent, as shown

Applying KVL around the base circuit loop

$$V = I_B R_b + V_{BE} + (I_B + I_C) R_e$$

Differentiating w.r.t $I_C$ and considering $V_{BE}$ to be independent of $I_C$, we get

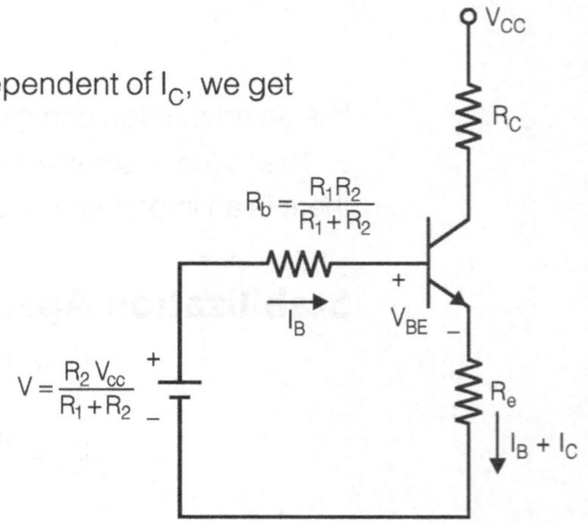


$$0 = \frac{\partial I_B}{\partial I_C} \cdot R_b + \frac{\partial I_B}{\partial I_C} R_e + R_e$$

$$\frac{\partial I_B}{\partial I_C}(R_b + R_e) = -R_e$$

$$\frac{\partial I_B}{\partial I_C} = \frac{-R_e}{R_b + R_e}$$

$$S = \frac{1+\beta}{1-\beta\frac{\partial I_B}{\partial I_C}} = \frac{1+\beta}{1+\beta\frac{R_e}{R_b+R_e}}$$

**From the above equation, we have**

(i) If $\frac{R_b}{R_e} << 1$, then Stability factor (S) $= (1+\beta)\cdot\frac{1}{(1+\beta)} = 1$

(ii) In the designing of the biasing circuit, $R_b$ value should be less. i.e. $R_1$ and $R_2$ will be taken less values. If suppose $R_1$ is very less in the circuit, then the life time of the battery will be less. So, we take $R_1 > R_2$ to overcome this disadvantage.

(iii) $R_e$ should be large. If it is consider, the negative feedback in the circuit increases which reduces the voltage gain. So keep a shunt capacitor parallel to $R_e$.

### Stability Factor S′

$$S' = \frac{\partial I_C}{\partial V_{BE}}$$

From the input loop, $V = I_B R_b + V_{BE} + (I_B + I_C) R_e$

Where $I_B = \frac{I_C}{\beta} - \frac{(1+\beta)}{\beta} I_{CO}$

$$V_{BE} = V + (R_b + R_e)\frac{(1+\beta)}{\beta} I_{CO} - \frac{R_b + (1+\beta)R_e}{\beta} I_C$$

d.w.r.t $V_{BE}$

$$1 = 0 + 0 - \frac{R_b + (1+\beta)R_e}{\beta}\frac{\partial I_C}{\partial V_{BE}}$$

$$S' = \frac{\partial I_C}{\partial V_{BE}} = \frac{-\beta}{R_b + (1+\beta)R_e}$$

$$\text{Stability factor (S)} = \frac{1+\beta}{1+\beta\frac{R_e}{R_b + R_e}} = \frac{(1+\beta)(R_b + R_e)}{R_b + (1+\beta)R_e}$$

$$\frac{S}{(1+\beta)(R_b + R_e)} = \frac{1}{R_b + (1+\beta)R_e}$$

$$S' = -\beta\frac{S}{(1+\beta)(R_b + R_e)}$$

$$\boxed{S' = \frac{S}{R_b + R_e}}$$

The lower the value of S, the lower is the value of S′,

### Stability Factor S″

$$S'' = \left.\frac{\partial I_C}{\partial \beta}\right|_{I_{co}\text{ and }V_{BE}\text{ constant}}$$

$$V_{BE} = V + (R_b + R_e)\frac{(1+\beta)}{\beta} I_{CO} - \frac{R_b + R_e(1+\beta)}{\beta} I_C = V + V_1 - \frac{R_b + R_e(1+\beta)}{\beta}\cdot I_C$$

Where $V_1 = (R_b + R_e) I_{co} \cdot \frac{\beta+1}{\beta}$ ; $I_C = \frac{\beta(V + V_1 - V_{BE})}{R_b + R_e(1+\beta)}$

$\beta R_e >> R_b$ ; $I_C = \frac{V + V_1 - V_{BE}}{R_e}$

From the above equation, $I_c$ is independent of β.

## COMPENSATION TECHNIQUES

- Up to now, we have seen the techniques of stabilizing the operating point of the transistor which makes use of resistors in the biasing circuits.
- But in compensation technique, we use temperature sensitive devices like diodes, thermistors, sensistors etc.

## 1.8 Bias Compensation by Diode

- The diode compensation circuit shown in Figure (1.10) offers stabilization against variations in $I_{co}$.
- Assume the diode D and transistor are made up of same material.
- Thus the effect of temperature is of the same order on both the diode and transistor.
- If there is an increase in temperature, the $I_{CO}$ and in turn, the $I_C$ of the transistor will increase.
- Also the leakage current $I_o$ through the diode D will increase, due to temperature change. Since $I_o$ will increase, the base current $I_B$ will decrease, this decrease in $I_B$ will lead to decrease in $I_C$, thereby trying to keep $I_C$ constant inspite of change in temperature.

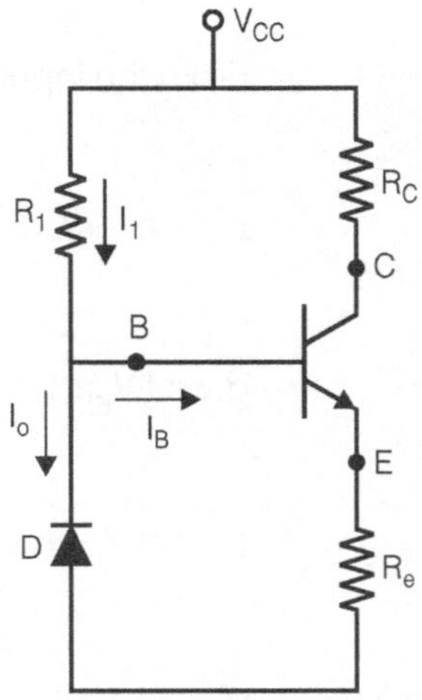


*Figure 1.10 Compensation through Diode*

## 1.9 Bias Compensation by Thermistor

- This is shown in Figure (1.11). Thermistors have negative temperature coefficient i.e. their resistance decreases with increase in temperature. If temperature increases, the resistance of $R_T$ decreases.
- So the value of $(R_2 \| R_T)$ also decreases. This makes to reduce base emitter voltage $V_{BE}$. Thus reducing the base current $I_B$ and in turn collector current $I_C$. Therefore thermistor provides the temperature compensation of the bias.

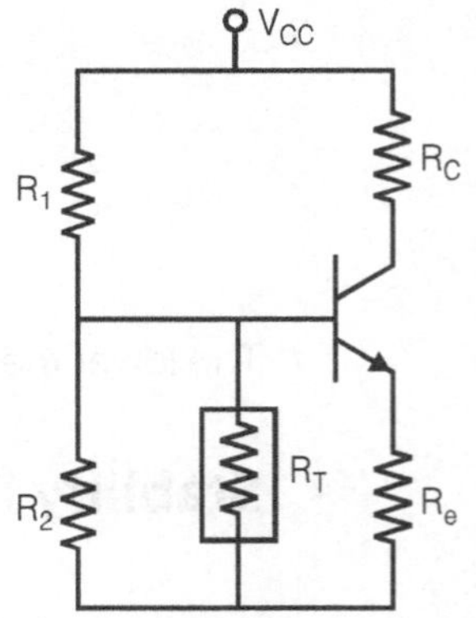


*Figure 1.11 Thermistor in parallel to $R_2$*

## 1.10 Bias Compensation by Sensistor

From the Figure (1.12), the resistance values $R_s$ and $R_2$ are so chosen that together they provide the correct bias at normal temperature.

- Sensistor have positive temperature coefficient i.e. their resistance increases with increase in temperature.
- If the temperature increases, the resistance of $R_s$ increases.
- As a result, the current flowing through $R_s$ decreases this in turn reduces the base current $I_B$, and in turn reduces the collector current $I_C$.
- In this way, use of sensistor provides the temperature compensation of the bias.

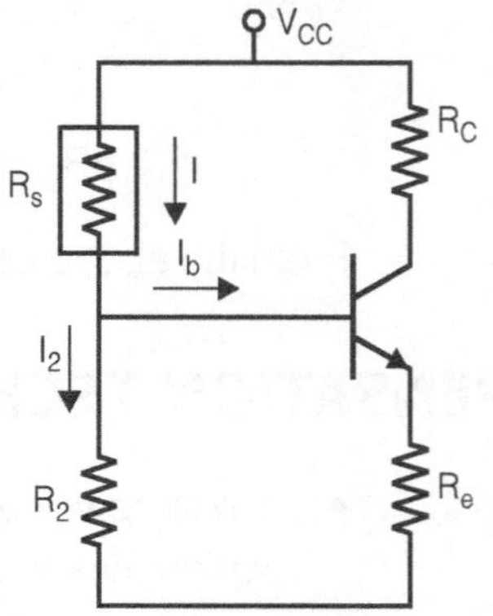


*Figure 1.12 Bias compensation by sensistor*

# 1.11 Thermal Run away

## THERMAL INSTABILITY

- Another important reason for instability of the bias (i.e. the operating point) is the change in temperature.
- We know that in a transistor, the reverse saturation current $I_{CO}$ changes with temperature infact $I_{CO}$ become double of every 10°C rise in temperature. The increase in the value of $I_{CO}$ will increase $I_C$, the collector current. The increase in $I_C$ will lead to more power loss in the collector-junction of the transistor. This in turn, will increase $I_{CO}$. This could become a spiralling action, sometimes leading to excessive temperature rise and consequent burning of the transistor. This concept is called as thermal run away.

## THERMAL RESISTANCE

Let $P_D$ = Power dissipation by the transistor

$T_j$ = Junction temperature

$T_A$ = Ambient temperature

The power is lost in the transistor which is the product of collector voltage and collector current. It is also dependent upon the temperature difference between the collector junction $T_j$ and ambient $T_A$.

$$\therefore \quad (T_j - T_A) \propto P_D$$

$$\text{or} \quad T_j - T_A = \theta P_D$$

θ, the constant of proportionality is known as the thermal resistance.

$$\therefore \quad \theta = \frac{T_j - T_A}{P_D}$$

The unit of θ, the thermal resistance is °C/Watt.

### Condition of Thermal Stability

The rate at which the heat is generated at the collector junction must not exceed the rate at which the heat can be dissipated i.e.

$$\boxed{\frac{\partial P_C}{\partial T_j} < \frac{\partial P_D}{\partial T_j}}$$

Where $P_c$ is heat generated at the collector junction.
Now, differentiating w.r.t. $T_j$, we get

$$1 = \theta \cdot \frac{\partial P_D}{\partial T_j}$$

$$\boxed{\frac{\partial P_C}{\partial T_j} < \frac{1}{\theta}}$$

This condition must be satisfied to prevent thermal runway. Now we can obtain a more practical expression for the above condition.

$$\frac{\partial P_c}{\partial T_j} = \frac{\partial P_C}{\partial I_C} \cdot \frac{\partial I_C}{\partial T_j}$$

$$P_C = V_{CC}I_C - I_C^2(R_C + R_e)$$

Taking derivative w.r.t. $I_C$

$$\frac{\partial P_c}{\partial I_c} = V_{cc} - 2I_c(R_c + R_e)$$

Stability factor S,

$$S = \frac{\partial I_C}{\partial I_{CO}} = \frac{\partial I_C}{\partial T_j} \times \frac{\partial T_j}{\partial I_{CO}}$$

We know that $I_{co}$ increases at the rate of 7% per°C

$$\frac{\partial I_{CO}}{\partial T_j} = 0.07\, I_{CO}$$

or

$$\frac{\partial T_j}{\partial I_{CO}} = \frac{1}{0.07\, I_{CO}}$$

$$\frac{\partial I_C}{\partial T_j} = S \times 0.07\, I_{CO}$$

$$\{V_{CC} - 2I_C(R_C + R_e)\} \times S \times 0.07 I_{CO} < \frac{1}{\theta}$$

## Conclusion

The following three conditions leads to thermal run-away:

(i) $\dfrac{\partial P_C}{\partial T_j} < \dfrac{\partial P_D}{\partial T_j}$

(ii) $\dfrac{\partial P_C}{\partial T_j} < \dfrac{1}{\theta}$

(ii) $\{V_{CC} - 2I_C(R_C + R_e)\} \times S \times 0.07 I_{CO} < \dfrac{1}{\theta}$

## SUMMARY

- The collector current of a transistor varies with $I_{CO}$, β and $V_{BE}$ which in turn are dependent on temperature variation.
- Stability factor S is given by

$$S = \frac{\partial I_c}{\partial I_{co}} = \frac{1+\beta}{1-\beta\dfrac{\partial I_b}{\partial I_c}}$$

- Fixed bias circuit is the simplest biasing arrangement and has a stability factor S = 1 + β.
- Collector to base bias circuit is an improvement over fixed bias circuit and has a stability factor

$$S = \frac{1+\beta}{1+\beta\dfrac{R_c}{R_b + R_c}}$$

- The self bias circuit consists of three resistors $R_1$, $R_2$, $R_e$, usually. The $R_E$ is bypassed with a large value capacitor $C_E$. The stability factor,

$$S = \frac{1+\beta}{1+\beta\dfrac{R_e}{R_b+R_e}}$$

for obtaining low value of S, it is necessary to choose a low value of $\frac{R_b}{R_e}$.
- Bias compensation can be achieved either by use of diode, thermistor or sensistor.
- Stability factor $S' = \frac{\partial I_C}{\partial V_{BE}}$ and stability factor $S'' = \frac{\partial I_C}{\partial \beta}$ can both be reduced if stability factor S is reduced.
- Condition for thermal stability or avoiding thermal runaway are

$$\Rightarrow \frac{\partial P_C}{\partial T_j} < \frac{\partial P_D}{\partial T_j} \text{ (or) } \frac{\partial P_c}{\partial T_j} < \frac{1}{\theta} \text{ (or) } [V_{CC} - 2I_C(R_c + R_e)] \times S \times 0.07 I_{CO} < \frac{1}{\theta}$$

where $\theta$ = thermal resistance.
- Best possible of choosing the operating point is $V_{CE} = \frac{V_{CC}}{2}$.

## GATE AND IES QUESTIONS

**Q.1** Introducing a resistor in the emitter of a common emitter amplifier stabilizes the dc operating point against variations in

(a) only the temperature
(b) only the $\beta$
(c) Both temperature and $\beta$
(d) none

**Exp. (c)**

$R_e$ resistor works as a negative feedback which gives stability of the circuit. Therefore it stabilizes against the variation of both temperature and $\beta$.

**Q.2** In the circuit of the figure, assume that the transistor is in the active region. It has a large $\beta$ and its base-emitter voltage is 0.7 V. The value of $I_c$ is

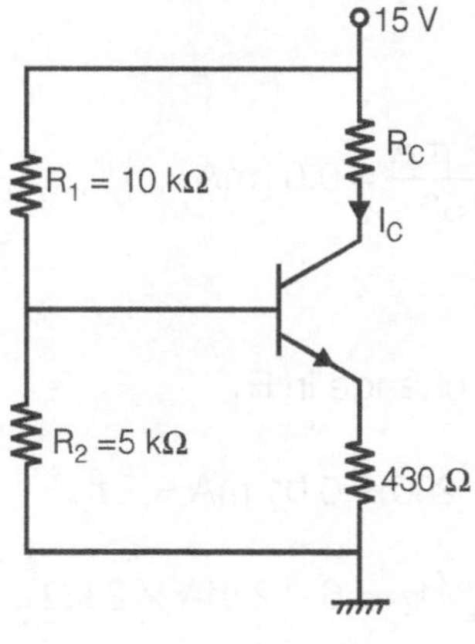


(a) Indeterminate since $R_c$ is not given
(b) 1 mA
(c) 5 mA
(d) 10 mA

**Exp.** (d)

The Thevenin equivalent is shown below:

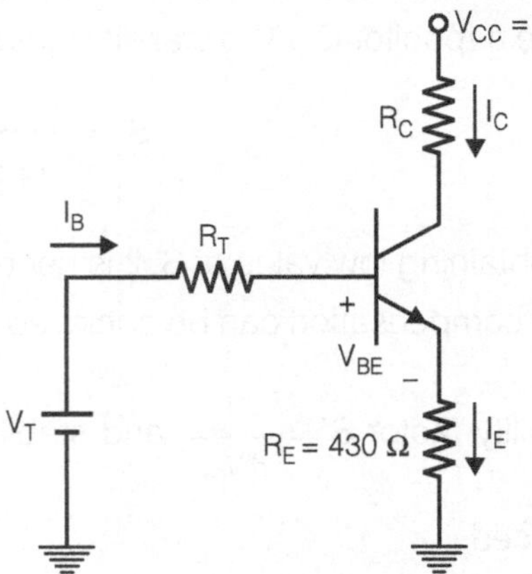


$$V_T = \frac{R_2}{R_1 + R_2} V_{cc}$$

$$= \frac{5}{10+5} \times 15 = 5\text{ V}$$

Since β is large

$$I_C \approx I_E,\ I_B \approx 0 \text{ and}$$

$$I_E = \frac{V_T - V_{BE}}{R_E}$$

$$= \frac{5 - 0.7}{0.430\text{ k}\Omega} = \frac{4.3}{0.430\text{ k}\Omega} = 10\text{ mA}$$

**Q.3** In the amplifier circuit shown in the figure, the values of $R_1$ and $R_2$ are such that the transistor is operating at $V_{CE} = 3$ V and $I_C = 1.5$ mA, when its β is 150. For a transistor with β of 200, the operating point ($V_{CE}$, $I_C$) is

(a) (2 V, 2 mA)

(b) (3 V, 2 mA)

(c) (4 V, 2 mA)

(d) (4 V, 1 mA)

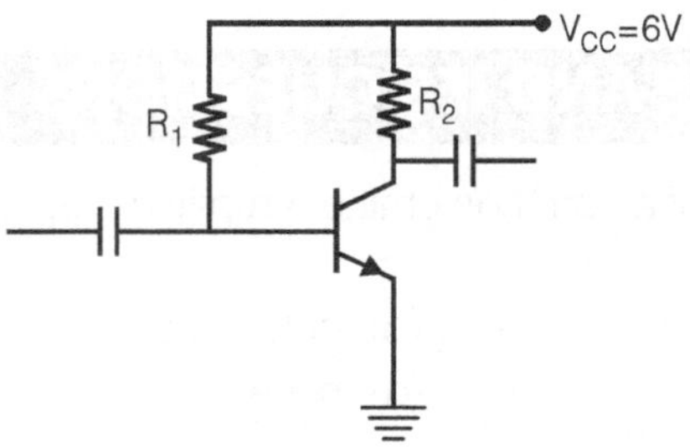


**Exp.** (a)

The DC equivalent circuit is shown below:

**First case:**

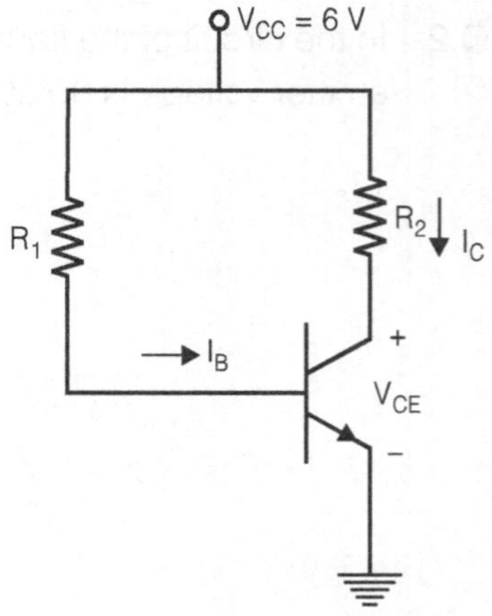


$$V_{CC} - I_{C_1} R_2 - V_{CE_1} = 0$$

$$6 - 1.5\text{ mA } R_2 - 3 = 0$$

$$R_2 = 2\text{ k}\Omega$$

$$I_{B_1} = \frac{I_{C_1}}{\beta_1} = \frac{1.5\text{ mA}}{150} = 0.01\text{ mA}$$

**Second case:**

$I_{B_2}$ will be equal to $I_{B_1}$ as there is no change in $R_1$

$$I_{C_2} = \beta_2 I_{B_2} = 200 \times 0.01\text{ mA} = 2\text{ mA}$$

$$V_{CE_2} = V_{CC} - I_{C_2} R_2 = 6 - 2\text{ mA} \times 2\text{ k}\Omega$$

$$= 2\text{ V}$$

**Q.4** Assuming $V_{CEsat} = 0.2$ V and $\beta = 50$, the minimum base current ($I_B$) required to drive the transistor in the figure to saturation is

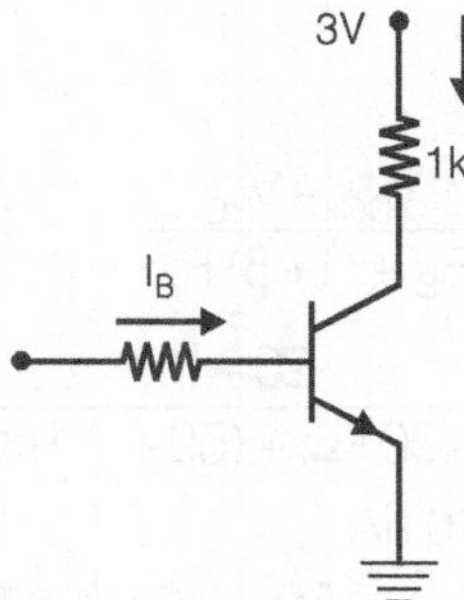


(a) 56 μA

(b) 140 mA

(c) 60 mA

(d) 3 mA

**Exp. (a )**

Applying KVL we get

$$V_{CC} - I_C R_C - V_{CE} = 0$$

$$I_C = \frac{V_{CC} - V_{CE}}{R_C} = \frac{3 - 0.2}{1k} = 2.8 \text{ mA} \; ; \; I_B = \frac{I_C}{\beta} = \frac{2.8 \text{ mA}}{50} = 56 \,\mu\text{A}$$

**Q.5** For an npn transistor connected as shown in the figure, $V_{BE} = 0.7$ V, given that reverse saturation current of the junction at room temperature 300 K is $10^{-13}$ A, the emitter current is

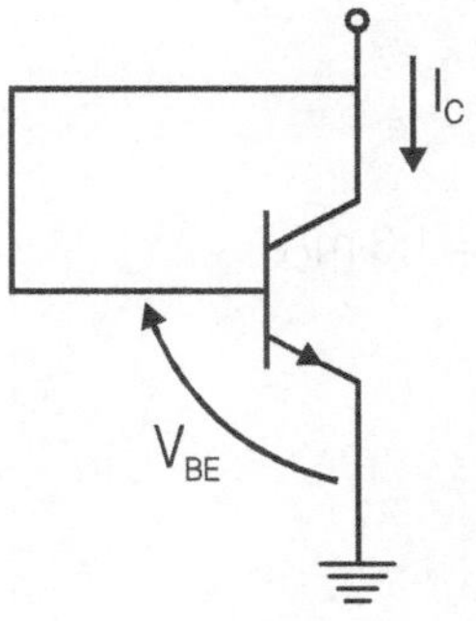


(a) 30 mA

(b) 39 mA

(c) 49 mA

(d) 20 mA

**Exp. (c)**

$$I_E = I_S\left[e^{V_{BE}/\eta V_T} - 1\right] = 10^{-13}\left[e^{0.7/1 \times 0.026} - 1\right] = 49 \text{ mA}$$

**Q.6** The circuit using a BJT with $\beta = 50$ and $V_{BE} = 0.7$ V is shown in the figure. The base current $I_B$ and collector voltage $V_C$ are respectively

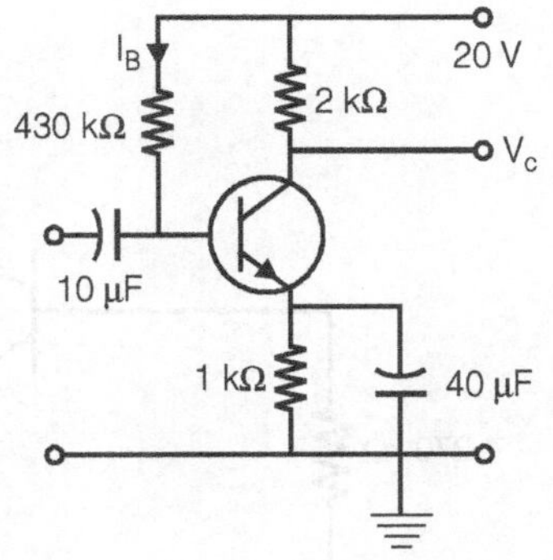


(a) 43 μA and 11.4 V

(b) 40 μA and 16 V

(c) 45 μA and 11 V

(d) 50 μA and 10 V

**Exp.** (b)

Under DC conditions

Using KVL we have

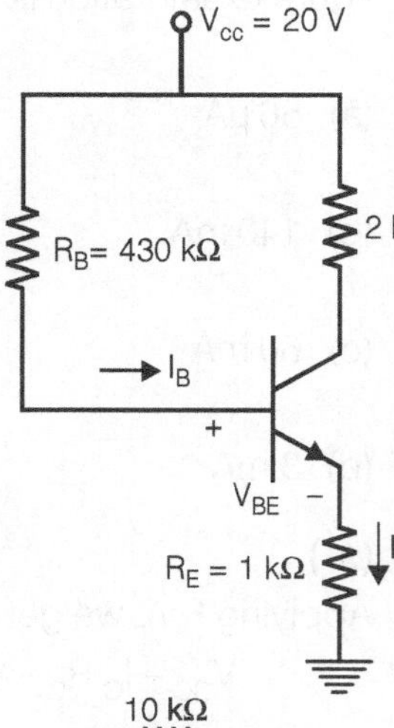


$$V_{CC} - R_B I_B - V_{BE} - R_E I_E = 0$$

$$V_{CC} - R_B I_B - V_{BE} - R_E (1 + \beta) I_B = 0$$

$$I_B = \frac{V_{CC} - V_{BE}}{R_B + (1+\beta) R_E}$$

$$= \frac{20 - 0.7}{430\text{ k}\Omega + (50+1)\,1\text{ k}\Omega}$$

$$= 40\ \mu A$$

$$I_C = \beta I_B = 50 \times 40\ \mu A = 2\text{ mA}$$

$$V_C = V_{CC} - I_C R_C$$

$$= 20 - 2\text{ mA} \times 2\text{ k}\Omega = 16\text{ V}$$

**Q.7** For the BJT circuit shown, assume that the $\beta$ of the transistor is very large and $V_{BE} = 0.7$ V. The mode of operation of BJT is

(a) cut-off

(b) saturation

(c) normal active

(d) reverse active

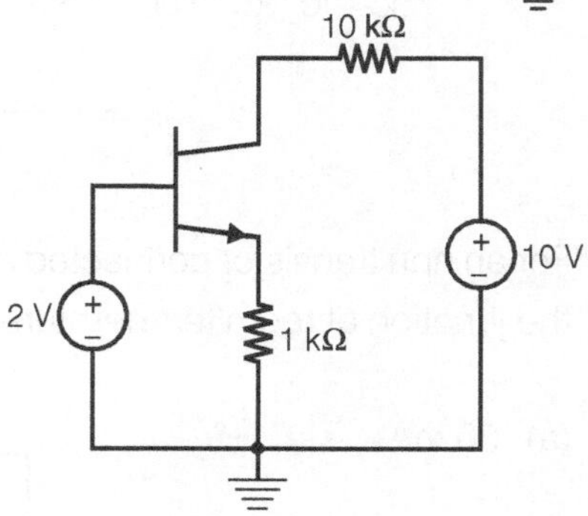


**Exp.** (b)

We assume that BJT is in active

$$I_E = \frac{2 - V_{BE}}{R_E} = \frac{2 - 0.7}{1\text{k}\Omega} = 1.3\text{ mA}$$

Since $\beta$ is very large, we have $I_E = I_C$, thus

$$I_C = 1.3\text{ mA}$$

Now applying KVL in collector-emitter loop

$$10 - 10\text{ k } I_C - V_{CE} - 1\text{k } I_C = 0$$

$$V_{CE} = -4.3\text{ V}$$

$$V_{BC} = V_{BE} - V_{CE}$$

$$= 0.7 - (-4.3) = 5\text{ V}$$

Since $V_{BC}$ is positive, thus transistor is in saturation.

**Q.8** The common emitter forward current gain of the transistor shown with $\beta_F = 100$

The transistor is operating in

(a) Saturation region

(b) Cutoff region

(c) Reverse active region

(d) Forward active region

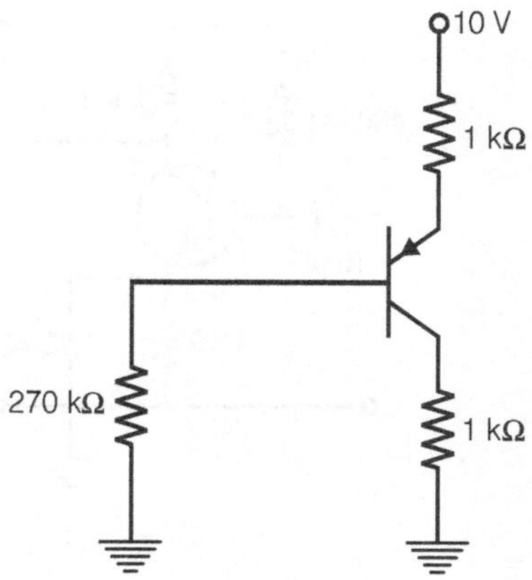

**Exp.** (d)

If transistor is in normal active region, base current can be calculated as

At input loop

$$10 - I_C(1 \times 10^3) - 0.7 - 270 \times 10^3 I_B = 0$$
$$1\,k \cdot \beta I_B + 270\,k\,I_B = 9.3 \text{ mA}$$
$$I_B(\beta\, 1\,k + 270\,k) = 9.3 \text{ mA}$$
$$I_B = \frac{9.3\text{ mA}}{270\,k + 100 \cdot 1k} = 0.025 \text{ mA} = 25\,\mu\text{A}$$

In saturation, base current is

$$10 - I_C(1\,k) - V_{CE\,sat} - I_E(1\,k) = 0$$
$$\frac{10}{2\,k} = (I_C)_{sat} \qquad (V_{CE\,sat} = 0\text{ V})$$
$$(I_C)_{sat} = 5 \text{ mA}$$
$$I_{B\,sat} = \frac{I_{C\,sat}}{\beta} = \frac{5}{100} = 0.050 \text{ mA} = 50\,\mu\text{A}$$

$I_B < (I_B)_{sat}$, so transistor is in forward active region.

**Q.9** Consider the circuit shown in figure. If the $\beta$ of the transistor is 30 and $I_{CBO}$ is 20 nA and the input voltage is + 5 V, the transistor would be operating in

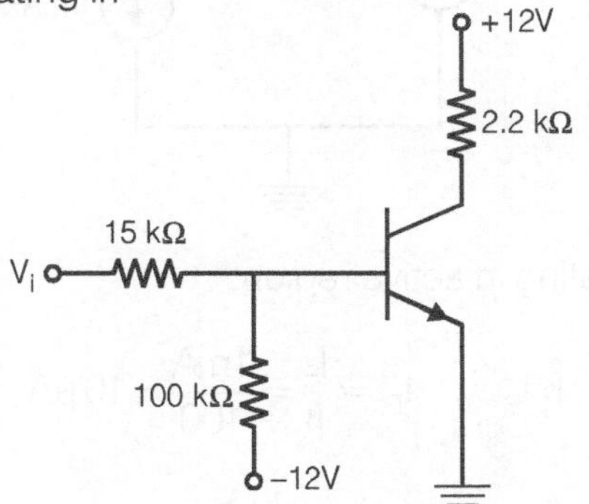


(a) saturation region

(b) active region

(c) breakdown region

(d) cut-off region

**Exp.** (b)

Assume that BJT is in active region, Thevenin equivalent circuit is obtained as

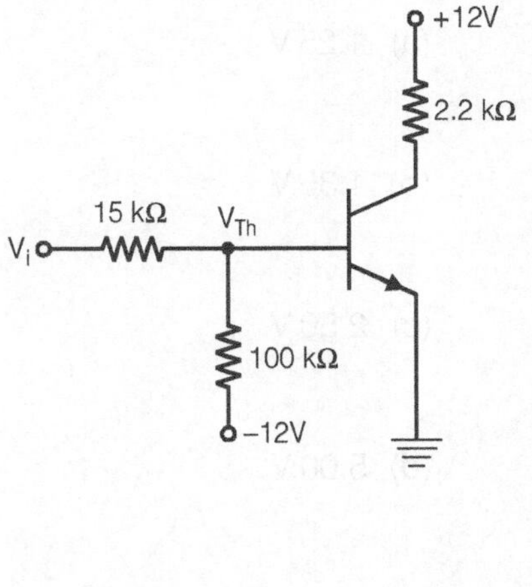


$$\frac{V_{Th} - V_i}{15} + \frac{V_{Th} - (-12)}{100} = 0$$
$$20\,V_{Th} - 20\,V_i + 3\,V_{Th} + 36 = 0$$
$$23\,V_{Th} = 20 \times 5 - 36 \qquad (V_i = 5\text{ V})$$
$$V_{Th} = 2.78 \text{ V}$$

Thevenin resistance $R_{Th} = 15\text{ k}\Omega \parallel 100\text{ k}\Omega = 13.04\text{ k}\Omega$

So the circuit is

Writing KVL for input loop

$$2.78 - R_{Th}\,I_B - 0.7 = 0$$
$$I_B = 0.159 \text{ mA}$$

Current in saturation is given as

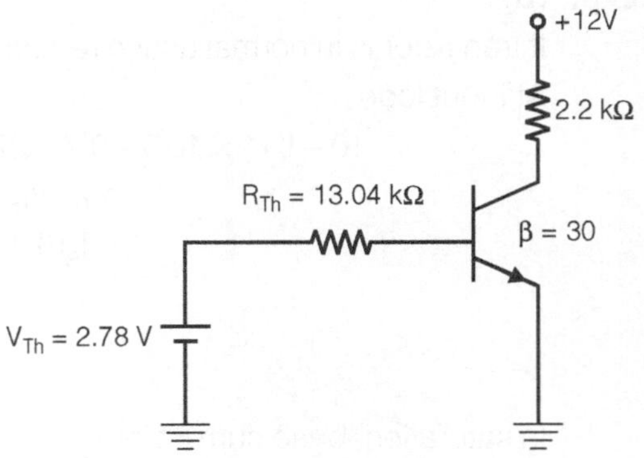


$$I_{B\,sat} = \frac{I_{C\,sat}}{\beta} \quad (V_{CE\,sat} = 0.2\text{ V})$$

$$I_{C\,sat} = \frac{12-0.2}{2.2\text{k}\Omega} = 5.36\text{ mA}$$

$$I_{B\,sat} = \frac{5.36\text{ mA}}{30} = 0.178\text{ mA}$$

Since $I_{B\,sat} > I_B$, It is active region.

**Q.10** The common emitter amplifier shown in the figure is biased using a 1 mA ideal current source. The approximate base current value is

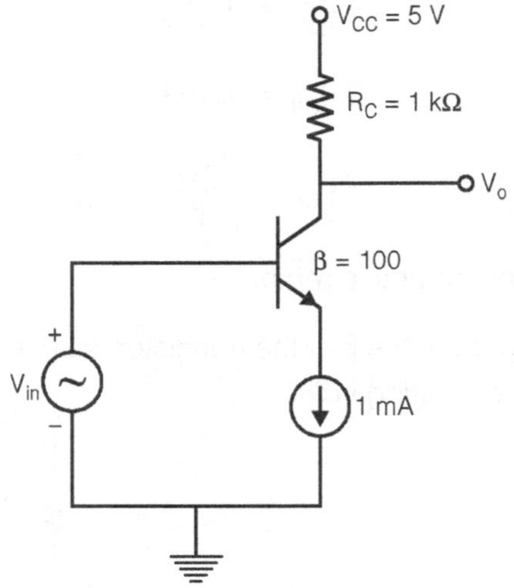


(a) 0 μA

(b) 10 μA

(c) 100 μA

(d) 1000 μA

**Exp.** **(b)**

Since the transistor is operating in active region.

$$I_E = \beta I_B \quad ; \quad I_B = \frac{I_E}{\beta} = \frac{1\text{mA}}{100} = 10\,\mu\text{A}$$

**Q.11** In the Schmitt trigger circuit shown in figure, if $V_{CE(sat)}$ = 0.1 V, the output logic low level ($V_{OL}$) is

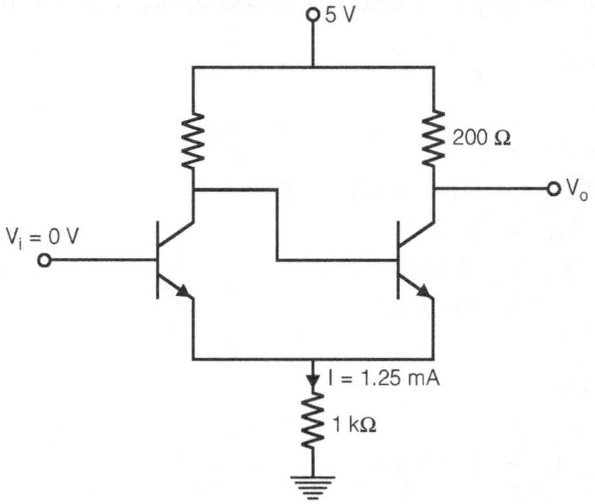


(a) 1.25 V

(b) 1.35 V

(c) 2.50 V

(d) 5.00 V

**Exp. (b)**

In the given circuit $V_i = 0$ V

So, transistor $Q_1$ cut-off and $Q_2$ saturation

$$5 - I_C R_C - 0.1 - 1.25 = 0 \quad (\because V_{CE\,sat} = 0.1\text{ V})$$

$$5 - I_C R_C = 1.35$$

$$V_o = 1.35\text{ V}$$

**Common Data For Q.12 (i), (ii) and (iii)**

For the circuit shown in figure, $I_E = 1$ mA, $\beta = 99$ and $V_{BE} = 0.7$ V

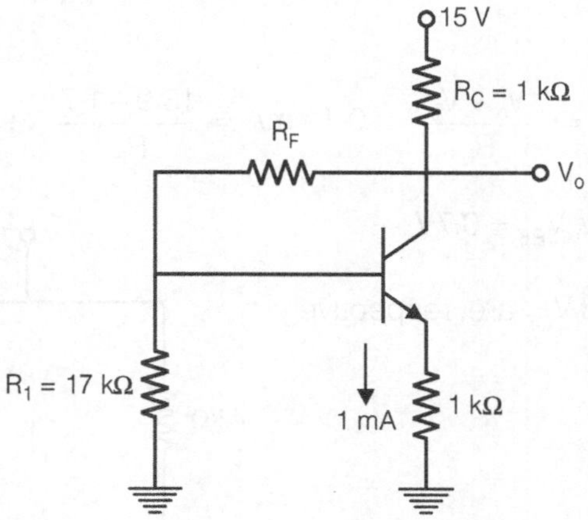


(i) The current through $R_C$ is

(a) 0.99 mA (b) 1.1 mA

(c) 1.2 0 mA (d) 1 mA

**Exp. (b)**

(i) Consider the transistor is in active region

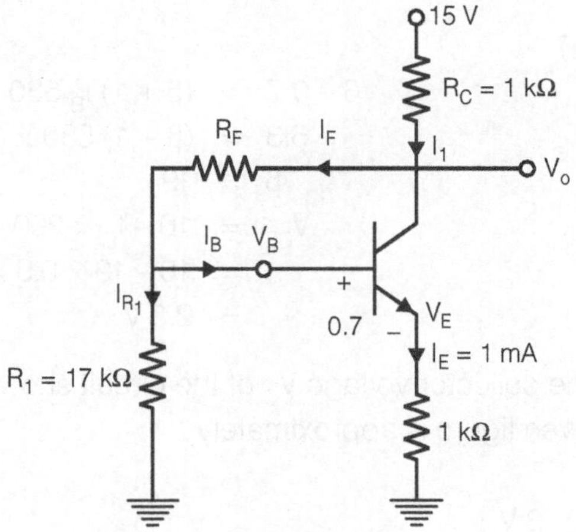


$$I_c = \alpha I_E\ ; \quad I_C = \frac{\beta}{(\beta+1)} I_E$$

$$I_C = \frac{99}{(99+1)} 1\text{mA}$$

$$I_C = 0.99\text{ mA}$$

$$V_{BE} = 0.7\text{ V}$$

$$V_E = I_E \times 1\text{ k}\Omega = 1\text{ V}$$

$$V_B - V_E = 0.7$$

$$V_B = 0.7 + 1 = 1.7\text{ V}$$

$$I_{R_1} = \frac{V_B}{17\text{ k}\Omega} = \frac{1.7}{17\text{ k}\Omega} = 100\,\mu\text{A} \quad ; \quad I_B = \frac{I_E}{\beta+1} = \frac{1\text{mA}}{(99+1)} = 10\,\mu\text{A}$$

By writing KCL at base

$$I_{R_F} = I_B + I_{R_1} = 10 + 100 = 110\,\mu\text{A}$$

$$I_1 = I_C + I_F$$

$$= 0.99\text{ mA} + 110\,\mu\text{A} = 1.1\text{ mA}$$

**(ii)** Output voltage $V_o$ will be

(a) 16.1 V (b) 14 V
(c) 13.9 V (d) None of these

**Exp. (c)**

$$V_o = 15 - I_1 R_C = 15 - (1.1 \text{ mA})(1 \text{ k}\Omega) = 13.9 \text{ V}$$

**(iii)** Value of resistance $R_F$ is

(a) 110.9 kΩ (b) 124.5 kΩ
(c) 130.90 kΩ (d) None of these

**Exp. (a)**

Current in $R_F$ $\quad I_F = \dfrac{V_o - V_B}{R_F}$ ; $0.11 \text{ mA} = \dfrac{13.9 - 1.7}{R_F}$ ; $R_F = 110.9 \text{ k}\Omega$

**Q.13** In the circuit shown below, $V_{BE} = 0.7\text{V}$

The β of the transistor and $V_{CE}$ are, respectively

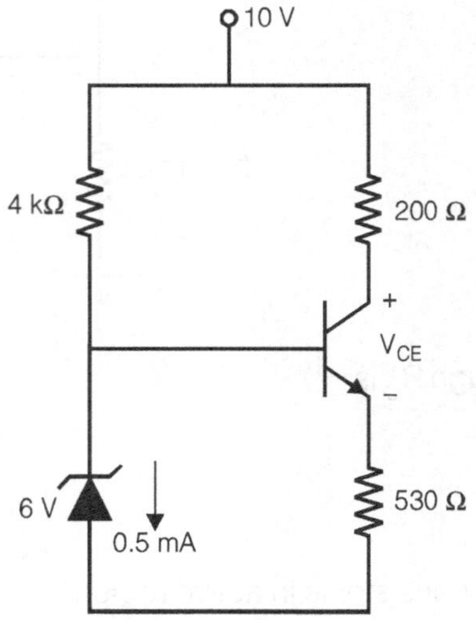


(a) 19 and 2.8 V

(b) 19 and 4.7 V

(c) 38 and 2.8 V

(d) 38 and 4.7 V

**Exp. (a)**

$$6 - 0.7 = (\beta + 1) I_B 530$$
$$5.3 = (\beta + 1) 530 \times 0.5$$
$$\beta = 19$$
$$V_{CE} = 10 - I_C \times 200 - 530 \times I_E = 10 - \beta I_B 200 - (\beta + 1) 530 I_B$$
$$= 10 - 19 \times 0.5 \times 200 \times 10^{-3} - 20 \times 530 \times 0.5 \times 10^{-3}$$
$$= 2.8 \text{ V}$$

**Q.14** The collector voltage $V_C$ of the circuit shown in the given figure is approximately

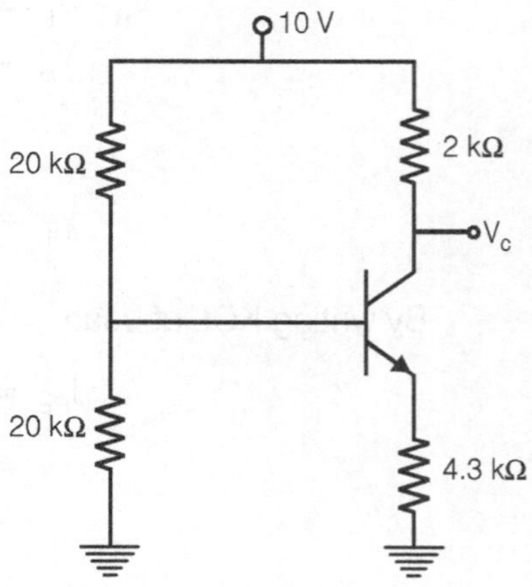


(a) 2 V

(b) 4.6 V

(c) 8 V

(d) 8.6 V

**Exp.** (c)

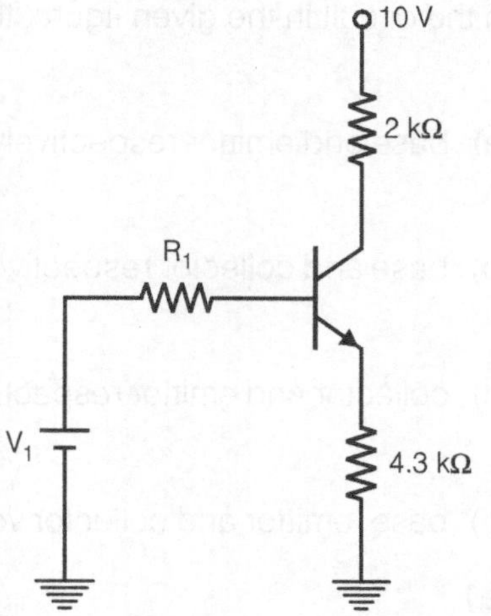


$$V_1 = \frac{20}{20+20} \times 10 = 5\text{ V}$$

$$R_1 = \frac{20 \times 20}{20+20} = 10\text{ k}\Omega$$

KVL at input loop $V_1 = R_1 I_B + V_{BE} + 4.3 I_E$

Let $\beta >> 1$ So $I_B \approx 0$ and $I_E \approx I_C$

$$V_1 = V_{BE} + 4.3 I_C$$

$$5 - 0.7 = 4.3 I_C$$

$$I_C = 1\text{ mA}$$

$$V_C = 10 - 2 \times 1$$

$$V_C = 8\text{ V}$$

**Q.15** Thermal run away will take place if the quiescent point is such that

(a) $V_{CE} > \frac{1}{2} V_{CC}$

(b) $V_{CE} < V_{CC}$

(c) $V_{CE} < 2 V_{CC}$

(d) $V_{CE} < \frac{1}{2} V_{CC}$

**Exp.** (a)

Thermal run-away will take place if $V_{CE} > \frac{1}{2} V_{CC}$.

**Q.16** A common emitter amplifier circuit is shown in the given figure

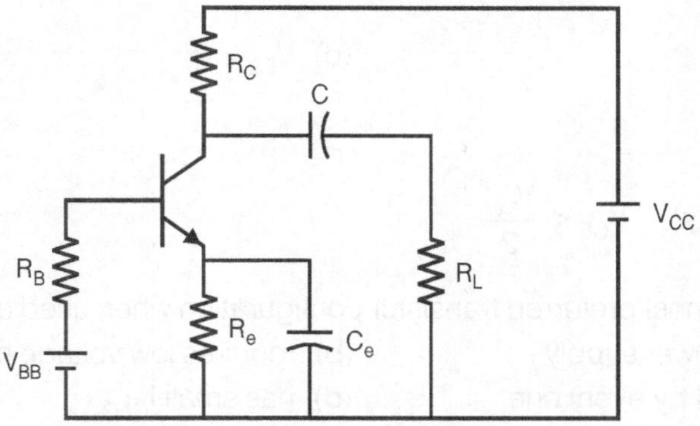


The slope of AC load line is

(a) $-\left[\frac{1}{R_L} + \frac{1}{R_C}\right]$

(b) $-\left[\frac{1}{R_L + R_C}\right]$

(c) $-\frac{1}{R_L}$

(d) $-\frac{1}{R_C}$

**Exp.** (a)

Slope of AC load line is $-\frac{1}{R_L'} = -\frac{1}{R_C \| R_L}$

**Q.17** In the circuit in the given figure, the approx voltages at the transistor

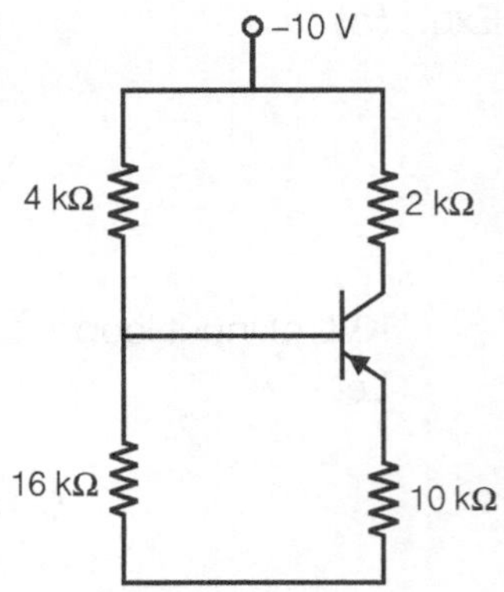


(a) base and emitter respectively are −8 V and −7.3 V

(b) base and collector respectively are −8 V and −5 V

(c) collector and emitter respectively are −8 V and −7.3 V

(d) base, emitter and collector voltage are −8 V, −7.3 V and −5 V

**Exp. (a)**

Let $I_B \approx 0$

Equivalent circuit is shown below:

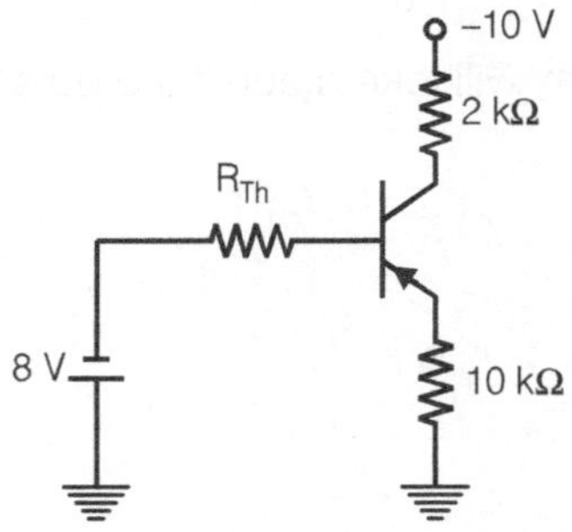


$$V_B = -8\text{ V}$$

$$V_E = V_{EB} + V_B$$

$$= 0.7 + (-8) = -7.3\text{ V}$$

**Q.18** The thermal runaway in a CE transistor amplifier can be prevented by biasing the transistor in such a manner that

(a) $V_{CE} > \frac{V_{CC}}{2}$ (b) $V_{CE} < \frac{V_{CC}}{2}$

(c) $V_{CE} = \frac{V_{CC}}{2}$ (d) $V_{CE} = 0$

**Exp. (b)**

To avoid thermal runaway $V_{CE} < \frac{V_{CC}}{2}$

**Q.19** CE configuration is the most preferred transistor configuration when used as a switch because it

(a) requires only one power supply (b) requires low voltage or current

(c) is easily understood by every one (d) has small $I_{CEO}$

**Exp. (b)**

CE configuration has high power gain requires low voltage or current.

**Q.20** The condition to be satisfied to prevent thermal runaway in a transistor amplifier where ($P_C$ = power dissipated at collector, $T_j$ = junction temperature, $T_A$ = ambient temperature, $\theta$ = thermal resistance) is

(a) $\frac{\partial P_C}{\partial T_j} > \frac{1}{\theta}$ (b) $\frac{\partial P_C}{\partial T_A} < \frac{1}{\theta}$

(c) $\frac{\partial P_C}{\partial T_j} < \frac{1}{\theta}$ (d) $\frac{\partial P_C}{\partial T_A} > \frac{1}{\theta}$

**Exp. (c)**

The condition to prevent thermal run-away is $\frac{\partial P_C}{\partial T_j} < \frac{1}{\theta}$.

**Q.21** In the circuit shown
The transistor is biased at

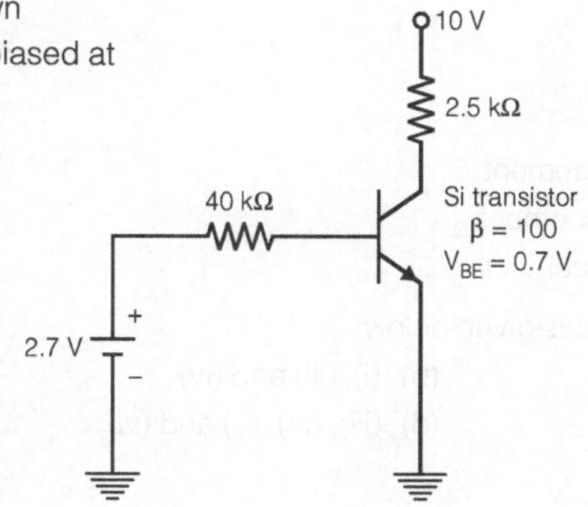


(a) 0 mA

(b) 5 mA

(c) 3.9 mA

(d) ∞

**Exp.** (c)

Let the transistor be in active region

$$2.7 - 0.7 = 40 \times I_b$$

$$I_b = \frac{1}{20} \text{ mA}$$

$$I_C = \beta I_b = 5 \text{ mA}$$

$$V_{CE} = 10 - 2.5 I_C = 10 - 2.5 \times 5 = -2.5 \text{ V}$$

$$V_{BE} + V_{CB} = V_{CE}$$

$$V_{CB} = -2.5 - 0.7 = -3.2 \text{ V}$$

$V_B > V_C$ that means it is in saturation ; $(I_C)_{sat} = \dfrac{V_{CC} - V_{CE\,sat}}{R_C} = \dfrac{10 - 0.2}{2.5\text{ k}\Omega} = 3.9 \text{ mA}$

**Q.22** Consider a silicon transistor connected a a common emitter amplifier as shown below. The quiescent collector voltage of the circuit is approx

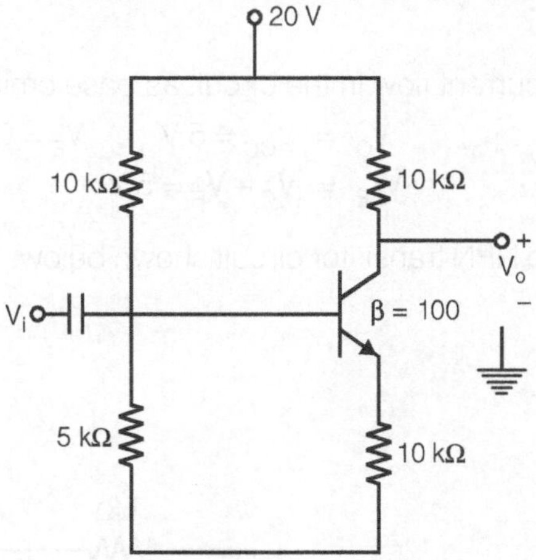


(a) 6.67 V

(b) 10 V

(c) 14 V

(d) 20 V

**Exp.** (c)

$$V_B = \frac{20 \times 5\text{ k}\Omega}{10\text{ k}\Omega + 5\text{ k}\Omega} = \frac{20}{3} \quad ; \quad \frac{20}{3} = 0.7 + 10 \times 10^3 I_C$$

$$I_C = 0.6 \text{ mA}$$

$$(V_C)_Q = 20 - 10 \times 10^3 I_C = 20 - 6 = 14 \text{ V}$$

**Q.23** Which of the following main properties of a bipolar junction transistor make it necessary for the transistor to have bias stabilization?

(i) Variation of $V_{BE}$ with temperature
(ii) Variation of $h_{FE}$ with temperature
(iii) Variation of $I_{CO}$ with temperature
(iv) Variation of $h_{FE}$ with transistor replacement
(v) Variation of $V_{BE}$ with transistor replacement
(vi) Variation of $I_{CO}$ with transistor replacement

Select the correct answer using the codes given below:

(a) (i), (ii) and (vi)
(b) (i), (iii) and (iv)
(c) (ii), (iii) and (v)
(d) (iii), (iv), (v) and (vi)

**Exp. (b)**

Variation of $V_{BE}$ and $I_{CO}$ with temperature and variation of $h_{FE}$ with transistor replacement.

**Q.24** Consider the following circuit

What is voltage difference between collector and emitter ($V_{CE}$) in the circuit?

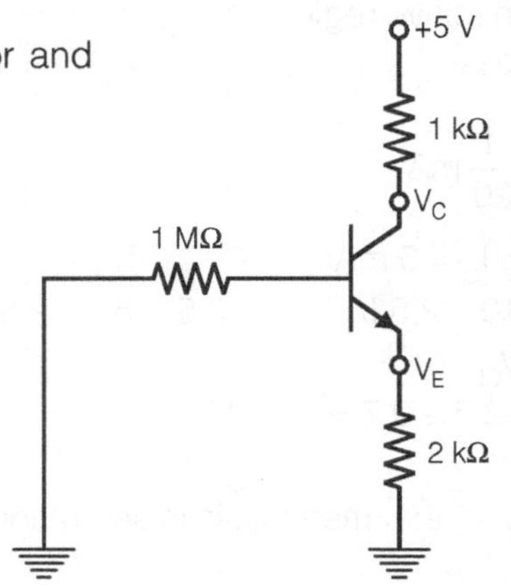


(a) $\frac{10}{3}$ V

(b) 0 V

(c) 5 V

(d) 3 V

**Exp. (c)**

There is no current flow in the circuit as base emitter junction is not FB

$$V_C = V_{CC} = 5\text{ V} \quad ; \quad V_E = 0$$
$$V_{CE} = V_C - V_E = 5\text{ V}$$

**Q.25** Consider the NPN transistor circuit shown below:

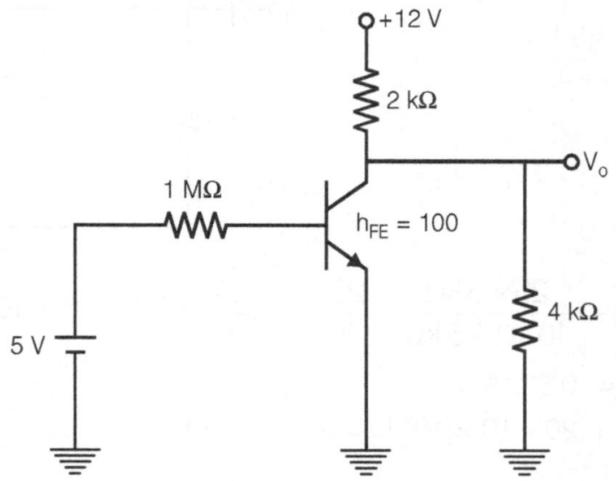

What is the output voltage $V_o$ in the above circuit?

(a) 0 V (b) 12 V
(c) 9 V (d) 7.5 V

**Exp. (d)**

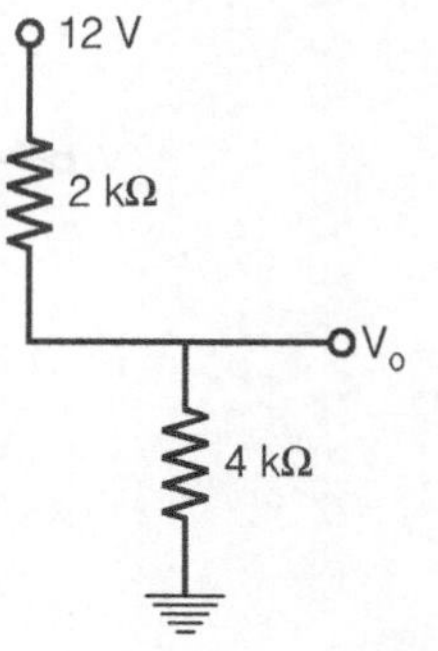


$$i_b = \frac{5}{1\text{M}\Omega} = 5\,\mu\text{A} \approx 0$$

Transistor is cutoff

$$V_o = \frac{12 \times 4\,\text{k}\Omega}{6\,\text{k}\Omega} = 8\text{ V} \rightarrow \text{ideal}$$

**Q.26** Consider the following statements:
The bias stability of an emitter bias amplifier circuit improves by
1. Decreasing the value of $R_B$
2. Increasing the value of $R_E$
3. Decreasing the value of $R_E$
4. Increasing the value of $R_B$
5. Increasing the value of $R_C$

Which of the above statements are correct?

(a) 1 and 2 (b) 2 and 3
(c) 3 and 4 (d) 4 and 5

**Exp. (a)**

$$\frac{R_B}{R_E} << 1$$

**Q.27** For the transistor circuit shown in the figure, when

(i) $V_{in} > 0$, transistor is OFF

(ii) $V_{in} \leq 0$, transistor is OFF

(iii) $I_B > \frac{I_C}{h_{FE}}$, transistor is ON

(iv) $I_B \leq \frac{I_C}{h_{FE}}$, transistor is ON

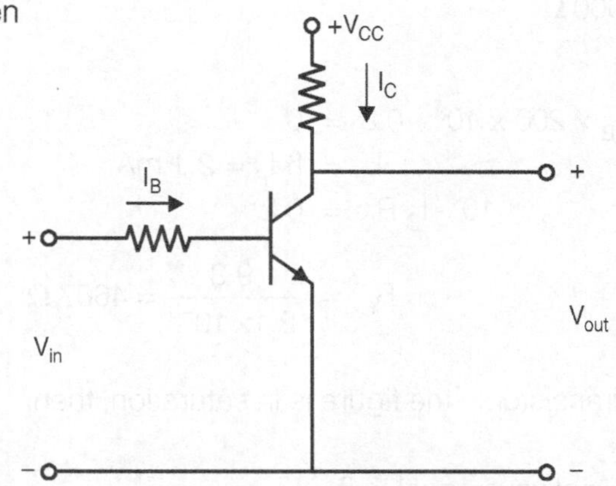


The correct statements are

(a) (i), (ii), (iii) and (iv) (b) (i) and (ii) only
(c) (ii) and (iii) only (d) (iii) and (iv) only

**Exp. (c)**

$V_{in} \leq 0$, transistor is OFF

$I_B > \frac{I_C}{h_{FE}}$, transistor is ON

**Q.28** In the circuit as shown below, the ratio of $V_o$ to $(V_1 - V_2)$ would approximately (neglecting constant due to $V_{CC}$) be

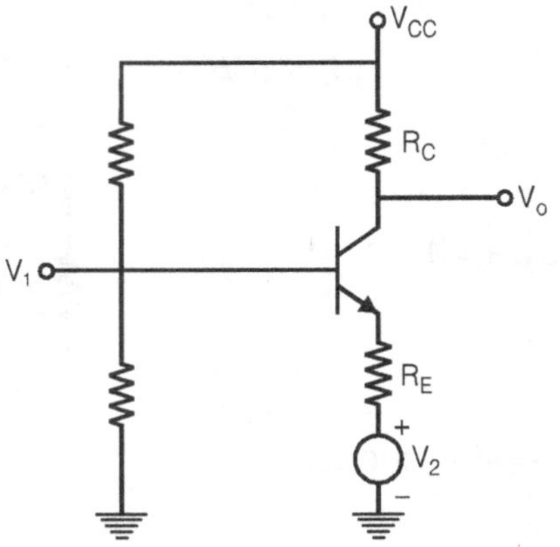


(a) $\frac{R_C}{R_E}$

(b) $\frac{R_E}{R_C}$

(c) $-\frac{R_C}{R_E}$

(d) $-\frac{R_E}{R_C}$

**Exp. (c)**

$$V_o = V_{CC} - I_C R_C = -I_C R_C \quad ; \quad V_1 - V_2 = I_C R_E$$

$$\frac{V_o}{V_1 - V_2} = -\frac{I_C R_C}{I_C R_E} = -\frac{R_C}{R_E}$$

**Q.29** A silicon transistor with $V_{BE\,sat} = 0.8$ V,

$\beta_{dc} = 100$ and $V_{CE\,sat} = 0.2$ V is used in the circuit shown below
What is the minimum value of $R_C$ for which transistor is in saturation?

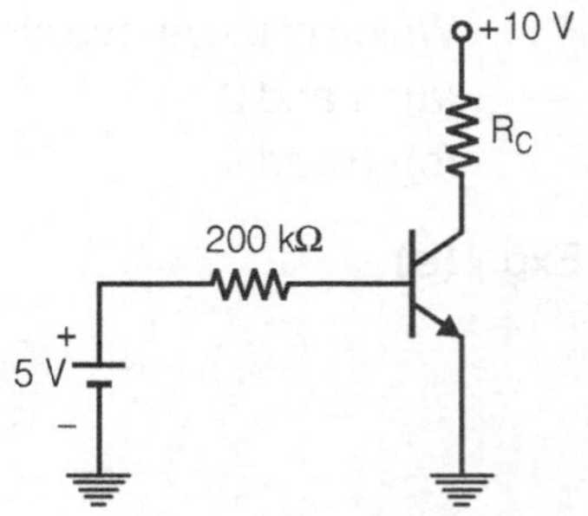


(a) 4286 Ω

(b) 4667 Ω

(c) 5000 Ω

(d) 1000 Ω

**Exp. (b)**

$$5 - I_B \times 200 \times 10^3 - 0.8 = 0$$

$$I_C = \beta I_B = 2.1 \text{ mA}$$

$$10 - I_C R_C = 0.2$$

$$R_C = \frac{9.8}{2.1 \times 10^{-3}} = 4667\ \Omega$$

**Q.30** If the transistor in the figure is in saturation, then

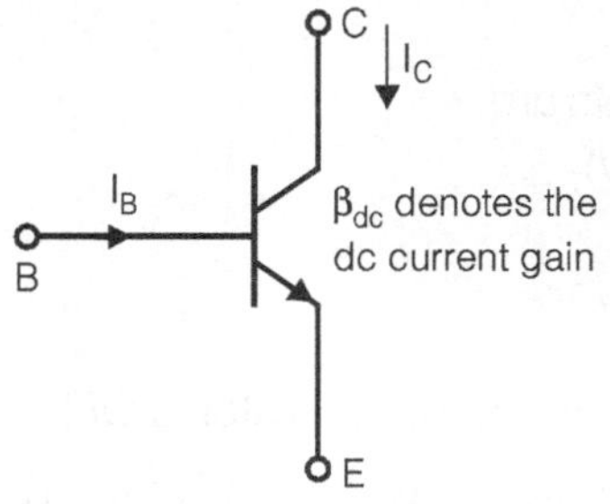


(a) $I_C$ is always equal to $\beta_{dc} I_B$

(b) $I_C$ is always equal to $-\beta_{dc} I_B$

(c) $I_C$ is greater than or equal to $\beta_{dc} I_B$

(d) $I_C$ is less than or equal to $\beta_{dc} I_B$

**Exp. (d)**

For the transistor to be in saturation

$$I_B > \frac{I_C}{\beta}$$

**Q.31** If for a silicon n-p-n transistor, the base-to-emitter voltage ($V_{BE}$) is 0.7 V and the collector to base voltage ($V_{CB}$) is 0.2 V, then the transistor is operating in the

(a) normal active mode
(b) saturation mode
(c) inverse active mode
(d) cutoff mode

**Exp. (a)**

$$V_{CE} = V_{BE} + V_{CB}$$
$$= 0.7 + 0.2$$
$$= 0.9\ V$$
$$(V_{CE})_{sat} = 0.2\ V$$

Therefore $V_{CE} > (V_{CE})_{sat}$

It is in normal active mode

## REVIEW QUESTIONS

1. What is the purpose of biasing a transistor? Indicate the methods of biasing and compare them.
2. Define the three different stability factors and give an expression that determine the variation of collector current in terms of stability factor.
3. Explain the effect of temperature on various parameters of a transistor amplifier circuit and discuss the methods of stabilization of the operating point.
4. Draw a fixed bias circuit and explain its drawbacks in detail.
5. Derive expressions for stability factors of a self bias circuit.
6. Explain how stability is improved in a self bias circuit.
7. Give a transistor circuit that employs a thermistor to compensate the variation of parameters with temperature and explain the circuit operation.
8. What is a sensistor? Explain how it can be employed in a transistor circuit to compensate the variation of operating point with temperature.
9. Distinguish between stabilization and compensation techniques. Discuss about compensation techniques in detail.
10. What do you understand by thermal run away? Derive the condition for avoiding thermal run-away in a transistor. Explain the term thermal resistance and derive the condition for thermal stability of BJT.

# CHAPTER 2

# SMALL SIGNAL ANALYSIS OF BJT

## About This Chapter

"If the signal applied is small, then it is possible to calculate the behaviour of transistor amplifier by mathematical means instead of graphical method".

In this chapter, we shall concentrate on

- Low frequency analysis
- High frequency analysis
- Frequency response of an amplifier

# 2 SMALL SIGNAL ANALYSIS OF BJT

## LOW FREQUENCY ANALYSIS OF BJT

## 2.1 Introduction

Assume in the figure (2.1), $J_E$ is forward bias and $J_C$ is reverse bias.
Where $J_E$ is emitter base junction and $J_C$ is collector base junction. The junctions of transistors possess capacitances. The impedance of a capacitance is inversely proportional to the frequency, $Z_c = \frac{1}{j\omega C} = \frac{1}{j2\pi fC}$.

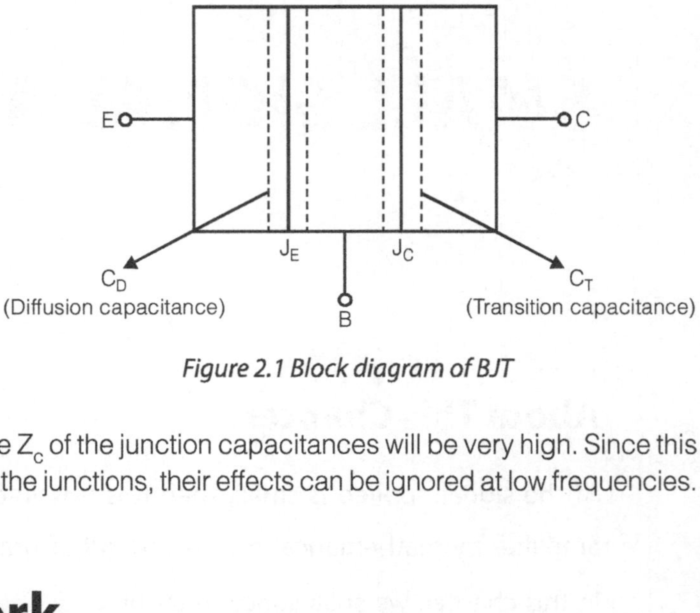

*Figure 2.1 Block diagram of BJT*

For low frequencies, the impedance $Z_c$ of the junction capacitances will be very high. Since this impedance appears in parallel with the junctions, their effects can be ignored at low frequencies.

## 2.2 Two Port Network

- In amplifiers, we bias the transistor at a desired operating point and allow small variations around this point by applying an ac signal.
- For this small variation, the behaviour of the transistor can be considered to be linear and therefore can be described by simpler equations.
- For circuit analysis, the transistor is considered to be a two port network. BJT is the best example of two port network.
- In h-parameter notation $V_1$ and $I_2$ are expressed as functions of $I_1$ and $V_2$ as follows:

$$V_1 = h_{11} I_1 + h_{12} V_2$$
$$I_2 = h_{21} I_1 + h_{22} V_2$$

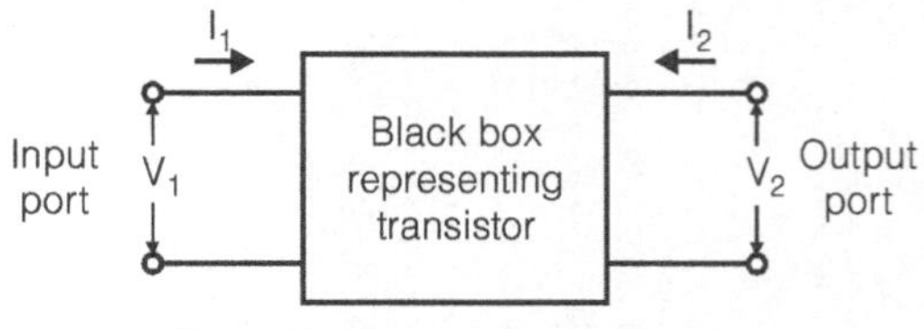

*Figure 2.2 BJT as a two port network*

## Case (a):

When output terminals are short circuited

$$\therefore \quad V_2 = 0$$

$$V_1 = h_{11} I_1$$

$$h_{11} = \frac{V_1}{I_1}$$

Therefore dimension of $h_{11}$ is voltage/current = ohms. Also, it represents input resistance when output is short circuited.

$$I_2 = h_{21} I_1$$

$$h_{21} = \frac{I_2}{I_1}$$

Therefore, $h_{21}$ (being ratio of two currents) is dimensionless. It represents (output current/input current) i.e. forward current gain with output short circuited.

## Case (b):

When input is open circuited

$$\therefore \quad I_1 = 0$$

$$V_1 = h_{12} V_2$$

$$h_{12} = \frac{V_1}{V_2}$$

Therefore, $h_{12}$ (being a ratio of two voltages) is dimensionless. It is reverse voltage gain

$$I_2 = h_{22} V_2$$

$$h_{22} = \frac{I_2}{V_2}$$

Therefore, $h_{22}$ has the dimension of mho (being a ratio of current/voltage). This represents output conductance, when input is open circuited.

The h-parameters in alphabetic notation

$$V_1 = h_i I_1 + h_r V_2$$

$$I_2 = h_f I_1 + h_o V_2$$

where, $h_i \rightarrow$ input resistance

$h_r \rightarrow$ reverse voltage gain

$h_f \rightarrow$ forward current gain

$h_o \rightarrow$ Output conductance

**Reasons for use of h-parameters:**

- They can be obtained from characteristic of transistor.
- h-parameters are easy to measure.
- They are more convenient to use in circuit analysis.

## 2.3 Hybrid Model of BJT

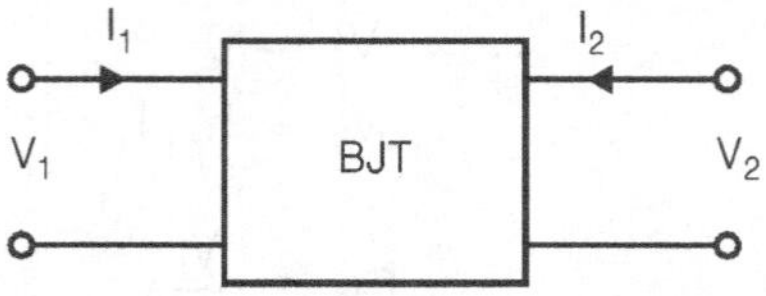


Replacing the rectangular box of BJT by Hybrid model

In case of transistors, another subscript (b, e or c) is added to designate the type of configuration. For example,

$h_{ic}$ = input resistance in CC configuration

$h_{fe}$ = short circuit forward current gain in CE configuration

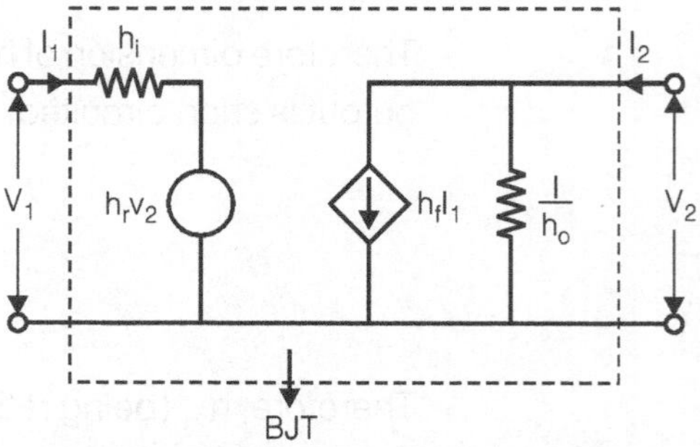


*Figure 2.3 H parameter model of BJT*

## 2.4 Hybrid Model in Different Configurations

Common emitter (CE)

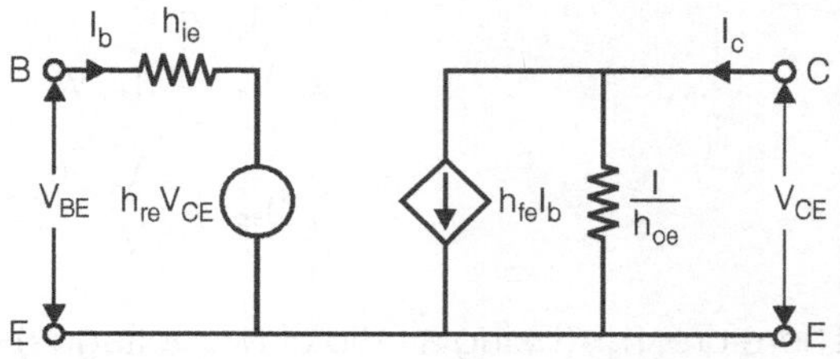


*Fig. 2.4 H parameter model for common emitter*

Common collector (CC)

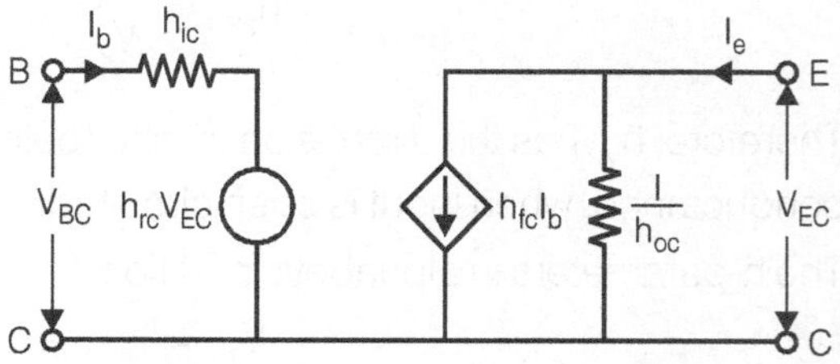


*Fig. 2.5 H parameter model for common collector*

Common Base (CB)

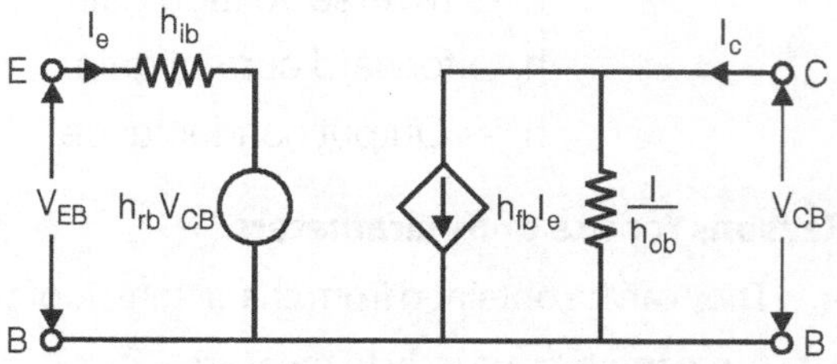


*Fig. 2.6 H parameter model for common base*

## 2.5 Typical Values of H-Parameters

| S.No. | H-Parameter | CE | CC | CB |
|---|---|---|---|---|
| 1. | $h_i$ | $1100\,\Omega$ | $1100\,\Omega$ | $22\,\Omega$ |
| 2. | $h_f$ | 50 | –51 | –0.98 |
| 3. | $h_r$ | $2.4\times10^{-4}$ | $\approx 1$ | $2.9\times10^{-4}$ |
| 4. | $h_o$ | $24\times10^{-6}$ A/V | $25\times10^{-6}$ A/V | $0.49\times10^{-6}$ A/V |

## 2.6 Conversion Formulas of H Parameters in Different Configuration

**CE to CB**

$$h_{ib} = \frac{h_{ie}}{1+h_{fe}} \; ; \; h_{fb} = -\frac{h_{fe}}{1+h_{fe}} \; ; \; h_{rb} = \frac{h_{ie}\cdot h_{oe}}{1+h_{fe}} - h_{re} \; ; \; h_{ob} = \frac{h_{oe}}{1+h_{fe}}$$

**CB to CE**

$$h_{ie} = \frac{h_{ib}}{1+h_{fb}} \; ; \; h_{fe} = -\frac{h_{fb}}{1+h_{fb}} \; ; \; h_{re} = \frac{h_{ib}\,h_{ob}}{1+h_{fb}} - h_{rb} \; ; \; h_{oe} = \frac{h_{ob}}{1+h_{fb}}$$

**CE to CC**

$$h_{ic} = h_{ie} \; ; \; h_{fc} = -(1+h_{fe}) \; ; \; h_{rc} = 1 \; ; \; h_{oc} = h_{oe}$$

## 2.7 Transistor Amplifier Analysis Using Exact Model

The characteristics of amplifiers are:

(i) Current gain $A_I$

(ii) Input impedance $Z_i$

(iii) Voltage gain $A_V$

(iv) Output impedance $Z_0$

(v) Voltage amplification $A_{VS}$

(vi) Current amplification $A_{IS}$

(vii) Effective input impedance $Z_i'$

(viii) Effective output impedance $Z_0'$

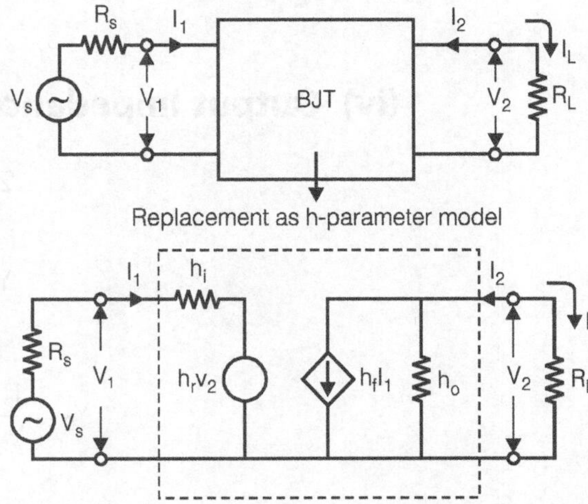


*Figure 2.7 Amplifier model with H parameters*

### (i) Current Gain $A_I$

$$A_I = \frac{I_L}{I_1} = -\frac{I_2}{I_1}$$

From h-parameters

$$I_2 = h_f I_1 + h_o V_2$$
$$V_2 = -I_2 R_L$$
$$I_2 = h_f I_1 - h_o I_2 R_L$$
$$I_2(1 + h_o R_L) = h_f I_1$$

$$\boxed{A_I = -\frac{I_2}{I_1} = \frac{-h_f}{1 + h_o R_L}}$$

### (ii) Input Impedance $Z_i$

$$Z_i = \frac{V_1}{I_1}$$
$$V_1 = h_i I_1 + h_r V_2$$
$$Z_i = \frac{V_1}{I_1} = h_i + h_r \frac{V_2}{I_1}$$
$$= h_i + h_r\left(\frac{-I_2}{I_1}\right)R_L = h_i + h_r A_I R_L$$

$$\boxed{Z_i = h_i - \frac{h_r h_f}{1 + h_o R_L} R_L}$$

### (iii) Voltage Gain $A_V$

$$A_V = \frac{V_2}{V_1} = -\frac{I_2 R_L}{V_1} = \frac{-\frac{I_2}{I_1} \cdot R_L}{V_1/I_1} = \frac{A_I R_L}{Z_i}$$

$$\boxed{A_V = \frac{A_I R_L}{Z_i}}$$

### (iv) Output Impedance ($Z_0$)

$$Z_0 = \frac{V_2}{I_2}$$
$$Y_0 = \frac{I_2}{V_2}$$
$$I_2 = h_f I_1 + h_0 V_2$$
$$Y_0 = \frac{I_2}{V_2} = \frac{h_f I_1 + h_0 V_2}{V_2}$$
$$Y_0 = h_f\left(\frac{I_1}{V_2}\right) + h_0$$

To calculate the output impedance:

(i) The output load $R_L$ should be open.

(ii) If the input is a voltage source, make it as short circuit, otherwise open circuit if it is a current source.

Taking input loop,

$$V_s = I_1(h_i + R_s) + h_r V_2$$

$$0 = I_1(h_i + R_s) + h_r V_2$$

$$\frac{I_1}{V_2} = \frac{-h_r}{R_s + h_i}$$

$$Y_o = \frac{-h_f h_r}{h_i + R_s} + h_o$$

$$Z_0 = \frac{1}{h_0 - \dfrac{h_f h_r}{h_i + R_s}}$$

## (v) Voltage Amplification

$$A_{VS} = \frac{V_2}{V_s} = \frac{V_2}{V_1} \cdot \frac{V_1}{V_s} = A_V \cdot \frac{V_1}{V_s}$$

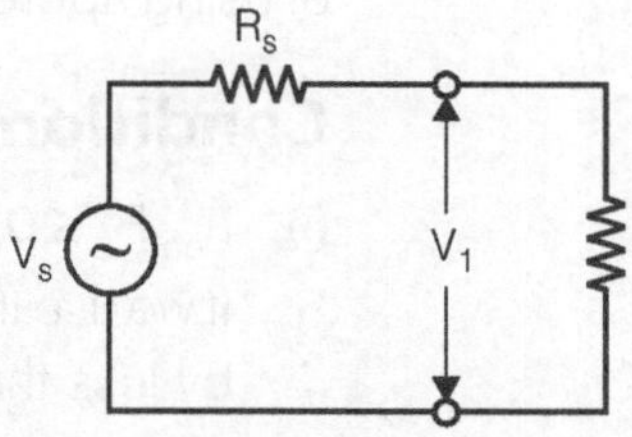


$$V_1 = \frac{R_i V_s}{R_i + R_s}$$

$$\frac{V_1}{V_s} = \frac{R_i}{R_i + R_s}$$

$$A_{VS} = A_V \frac{R_i}{R_i + R_s}$$

## (vi) Current Amplification

$$A_{IS} = \frac{I_L}{I_s} = \frac{I_L}{I_1} \cdot \frac{I_1}{I_s} = A_I \cdot \frac{I_1}{I_s}$$

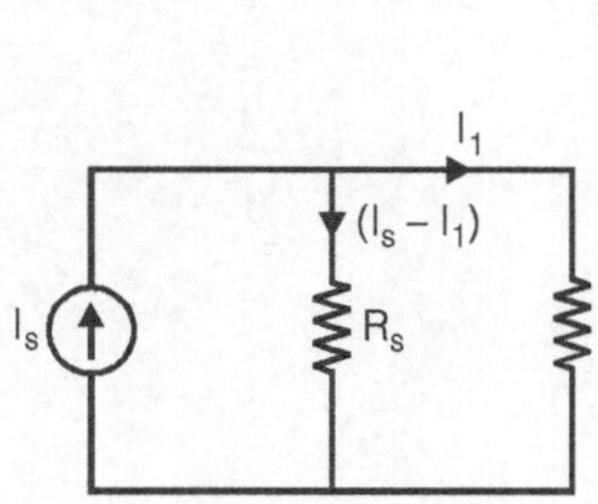


$$(I_s - I_1)R_s = I_1 R_i$$

$$I_s R_s = I_1(R_i + R_s)$$

$$\frac{I_1}{I_s} = \frac{R_s}{R_i + R_s}$$

$$A_{IS} = A_I \cdot \frac{R_s}{R_i + R_s}$$

### (vii) Effective Input and Output Impedance

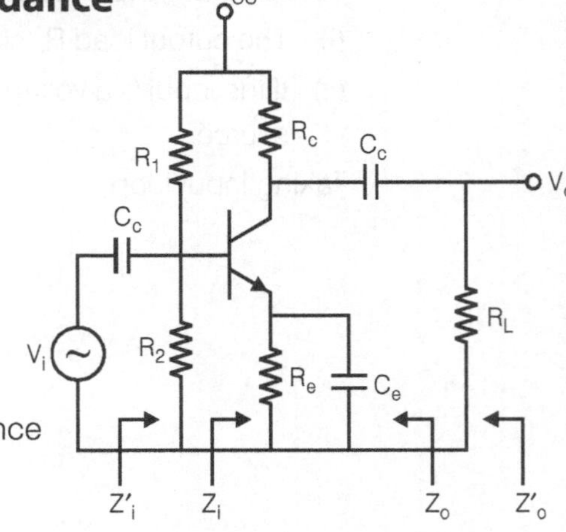


$$Z_i = h_{ie}$$

$$Z_i' = h_{ie} || R_1 || R_2 \rightarrow \text{Effective input impedance}$$

$$Z_0 = \frac{1}{h_{oe}}$$

$$Z_0' = \frac{1}{h_{oe}} || R_C || R_L \rightarrow \text{Effective output impedance}$$

## 2.8 Transistor Amplifier Analysis Using Approximate Model

In most practical cases it is justifiable to obtain approximate values of $A_I$, $A_V$, $R_i$ and $R_o$. The two of the four h-parameters, $h_{ie}$ and $h_{fe}$ are sufficient for approximate analysis. The main advantage of using approximate model is the ease of calculation.

*Simplified model of H parameter*

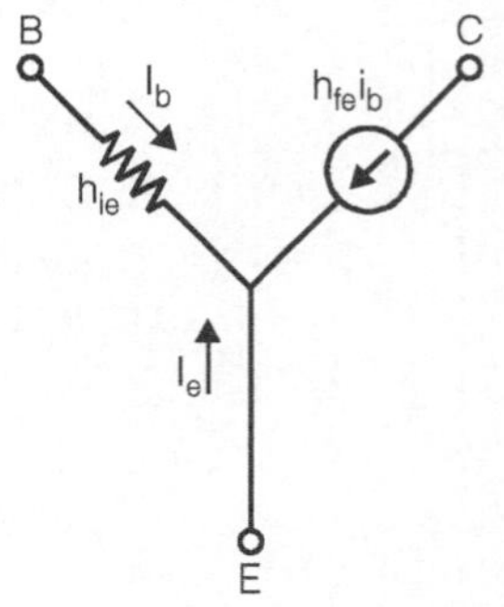


### Conditions

(i) $h_{oe} R_L \leq 0.1$

If we use this condition, the error will decrease by less than 10%.

(ii) $h_{re}$ and $h_{oe}$ are neglected

(iii) It is valid for common emitter mode only.

### CE Unbypass Amplifier

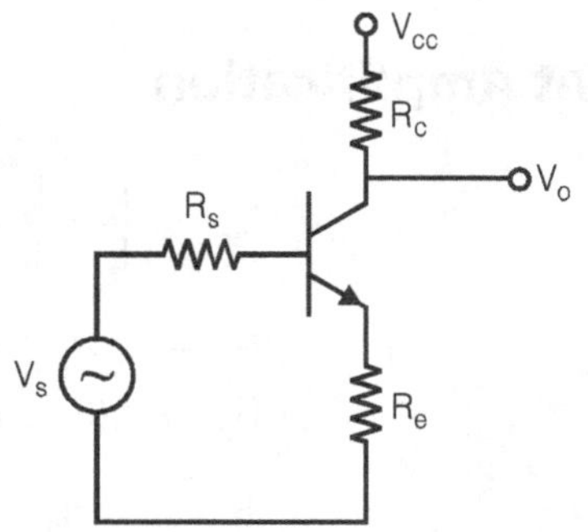


*Fig. 2.8 Common emitter amplifier with unbypass*

The approximate model is given by

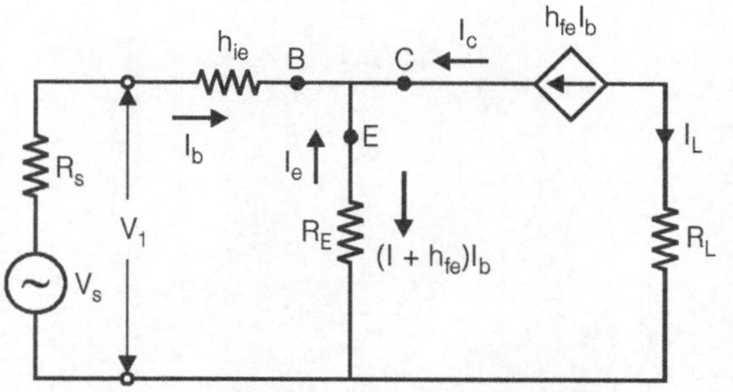


(i) $A_I = \frac{I_L}{I_b} = \frac{-I_c}{I_b} = \frac{-h_{fe} I_b}{I_b} = -h_{fe}$

(ii) $Z_i = \frac{V_1}{I_b} = \frac{h_{ie} I_b + (1+h_{fe}) I_b R_E}{I_b}$

$$Z_i = h_{ie} + (1 + h_{fe})R_E$$

(iii) $A_V = \frac{A_I R_L}{Z_i} = \frac{-h_{fe} R_L}{h_{ie} + (1+h_{fe})R_E}$

(iv) $Z_0 = \frac{V_2}{I_C} = \frac{V_2}{0} = \infty$

## CE Bypass Amplifier

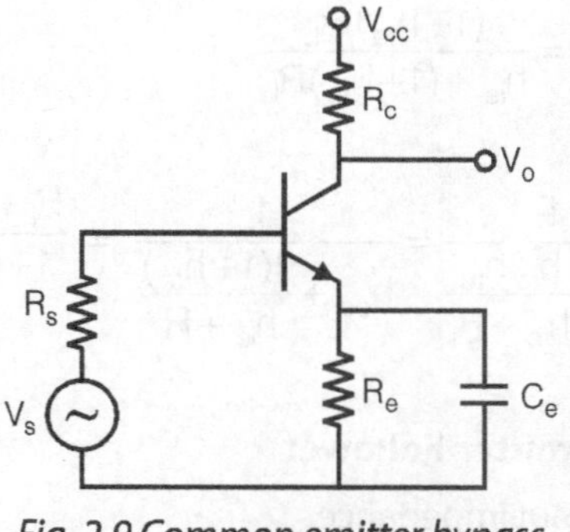


*Fig. 2.9 Common emitter bypass*

The approximate model is given by

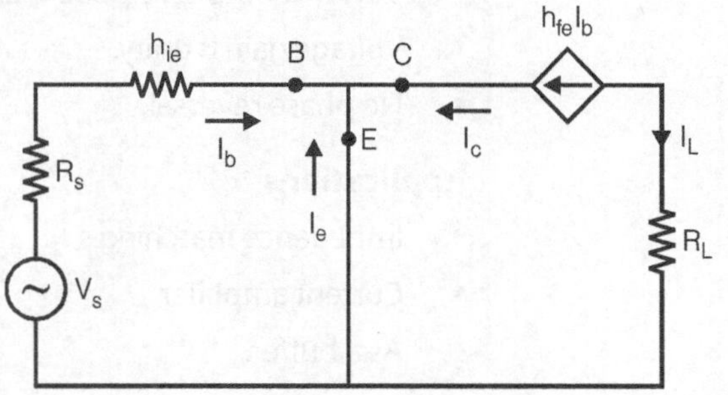


(i) $A_I = -h_{fe}$

(ii) $Z_i = h_{ie}$

(iii) $A_V = \frac{-h_{fe} R_L}{h_{ie}}$

(iv) $Z_o = \infty$

## Conclusions

- In CE unbypass amplifier voltage gain $A_V$ is less and input impedance is high.
- In CE bypass amplifier voltage gain $A_V$ is more and input impedance is less.

## Common Collector Amplifier

Common collector is also called as emitter follower. As the voltage gain is close to unity, any change in base voltage appears across the load at the emitter. Thus emitter follows the potential variations at the base and hence called as emitter follower.

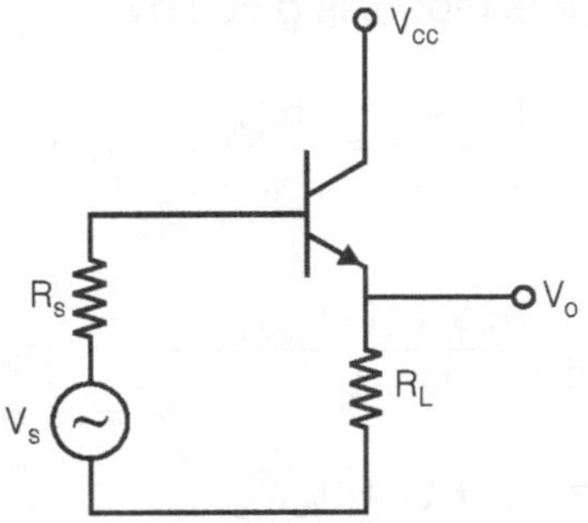


*Fig. 2.10 Common collector amplifier*

The approximate model is given by

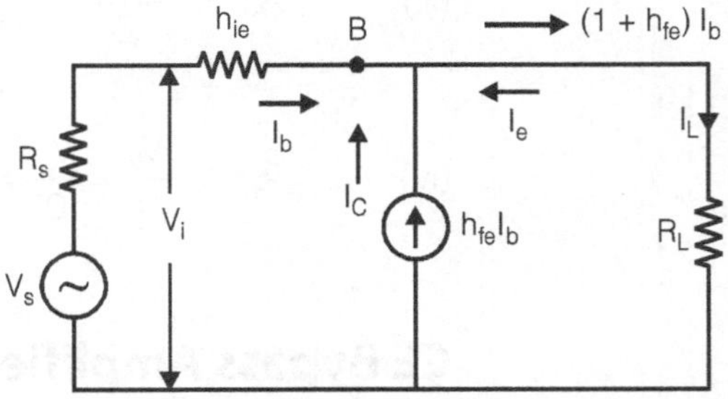


(i) $A_I = \frac{I_L}{I_b} = \frac{-I_e}{I_b} = \frac{(1+h_{fe})I_b}{I_b} = (1+h_{fe})$

(ii) $Z_i = \frac{V_i}{I_b} = \frac{h_{ie}I_b + (1+h_{fe})I_b R_L}{I_b} = h_{ie} + (1+h_{fe})R_L$

(iii) $A_V = \frac{A_I R_L}{Z_i} = \frac{(1+h_{fe})R_L}{h_{ie} + (1+h_{fe})R_L}$

(iv) $Z_0 = \frac{1}{h_{oc} - \frac{h_{rc}h_{fc}}{h_{ic}+R_i}} = \frac{1}{h_{oe} + \frac{1(1+h_{fe})}{h_{ie}+R_i}} = \frac{R_i + h_{ie}}{1+h_{fe}}$

**Properties of Emitter Follower**

- Very high input impedance.
- Very low output impedance.
- Voltage gain is unity.
- No phase reversal.

**Applications**

- Impedance matching.
- Current amplifier.
- As a buffer.

## Common Base Amplifier

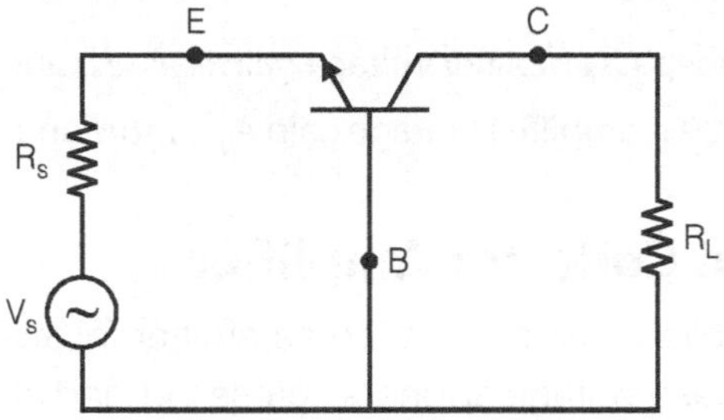


*Fig. 2.11 Common base amplifier*

The approximate model is given by

(i) $A_I = \frac{I_L}{I_e} = -\frac{I_C}{I_e} = -\frac{h_{fe} I_b}{-(1+h_{fe})I_b} = \frac{h_{fe}}{1+h_{fe}}$

(ii) $R_i = \frac{h_{ie}}{1+h_{fe}}$

(iii) $A_V = \frac{h_{fe} R_L}{h_{ie}}$

(iv) $Z_0 = \infty$

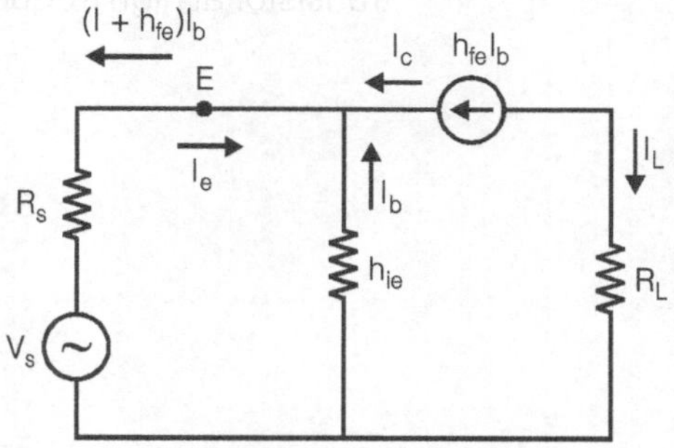


**Conclusions**

- CB is used when we want ideal current source at the output.
- CB is widely used in wideband amplification.

# HIGH FREQUENCY ANALYSIS OF BJT

## 2.9 Introduction

If the transistor is used for high frequency application, the time taken is comparable with the time period of the signal. Therefore, the h-parameter model discussed earlier is valid only for low frequencies. To study the high frequency behaviour, various junction capacitances have to be taken into account. The hybrid-π model is used for this purpose.

## 2.10 Hybrid-π Model

Assume a transistor block diagram as shown below:

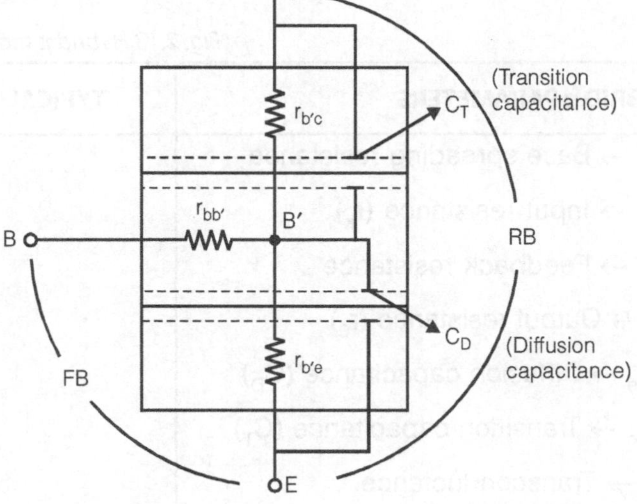

The hybrid-π equivalent circuit shown in Figure (2.12) is used to analyse the behaviour of a transistor at high frequencies.

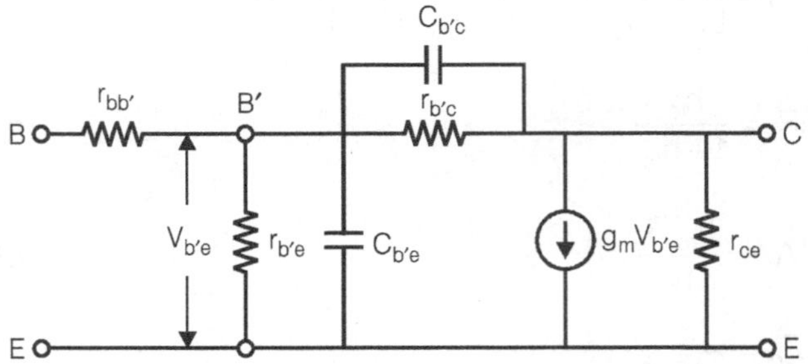


*Fig. 2.12 Hybrid π model*

The internal node B′ is imaginary and it is not physically accessible. $r_{bb'}$ is called as base spreading resistance between B and B′. The change in the collector current is proportional to change in $V_{b'e}$. The constant of proportionality is called transconductance ($g_m$). The physical significance of each component in the hybrid-π model is discussed below.

## 2.11 Typical Values of Hybrid-π Parameters

Hybrid-π model can also be represented with BJT symbol as shown in the Figure (2.13).

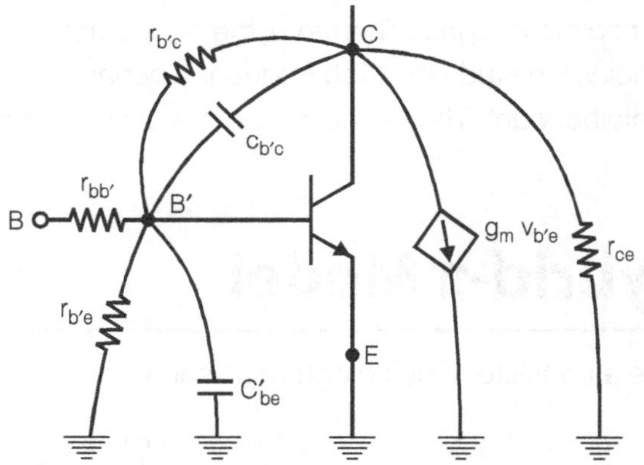


*Fig. 2.13 Hybrid π model*

| HYBRID π PARAMETERS | TYPICAL VALUES |
|---|---|
| $r_{bb}'$ → Base spreading resistance | 100 Ω |
| $r_{b'e}$ → Input resistance ($r_\pi$) | 1 kΩ |
| $r_{b'c}$ → Feedback resistance | 4 MΩ |
| $r_{ce}$ → Output resistance ($r_0$) | 80 kΩ |
| $C_{b'e}$ → Diffusion capacitance ($C_D$) | 100 pf |
| $C_{b'c}$ → Transition capacitance ($C_T$) | 3 pf |
| $g_m$ → Transconductance | 50 mA/V |

# 2.12 Expressions for Hybrid-π Parameters

## (a) Transconductance $g_m$

Transconductance is the ratio of change in collector current due to a small change in base-emitter voltage of a transistor.

$$\therefore \quad g_m = \frac{\partial I_c}{\partial V_{B'E}}$$

$$(\text{Assume } V_{B'E} = V_E)$$

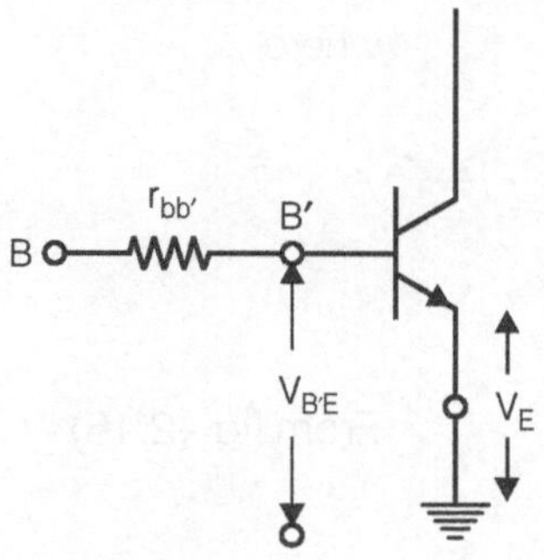


*Fig. 2.14 BJT symbol representation*

We know

$$I_c = \alpha I_E$$

$$\partial I_c = \alpha \partial I_E$$

$$g_m = \alpha \frac{\partial I_E}{\partial V_E} = \frac{\alpha I_E}{V_T}$$

$$\left(\text{where } \frac{\partial I_E}{\partial V_E} = \frac{I_E}{V_T}\right)$$

$$\boxed{g_m = \frac{I_c}{V_T}}$$

where $$g_m = \frac{I_c}{T} \times 11{,}600$$

## (b) Input Resistance $r_{b'e}$

To convert high frequency model into low frequency the conditions are:

1. $C_{b'e}$ should be open
2. $C_{b'c}$ should be open

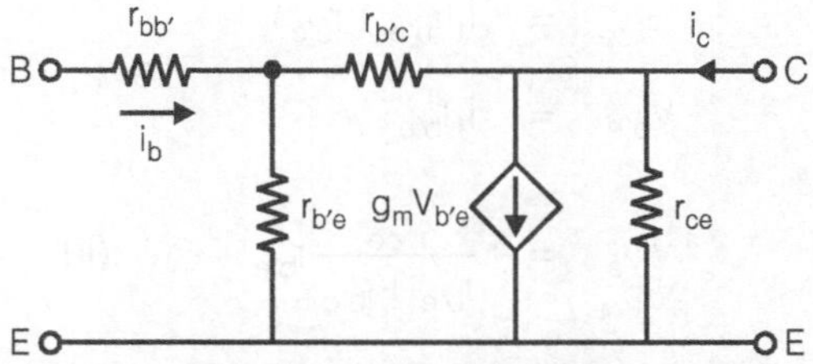


*Fig. 2.15 Hybrid π model in low frequency*

Actual low frequency model

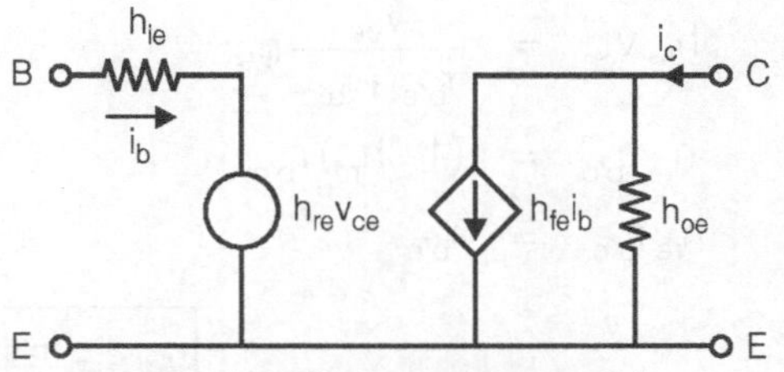


*Fig. 2.16 H parameter model*

To get the input resistance $r_{b'e}$, output voltage $V_{CE}$ should be neglected ($V_{CE}$ should be short circuit)

Assuming fig. (2.15)

As $V_{CE}$ is short circuit

$$i_c = g_m V_{b'e}$$

where $$V_{b'e} = i_b r_{b'e}$$

$$i_c = g_m i_b r_{b'e}$$

$$\frac{i_c}{i_b} = g_m r_{b'e} \quad \text{...(i)}$$

From fig. (2.16)

$$i_c = h_{fe} i_b$$

$$\frac{i_c}{i_b} = h_{fe} \quad \text{...(ii)}$$

From (i) and (ii) $$h_{fe} = g_m r_{b'e}$$

$$\boxed{r_{b'e} = \frac{h_{fe}}{g_m}}$$

## (c) Feedback Resistance $r_{b'c}$

To get the feedback resistance $r_{b'c}$ input parameter $i_b$ should be zero

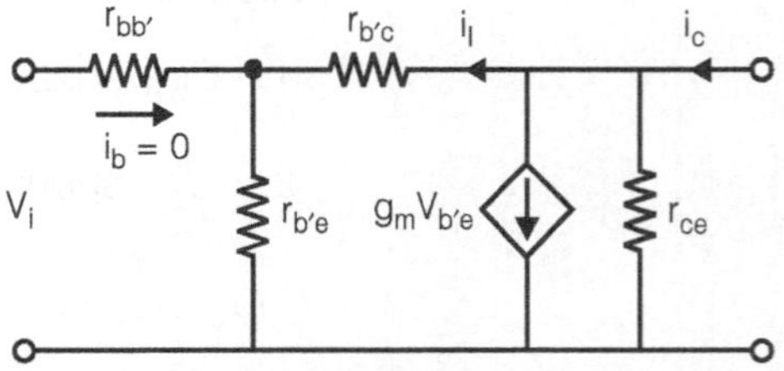


*Fig. 2.17 Hybrid π model*

From h parameter model

$$V_i = h_{re} V_{ce} \quad \text{...(i)}$$

According to the above figure

$$V_{ce} = i_1(r_{b'e} + r_{b'c})$$

$$V_{b'e} = i_1 r_{b'e}$$

$$V_{b'e} = \frac{V_{ce}}{r_{b'e} + r_{b'c}} r_{b'e} \quad \text{...(ii)}$$

we have $$V_i = V_{b'e}$$

From (i) and (ii)

$$h_{re} V_{ce} = \frac{V_{ce}}{r_{b'e} + r_{b'c}} r_{b'e}$$

$$h_{re} r_{b'c} = (1 - h_{re}) r_{b'e}$$

here $$h_{re} r_{b'c} = r_{b'e}$$

$$\boxed{r_{b'c} = \frac{r_{b'e}}{h_{re}}}$$

## (d) Base Spreading Resistance $r_{bb'}$

It is the input parameter, so $V_{CE} = 0$

The equivalent model of hybrid $\pi$ in low frequency analysis by applying condition $V_{CE} = 0$.

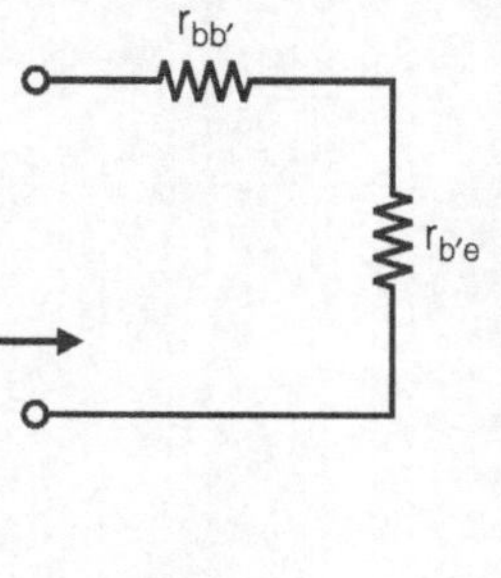


$$z_i = r_{bb'} + r_{b'e} \quad \text{...(i)}$$

According to h-parameter model

$$z_i = h_{ie} \quad \text{...(ii)}$$

According to (i) and (ii)

$$h_{ie} = r_{bb'} + r_{b'e}$$

$$\boxed{r_{bb'} = h_{ie} - r_{b'e}}$$

## (e) Output resistance $r_{ce}$

It is output parameter, so $I_b = 0$

According to hybrid $\pi$ model

$$I_c = \frac{V_{CE}}{r_{ce}} + g_m V_{b'e} + \frac{V_{CE}}{r_{b'e} + r_{b'c}}$$

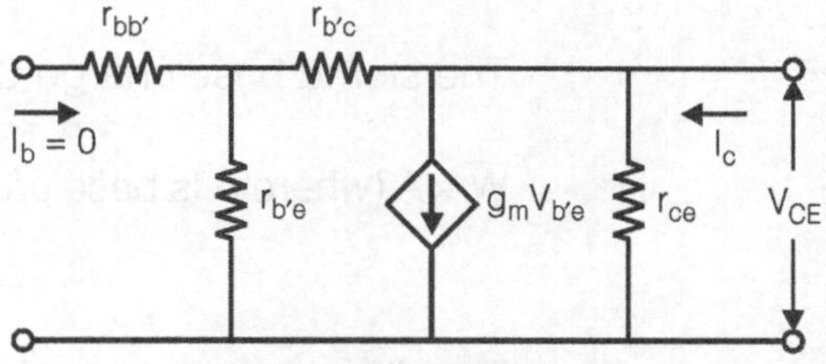


*Fig. 2.18 Hybrid $\pi$ model*

According to h-parameter model

$$h_{oe} = \frac{I_c}{V_{CE}}$$

$$I_c = \frac{V_{CE}}{r_{ce}} + \frac{g_m V_{CE} r_{b'e}}{r_{b'e} + r_{b'c}} + \frac{V_{CE}}{r_{b'e} + r_{b'c}}$$

$$\left(\text{where} \quad V_{b'e} = \frac{V_{CE} r_{b'e}}{r_{b'e} + r_{b'c}}\right)$$

$$\frac{I_c}{V_{CE}} = \frac{1}{r_{ce}} + \frac{h_{fe}}{r_{b'e} + r_{b'c}} + \frac{1}{r_{b'e} + r_{b'c}}$$

$$h_{oe} = \frac{1}{r_{ce}} + \frac{h_{fe} + 1}{r_{b'e} + r_{b'c}} \quad \text{where } (h_{fe} >> 1 \,\&\, r_{b'c} >> r_{b'e})$$

$$h_{oe} = g_{ce} + h_{fe}\, g_{b'c}$$

$$\boxed{g_{ce} = h_{oe} - h_{fe}\, g_{b'c}}$$

$$r_{ce} = \frac{1}{g_{ce}}$$

## (f) Capacitive Effect in Hybrid $\pi$ Model

$$C_T = C_e + C_c$$

$C_e$ ↓ Diffusion capacitance; $C_c$ → Transition capacitance

where $C_e >> C_c$

## Expression for $C_{b'e}$

Assume carrier concentration versus distance graph for a pnp transistor shown below:

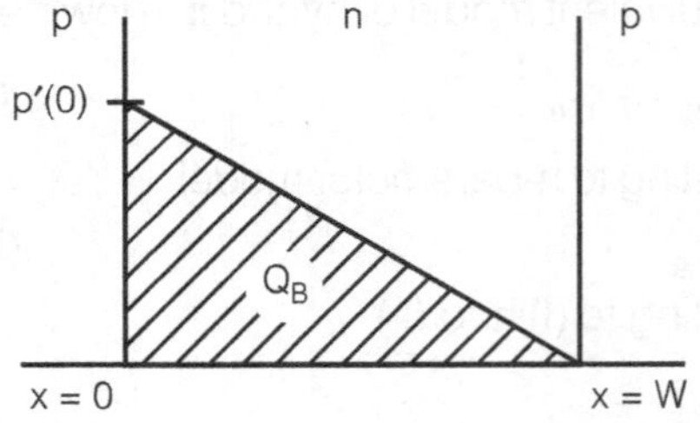


*Fig. 2.19 Carrier concentration versus distance*

**$Q_B$ is called as stored base charge**

If $W << L_B$, then p varies almost linearly from the value P'(0) at the emitter to zero at the collector.

The stored base charge $Q_B$ is the average concentration $\frac{P'(0)}{2}$ times the volume of the base $W \times A$ (where A is base cross-sectional area) times the electronic charge e; that is

$$Q_B = \frac{1}{2} P'(0)\, A W e \quad \text{...(i)}$$

The diffusion current is given by

$$I = -AeD_B \frac{dp}{dx} = AeD_B \frac{P'(0)}{W} \quad \text{...(ii)}$$

**Combining (i) and (ii)**

$$Q_B = \frac{IW^2}{2D_B}$$

The emitter diffusion capacitance $C_{b'e}$ is given by the rate of change of $Q_B$ with respect to emitter voltage V i.e.

$$C_{b'e} = \frac{\partial Q_B}{\partial V} = \frac{W^2}{2D_B} \frac{dI}{dV} = \frac{W^2}{2D_B} \frac{1}{r_e}$$

where $r_e = \frac{dV}{dI}$. The emitter junction incremental resistance.

We know that,

$$r_e = \frac{V_T}{I_E}$$

where $V_T$ is the volt equivalent of temperature and is defined as

$$V_T = \frac{T}{11,600}$$

Substituting $$r_e = \frac{V_T}{I_E}$$

$$\boxed{C_{b'e} = \frac{W^2}{2D_B} \frac{I_E}{V_T}}$$

The above equation indicates that the diffusion capacitance is proportional to the emitter bias current $I_E$. Since $D_B$ is inversely proportional to T, and $V_T$ is proportional to T, the diffusion capacitance $C_{b'e}$ is almost independent of temperature.

**Dependence of $f_\alpha$ upon Base Width or Transit time:**

$$f_\alpha = \frac{1}{2\pi r_e C_{b'e}} = \frac{D_B}{\pi W^2}$$

where,
$$C_{b'e} = \frac{W^2}{2D_B} \cdot \frac{1}{r_e}$$

or
$$\omega_\alpha = 2\pi f_\alpha = \frac{2D_B}{W^2}$$

The above equation indicates that the alpha cutoff frequency varies inversely as the square of the base width W.

We have,
$$I = Q_B \omega_\alpha \quad \text{...(i)}$$

$t_B$ is the base transit time, defined as the time taken by the charge carriers to cross the base region, that means base transit time $t_B$ an amount of charge reaching the collector is equal to base charge $Q_B$.

The resulting current is given by

$$I = \frac{Q_B}{t_B} \quad \text{...(ii)}$$

From (i) and (ii)
$$\boxed{\omega_\alpha = \frac{1}{t_B}}$$

The above equation gives relation between the alpha cutoff frequency and transit time. It indicates that the alpha cutoff frequency is proportional to the reciprocal of the transit time $t_B$.

**Note:**

**If $V_{CE}$ increases what happens to diffusion capacitance?**

**Reason:**

As $V_{CE}$ increases, the physical width of base reduces, therefore $C_D$ decreases

As $V_{CE}\uparrow$, $W_B\downarrow$, $C_D\downarrow$

## Expression for $C_{b'c}$

$$C_{b'c} \propto \frac{1}{(V_{CE})^n}$$

where
$$n = \frac{1}{2} \text{ for alloy type}$$

$$n = \frac{1}{3} \text{ for grown junction type}$$

# 2.13 Variation of Hybrid π Parameters With $I_c$, $V_{CE}$ and Temperature

1. $$g_m = \frac{I_c}{V_T}$$

where, $$g_m \propto I_c$$

$$g_m \propto \frac{1}{T}$$

2. $$r_{b'e} = \frac{h_{fe}}{g_m}$$

$$r_{b'e} \propto T$$

$$r_{b'e} \propto \frac{1}{I_c}$$

3. $C_{b'e}$ (Diffusion capacitance)

   **As $V_{CE}$ increases, width of the base decreases (early effect). Therefore $C_{b'e}$ decreases.**

4. $C_{b'c}$ (Transition capacitance)

$$C_{b'c} \propto \frac{1}{(V_{CE})^n}$$

   **As $V_{CE}$ increases $c_{b'c}$ decreases**

   $C_{b'c}$ is independent of $I_c$ and T

5. $r_{bb'}$

   As $I_c$ increases, $r_{bb'}$ decreases

   Temperature increases, $r_{bb'}$ increases.

# 2.14 Common Emitter Short Circuit Current Gain

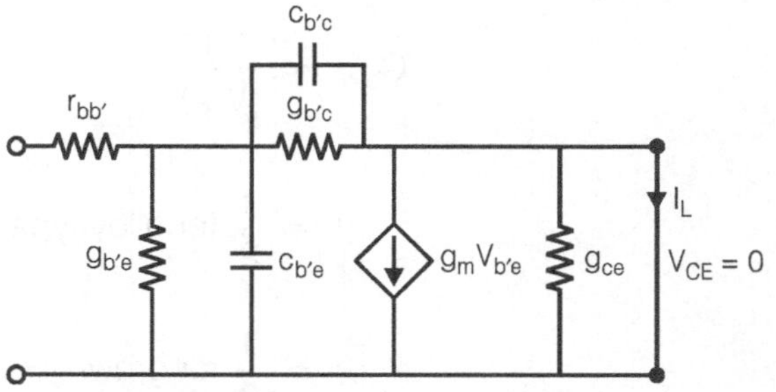


*Fig. 2.20 Common emitter short circuit current gain*

Equivalent circuit is given by

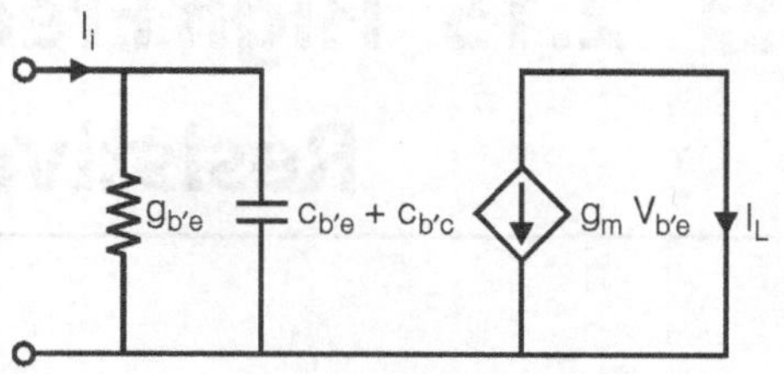


Assume

$r_{bb'} \to 0$

$g_{b'c} \to \infty$

Current gain $\quad A_I = \dfrac{I_L}{I_i}$

$$I_L = -g_m V_{b'e}$$

$$I_i = V_{b'e}(g_{b'e} + j\omega(C_{b'e} + C_{b'c}))$$

where $(C_{b'e} >> C_{b'c})$

$$A_I = \frac{-g_m}{g_{b'e} + j\omega C_{b'e}} = \frac{-g_m}{\frac{g_m}{h_{fe}} + j2\pi f \frac{g_m}{2\pi f_T}}$$

$$\therefore\ C_{b'e} = \frac{g_m}{2\pi f_T}$$

$$A_I = \frac{-g_m}{\frac{g_m}{h_{fe}}\left(1 + j\frac{2\pi f}{2\pi}\frac{h_{fe}}{f_T}\right)} = \frac{-h_{fe}}{1 + j\frac{f}{\left(\frac{f_T}{h_{fe}}\right)}}$$

where $\quad f_\beta = \dfrac{f_T}{h_{fe}}$

$$|A_I| = \frac{h_{fe}}{\sqrt{1 + \left(\frac{f}{f_\beta}\right)^2}}$$

when $f = f_\beta$

$$|A_I| = \frac{h_{fe}}{\sqrt{2}}$$

$$1 = \frac{h_{fe}}{\sqrt{1 + \left(\frac{f_T}{f_\beta}\right)^2}}$$

$$\frac{h_{fe}}{\left(\frac{f_T}{f_\beta}\right)} = 1$$

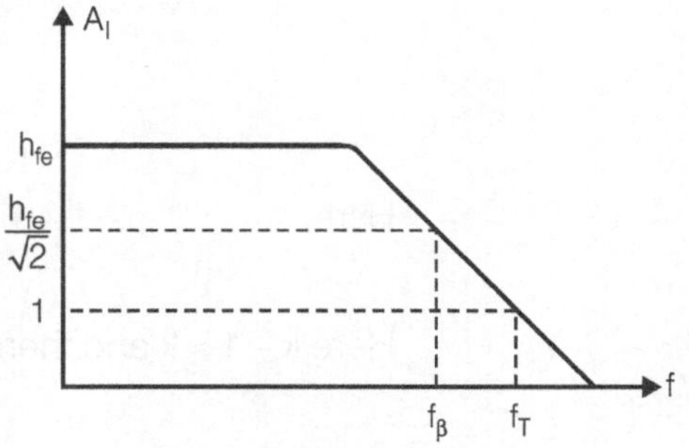


*Fig. 2.21 Frequency response*

$$f_T = h_{fe}\, f_\beta$$

where

**$f_\beta$ (β cutoff frequency):**

"It is the frequency at which the short circuit common emitter current gain reduces by 3 dB less".

**$f_T$ (Unit gain bandwidth frequency):**

"It is the frequency at which the gain reaches to unity."

**$f_\alpha$ (α cutoff frequency):**

"It is the frequency at which the short circuit common base current gain reduces by 3 dB of its value.

$$f_\alpha = (1+\beta)\, f_\beta = \left(\frac{1}{1-\alpha}\right) f_\beta$$

# 2.15 High Frequency Current Gain with Resistive Load

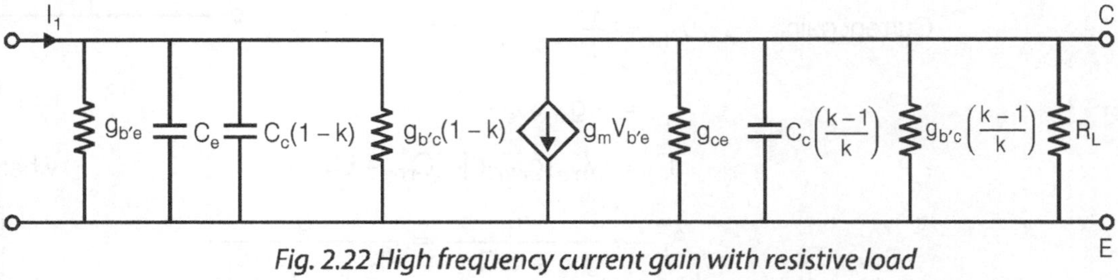


*Fig. 2.22 High frequency current gain with resistive load*

Approximate equivalent circuit for calculation of response of a transistor amplifier stage with a resistive load.

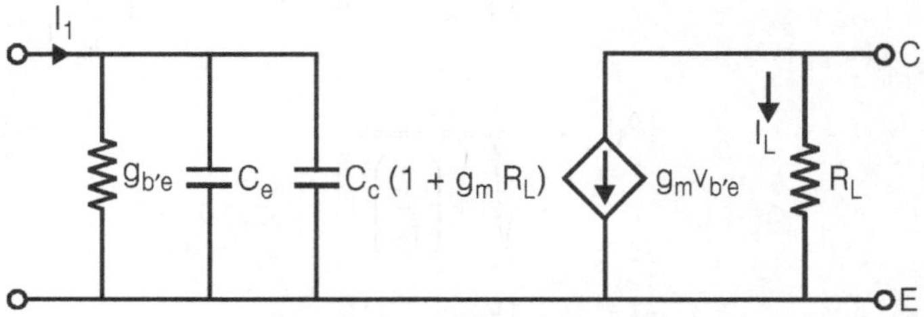


*Fig. 2.23 Simplified circuit*

Here,

$$k = \frac{V_{ce}}{V_{b'e}} = -g_m R_L$$

Here $k - 1 \approx k$ and therefore

$$g_{b'c}\left(\frac{k-1}{k}\right) = g_{b'c}$$

$g_{b'c}$ and $g_{ce}$ are neglected with comparison to $R_L$.

Here 3 dB frequency is now $f_2$ (rather than $f_\beta$), where

$$f_2 = \frac{1}{2\pi r_{b'e} C} = \frac{g_{b'e}}{2\pi C}$$

where

$$C = C_e + C_c (1 + g_m R_L)$$

# FREQUENCY RESPONSE OF AN AMPLIFIER

Assume CE amplifier circuit

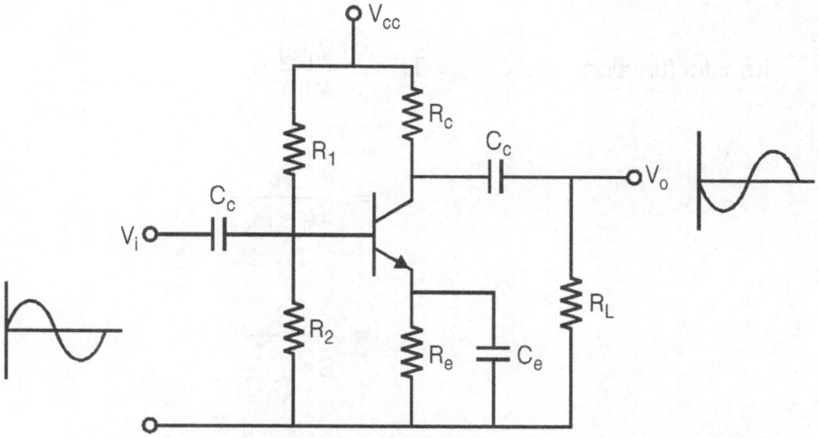


*Fig. 2.24 Common emitter amplifier*

The frequency response of the above circuit is

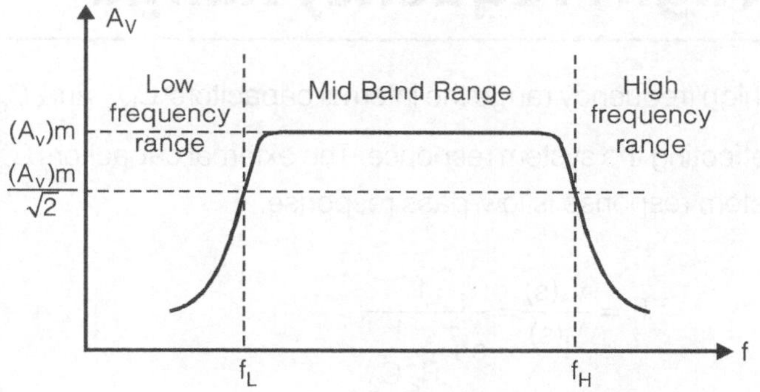


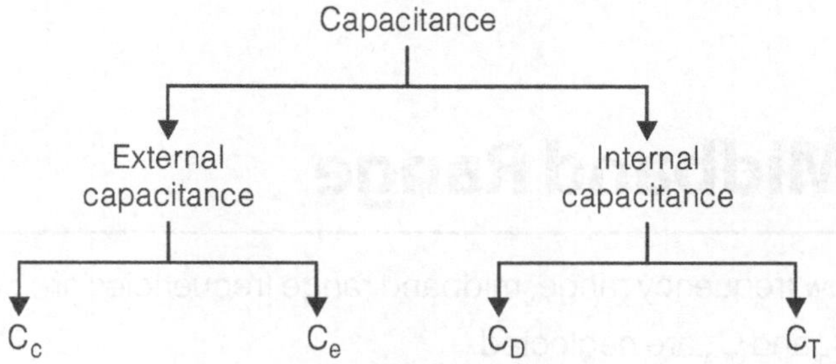


where,

$C_c \rightarrow$ Coupling capacitor

$C_e \rightarrow$ Bypass capacitor

$C_D \rightarrow$ Diffusion capacitance

$C_T \rightarrow$ Transition capacitance

## 2.16 Low Frequency Range

At low frequency range, the external capacitors $C_c$ and $C_e$ are effecting the system response, but the internal capacitors $C_D$ and $C_T$ are neglected. The system response is high pass response.

Transfer function

$$T.F. = \frac{V_0(s)}{V_i(s)}$$

$$= \frac{R_1}{R_1 - jX_{C_1}}$$

$$= \frac{S}{S + \frac{1}{R_1 C_1}}$$

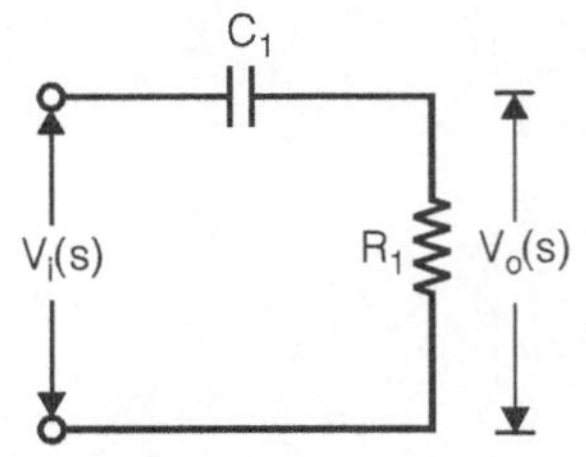

*Fig. 2.25 High pass circuit*

One zero and one pole is considered.

## 2.17 High Frequency Range

At high frequency range the internal capacitors $C_{b'e}$ and $C_{b'c}$ are acting like short circuit, which is effecting the system response. The external capacitors $C_c$ and $C_e$ effect will be neglected. The system response is low-pass response.

$$T.F. = \frac{V_0(s)}{V_i(s)} = \frac{1}{s + \frac{1}{R_2 C_2}}$$

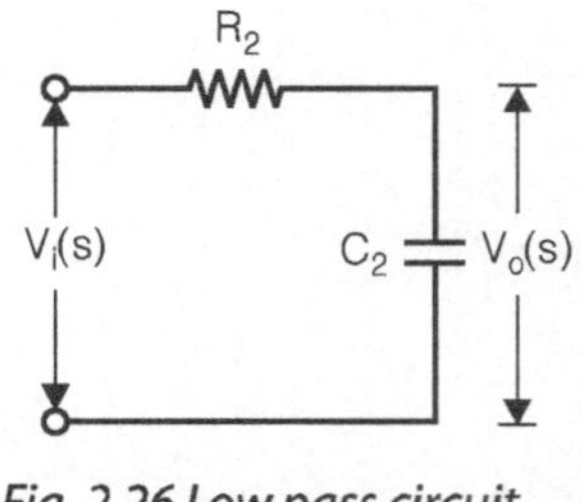

*Fig. 2.26 Low pass circuit*

Only one pole is considered.

## 2.18 Midband Range

At low frequency range, midband range frequencies are considered as high frequency range.

$\therefore$ $C_c$ and $C_e$ are neglected

At high frequency range, midband range frequencies are considered as low frequency range.

$\therefore$ $C_{b'e}$ and $C_{b'c}$ are neglected

Therefore the gain is independent of frequency, so the gain remains constant in midband range.

# 2.19 Application of Bodeplot in Frequency Response of Amplifier

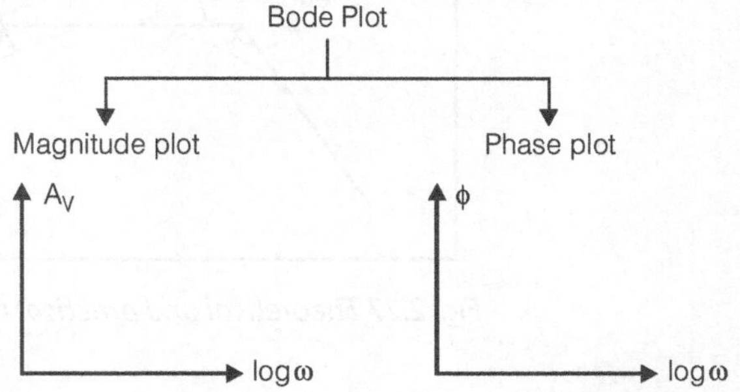


### Magnitude Analysis

For example $G(s) = \dfrac{ks}{1+sT}$

**G(s) = constant k**

$$G(s) = k$$

$$|G(j\omega)|_{dB} = 20 \log k$$

If $k > 1 \Rightarrow +\text{ dB}$

$k = 1 \Rightarrow 0\text{ dB}$

$k < 1 \Rightarrow -\text{ dB}$

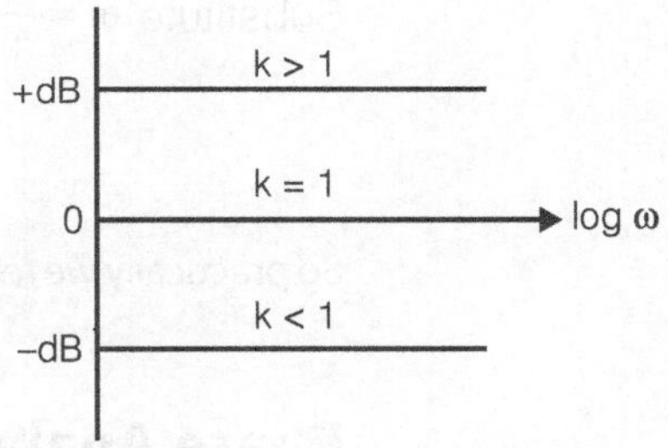


**G(s) = s (one zero)**

$$G(j\omega) = (j\omega)$$

$$|G(j\omega)|_{dB} = 20 \log \omega$$

$\omega = 0.1 \Rightarrow -20\text{ dB}$

$\omega = 1 \Rightarrow 0\text{ dB}$

$\omega = 10 \Rightarrow +20\text{ dB}$

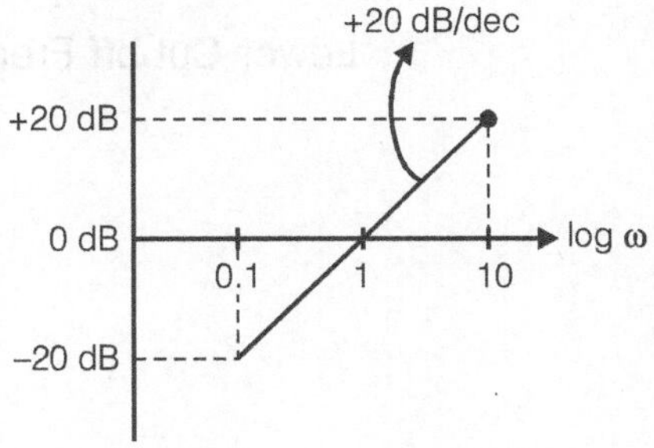


$G(s) = \dfrac{1}{1+sT}$ **(one pole)**

$$|G(j\omega)|_{dB} = 20 \log\sqrt{1+\omega^2T^2}$$

$\omega T << 1 \Rightarrow |G(j\omega)| = 0\text{ dB}$ (low frequencies)

$\omega T >> 1 \Rightarrow |G(j\omega)| = -20 \log \omega T$ (high frequencies)

$\omega = \dfrac{1}{T} \Rightarrow |G(j\omega)| = 0\text{ dB}$

$\omega = \dfrac{10}{T} \Rightarrow |G(j\omega)| = -20\text{ dB}$

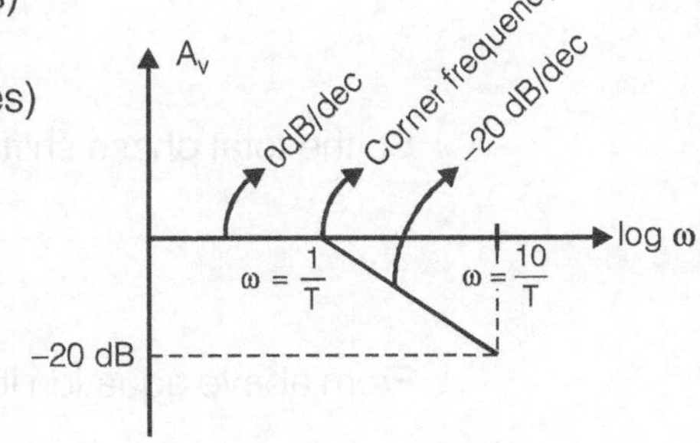

The actual theoretical and practical frequency response is given by

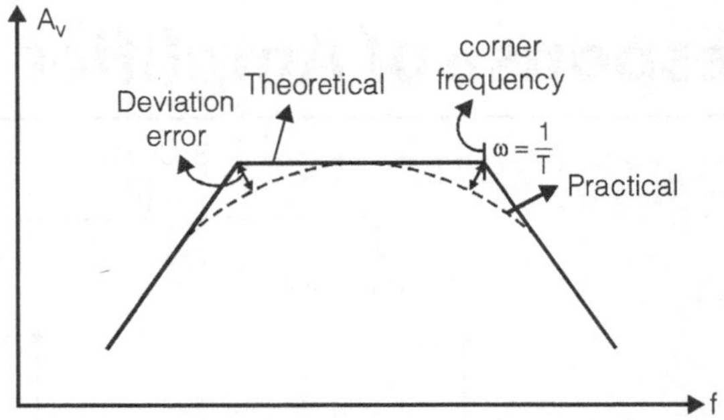


*Fig. 2.27 Theoretical and practical frequency response*

When $\omega = \frac{1}{T}$ (corner frequency) we can find the loss

$$G(j\omega) = -20\log\sqrt{1+\omega^2T^2}$$

Substitute $\omega = \frac{1}{T}$ in above equation

$$G(j\omega) = -20\log\sqrt{1+1} = -3\text{ dB}$$

So practically we reduce the midband gain less than 3 dB to find the lower and higher cutoff frequencies.

## Phase Analysis

Lower Cut off Frequency

$$(A_V)_l = \frac{(A_V)_m}{1-j\left(\frac{f_L}{f}\right)}$$

$$|(A_V)_l| = \frac{|(A_V)|_m}{\sqrt{1+\left(\frac{f_L}{f}\right)^2}}$$

$$\theta_l = \tan^{-1}\left(\frac{f_L}{f}\right)$$

So the total phase shift of the voltage gain $A_{Vl}$ in the low frequency range is given by

$$\theta_l' = 180° + \theta_l = 180° + \tan^{-1}\left(\frac{f_L}{f}\right)$$

From above equation it is obvious that with the decrease in frequency, $\theta_l$ increases and at $f = f_l$

$$\boxed{\theta_l' = 180° + 45° = 225°}$$

Higher Cut off Frequency

$$(A_V)_h = \frac{(A_V)_m}{1+j\left(\frac{f}{f_h}\right)}$$

$$|(A_V)_h| = \frac{(A_V)_m}{\sqrt{1+\left(\frac{f}{f_h}\right)^2}}$$

So, $\theta_h' = 180° - \theta_h$

$$= 180° - \tan^{-1}\left(\frac{f}{f_h}\right)$$

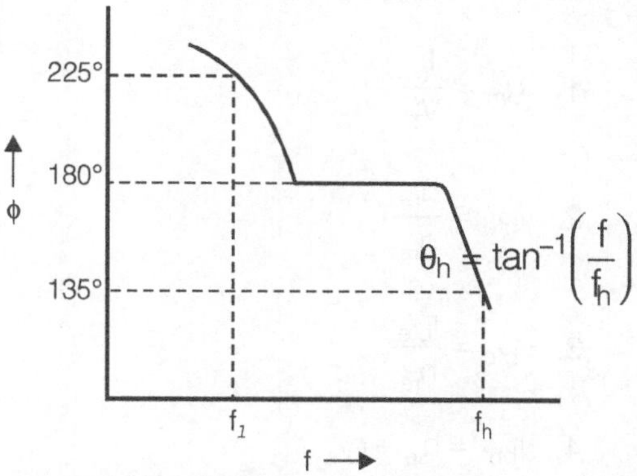

*Fig. 2.28 Phase relationship in a RC coupled amplfier*

At $f = f_2$, $\boxed{\theta_h' = 180° - 45° = 135°}$

$\theta_h'$ decreases with the increase in input frequency f, the variation of total phase shift between input and output voltages with frequency is shown in Figure (2.28).

## SUMMARY

### LOW FREQUENCY ANALYSIS

The h-parameter equations are

$$V_1 = h_i I_1 + h_r V_2$$
$$I_2 = h_f I_1 + h_o V_2$$

### Exact Model

(i) $A_I = \frac{-h_f}{1+h_o R_L}$

(ii) $R_i = h_i - \frac{h_r h_f}{1+h_o R_L} R_L$

(iii) $A_V = \frac{A_I R_L}{R_i}$

(iv) $R_o = \frac{1}{Y_o} = \frac{1}{h_o - \frac{h_f h_r}{h_i + R_s}}$

Exact model formulas are valid for all the configurations (CE, CB, CC)

### Approximate Model Conditions

1. $h_{oe} R_L \leq 0.1$
2. $h_{re}$ and $h_{oe}$ are neglected
3. It is valid for CE mode only

## HIGH FREQUENCY ANALYSIS

1. $g_m = \frac{I_c}{V_T}$

2. $r_{b'e} = \frac{h_{fe}}{g_m}$

3. $r_{b'c} = \frac{r_{b'e}}{h_{re}}$

4. $r_{bb'} = h_{ie} - r_{b'e}$

5. $r_{ce} = \frac{1}{g_{ce}}$ where $g_{ce} = h_{oe} - h_{fe}\, g_{b'c}$

6. $C_{b'e} = \frac{W^2}{2D_B} \cdot \frac{1}{r_e}$ or $C_{b'e} = \frac{W^2}{2D_B} \cdot g_m$

7. $C_{b'e} = \frac{g_m}{2\pi f_T}$

8. $f_\alpha = \frac{1}{2\pi r_e C_{b'e}} = \frac{D_B}{\pi W^2}$

9. $\omega_\alpha = 2\pi f_\alpha = \frac{2D_B}{W^2}$

10. $\omega_\alpha = \frac{1}{t_B}$ where $t_B$ = Base transit time

11. $C_{b'c} \propto \frac{1}{(V_{ce})^n}$ where $n = \frac{1}{2}$ for alloy type

    $n = \frac{1}{3}$ for grown junction type

12. $f_T = h_{fe} f_\beta$

13. $f_\alpha = (1+\beta) f_\beta = \frac{1}{1-\alpha} f_\beta$

## FREQUENCY RESPONSE OF AN AMPLIFIER

1. In low frequency range, the system response is high pass response.
2. In high frequency range, the system response is low-pass response.
3. The phase angle at lower cut off frequency is 225°.
4. The phase angle at higher cut off frequency is 135°.

## GATE AND IES QUESTIONS

**Q.1** The transconductance $g_m$ of the transistor shown in figure is 10 mS. The value of the input resistance $R_{IN}$ is

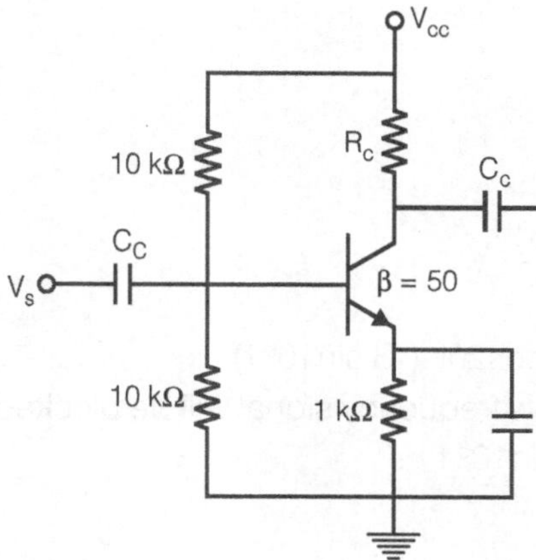


(a) 10.0 kΩ
(b) 8.3 kΩ
(c) 5.0 kΩ
(d) 2.5 kΩ

**Exp. (d)**

Input resistance $R_{IN} = R_i || 10k || 10k$

$$R_i = \frac{h_{fe}}{g_m} = \frac{50}{10\,\text{mS}}$$

$$= 5\text{ k}\Omega$$

$$R_{IN} = 5\text{ k} || 5\text{ k}$$

$$= 2.5\text{ k}\Omega$$

**Q.2** The current $i_b$ through the base of a silicon npn transistor is $1 + 0.1\cos(10000\,\pi t)$mA. At 300 K, the $r_\pi$ in the small signal model of transistor is

(a) 250 Ω
(b) 27.5 Ω
(c) 25 Ω
(d) 22.5 Ω

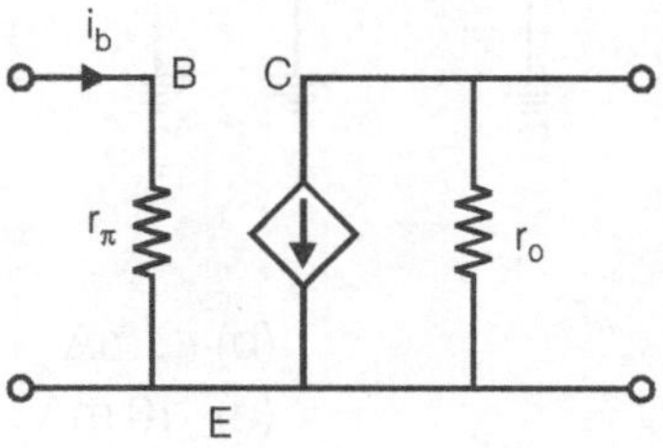


**Exp. (c)**

$$r_\pi = \frac{h_{fe}}{g_m} = \frac{h_{fe}}{I_c} \times V_T = \frac{h_{fe}}{h_{fe} I_b} \times V_T = \frac{25\text{ mV}}{1\text{mA}} = 25\ \Omega$$

**Q.3** A small signal source $V_i(t) = A\cos 20t + B\sin 10^6 t$ is applied to a transistor amplifier as shown below. The transistor has $\beta = 150$ and $h_{ie} = 3\text{ k}\Omega$, which expression best approximate $V_0(t)$

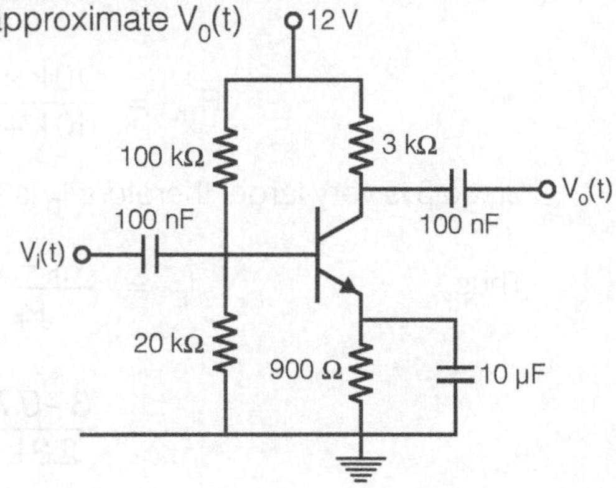


(a) $V_0(t) = -1500\,(A\cos 20t + B\sin 10^6 t)$
(b) $V_0(t) = -150\,(A\cos 20t + B\sin 10^6 t)$
(c) $V_0(t) = -1500\,B\sin 10^6 t$
(d) $V_0(t) = -150\,B\sin 10^6 t$

**Exp. (d)**

The output voltage is

$$V_0 = A_v V_i$$

$$= \frac{-h_{fe} R_c V_i}{h_{ie}}$$

Here $R_C = 3\ k\Omega$ and $h_{ie} = 3\ k\Omega$

$$V_0 \approx -\frac{150 \times 3\,k}{3\,k} V_i$$

$$\approx -150\ (A \cos 20t + B \sin 10^6\ t)$$

Since coupling capacitor is large so low frequency signal will be blocked and best approximation is

$$V_0 \approx -150\ B \sin 10^6\ t$$

**Linked Answer-Questions (4) and (5):**

In the following transistor circuit, $V_{BE} = 0.7$ V, $r_e = 25\ mV/I_E$ and the capacitance are very large.

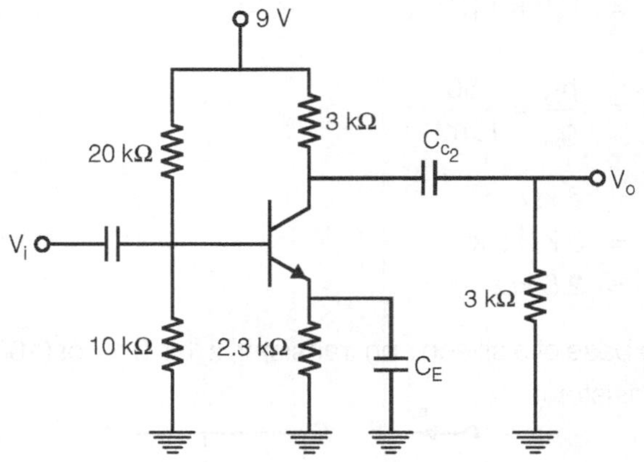


**Q.4** The value of DC current $I_E$ is

(a) 1 mA
(b) 2 mA
(c) 5 mA
(d) 10 mA

**Exp. (a)**

The Thevenin equivalent circuit for DC model is

$$V_{Th} = \frac{10}{10+20} \times 9 = 3\ V$$

$$R_{Th} = \frac{10\,k \times 20k}{10\,k + 20k} = 6.67\ k\Omega$$

Since β is very large, therefore $I_B$ is small and can be ignored.

Thus,

$$I_E = \frac{V_{Th} - V_{BE}}{R_E}$$

$$= \frac{3 - 0.7}{2.3\,k} = 1 mA$$

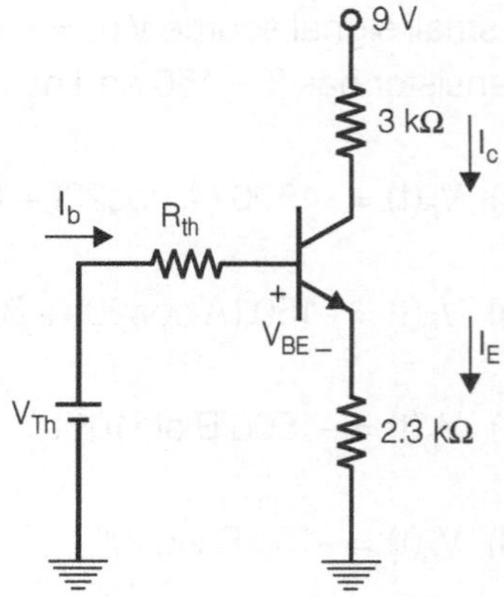

**Q.5** The mid band voltage gain of the amplifier is approximately

(a) –180 (b) –120
(c) –90 (d) –60

**Exp. (d)**

The small signal model is

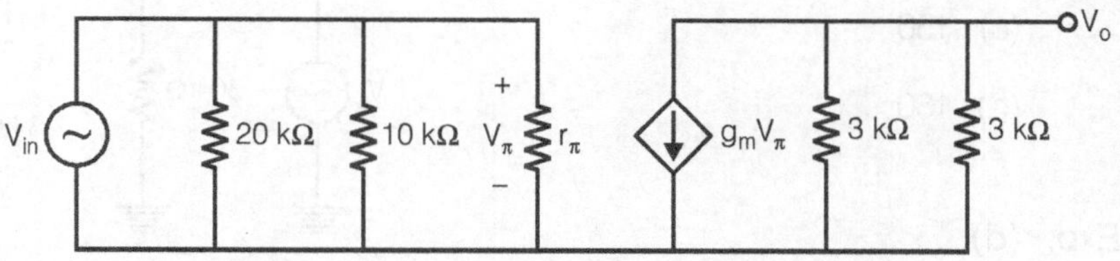


$$g_m = \frac{|I_c|}{V_T} = \frac{1\,mA}{25\,mV} = \frac{1}{25}\,A/V$$

$$V_o = -g_m V_\pi \times (3k \,||\, 3k)$$

$$= -\frac{1}{25} V_{in}(1.5k) = -60\,V_{in}$$

i.e. $\dfrac{V_o}{V_{in}} = -60$

**Q.6** An npn BJT has $g_m = 38$ mA/V, $C_\mu = 10^{-14}$ F, $C_\pi = 4 \times 10^{-13}$ F and DC current gain $\beta_0 = 90$. For this transistor $f_T$ and $f_\beta$ are

(a) $f_T = 1.64 \times 10^8$ Hz and $f_\beta = 1.47 \times 10^{10}$ Hz
(b) $f_T = 1.47 \times 10^{10}$ Hz and $f_\beta = 1.64 \times 10^8$ Hz
(c) $f_T = 1.33 \times 10^{12}$ Hz and $f_\beta = 1.47 \times 10^{10}$ Hz
(d) $f_T = 1.47 \times 10^{10}$ Hz and $f_\beta = 1.33 \times 10^{12}$ Hz

**Exp. (b)**

$$f_T = \frac{g_m}{2\pi(C_\mu + C_\pi)} = \frac{38 \times 10^{-3}}{2\pi(10^{-14} + 4 \times 10^{-13})}$$

$$f_T = 1.47 \times 10^{10}\text{ Hz}$$

$$f_\beta = \frac{f_T}{\beta} = \frac{1.47 \times 10^{10}}{90} = 1.64 \times 10^8\text{ Hz}$$

**Q.7** The current gain of a bipolar transistor drops at high frequencies because of

(a) transistor capacitances
(b) high current effects in the base
(c) parasitic inductive elements
(d) the early effect

**Exp. (a)**

At high frequency

$$A_I = \frac{-g_m}{g_{b'e} + j\omega C}$$

$$A_I \propto \frac{1}{\text{capacitance}}$$

Thus due to the transistor capacitance current gain of a bipolar transistor drops.

**Q.8** An amplifier circuit is shown in the given figure The voltage gain $V_0/V_s$ is

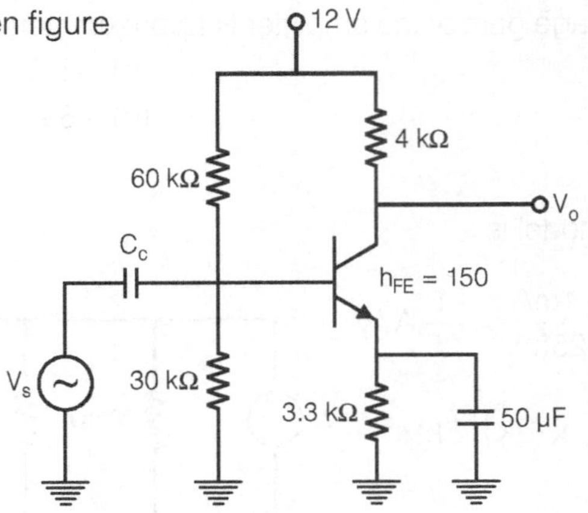


(a) $\frac{4}{3.33}$

(b) 100

(c) 150

(d) 160

**Exp. (d)**

The amplifier equivalent model is

$$V_0 = -g_m V_\pi \times 4\text{ k}$$

where

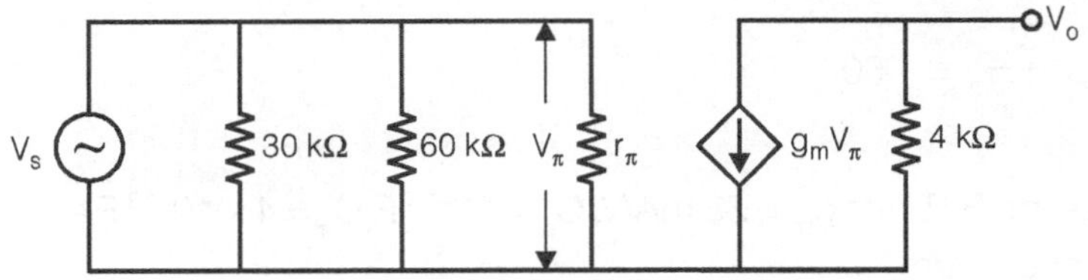


$$V_\pi = V_s$$

$$V_0 = -g_m V_s \times 4\text{ k}$$

$$\frac{V_0}{V_s} = -g_m \times 4\text{ k}$$

$$g_m = \frac{|I_C|}{V_T}$$

$I_c$ is a DC current.

The DC equivalent of the circuit is

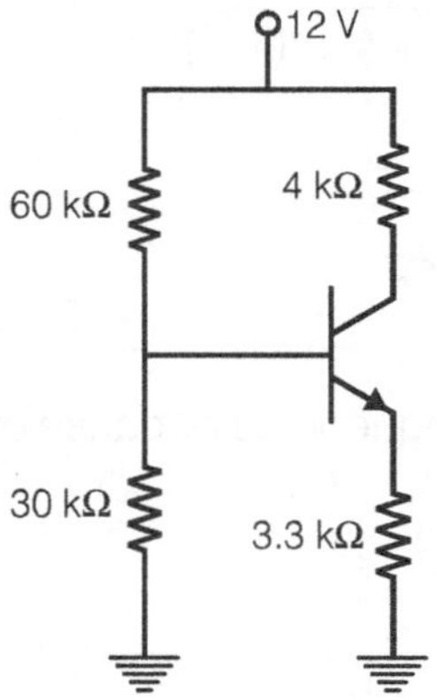


⇒

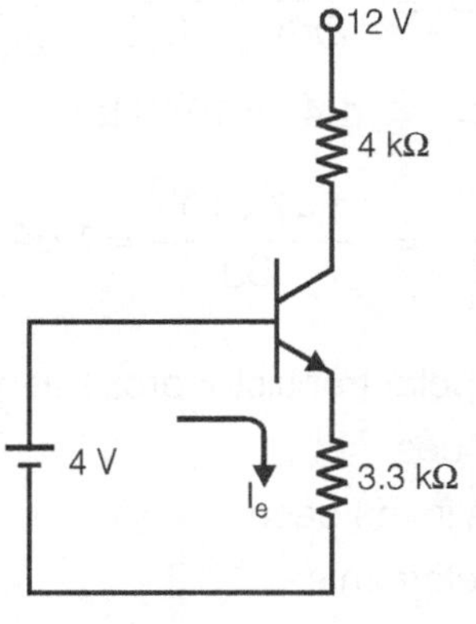


$$I_e = \frac{4-0.7}{3.3} = 1\text{mA} \quad ; \quad I_c = I_e$$

$$g_m = \frac{1\text{mA}}{25\text{ mV}} \qquad \text{(where } V_T = 25\text{ mV)}$$

$$A_v = \frac{V_0}{V_s} = -\frac{1\text{mA}}{0.025} \times 4\text{ k} = -160$$

$$|A_v| = 160$$

**Q.9** In the circuit shown below, if $R_1 > R_P$ and the impulses can completely saturated transistor $Q_1$, then the output voltage $V_0$ will be

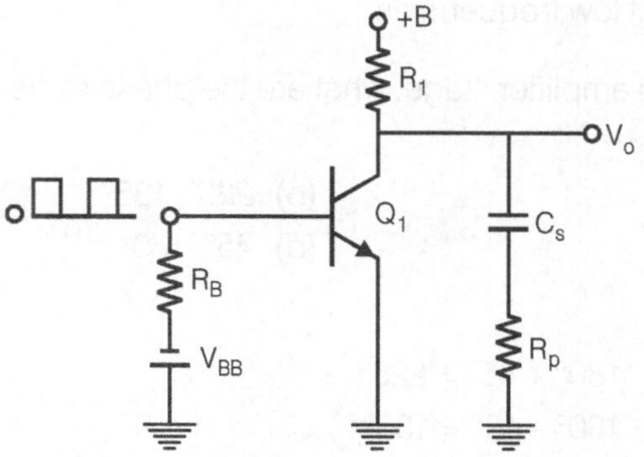


(a) 
(b) 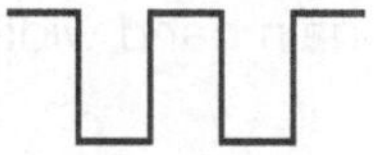
(c) 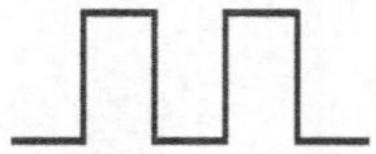
(d) 

**Exp. (d)**
This is CE amplifier, that means 180° phase shift property, but due to capacitive effect output will be distorted from input.

**Q.10** A common emitter transistor amplifier has a collector load of 10 kΩ. If its $h_{fe}$ = 100 and $h_{ie}$ = 2 kΩ ($h_{re} \approx h_{oe} \approx 0$). The voltage amplification of the amplifier is nearly equal to
(a) 500 (b) 200
(c) 100 (d) 50

**Exp. (a)**

$$A_V = \frac{A_I R_L}{R_i} = \frac{100 \times 10\ \text{k}\Omega}{2\ \text{k}\Omega} = 500$$

**Q.11** In an RC coupled amplifier, the gain decreases in the frequency response due to the
(a) coupling capacitor at low frequency and bypass capacitor at high frequency.
(b) coupling capacitor at high frequency and bypass capacitor at low frequency.
(c) coupling junction capacitance at low frequency and coupling capacitor at high frequency.
(d) device junction capacitor at high frequency and coupling capacitor at low frequency.

**Exp. (d)**
At low frequency range, external capacitors i.e. $C_c$ and $C_e$ creates a problem and at high frequency range, internal capacitors i.e. $C_{b'e}$ and $C_{b'c}$ create a problem.

**Q.12** If the coupling capacitors of a CE transistor amplifier is shorted, which one of the following graphs will represent the frequency response curve of the amplifier?

(a) 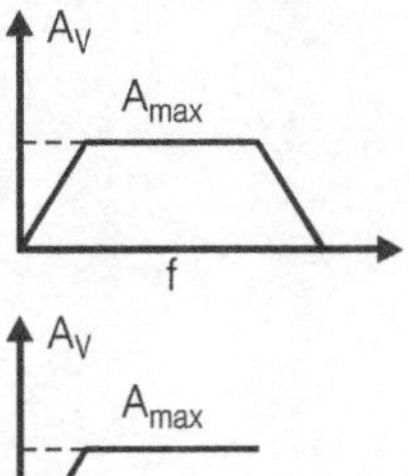

(b) 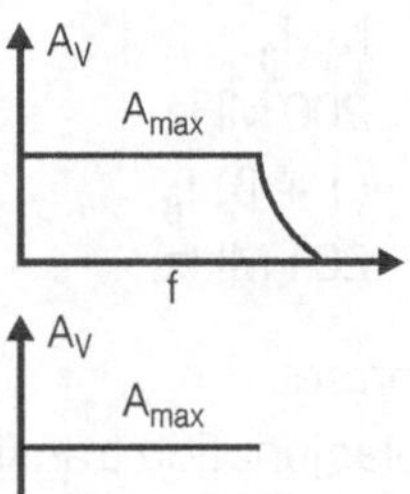

(c) 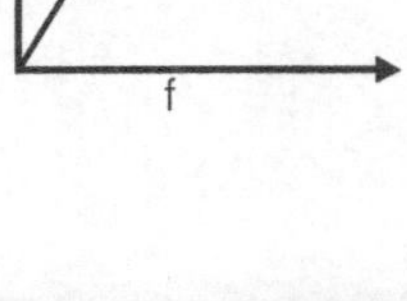

(d) 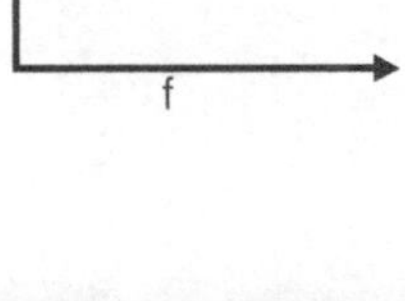

**Exp. (b)**

When coupling capacitors of CE is shorted. External capacitors are neglected at low frequencies therefore there will be no attenuation at low frequencies.

**Q.13** In a single stage RC coupled amplifier stage, what are the phase shifts introduced at lower and upper 3 dB frequencies, respectively?

(a) 45°, 225°
(b) 225°, 135°
(c) 90°, 180°
(d) 45°, 180°

**Exp. (b)**

$$\theta_L = 180° + 45° = 225°$$
$$\theta_h = 180° - 45° = 135°$$

**Q.14** The common emitter current gain band width product of a transistor ($f_T$) is defined as the frequency at which

(a) Alpha of the transistor falls by 3 dB
(b) Beta of the transistor falls by 3 dB
(c) Beta of the transistor falls to unity
(d) Power gain of the transistor falls to unity

**Exp. (c)**

$f_T$ is the frequency at which the common emitter short circuit current gain reaches to unity.

**Q.15** A BJT is to be used in a high frequency circuit in common emitter amplifier. For a higher cut off frequency ($C_\mu$, $C_\pi$ and $r_0$ have their usual meanings)

(a) $C_\mu$ should be as small as possible
(b) $r_0$ and $C_\mu$ should be as large as possible
(c) $C_\pi$ and $C_\mu$ should be as large as possible
(d) $r_0$, $C_\pi$ and $C_\mu$ should be as large as possible

**Exp. (a)**

$$f_\beta = \frac{f_T}{h_{fe}} = \frac{1}{h_{fe}} \frac{g_m}{2\pi(C_\pi + C_\mu)}$$

$C_\mu$ should be small.

**Q.16** An npn transistor has a beta cut off frequency $f_\beta$ of 1 MHz and common emitter short circuit low frequency current gain $\beta$ of 200. Its unity gain frequency $f_T$ and the alpha cut off frequency $f_\alpha$ respectively are

(a) 200 MHz, 201 MHz
(b) 200 MHz, 199 MHz
(c) 199 MHz, 200 MHz
(d) 201 MHz, 200 MHz

**Exp. (a)**

$$f_T = h_{fe} f_\beta = 200 \text{ MHz}$$
$$f_\alpha = (1 + \beta) f_\beta = 201 \text{ MHz}$$

**Q.17** Which of the following are correct.

$\alpha$ cut off frequency of a bipolar junction transistor increases with

(A) The increase in band width
(B) The increase in emitter width
(C) The increase in collector width
(D) The decrease in base width
(a) All
(b) B, C
(c) D
(d) B, C, D

**Exp. (c)**

$$f_\alpha \uparrow \alpha\, f_\beta \uparrow \alpha\, f_T \uparrow \alpha\, g_m \uparrow \alpha\, I_c \uparrow$$

$$I_b \downarrow w_B \downarrow$$

**Q.18** A bipolar transistor is operating in the active region with a collector current of 1 mA, assuming that the $\beta$ of the transistor is 100 and the thermal voltage ($V_T$) is 25 mV, the transconductance $g_m$ and the input resistance ($r_\pi$) of the transistor in the common emitter configuration, are
(a) $g_m$ = 25 mA/V and $r_\pi$ = 15.65 kΩ
(b) $g_m$ = 40 mA/V and $r_\pi$ = 4.0 kΩ
(c) $g_m$ = 25 mA/V and $r_\pi$ = 2.5 kΩ
(d) $g_m$ = 40 mA/V and $r_\pi$ = 2.5 kΩ

**Exp. (d)**

$$g_m = \frac{I_c}{V_T} = 40 \text{ mA/V}$$

and

$$r_\pi = \frac{\beta}{g_m} = \frac{100}{40 \times 10^{-3}} = 2.5 \text{ k}\Omega$$

**Common Data Questions for (19 and 20):**
Consider the common emitter amplifier shown below with the following circuit parameters $\beta$ = 100, $g_m$ = 0.3861 A/V, $r_0 = \infty$, $r_\pi$ = 259 Ω, $R_s$ = 1 kΩ, $R_B$ = 93 kΩ, $R_C$ = 250 Ω, $R_L$ = 1 kΩ, $C_1 = \infty$, $C_2$ = 4.7 μF

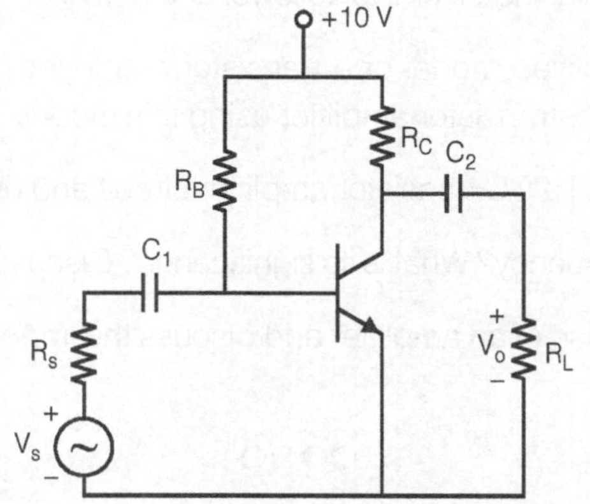


**Q.19** The resistance seen by the source $V_s$ is
(a) 250 Ω
(b) 1258 Ω
(c) 93 kΩ
(d) ∞

Exp. (b)

The amplifier model is

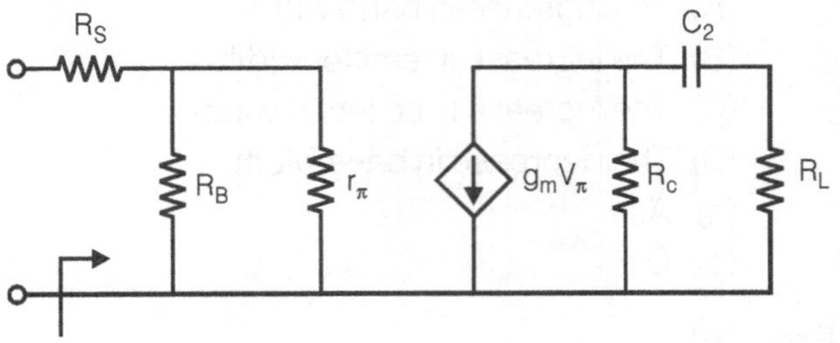


$$Z_{in} = R_s + R_B \,||\, r_\pi$$
$$= 1\text{ k}\Omega + 93\text{ k} \,||\, 259$$
$$\approx 1258\ \Omega$$

Q.20 The lower cut off frequency due to $C_2$ is

(a) 33.9 Hz
(b) 27.1 Hz
(c) 13.6 Hz
(d) 16.9 Hz

Exp. (b)

$$f_L = \frac{1}{2\pi(R_L + R_C)C_2} = \frac{1}{2\pi(1\text{k} + 250\ \Omega)4.7\ \mu\text{F}}$$
$$= 27.1\text{ Hz}$$

## REVIEW QUESTIONS

1. Draw the high frequency model of one stage RC coupled amplifier and derive an expression for upper 3 dB frequency.
2. Discuss the effect of emitter by pass capacitor on the low frequency response of a CE amplifier.
3. Draw the hybrid $\pi$ equivalent circuit of a transistor in CE configuration and explain the various parameters in it. In terms of h parameters derive the equations for the various components of this circuit.
4. Among the three configuration of BJT, CE configuration is widely used. Justify this statement and give its applications.
5. Draw a circuit diagram for a CE transistor amplifier and derive expressions for voltage and current gain, input and output impedance.
6. Draw the circuit diagram of Darlington emitter follower and derive the expression for input impedance.
7. When and why is the simplified model of a transistor employed? Give its circuit and derive the amplifier parameters of a CE transistor amplifier using this model.
8. Draw the simplified model of a CC transistor amplifier circuit and derive its parameters.
9. What is unity cross over frequency? What is its significance? Discuss about gain bandwidth product.
10. Draw the frequency response of an amplifier and discuss the importance of –3dB.

OOOO

# CHAPTER 3

# FEEDBACK AMPLIFIERS

## About This Chapter

The concepts introduced in this chapter are:

- Various types of amplifiers.
- Need for feedback in an amplifier.
- Types of feedback amplifiers.
  - Voltage series
  - Voltage shunt
  - Current series
  - Current shunt
- Analysis of negative feedback amplifiers.

# 3 FEEDBACK AMPLIFIERS

## 3.1 Introduction

- The basic characteristics of an amplifier are input impedance, output impedance, voltage gain, bandwidth etc.
- Suppose if we want to change the characteristics of a basic amplifier for certain application, we have to use a technique called as feedback.
- When a part or a fraction of the output fedback to the input, feedback is said to be exist.
- Feedback is classified in to two types:
  (i) Positive feedback
  (ii) Negative feedback
- When the net effect of the feedback increases the magnitude of the input signal, it is called as positive or direct or regenerative feedback.
- When the net effect of the feedback reduces the magnitude of the input signal, it is called as negative or inverse or degenerative feedback.

## 3.2 Limitations of Basic Amplifier

The main limitations of basic amplifier are:

**(i)** Instability of ac gain.
**(ii)** The input impedance is low.
**(iii)** The output impedance is high.
**(iv)** Noise and distortion are more.
**(v)** Bandwidth is not sufficiently large.

## 3.3 Feedback Principle

Fig. (3.1) shows a block diagram of a feedback system. 'A' represents the main amplifier and 'β' represents feedback factor.

Where
$$\beta = \frac{\text{Voltage feedback to the input side}}{\text{output voltage}}$$

From the block diagram, we can notice that the net voltage input to the main amplifier is

$$V_i = V_s - \beta V_o$$

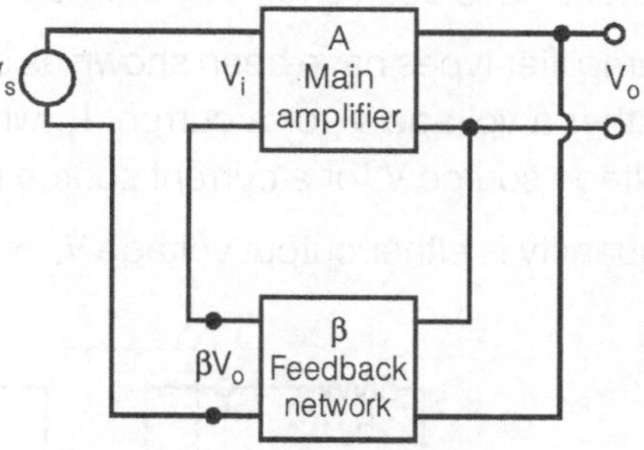


*Fig. 3.1 Block diagram of a feedback system*

Since the main amplifier has a gain of A, so the output $V_o$ will be

$$V_o = AV_i = A(V_s - \beta V_o)$$

(or) $$V_o = AV_s - A \cdot \beta \cdot V_o$$

(or) $$V_o(1 + A\beta) = A \cdot V_s$$

The overall feedback amplifier system has an input of $V_s$ and output $V_o$. Therefore gain of the amplifier with feedback is obtained from above equation

$$A_f = \text{Gain with feedback} = \frac{V_o}{V_s} \quad \text{(or)} \quad \boxed{A_f = \frac{A}{1 + A\beta}}$$

The gain of the amplifier is changed by the factor $(1 + A\beta)$. Three cases can arise:

- **If $|A_f| < |A|$, the type of feedback is negative (or) degenerative. Negative feedback reduces gain, but also reduces distortion and noise and increases bandwidth and stability. A more detailed analysis is presented later in the chapter.**
- **If $|A_f| > |A|$, the type of feedback is positive (or) regenerative. Positive feedback increases gain, increases distortion, reduces bandwidth and stability.**
- **If $(1 + A\beta) = 0$, then the gain becomes infinite and output can be obtained with no external input and such circuit becomes an oscillator.**

In this chapter we shall restrict our study to case (i) i.e. negative feedback.

## 3.4 Classification of Amplifiers

For understanding the concepts of feedback, amplifiers have been classified into four broad categories:

(i) Voltage amplifiers

(ii) Current amplifiers

(iii) Transconductance amplifiers

(iv) Transresistance amplifiers

The relative magnitude of input impedance as compared to the source impedance and that of output impedance as compared to load impedance are the factors which determine the category of the amplifier.

Figures. 3.2 (a) to 3.2 (d) show the equivalent circuits of the four types of amplifiers. For each of the types, the relative magnitudes of input and output impedances as compared to source and load impedances have been shown by the side of each diagram.

Each of the amplifier types have been shown as a four terminal (i.e. two port) network. The input quantity is either a voltage $V_1$ or a current $I_1$, which in turn is obtained from the input source, namely, a voltage source $V_s$ or a current source $I_s$, respectively.

The output quantity is either output voltage $V_o$ or output current $I_o$ (= $I_L$)

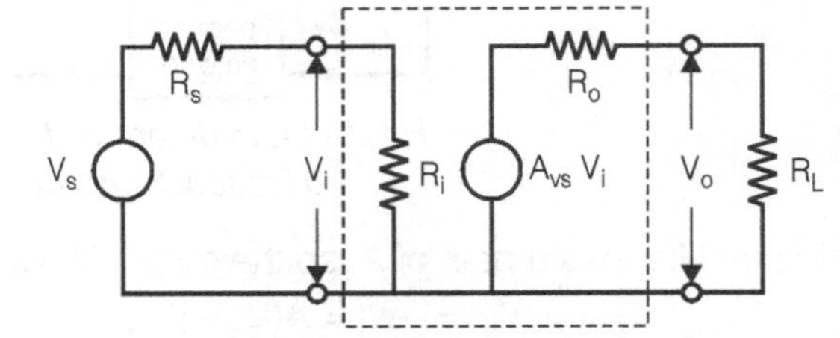


Condition
$R_i >> R_s$
$R_o << R_L$

*Fig. 3.2 (a) Equivalent circuit of a voltage amplifier*

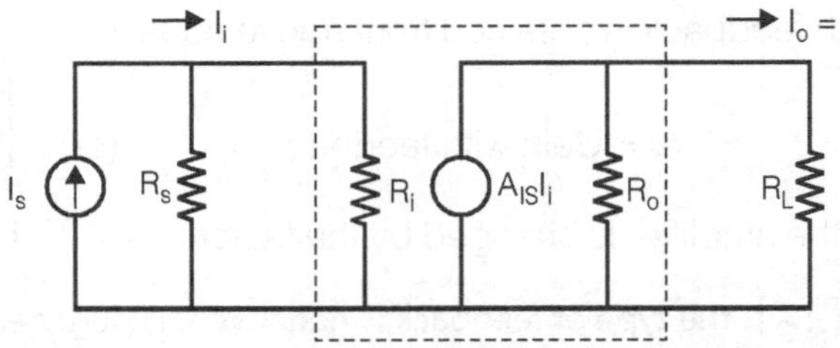


Condition
$R_i << R_s$
$R_o >> R_L$

*Fig. 3.2 (b) Equivalent circuit of a current amplifier*

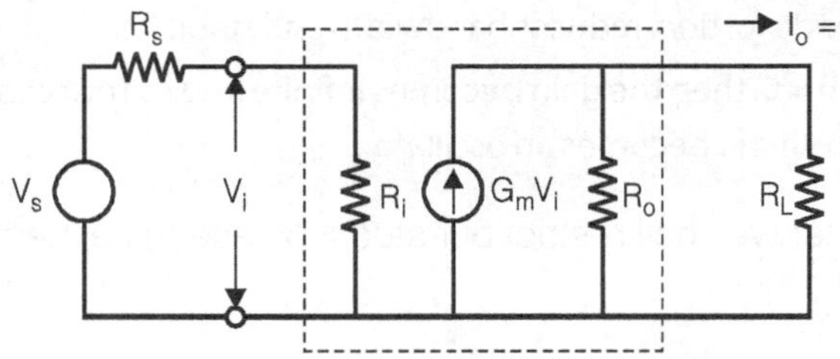


Condition
$R_i >> R_s$
$R_o >> R_L$

*Fig. 3.2 (c) Equivalent circuit of a transconductance amplifier*

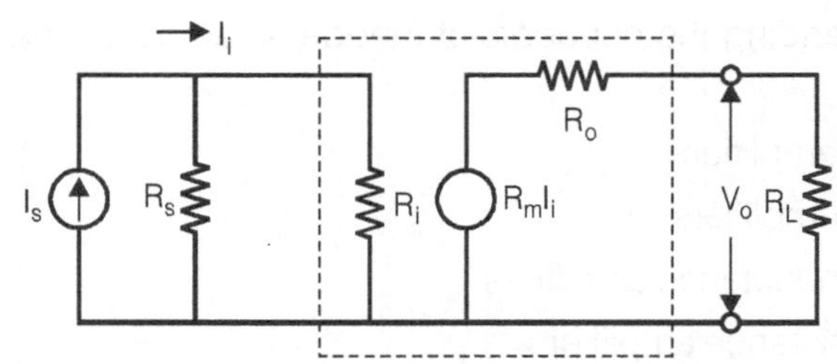


Condition
$R_i << R_s$
$R_o << R_L$

*Fig. 3.2 (d) Equivalent circuit of a transresistance amplifier*

## VOLTAGE AMPLIFIER

From Figure 3.2 (a), we find that the input resistance $R_i >> R_s$, then $V_i \approx V_s$.
Also, if load resistance $R_L >> R_o$,

$$\text{then} \quad V_o = A_v V_i \simeq A_v \cdot V_s$$

$$\therefore \quad A_v = \frac{V_o}{V_s}$$

**An ideal voltage amplifier should have infinite input resistance (i.e. $R_i = \infty$) and zero output resistance (i.e. $R_o = 0$).**

In a similar way, the other types amplifiers can also be understood.
Table (3.1) brings out the behaviour or characteristics of the four types of amplifiers.

***TABLE (3.1) IDEAL AMPLIFIER CHARACTERISTICS***

| | Voltage amplifier | Current amplifier | Transconductance amplifier | Transresistance amplfier |
|---|---|---|---|---|
| Ideal value of $R_i$ | $\infty$ | 0 | $\infty$ | 0 |
| Ideal value of $R_o$ | 0 | $\infty$ | $\infty$ | 0 |
| Input quantity | $V_s$ | $I_s$ | $V_s$ | $I_s$ |
| Output Quantity | $V_o$ | $I_L$ | $I_L$ | $V_o$ |
| Transfer characteristic | $A_{vs} = \frac{V_s}{V_o}$ | $A_{IS} = \frac{I_L}{I_s}$ | $G_m = \frac{I_L}{V_s}$ | $R_m = \frac{V_o}{I_s}$ |

# 3.5 General Theory of Feedback

We have seen in section 3.3 that in a feedback system, a port of the output signal is combined with the external input signal. In this section, we present a generalised block diagram (see Fig. 3.3.) of a feedback system.

The output quantity (either voltage or current) is sampled by a suitable sampling network and fed to the feedback network. The output of feedback network is combined with external (source) signal through a mixer network, and fed to the main amplifier. We now examine the different varieties possible in each of these blocks.

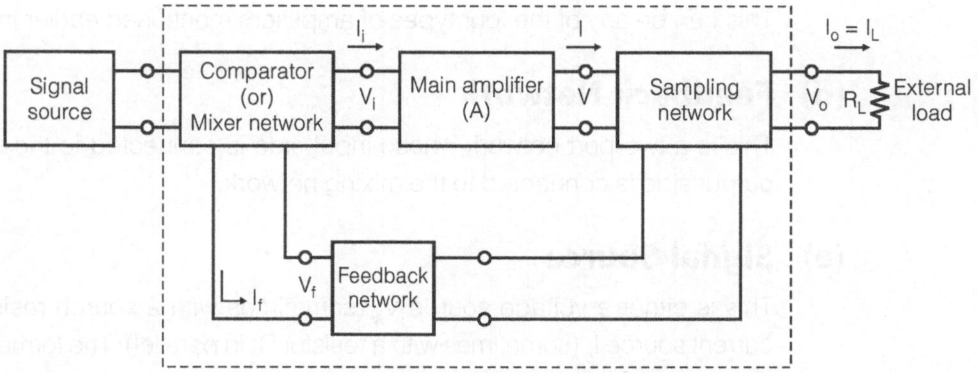


*Fig. 3.3 Block diagram of a feedback system*

## (a) Sampling Network

Mainly there are two types of sampling network, namely, voltage sampling and current sampling.

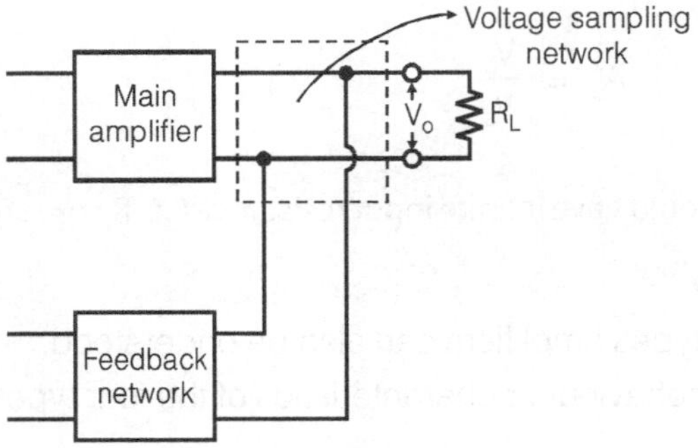


*Fig. 3.4 (a) Voltage sampling*

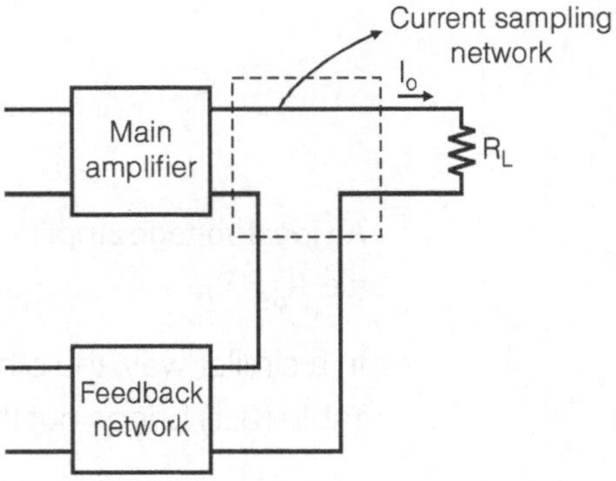


*Fig. 3.4 (b) Current sampling*

In Figure 3.4 (a), the output voltage is sampled, where as in Figure 3.4 (b), the output current is sampled. The voltage sampling is also known as node sampling and current sampling is also known as loop sampling.

## (b) Mixer Network

This is also known as comparator network. There are two types of mixer network, namely, series mixing [Figure 3.5 (a)] and shunt mixing [Figure 3.5 (b)].

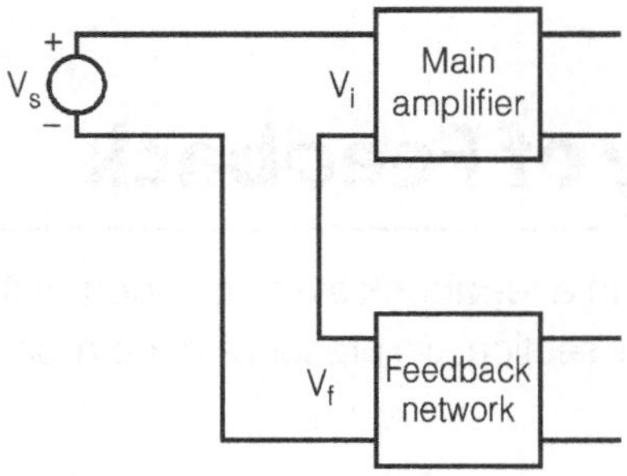


*Fig. 3.5 (a) Series mixing*
*($V_i = V_s - V_f$)*

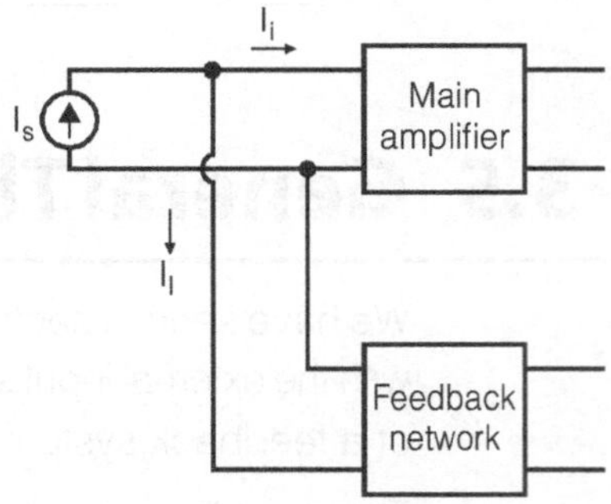


*Fig. 3.5 (b) Shunt mixing*
*($I_i = I_s - I_f$)*

## (c) The Main Amplifier

This can be any of the four types of amplifiers mentioned earlier in section 3.4.

## (d) Feedback Network

This is a two-port network whose input side is connected to the sampling network and output side is connected to the mixing network.

## (e) Signal Source

This is either a voltage source $V_s$ (sometimes with a source resistor $R_s$ in series), or a current source $I_s$ (sometimes with a resistor $R_s$ in parallel). The former is known as Thevenin source and the latter as Norton source.

## 3.6 Concept of Positive Feedback and Negative Feedback

Consider a basic block diagram of feedback system, with basic amplifier A, shown in Figure (3.6).

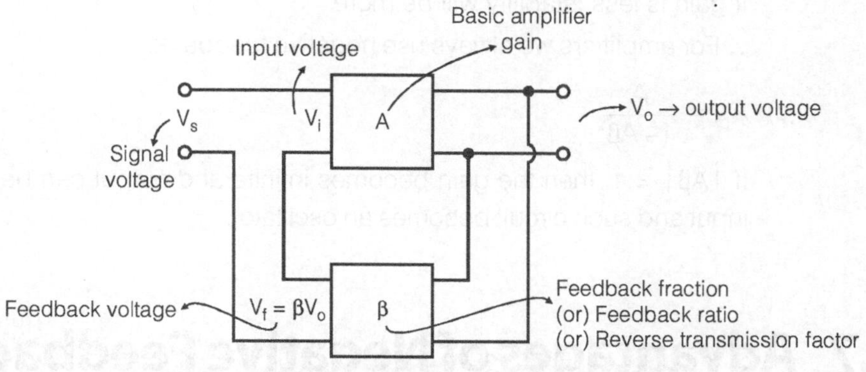


*Fig. 3.6 Basic block diagram of feedback system*

### Analysis

From the above Figure 3.6, we can define

$$V_i = V_s \pm V_f \qquad ...(i)$$

When it is considered as positive feedback

$$V_i = V_s + V_f \qquad ...(ii)$$

When it is considered as negative feedback

$$V_i = V_s - V_f \qquad ...(iii)$$

### Positive Feedback

$$V_o = AV_i$$

Substitute equation (ii) in above

$$V_o = A(V_s + V_f) = A(V_s + \beta V_o)$$

$$V_o = AV_s + A\beta V_o$$

$$V_o(1 - A\beta) = AV_s$$

$$A_{pf} \text{ (positive feedback gain)} = \frac{A}{1 - A\beta}$$

### Negative Feedback

$$V_o = AV_i$$

Substitute equation (iii) in above

$$V_o = A(V_s - V_f)$$

$$V_o = A(V_s - \beta V_o)$$

$$V_o = AV_s - A\beta V_o$$

$$V_o(1 + A\beta) = AV_s$$

$$A_{nf} \text{(negative feedback gain)} = \frac{A}{1 + A\beta}$$

### Conclusions

(a) $A_{pf} > A > A_{nf}$

(b) In amplifier design there are two criteria
   (i) Stability
   (ii) gain
   If gain is less, stability will be more.
   ∴ For amplifiers we always use negative feedback.

(c) $A_{pf} = \dfrac{A}{1-A\beta}$

   If $|A\beta| = 1$, then the gain becomes infinite and output can be obtained with no external input and such circuit becomes an oscillator.

## 3.7 Advantages of Negative Feedback Amplifiers

### (a) Stability of AC Gain

Consider negative feedback gain

$$A_{nf} = \frac{A}{1+A\beta}$$

If $A\beta >> 1$

$$A_{nf} = \frac{A}{A\beta} = \frac{1}{\beta}$$

Where $\beta$ in circuit analysis, it is a passive component resistor.

Negative feedback gain is depending only on $\beta$.

∴ Gain is stabilized.

### Practical Consideration

Suppose there is a small change in the internal resistance of an amplifier, therefore fractional change of gain with feedback is given by

Consider negative feedback gain,

$$A_f = \frac{A}{1+A\beta}$$

d.w.r.t.A

$$\frac{\partial A_f}{\partial A} = \frac{1+A\beta-A\beta}{(1+A\beta)^2}$$

$$\frac{\partial A_f}{A_f} = \frac{1}{(1+A\beta)^2}\frac{\partial A}{A_f}$$

$$\frac{\partial A_f}{A_f} = \frac{1}{(1+A\beta)^2}\frac{\partial A}{A}\times(1+A\beta)$$

$$\boxed{\frac{\partial A_f}{A_f} = \frac{1}{(1+A\beta)}\frac{\partial A}{A}}$$

Where $\frac{\partial A_f}{A_f}$ → Fractional change of gain with feedback

$\frac{\partial A}{A}$ → Fractional change of gain without feedback

$\frac{1}{1+A\beta}$ → Sensitivity (S)

$1 + A\beta$ → Desensitivity (D)

In practical consideration also, stability is improved.

## (b) Increase in Input Impedance

Consider the block diagram of feedback system

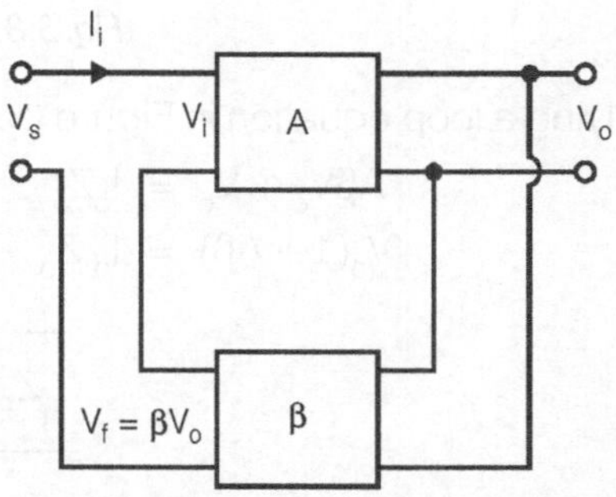


*Fig. 3.7 Block diagram of feedback system*

From the Figure (3.7)

$$Z_{if}(\text{Input impedance with feedback}) = \frac{V_s}{I_i} \quad \text{...(i)}$$

$$Z_i(\text{Input impedance without feedback}) = \frac{V_i}{I_i} \quad \text{...(ii)}$$

$$V_i = V_s - V_f = V_s - \beta V_o = V_s - \beta A V_i$$

$$V_i = V_s - A\beta V_i$$

$$V_i(1 + A\beta) = V_s$$

Dividing both sides with $I_i$

$$\frac{V_i}{I_i}(1 + A\beta) = \frac{V_s}{I_i}$$

From equation (i) and (ii)

$$\boxed{Z_{if} = Z_i(1 + A\beta)}$$

## (c) Decrease in Output Impedance

To calculate output impedance, two conditions should be followed.

(i) $R_L$ should be open i.e. $R_L = \infty$

(ii) If the input is a voltage source, make it as short circuit, otherwise make it as open circuit, if it is a current source.

The above conditions are applied to Figure (3.7), then

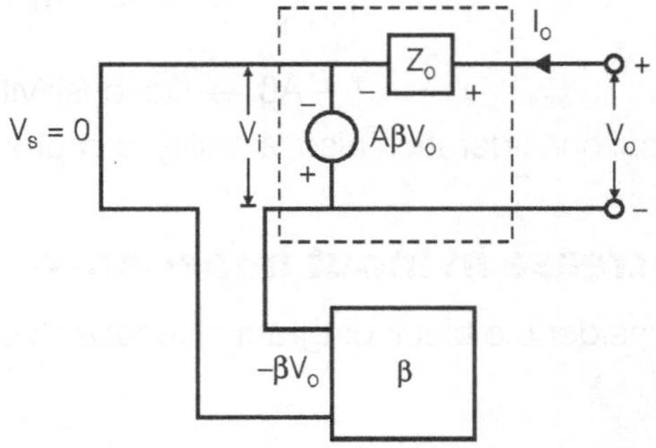


*Fig. 3.8 Internal model of an amplifier*

Applying a loop equation in Figure (3.8), we have

$$A\beta V_o + V_o = I_o Z_o$$

$$V_o(1 + A\beta) = I_o Z_o$$

$$Z_{of} = \frac{V_o}{I_o} = \frac{Z_o}{1 + A\beta}$$

Where $Z_{of} \rightarrow$ Output impedance with feedback

$Z_o \rightarrow$ Output impedance with out feedback

## (d) Reduction in Distortion and Noise

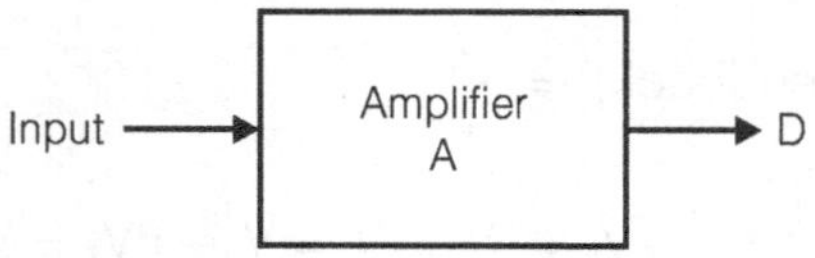


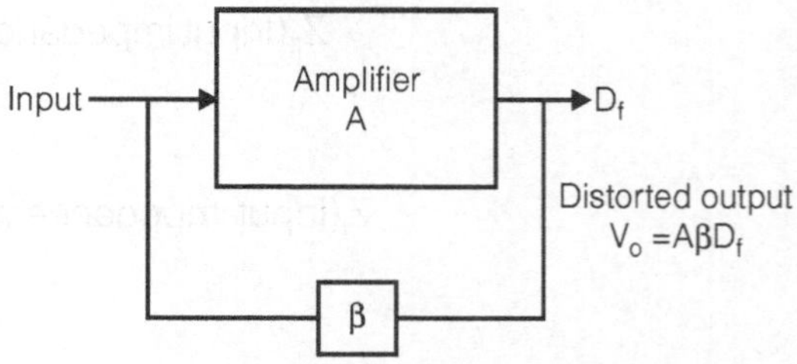


*Fig. 3.9 (a) Without feedback with distortion D* *Fig. 3.9 (b) With feedback with distortion $D_f$*

We have, Net distortion = Original distortion + distorted output

$$D_f = D - A\beta D_f$$

$$D_f(1 + A\beta) = D$$

$$D_f = \frac{D}{1 + A\beta}$$

Similarly

$$N_f = \frac{N}{1 + A\beta}$$

## (e) Increase in Bandwidth

Consider the frequency response of an amplifier is show below:

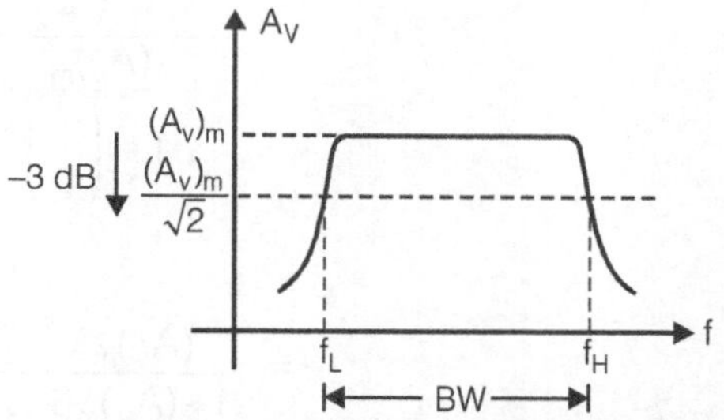


*Fig. 3.10 Frequency response of an amplifier*

## (i) Lower Cut off Frequency

$$(A_v)_l \text{ (Lower cut off frequency gain)} = \frac{(A_v)_m}{1-j\left(\frac{f_L}{f}\right)}$$

$$(A_v)_{lf} \text{ (Lower cut off frequency gain with feedback)} = \frac{(A_v)_l}{1+(A_v)_l\beta} = \frac{\frac{(A_v)_m}{1-j\left(\frac{f_L}{f}\right)}}{1+\frac{(A_v)_m}{1-j\left(\frac{f_L}{f}\right)}\beta}$$

$$= \frac{(A_v)_m}{1+(A_v)_m\beta - j\left(\frac{f_L}{f}\right)} = \frac{(A_v)_m}{1+(A_v)_m\beta}\cdot\frac{1}{1-j\frac{f_L}{f(1+(A_v)_m\beta)}} = \frac{(A_v)_{mf}}{1-j\left(\frac{f_L'}{f}\right)}$$

Where $(A_v)_{mf} = \frac{(A_v)_m}{1+(A_v)_m\beta}$

$$f_L' = \frac{f_L}{1+(A_v)_m\beta}$$

∴ Lower cut off frequency is reduced by a factor $1 + (A_V)_m\beta$ with negative feedback.

## (ii) Higher Cut off Frequency

$$(A_v)_h \text{ (Higher cut off frequency gain)} = \frac{(A_v)_m}{1+j\left(\frac{f}{f_h}\right)}$$

$$(A_v)_{hf} \text{ (Higher cut off frequency gain with feedback)} = \frac{(A_v)_h}{1+(A_v)_h\beta}$$

$$(A_v)_{hf} = \frac{\dfrac{(A_v)_m}{1+j\left(\dfrac{f}{f_h}\right)}}{1+\dfrac{(A_v)_m}{1+j\left(\dfrac{f}{f_h}\right)}\cdot\beta} = \frac{(A_v)m}{1+(A_v)_m\beta+j\left(\dfrac{f}{f_h}\right)}$$

$$= \frac{(A_v)_m}{1+(A_v)_m\beta}\cdot\frac{1}{1+j\dfrac{f}{f_h(1+(A_v)_m\beta)}} = \frac{(A_v)_{mf}}{1+j\left(\dfrac{f}{f_h'}\right)}$$

Where $$f_h' = f_h\,(1+(A_v)_m\beta)$$

∴ Higher cut off frequency is increased by a factor $(1+(A_v)_m\beta)$ with negative feedback.

Bandwidth is defined as difference between higher cut off frequency and lower cut off frequency, therefore bandwidth is increased with negative feedback.

## 3.8 Topologies of Feedback

In a feedback amplifier, sample the output voltage or current by means of sampling or feedback network and applied at the input of the amplifier. Based on the quantity to be amplified. The amplifiers are classified into four types such as:

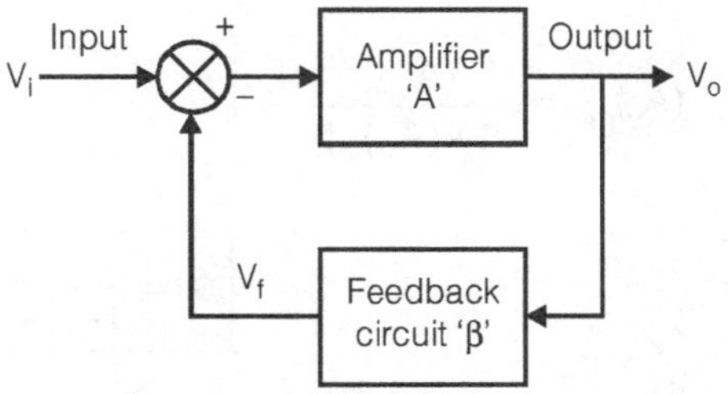


*Fig. 3.11 Block diagram representation of feedback amplifier*

(i) Voltage amplifier
(ii) Current amplifier
(iii) Transconductance amplifier
(iv) Transresistance amplifier

### (i) Voltage Amplifier

In voltage amplifier, voltage $V_i$ acts as input and amplified voltage acts as output. The Figure (3.12) shows Thevenins equivalent circuit of a voltage amplifier.

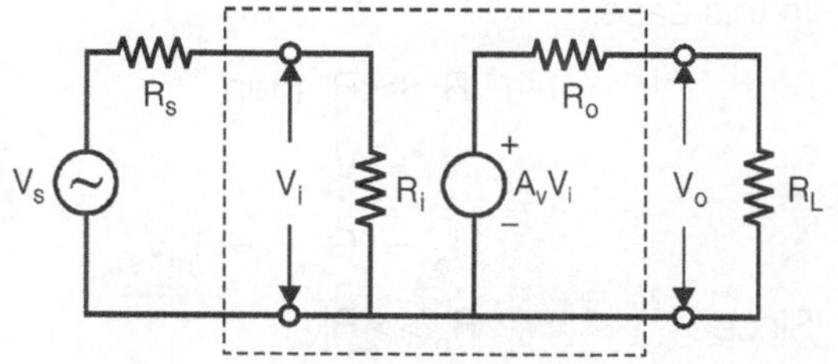


*Fig. 3.12 Voltage amplifier*

In this circuit usually, $R_i >> R_s$ then $V_i \approx V_s$ where $R_i$, $R_s$ are input and source resistance and also if $R_L >> R_o$, thus

$$V_o = A_v V_i \quad \text{(or)} \quad V_o \approx A_v V_s$$

Therefore the output voltage is directly proportional to input voltage and the proportionality factor is independent of the magnitudes of the source and load resistances, such a circuit is called as "voltage amplifier". An ideal voltage amplifier must have "Infinite input resistance" $R_i$ and zero "output resistance" $R_o$.

Voltage amplifier can also be called as voltage controlled voltage source.

## (ii) Current Amplifier

The current amplifier circuit shown in Figure (3.13).

In this amplifier output quantity is a current, thus the feedback signal is in current form, hence $R_i << R_s$ then $I_i \approx I_s$ and also $R_o >> R_L$.

Then $$I_L \approx A_I I_i \quad \text{(or)} \quad A_I = \frac{I_L}{I_s}$$

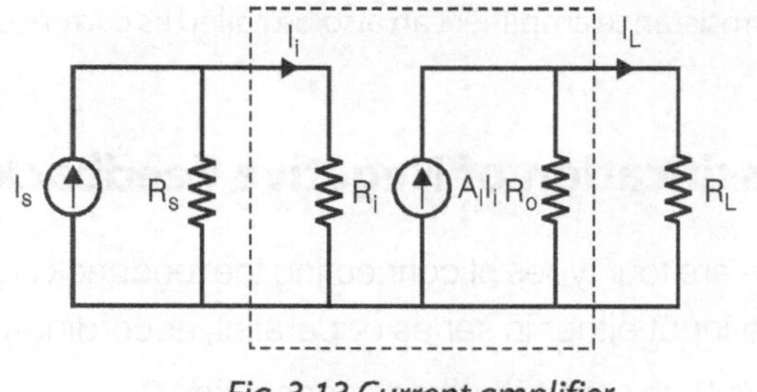


*Fig. 3.13 Current amplifier*

For an ideal current amplifier have "zero input impedance and infinite output impedance" i.e. $R_i = 0$ and $R_o = \infty$.

Current amplifier can also be called as current controlled current source.

## (iii) Transconductance Amplifier

In transconductance amplifier (Figure 3.14) as the name implies the voltage is given as input and current is taken as output and the output current is directly proportional to input voltage

i.e. $$I_o \propto V_i$$

$\therefore$ $$G_m = \frac{I_o}{V_i}$$

In this case

$R_i >> R_s$ then

$V_i \approx V_s$

$\therefore \quad I_o = G_m V_I \approx G_m V_s$

Since $\quad R_o >> R_L$

It drives a low resistance load.

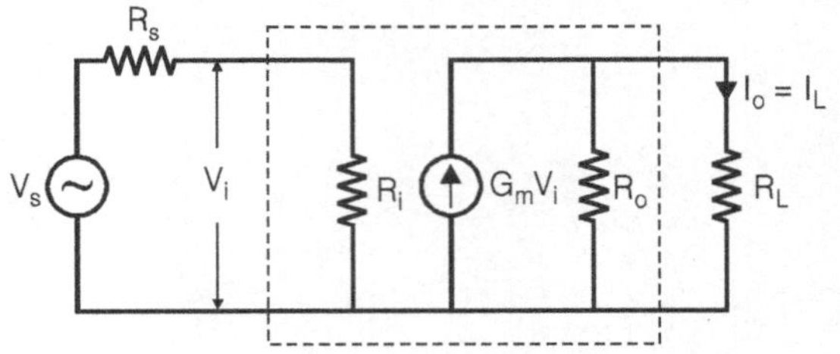


*Fig. 3.14 Transconductance amplifier*

Transconductance amplifier can also be called as voltage controlled current source.

## (iv) Transresistance Amplifier

In transresistance amplifier (Figure 3.15) current source acts as input and voltage is taken as output.

Input is represented as Nortons equivalent circuit and the output circuit is Thevenins equivalent circuit.

In this case $R_s >> R_i$, then

$I_i \approx I_s$

and If $R_o << R_L$

Then $V_o = R_m I_i \approx R_m I_s$

$$R_m = \frac{V_o}{I_i}$$

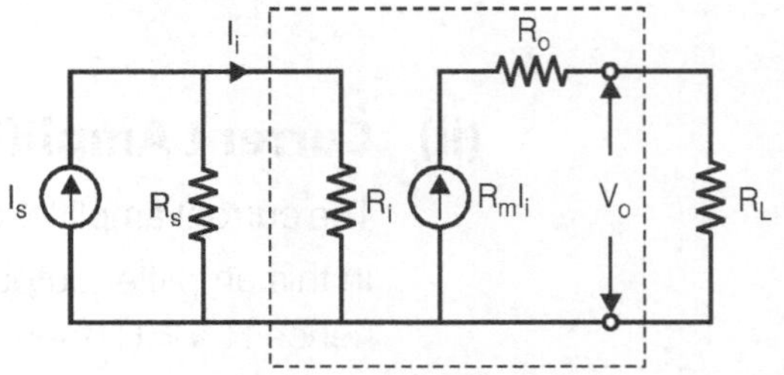


*Fig. 3.15 Transresistance amplifier*

Transresistance amplifier can also be called as current controlled voltage source

## Classification of Negative Feedback Amplifier

There are four types of connecting the feedback signal. Both current or voltage can be feedback to the input either in series or parallel, accordingly it can be classified as:

**(a)** Voltage series feedback connection

**(b)** Voltage shunt feedback connection

**(c)** Current series feedback connection

**(d)** Current shunt feedback connection

In the classification, the first term "voltage" refers to connecting output voltage as input to the feedback network. Similarly the term "current" refers to connecting output current as the input to the feedback network.

The second term "series" refers to connecting the feedback signal in series to the input signal and term "shunt" refers to connecting the feedback signal in shunt with an input source.

For example voltage shunt feedback means, output voltage taken as feedback signal, which is connected in shunt with the input.

## 3.9 Voltage Series Feedback

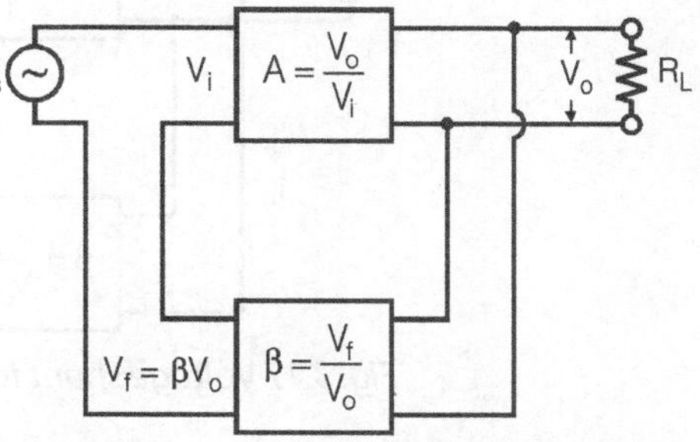


*Fig. 3.16 Voltage series feedback connection*

Figure (3.16) shows the block diagram of voltage series feedback. A sample of output voltage i.e., $V_f = \beta V_o$ is connected in series opposition to the input voltage signal $V_s$. The amplifier has a gain of

$$A = \frac{V_o}{V_i} = \text{gain of the amplifier with out feedback}$$

If feedback is connected then

$$V_s = V_i + V_f \quad \text{or} \quad V_i = V_s - V_f$$
$$V_s = V_i + \beta V_o = V_i + \beta A V_i = (1 + A\beta)V_i$$

$$A_{vf} = \frac{V_o}{V_s} = \frac{V_i A}{(1+A\beta)V_i} = \frac{A}{1+A\beta}$$

Voltage series feedback can also be named as series-shunt feedback.

## 3.10 Voltage Shunt Feedback

Figure (3.17) shows the block diagram for voltage shunt feedback connection. In this, the feedback gain β has the units Siemens. In this case the output voltage is fed in parallel with the input current. Thus it is called as voltage shunt feedback. This configuration acts as transresistance amplifier.

Let $$A = \frac{V_o}{I_i} = \text{gain of amplifier without feedback and } \beta = \frac{I_f}{V_o}$$

$$I_s = I_i + I_f = I_i + \beta V_o = I_i + \beta A\, I_i = I_i(1 + A\beta)$$

$$A_f = \frac{V_o}{I_s} = \text{gain of amplifier with feedback}$$

$$A_f = \frac{V_o}{I_i + \beta A I_i} = \frac{A I_i}{(1+A\beta)I_i} = \frac{A}{1+A\beta}$$

Thus $$\boxed{A_f = \frac{A}{1+A\beta}}$$

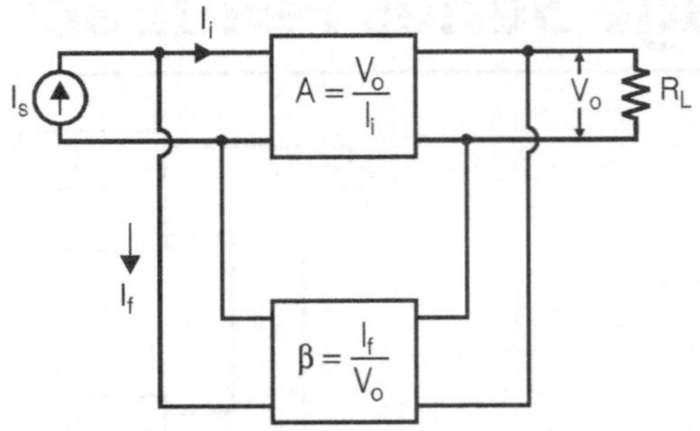


*Fig. 3.17 Voltage shunt feedback connection*

This confirms that the amplifier gain without feedback is reduced by a factor of (1 + βA) because of the feedback.

## Input Impedance

The input impedance with feedback is given by

$$Z_{if} = \frac{V_i}{I_s} = \frac{V_i}{I_i + I_f} = \frac{V_i}{I_i + \beta V_o} = \frac{V_i}{I_i + \beta A I_i} = \frac{V_i}{I_i(1 + A\beta)}$$

$$Z_{if} = \frac{Z_i}{1 + A\beta}$$

Where $Z_i = \frac{V_i}{I_i}$ = Input impedance of amplifier without feedback.

Thus the input impedance gets reduced by a factor (1 + βA). This is true for both voltage shunt and current shunt feedback connections.

## Output Impedance

From the figure 3.18

$$V_o = I_o R_o + AI_i$$

We know $I_i = I_s - I_f$

Thus $I_s = 0$

Then $I_i = -I_f = -\beta V_o$

∴ $V_o = I_o R_o - AI_f$

$$= I_o R_o - A\beta V_o$$

$$V_o(1 + A\beta) = I_o R_o$$

$$R_{of} = \frac{V_o}{I_o} = \frac{R_o}{1 + A\beta}$$

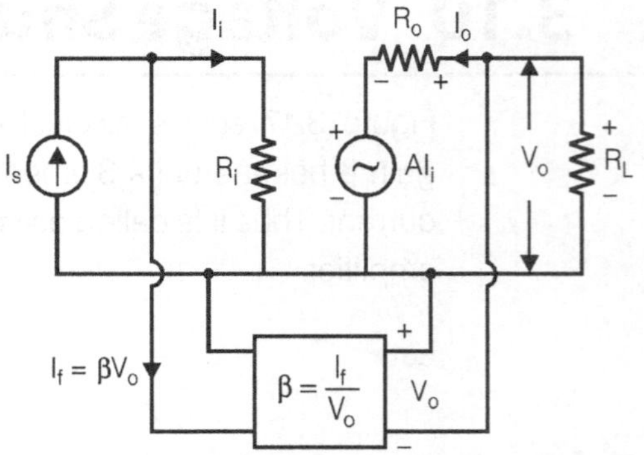


*Fig. 3.18 Equivalent circuit of voltage shunt feedback*

The above equation shows that output impedance of voltage shunt feedback is reduced by the desensitivity factor of (1 + Aβ) from the output impedance of amplifier without feedback.

Voltage shunt feedback can also be named as shunt-shunt feedback.

# 3.11 Current Series Feedback

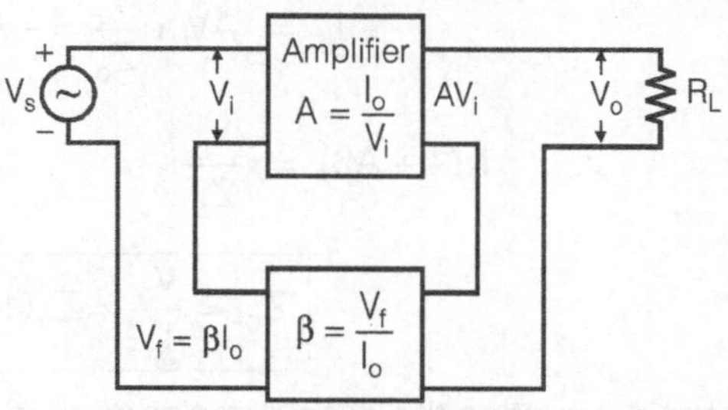


*Fig. 3.19 Current series feedback connection*

Figure (3.19) shows the block diagram of current series feedback. In this, input voltage is proportional to output current is fed back negatively to the input in series to the voltage source. It acts as "transconductance amplifier".

Let gain of amplifier with out feedback is

$$A = \frac{I_o}{V_i} \quad \text{and} \quad \beta = \frac{V_f}{I_o}$$

We have, $V_s = V_i + V_f$

The gain of the amplifier with feedback is given by

$$A_f = \frac{I_o}{V_s} = \frac{I_o}{V_i + V_f} = \frac{AV_i}{V_i + \beta I_o} = \frac{AV_i}{V_i + \beta AV_i} = \frac{A}{1+\beta A}$$

$$\therefore \quad \boxed{A_f = \frac{A}{1+A\beta}}$$

The above equation shows that the gain is reduced by the factor (1 + βA) when feedback is provided.

## Input Impedance

From Figure 3.20

$$V_s = I_i R_i + V_f$$
$$= I_i R_i + \beta I_o$$
$$= I_i R_i + \beta AV_i$$

Where $I_o = AV_i$

$$\therefore \quad V_s = I_i R_i + A\beta I_i R_i$$
$$(V_i = I_i R_i)$$
$$V_s = I_i R_i (1 + A\beta)$$

$$\boxed{R_{if} = R_i(1+A\beta)}$$

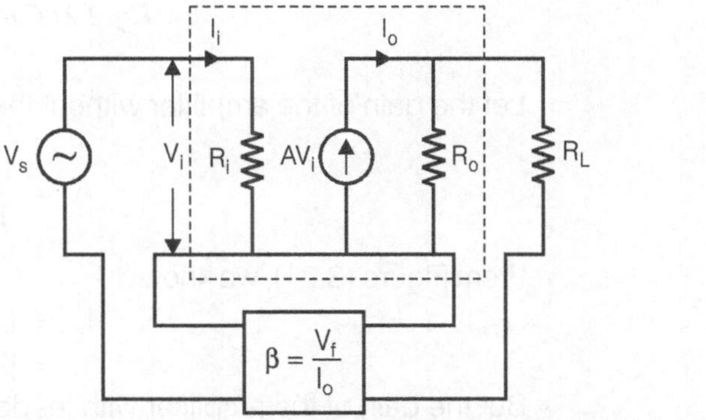


*Fig. 3.20 Equivalent circuit of current series feedback connection*

Thus the input impedance gets increased by the factor (1 + βA) in current series feedback connection.

## Output Impedance

Figure (3.20) shows the current series feedback with basic amplifier represented by its circuit model. To get the output impedance assume source voltage shorted i.e. $V_s = 0$, resulting in a current $I_o$ in the circuit.

$$V_s = V_i + V_f$$

If $V_s = 0$, Then $V_i = -V_f$

$$I_o = AV_i + \frac{V_o}{Z_o} = -AV_f + \frac{V_o}{Z_o} = -A\beta I_o + \frac{V_o}{Z_o}$$

$$\therefore \quad I_o(1 + A\beta) = \frac{V_o}{Z_o}$$

Hence
$$Z_{of} = \frac{V_o}{I_o} = Z_o(1 + A\beta)$$

From the above equation the output impedance is increased by a factor of $(1 + A\beta)$.

Current series feedback can also be named as series-series feedback.

## 3.12 Current Shunt Feedback

The block diagram of current shunt feedback connection is shown in the Figure (3.21). The output current is fed in parallel with the input circuit. Thus it is called as current shunt feedback configuration. It behaves like a "current amplifier".

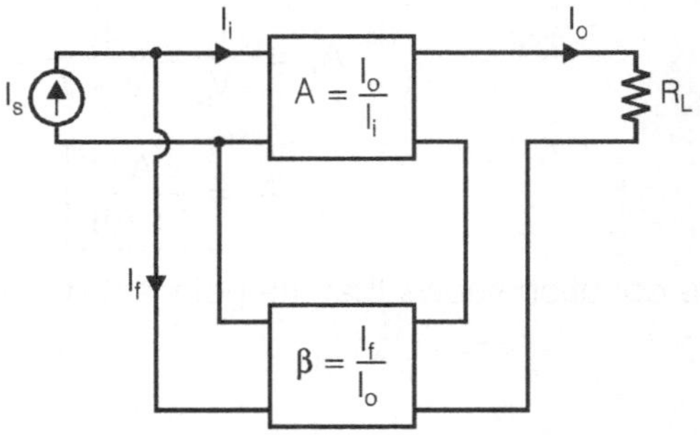


*Fig. 3.21 Current shunt feedback connection*

Let the gain of the amplifier without feedback is

$$A = \frac{I_o}{I_i} \quad \text{and} \quad \beta = \frac{I_f}{I_o}$$

From Figure (3.21), we know

$$I_s = I_i + I_f$$

$$I_f = \beta I_o \quad \text{and} \quad I_o = AI_i$$

But the gain of the amplifier with feedback

$$A_f = \frac{I_o}{I_s} = \frac{I_i A}{I_i + I_f} = \frac{I_i A}{I_i + \beta I_o} = \frac{I_i A}{I_i + \beta A I_i}$$

Thus
$$A_f = \frac{A}{1 + \beta A}$$

Again the gain without feedback is seen to be reduced by a factor $(1 + \beta A)$, when negative current shunt feedback is provided.

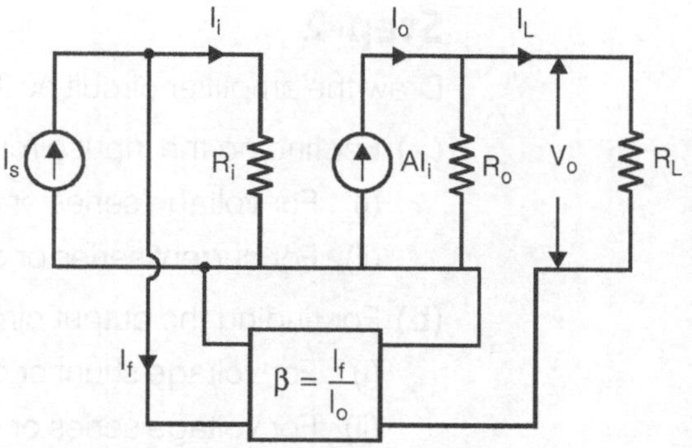

*Fig. 3.22 Equivalent circuit of current shunt*

## Input Impedance

From the Figure (3.22)

$$I_s = I_i + I_f = \frac{V_i}{R_i} + \beta I_o = \frac{V_i}{R_i} + \beta A I_i$$

$$I_s = \frac{V_i}{R_i} + \frac{\beta A V_i}{R_i}$$

$$= \frac{V_i}{R_i}(1 + A\beta)$$

$$R_{if} = \frac{V_i}{I_s} = \frac{R_i}{1 + A\beta}$$ Input resistance of the amplifier with feedback.

Thus the input impedance gets decreased by the factor $(1 + A\beta)$.

## Output Impedance

We know

$$I_s = I_i + I_f$$

(or) $$I_i = I_s - I_f = -I_f \quad [\because I_s = 0]$$

$$I_o = AI_i + \frac{V_o}{R_o} = \frac{V_o}{R_o} - AI_f = \frac{V_o}{R_o} - A\beta I_o$$

$$I_o(1 + A\beta) = \frac{V_o}{R_o}$$

$\therefore$ $$R_{of} = \frac{V_o}{I_o} = R_o(1 + A\beta)$$

Thus the output impedance increased by $[1 + A\beta]$.

Current shunt feedback can also be named as shunt-series feedback.

# 3.13 Method of Analysis of a Feedback Amplifier

The procedure for the analysis of a feedback amplifier is given below:

### Step-1

Identify the topology of feedback by following questions:

(i) Is the sampled output signal a voltage or a current.

(ii) Is the feedback signal being mixed in series or in shunt with external input signal.

Depending on the above there would be four different topologies, as given below:

- Voltage Series
- Voltage shunt.
- Current series.
- Current shunt.

### Step-2

Draw the amplifier circuit as it were without feedback, by using the following analysis:

**(a)** For finding the input circuit, modify the output side as follows:
   (i) For voltage series or voltage shunt topologies, short the output node to ground
   (ii) For current series or current shunt topologies, break/open the output loop.

**(b)** For finding the output circuit, modify the input side as follows:
   (i) For voltage shunt or current shunt topologies, short the input node.
   (ii) For voltage series or current series topologies, break/open the input loop.

### Step-3

The input signal source has to be converted to:
   (i) A Thevenin source i.e. a voltage source for voltage series and current series topologies.
   (ii) A Norton source i.e. current source for voltage shunt and current shunt topologies.

### Step-4

Mark out the feedback signal (i.e. $V_f$ or $I_f$) on the circuit diagram.

### Step-5

Mark out the output signal (i.e. $V_o$ or $I_o$) on the circuit diagram.

### Step-6

Measure the feedback factor

$$\beta = \frac{\text{feedback signal}}{\text{output signal}}$$

### Step-7

Draw the equivalent circuit of the basic amplifier (without feedback), by replacing each active device by its proper model (for example, h-parameter model for BJT). From this equivalent circuit, calculate A.

### Step-8

From the values of β obtained from **step-6** and value of A from **step-7** above,

Calculate $$D = 1 + A\beta$$

### Step-9

Calculate $$A_f = \frac{A}{D}$$

### Step-10

Similarly calculate $R_{if}$ and $R_{of}$ as below:

| | Voltage series | Voltage shunt | Current Series | Current shunt |
|---|---|---|---|---|
| $R_{if}$ | $R_i(1+A\beta)$ | $\frac{R_i}{(1+A\beta)}$ | $R_i(1+A\beta)$ | $\frac{R_i}{1+A\beta}$ |
| $R_{of}$ | $\frac{R_o}{(1+A\beta)}$ | $\frac{R_o}{(1+A\beta)}$ | $R_o(1+A\beta)$ | $R_o(1+A\beta)$ |

## SUMMARY

- Feedback is classified into two types:
  (i) Positive feedback
  (ii) Negative feedback
- If $|A_f| < |A|$, the type of feedback is negative (or) degenerative.
- If $|A_f| > |A|$, the type of feedback is positive (or) regenerative.
- If $(1 + A\beta) = 0$, then the gain becomes infinite and output can be obtained with no external input and such circuit becomes an oscillator.
- There are four types of basic amplifiers based on input and output impedances:
  (i) Voltage amplifier
  (ii) Current amplifier
  (iii) Transconductance amplifier
  (iv) Transresistance amplifier
- For voltage amplifier $Z_i = \infty$ and $Z_o = 0$
- For current amplifier $Z_i = 0$ and $Z_o = \infty$
- For transconductance amplifier $Z_i = \infty$ and $Z_o = \infty$
- For transresistance amplifier $Z_i = 0$ and $Z_o = 0$
- The stability of gain with negative feedback is

$$\frac{\partial A_f}{A_f} = \frac{1}{1+A\beta}\frac{\partial A}{A}$$

- The topologies of negative feedback are:
  (i) Voltage series
  (ii) Voltage shunt
  (iii) Current series
  (iv) Current shunt
- In voltage series topology:
  (i) Voltage amplifier is employed
  (ii) $Z_{if} = Z_i(1 + A\beta)$
  (iii) $Z_{of} = \dfrac{Z_o}{(1+A\beta)}$
  (iv) $A_f = \dfrac{A}{1+A\beta}$
  (v) Voltage series can also called as series shunt topology.
- In voltage shunt topology:
  (i) Transresistance amplifier is employed
  (ii) $Z_{if} = \dfrac{Z_i}{(1+A\beta)}$

(iii) $Z_{of} = \dfrac{Z_o}{(1+A\beta)}$

(iv) $A_f = \dfrac{A}{1+A\beta}$

(v) Voltage shunt can also called as shunt-shunt topology.

- In current series topology:
  (i) Transconductance amplifier is employed
  (ii) $Z_{if} = Z_i(1 + A\beta)$
  (iii) $Z_{of} = Z_o(1 + A\beta)$
  (iv) $A_f = \dfrac{A}{1+A\beta}$
  (v) Current series can also called as series-series topology.
- In current shunt topology:
  (i) Current amplifier is employed
  (ii) $Z_{if} = \dfrac{Z_i}{(1+A\beta)}$
  (iii) $Z_{of} = Z_o(1 + A\beta)$
  (iv) $A_f = \dfrac{A}{1+A\beta}$
  (v) Current shunt can also called as shunt series topology.

## GATE AND IES QUESTIONS

**Q.1** Negative feedback in an amplifier
(a) reduces gain
(b) increases frequency and phase distortions
(c) reduces bandwidth
(d) increases noise

**Exp. (a)**

$$A_f = \frac{A}{1+A\beta}$$

Always negative feedback gain will be less than basic amplifier gain.

**Q.2** An amplifier has an open-loop gain of 100, an input impedance of 1 kΩ, and an output impedance of 100 Ω. A feedback network with a feedback factor of 0.99 is connected to the amplifier in a voltage series feedback mode. The new input and output impedances, respectively, are
(a) 10 Ω and 1 Ω
(b) 10 Ω and 10 Ω
(c) 100 kΩ and 1 Ω
(d) 100 kΩ and 1 kΩ

**Exp. (c)**

Given

$$A = 100$$
$$Z_i = 1\ k\Omega$$
$$Z_o = 100\ \Omega$$
$$\beta = 0.99$$

In voltage series, $Z_{if} = Z_i(1 + A\beta) = 1\ k\Omega\,(1 + 100 \times 0.99) = 1\ k\Omega\,(1 + 99) = 100\ k\Omega$

$$Z_{of} = \frac{Z_o}{1+A\beta} = \frac{100\ \Omega}{(1+100\times 0.99)} = \frac{100}{100} = 1\Omega$$

**Q.3** In a negative feedback amplifier using voltage-series.

(a) $R_i$ decreases and $R_o$ decreases
(b) $R_i$ decreases and $R_o$ increases
(c) $R_i$ increases and $R_o$ decreases
(d) $R_i$ increases and $R_o$ increases

**Exp. (c)**

In voltage series $R_{if} = R_i(1 + A\beta)$ i.e. $R_{if}$ increases

$$R_{of} = \frac{R_o}{(1+A\beta)}$$ i.e. $R_{of}$ decreases

**Q.4** An amplifier without feedback has a voltage gain of 50, input resistance of 1 kΩ and output resistance of 2.5 kΩ. The input resistance of the current-shunt negative feedback amplifier using the above amplifier with a feedback factor of 0.2 is

(a) 1/11 kΩ
(b) 1/5 kΩ
(c) 5 kΩ
(d) 11 kΩ

**Exp. (a)**

We have $R_i = 1\ k\Omega$, $\beta = 0.2$, $A = 50$ ; Thus $R_{if} = \frac{R_i}{1+A\beta} = \frac{1k\Omega}{1+50\times 0.2} = \frac{1}{11}k\Omega$

**Q.5** The action of a JFET in its equivalent circuit can best be represented as a

(a) Current Controlled Current Source
(b) Current Controlled Voltage Source
(c) Voltage Controlled Voltage Source
(d) Voltage Controlled Current Source

**Exp. (d)**

JFET is a voltage controlled current source

i.e. $i_d \propto V_{GS}$

**Q.6** An ideal op-amp is an ideal

(a) voltage controlled current source
(b) voltage controlled voltage source
(c) current controlled current source
(d) current controlled voltage source

**Exp. (b)**

An ideal op-amp is a voltage controlled voltage source

i.e. $V_o \propto V_d$

**Q.7** Voltage series feedback (also called series-shunt feedback) results in

(a) increase in both input and output impedances
(b) decrease in both input and output impedances
(c) increase in input impedance and decrease in output impedance
(d) decrease in input impedance and increase in output impedance

**Exp. (c)**

Voltage series feedback results in increase in input impedance and decrease in output impedance.

**Q.8** The effect of current shunt feedback in an amplifier is to

(a) increase the input resistance and decrease the output resistance
(b) increase both input and output resistances
(c) decrease both input and output resistances.
(d) decrease the input resistance and increase the output resistance

**Exp.** (d)

In current shunt $R_{if} = \frac{R_i}{1+A\beta}$

$R_{of} = R_o(1 + A\beta)$

**Q.9** The input impedance ($Z_i$) and the output impedance ($Z_o$) of an ideal trans-conductance (voltage controlled current source) amplifier are

(a) $Z_i = 0, Z_0 = 0$
(b) $Z_i = 0, Z_0 = \infty$
(c) $Z_i = \infty, Z_0 = 0$
(d) $Z_i = \infty, Z_0 = \infty$

**Exp.** (d)

For transconductance amplifier

$Z_i = \infty, \quad Z_o = \infty$

**Q.10** The feedback used in the circuit shown in figure can be classified as

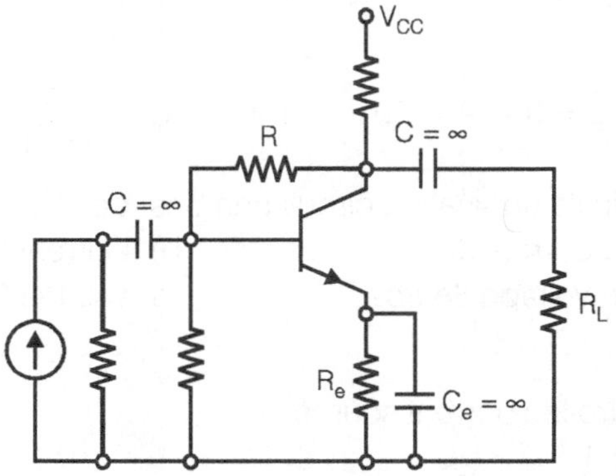


(a) shunt-series feedback
(b) shunt-shunt feedback
(c) series-shunt feedback
(d) series-series feedback

**Exp.** (b)

It is voltage shunt also called as shunt-shunt feedback.

**Q.11** The nature of feedback in the opamp circuit shown is

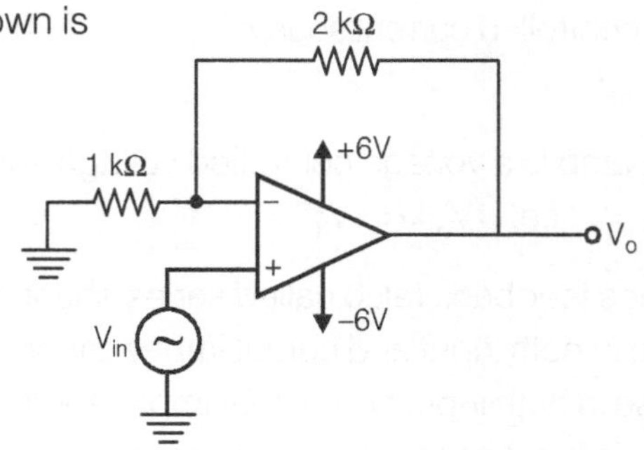


(a) Current – Current feedback
(b) Voltage – Voltage feedback
(c) Current – Voltage feedback
(d) Voltage – Current feedback

**Exp.** (d)

The nature of feedback is voltage shunt also called as voltage current feedback.

**Q.12 Assertion (A):** A large negative feedback is deliberately introduced in an amplifier to make its gain independent of the variation of parameters of the active device and other circuit components.

**Reason (R):** A large negative feedback results in a high value of return difference compared to unity, which makes the feedback gain inversely proportional to the feedback factor.

(a) Both A and R are true and R is the correct explanation of A
(b) Both A and R are true but R is NOT the correct explanation of A
(c) A is true but R is false
(d) A is false but R is true

**Exp. (a)**

$$A_{vf} = \frac{A_v}{1+A_v\beta}$$

When $\beta >> 1, \quad A_{vf} \approx \frac{A_v}{A_v\beta} = \frac{1}{\beta}$

Thus, gain $A_{vf}$ is independent of $A_v$

Return difference $D = 1 + A_v\beta$

**Q.13** In a feedback series regulator circuit, the output voltage is regulated by controlling the
(a) magnitude of the input voltage
(b) gain of the feedback transistor
(c) reference voltage
(d) voltage drop across the series pass transistor

**Exp. (d)**

In a feedback series regulator circuit. The output voltage is regulated by controlling the voltage drop across the series pass transistor.

**Q.14** The given circuit has a feedback factor of

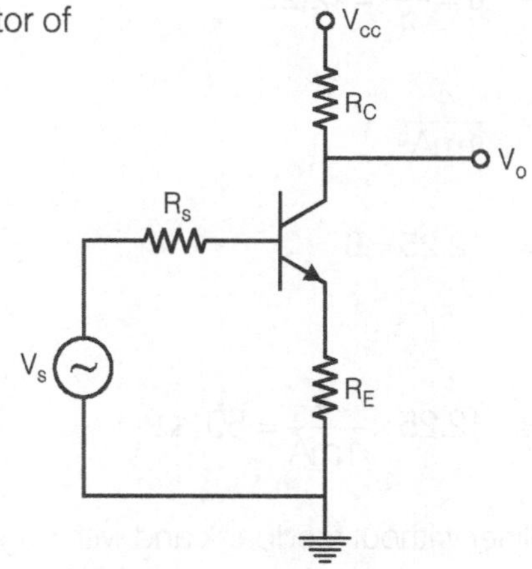


(a) $-\frac{R_c}{Rs}$

(b) $-\frac{R_E}{R_c}$

(c) $-\frac{R_E}{R_s}$

(d) $-\frac{R_c}{R_E}$

**Exp. (b)**

The equivalent in ac analysis

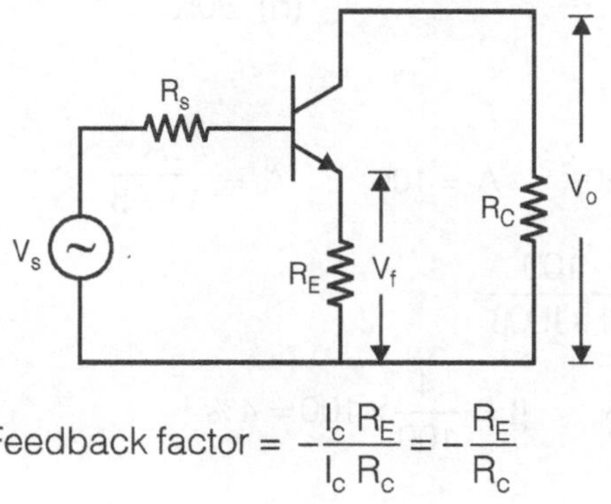


$$\beta = \text{Feedback factor} = -\frac{I_c R_E}{I_c R_c} = -\frac{R_E}{R_c}$$

**Q.15** Consider the following statements regarding the bootstrap biasing arrangement for a BJT emitter follower:

1. The input impedance is very high.
2. The voltage gain is exactly equal to one.
3. The output impedance is equal to zero.

Which of these statements is correct?

(a) 1 alone (b) 2 alone
(c) 3 alone (d) 1 and 3

**Exp. (d)**

In the bootstrap biasing arrangement, voltage gain $A_v \to 1$, but not exactly equal to 1.

**Q.16** If a common emitter amplifier with an emitter resistance $R_e$ has an overall transconductance gain of –1 mA/V, a voltage gain of –4 and desensitivity of 50, then the value of the emitter resistance $R_e$ would be

(a) 50 kΩ (b) 0.98 kΩ
(c) 50 MΩ (d) 0.98 MΩ

**Exp. (a)**

$$\frac{I_c}{V_i} = g_m = -1\,\text{mA/V}$$

$$\frac{V_o}{V_i} = A_v = -4$$

$$1 + A\beta = 50$$

$$1 - 4\beta = 50$$

$$\frac{V_f}{V_o} = \beta = \frac{49}{4} = 12.25$$

$$\frac{V_o}{I_c} = \frac{4}{1\text{mA}}$$

$$\frac{I_c R_e}{I_c \times \dfrac{4}{1\text{mA}}} = 12.25 = \beta$$

$$Re = 12.25 \times \frac{4}{1\text{mA}} = 50\,\text{k}\Omega$$

**Q.17** The voltage gain of an amplifier without feedback and with negative feedback respectively are 100 and 20. The percentage of negative feedback (β) would be

(a) 4% (b) 5%
(c) 20% (d) 80%

**Exp. (a)**

$$A_f = 20 \;;\; A = 100 \;;\; A_f = \frac{A}{1+A\beta}$$

$$20 = \frac{100}{1+100\beta}$$

$$1 + 100\beta = 5 \;;\; \beta = \frac{4}{100} \times 100 = 4\%$$

**Q.18 Assertion (A):** In an amplifier with negative feedback, the gain-stability is improved by the factor $1 + A\beta$ where A is the magnitude of gain and $\beta$ is the feedback factor.

**Reason (R):** The relation between gain with feedback $A_f$ and gain without feedback A, is $A_f = \frac{A}{1+A\beta}$.

Thus $\frac{dA_f}{A_f} = \frac{1}{1+A\beta}\left|\frac{dA}{A}\right|$

(a) Both A and R are true and R is the correct explanation of A
(b) Both A and R are true but R is NOT the correct explanation of A
(c) A is true but R is false
(d) A is false but R is true

**Exp. (a)**

The gain stability relation with negative feedback is

$$\frac{\partial A_f}{A_f} = \frac{1}{1+A\beta}\left|\frac{dA}{A}\right|$$

**Q.19** The biasing shown in the below circuit is

(a) Emitter bias
(b) Self bias
(c) Potential divider bias
(d) Bootstrap bias

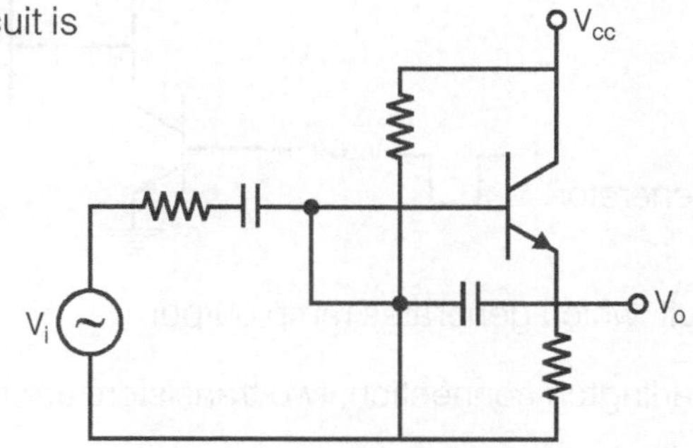


**Exp. (d)**

This arrangement is bootstrap bias.

**Q.20** In a BJT amplifier with the introduction of feedback, the input impedance is reduced, output impedance is increased, bandwidth is increased and distortion is reduced. The feedback is

(a) Voltage series
(b) Current series
(c) Voltage shunt
(d) Current shunt

**Exp. (d)**

It is a current shunt.

**Q.21 Assertion (A):** If the emitter bypass capacitor of an R-C coupled CE amplifier gets disconnected, its voltage gain increases.

**Reason (R):** The unbypassed emitter resistor gives negative feedback.

(a) Both A and R are true and R is the correct explanation of A
(b) Both A and R are true but R is NOT the correct explanation of A
(c) A is true but R is false
(d) A is false but R is true

**Exp. (d)**

The unbypassed emitter resistor gives negative feedback which reduces the voltage gain.

**Q.22** Consider the following circuits:

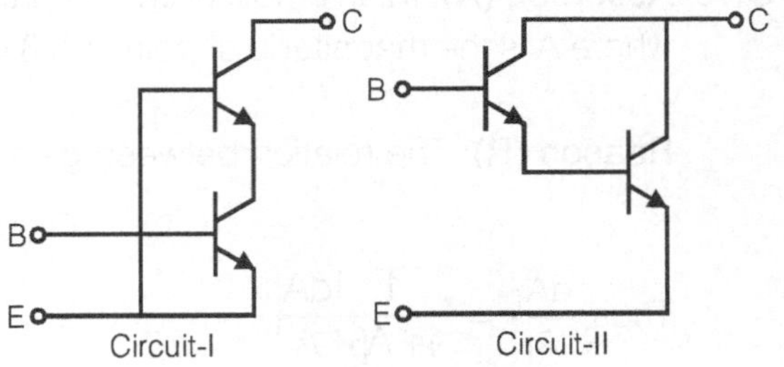


Which one of the following statements is correct?

(a) Circuit 1 is parallel connection and Circuit 2 is Darlington connection
(b) Circuit 1 is cascode connection and Circuit 2 is Darlington connection
(c) Circuit 1 is Darlington connection and Circuit 2 is cascode connection
(d) Circuit 1 is cascode connection and Circuit 2 is parallel connection

**Exp. (b)**

Circuit-1 is cascode (CE – CB)
Circuit-2 is Darlington connection (CC – CC).

**Q.23** Consider the following circuit:

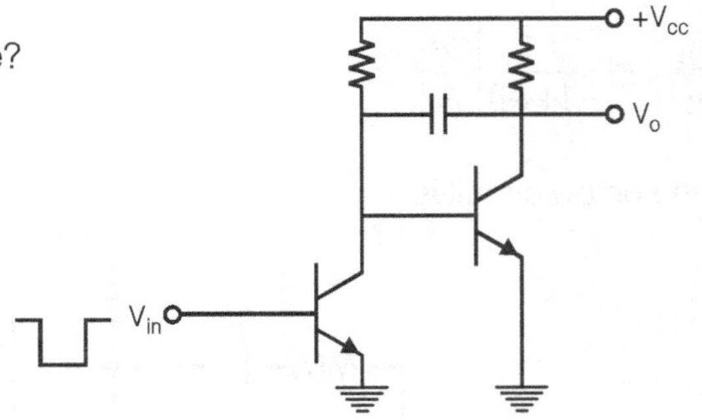


What is the circuit shown above?

(a) Miller sweep
(b) Bootstrap sweep
(c) Schmitt trigger
(d) Triangular wave generator

**Exp. (a)**

It is Miller sweep circuit, which generates ramp output.

**Q.24 Assertion (A):** In a Darlington connection, two transistors are connected in cascade in common emitter configuration.

**Reason (R):** The Darlington connection aims at making the current gain very high, almost equal to the product of beta of individual transistors.

(a) Both A and R are true and R is the correct explanation of A
(b) Both A and R are true but R is NOT the correct explanation of A
(c) A is true but R is false
(d) A is false but R is true

**Exp. (d)**

In a Darlington connection, two transistors are connected in cascade in common collector configuration.

**Q.25** An amplifier has an open loop gain of 1000 ± 10. Negative feedback is provided such that the gain variation remains within 0.1%. What is the amount of feedback $\beta_F$?

(a) 1/10 (b) 1/9
(c) 9/100 (d) 9/1000

**Exp. (d)**

$$\frac{\partial A_f}{A_f} = \frac{1}{1+A\beta}\frac{\partial A}{A} \quad ; \quad \frac{0.1}{100} = \frac{10}{1000}\left[\frac{1}{1+1000\beta}\right]$$

$$1 + 1000\beta = 10$$

$$\beta = \frac{9}{1000}$$

**Q.26** Cascode amplifiers when compared with a simple common-emitter amplifier provide which of the following?
(a) Higher voltage gain and same bandwidth
(b) Same voltage gain but higher bandwidth
(c) No change in either voltage gain or bandwidth
(d) Voltage gain less than one but bandwidth equal to $f_T$

**Exp.** (b)
Cascode improves the bandwidth of CE.

**Q.27** Which of the following features are offered by a bipolar junction transistor amplifier in Darlington connection?
1. High voltage gain 2. High input impedance 3. High current gain
Select the correct answer using the code given below:
(a) 1 and 2 only (b) 2 and 3 only
(c) 1 and 3 only (d) 1, 2 and 3

**Exp.** (b)
In Darlington connection it will have high $Z_i$ and $A_I$ but voltage gain is unity.

**Q.28** In a negative feedback amplifier, when is the input impedance increased?
(a) If the signal sampled is a voltage (b) If the signal sampled is a current
(c) If the feedback signal is a voltage (d) If the feedback signal is a current

**Exp.** (c)
The input impedance is increased if the feedback signal is a voltage.

**Q.29** A feedback amplifier has an open loop gain of –100. If 4% of the output is feedback in a degenerative loop. What is the closed loop gain of the amplifier?
(a) –33.3 (b) –25
(c) –20 (d) +25

**Exp.** (c)
Open loop gain $A = -100$

Negative feedback $\beta = \frac{-4}{100}$

Closed loop gain $$A_f = \frac{A}{1+A\beta} = \frac{-100}{1+(-100)\left(\frac{-4}{100}\right)} = -20$$

**Q.30** Which of the following describe the correct properties of an emitter follower circuit?
1. It is a voltage series feedback circuit.
2. It is a current series feedback circuit.
3. Its voltage gain is less than unity.
4. Its output impedance is very low.

Select the correct answers from the codes given below :
(a) 1, 3 and 4 (b) 2, 3 and 4
(c) 2 and 3 only (d) 2 and 4 only

**Exp.** (a)
Emitter follower properties are
- It is a voltage series feedback circuit.
- Its voltage gain is less than unity.
- Its output impedance is very low.

## REVIEW QUESTIONS

1. What is feedback? Why is it employed in amplifier circuit? How many types of feedback are possible? Discuss.
2. What is negative feedback? Discuss how it can improve stability in an amplifier.
3. Enumerate the advantages and disadvantages with negative feedback in an amplifier.
4. Derive an expression for the transfer gain of a feedback amplifier.
5. What are the various types of amplifiers based on the output and input parameters? Give their equivalent circuits and ideal characteristics.
6. Derive an expression for the input impedance with feedback of a voltage shunt feedback amplifier.
7. Derive an expression for output impedance with feedback of a current shunt feedback amplifier.
8. Draw a feedback amplifier in block diagram form and explain each block in detail giving its functions.
9. What are the four topologies of a feedback amplifier?
10. Define desensitivity factor D. Give its significance in negative feedback amplifiers.

OOOO

# CHAPTER 4

# OSCILLATORS

## About This Chapter

The concepts introduced in this chapter are:

- Positive feedback in amplifier circuits.
- Condition and frequency of sinusoidal oscillations.
- Various types of oscillator circuits:
  - RC phase shift oscillator.
  - Wien bridge oscillator.
  - Hartley oscillator.
  - Colpitts oscillator.
  - Clapp oscillator.
- Study of crystal oscillator for maintaining high frequency stability.

# 4 OSCILLATORS

## 4.1 Introduction

An oscillator is a device which generates an alternating voltage. This may also be defined as a circuit which generates an ac output signal without requiring any externally applied input signal. The oscillator converts dc energy into ac energy at a required frequency. So the operation of an oscillator is opposite to that of a rectifier which converts ac power into dc power.

The function of an oscillator circuit is similar to that of an amplifier circuit. The main difference between an amplifier and an oscillator Figure (4.1) can be stated as follows:

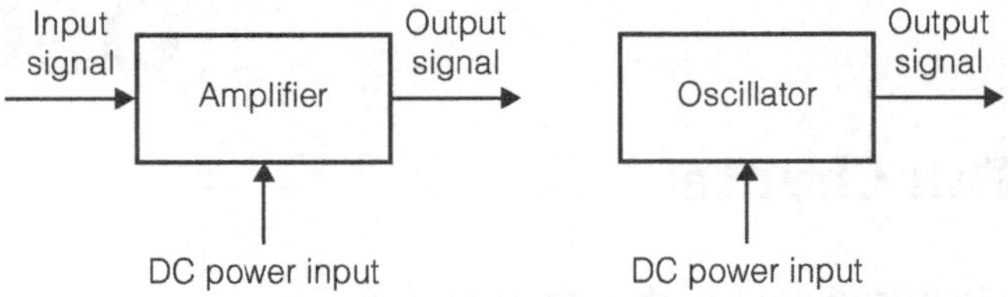


*Figure 4.1*

All electronic communication systems like TV, radio, computers, and industrial instrumentation systems required one or more of the different types of waveform viz, sinusoidal, square wave, pulses or triangular wave of specified frequency and amplitude. These signals are generated by electronic circuits known as "oscillators" or "waveform generators".

The condition of oscillation and frequency of oscillation depends on the principle known as Barkhausen criterion. The stability of both amplitude and frequency is of prime concern since the device employed here is a transistor whose parameters are not constant. This concept of stability is also considered in this chapter.

## 4.2 Types of Oscillators

There are various ways of classifying oscillators. For example depending upon the frequency region, it could be audio frequency oscillators, radio frequency oscillators etc. Sinusoidal oscillators, or sawtooth oscillator or square wave generator etc. are depending upon the components used for frequency selection. The classification of sinusoidal oscillators could be RC oscillators, LC oscillators and crystal oscillators.

**Oscillators may be classified in a variety of ways:**

**(a) According to the nature of waveform generated**

(i) Sinusoidal or harmonic oscillators. The oscillators which produce sinusoidal or nearly a sinusoidal waveform of a definite frequency are known as sinusoidal oscillators.

(ii) Relaxation oscillators or non sinusoidal oscillators. The oscillators which produce square waves, triangular waves, pulses or sawtooth waves are known as relaxation oscillators.

**(b) According to the frequency generated**

(i) Audio frequency oscillators. The oscillators which generate signals in the audio frequency range are known as audio frequency oscillators.

(ii) Radio frequency oscillators. The oscillators which generate signals in the radio frequency range are called radio frequency oscillators.

The classification of such oscillators is shown below:

| Oscillator | Frequency range |
|---|---|
| Audio frequency (A.F) oscillators | 20 Hz-20 kHz |
| Radio frequency (R.F) oscillators | 30 kHz-30 MHz |
| Very high frequency (V.H.F) oscillators | 30 MHz-300 MHz |
| Ultra high frequency (U.H.F) oscillators | 300 MHz-3 GHz |
| Micro wave oscillators | 3 GHz-Several GHz |

Sinusoidal oscillators may be further subdivided into following categories:

**(i) RC Oscillators**

Oscillators using RC network. The examples are phase shift oscillator and Wien-bridge oscillator. Practical considerations limit the use of RC oscillators in audio frequency range.

**(ii) LC Feedback Oscillators**

Oscillators using tuned circuits are known as LC feedback oscillators. Practically available values of L and C confine the application of these oscillators for the generation of radio frequency signals. Hartley, Colpitts, Clapp etc. are the examples of such oscillators.

**(iii) Negative Resistance Oscillators**

The oscillators which use a tunnel diode (or) UJT that possesses a current voltage characteristic curve of negative slope with in some range of operation are known as negative resistance oscillators. The examples are tunnel diode oscillator and UJT relaxation oscillator.

**(iv) Crystal Oscillators**

When piezo electric crystals are used in place of LC circuit for higher frequency stability, the oscillators are called as crystal oscillators.

## 4.3 Essentials of Transistor Oscillator

**Oscillator must have the following three elements:**

1. Oscillatory circuit.
2. Amplifier
3. Feedback network

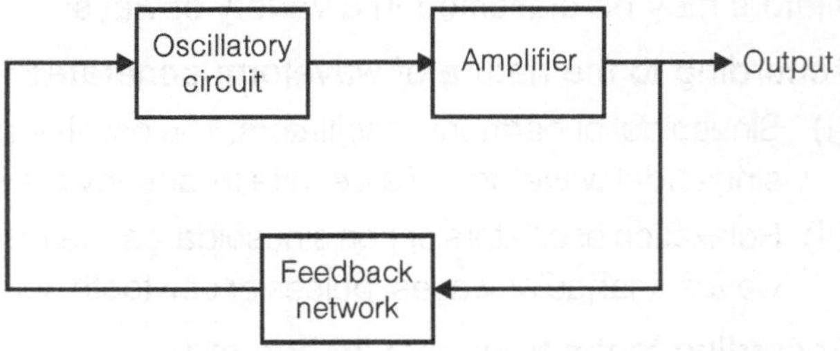


*Figure 4.2 Block diagram of an oscillator*

The electronic amplifier receives dc power from the battery or dc power supply and converts it into ac power for supply to the tank circuit. The oscillations occurring in the tank circuit are applied to the input of the electronic amplifier. Because of the amplifying properties of the amplifier, we get increased output of these oscillations. This amplified output of oscillations is because of dc power supplied from the external source (a battery or power supply). The output of the amplifier can be supplied to the tank circuit to meet the losses.

The feedback network supplies a part of output power to the tank or oscillatory circuit in correct phase to aid the oscillations. In other words the feedback circuit provides positive feedback.

The oscillatory circuit or element, also called the tank circuit, consists of an inductive coil of inductance L, connected in parallel with a capacitor of capacitance C. The frequency of oscillation in the circuit depends upon the values of L and C. The frequency of oscillation is the resonant or natural frequency and is given by the equation

$$f = \frac{1}{2\pi\sqrt{LC}} \text{ Hz}$$

Where L is inductance of coil in Henrys, and C is the capacitance of capacitor in Farads.

## 4.4 Barkhausen Criterion

Consider an inverting amplifier (i.e. output of the amplifier is 180° out of phase with input) with an open loop gain 'A' and the feedback network with feedback factor 'β' (always $\beta > 1$). In an oscillator circuit the input voltage $V_i$ to the amplifier is derived from its output of the amplifier through feedback network.

To produce the oscillation, the feedback must be positive i.e. feedback voltage should be in phase with input voltage $V_i$. Thus the feedback network needs to produce another phase shift of 180°. This ensure the total phase shift with in the loop is 360°, thus it produces the oscillation.

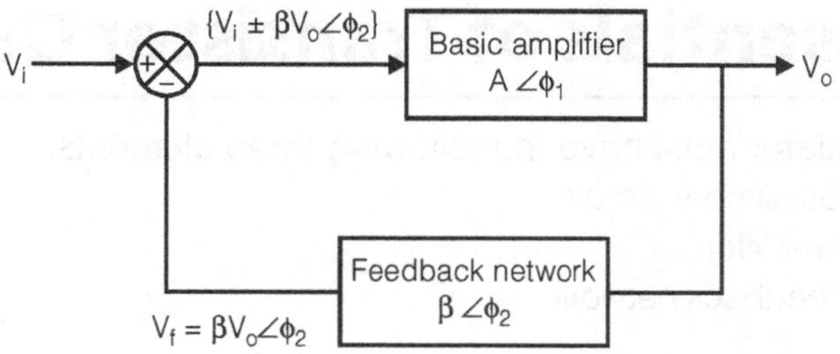


*Figure 4.4 Block diagram of feedback system*

Let $\phi_1 \rightarrow$ Phase of the basic amplifier

$\phi_2 \rightarrow$ Phase of the feedback network

$V_o \rightarrow$ Output voltage of the amplifier

$V_f = \beta V_o \angle\phi_2 \rightarrow$ Feedback voltage of the amplifier

From Figure (4.4) $V_o = \{V_i \pm \beta V_o \angle\phi_2\} A\angle\phi_1$

For oscillators, we use positive feedback

$$V_o = \{V_i + \beta V_o \angle\phi_2\} A\angle\phi_1$$

$$V_o\left[1 - A\beta\underline{|\phi_1 + \phi_2}\right] = AV_i\angle\phi_1$$

$$V_o = \frac{AV_i \angle\phi_1}{1 - A\beta\underline{|\phi_1 + \phi_2}}$$

Oscillations are produced when

$$A_f = \frac{V_o}{V_i} = \infty$$

This is possible when

(i) $|A\beta| = 1$

(ii) $\underline{|\phi_1 + \phi_2} = 0°$ or multiples of $2\pi$.

To generate oscillation $|A\beta| > 1$. To maintain oscillation the condition $|A\beta| = 1$. The Barkhausen criterion, defines two basic requirements for oscillations.

- **Total phase shift in the closed loop is 0° or 360°.**
- **The magnitude of loop gain i.e. $|A\beta| = 1$.**

If the above two conditions are satisfied, the feedback amplifier will produce an oscillation with out applying any external input.

# 4.5 RC Phase Shift Oscillator

## PRINCIPLE OF PHASE SHIFT OSCILLATOR

For low frequencies (i.e. audio frequencies) RC oscillators are more suitable. Tuned circuit is not an essential requirement for oscillation. The essential requirement is that there must be a 180° phase shift around the feedback network and loop gain should be greater than unity. The 180° phase shift in feedback signal can be achieved by a suitable RC network consisting of three RC sections as shown below:

## Circuit Arrangement

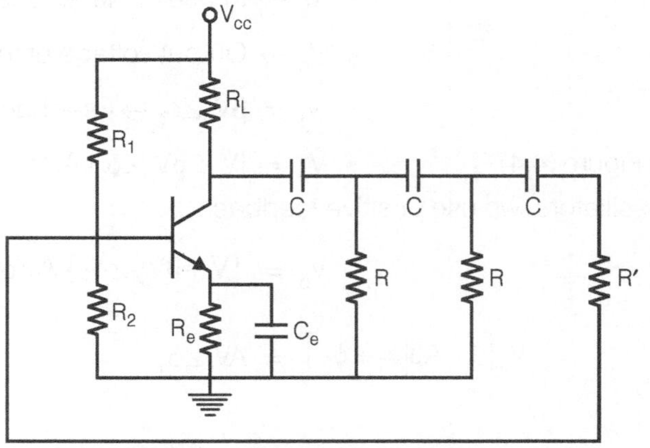


*Figure 4.5 RC phase shift oscillator*

The circuit arrangement of a phase shift oscillator using NPN transistor in common emitter configuration is shown in Figure (4.5). Here $R_1$ and $R_2$ provides biasing to the circuit. $R_L$ is the load which controls the collector voltage. $C_e$ is a bypass capacitor which prevents ac signal degeneration. The output of the amplifier goes to a feedback network which consists of three identical RC sections. It should be noted that the last section contains a resistance $R' = R - h_{ie}$. Since this resistor is connected with the base of the transistor, the input resistance $h_{ie}$ of the transistor is added to it to give a total resistance R.

## Circuit Diagram

From the Figure (4.5) RC network produces a phase shift of 180° between input and output voltages. Since CE amplifier produces a phase of 180°, the total phase change becomes 360° or 0° which is the essential requirement of sustained oscillations. The RC phase shift network serve as frequency determining circuit. These oscillators are used for audio frequency ranges. LC tuned circuits at low frequencies become to much bulky and expensive more over they suffer from frequency instability and poor waveform.

## Frequency of Oscillations

The equivalent circuit of Figure (4.5) is shown in Figure (4.6).

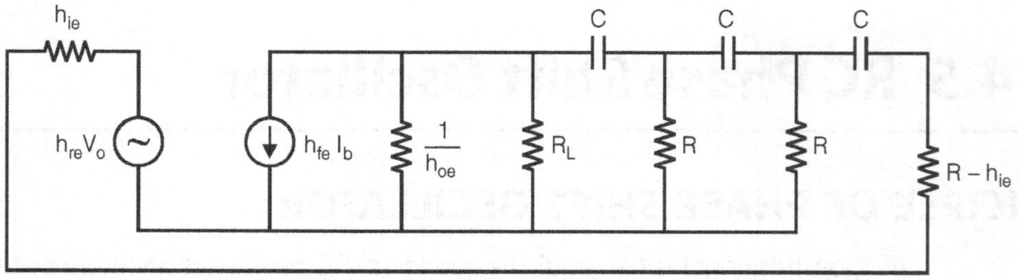


*Figure 4.6 Equivalent circuit of RC phase shift oscillator*

The above circuit of Figure (4.6) can be simplified by the following assumptions.

- $h_{re}$ of the transistor is usually small and hence $h_{re} V_o$ is eliminated from the circuit.
- $1/h_{oe}$ is much larger than $R_L$, its effect can be neglected.
- The current source is replaced by voltage source using Thevenin's source.

Equivalent circuit is shown in Figure (4.7).

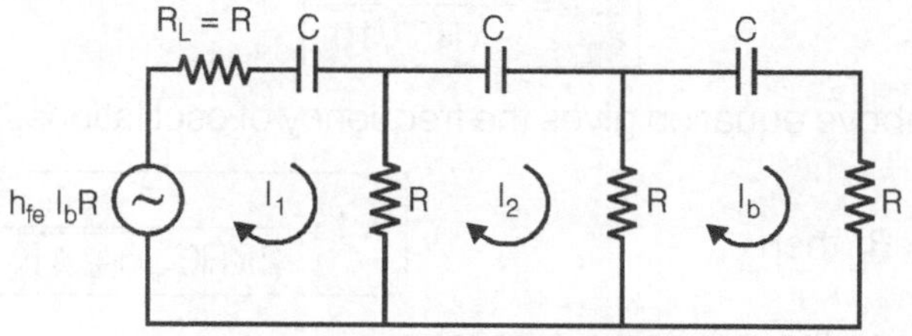


*Figure 4.7 Simplified circuit*

Applying Kirchhoffs voltage law equations for three loops, we have

$$\left[2R + \frac{1}{j\omega C}\right] I_1 - RI_2 + h_{fe}\, I_b R = 0 \quad \text{...(i)}$$

$$-RI_1 + \left[2R + \frac{1}{j\omega C}\right] I_2 - RI_b = 0 \quad \text{...(ii)}$$

$$0 - RI_2 + \left[2R + \frac{1}{j\omega C}\right] I_b = 0 \quad \text{...(iii)}$$

The currents $I_1$, $I_2$ and $I_b$ are non-vanishing, the determinant of the coefficients of $I_1$, $I_2$ and $I_b$ must be zero putting $X_C = \frac{1}{\omega C}$, we get

$$\begin{vmatrix} (2R - jX_C) & (-R) & h_{fe}R \\ (-R) & (2R - jX_C) & (-R) \\ 0 & (-R) & (2R - jX_C) \end{vmatrix} = 0$$

After the simplification

We have

Equating the imaginary part, we get

$$X_C^3 = 10R^2 X_C$$

$$\text{(or)} \quad X_C^2 = 10R^2$$

$$\text{(or)} \quad X_C = \sqrt{10} \cdot R$$

$$\text{(or)} \quad \frac{1}{\omega C} = \sqrt{10} \cdot R$$

$$\text{(or)} \quad \frac{1}{\omega} = \sqrt{10} \cdot RC$$

$$\text{(or)} \quad \omega = \frac{1}{\sqrt{10}\, RC}$$

$$f = \frac{1}{2\pi RC\sqrt{10}}$$

$$\therefore f = \frac{1}{2\pi RC\sqrt{10}}$$

The above equation gives the frequency of oscillations.

If $R \neq R_L$ then $$\therefore f = \frac{1}{2\pi RC\sqrt{6+4K}}$$ where $K = \frac{R_L}{R}$

Equating the real part, we have

$$4R^3 - 6RX_C^2 + h_{fe}R^3 = 0$$

(or) $4R^3 - 6R \times (10R^2) + h_{fe} \times R^3 = 0$

(or) $4 - 60 + h_{fe} = 0$

(or) $$h_{fe} = 56$$

Thus for sustained oscillations, the value of $h_{fe}$ of the transistor should be 56.

If $R_L \neq R$ then

$$h_{fe} = \frac{29}{K} + 4K + 23$$

To get minimum value of $h_{fe}$ for sustained oscillation differentiate the above equation w.r.t.k and equated to zero

$$\frac{\partial h_{fe}}{\partial K} = \frac{\partial}{\partial K}\left[\frac{29}{K} + 23 + 4K\right] = 4 - \frac{29}{K^2} = 0$$

(or) $$K^2 = \frac{29}{4} \text{ or } K = 2.693$$

$$\therefore \quad h_{fe} = \frac{29}{2.693} + 23 + 4(2.693) = 44.5$$

Thus for producing oscillations minimum $h_{fe}$ required is 44.5.

## Points to Remember

1. Transistor (180° phase shift) + 3 sets of RC (180° phase shift) gives 360°.
2. Frequency of oscillation

$$f = \frac{1}{2\pi RC\sqrt{6+4K}}$$ where $K = \frac{R_C}{R}$

If $K = 1$; $f = \frac{1}{2\pi RC\sqrt{10}}$

If $K < 1$; $f = \frac{1}{2\pi RC\sqrt{6}}$

3. Minimum value of $h_{fe}$ of the transistor is 44.5.
4. Gain of the amplifier in RC phase shift oscillator should be 29. The RC network introduces a phase shift of 180° and $\beta = \frac{1}{29}$.

## 4.6 Wien Bridge Oscillator

Wien bridge is a audio frequency RC oscillator. The merit of this oscillator is that the frequency may be varied in the range of 20 Hz to about 20 kHz, whereas in RC oscillators, the frequency can not be varied. The oscillator consists of two stages of RC coupled amplifier and a feedback network. The block diagram is shown in Figure (4.8).

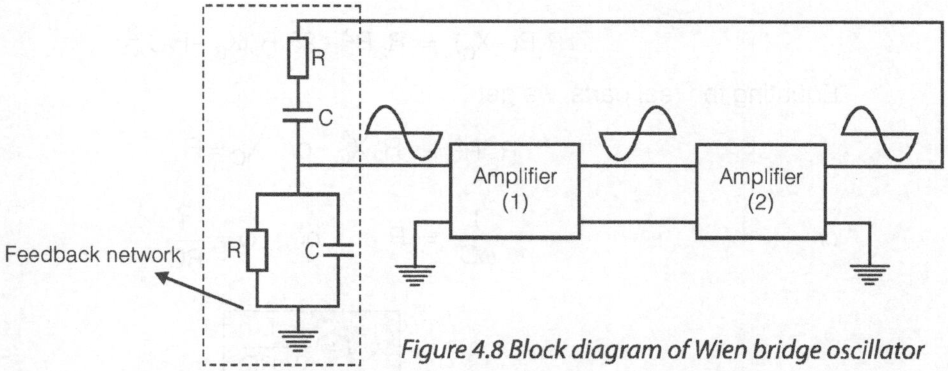


*Figure 4.8 Block diagram of Wien bridge oscillator*

The voltage across the parallel combination of R and C is fed to the input of amplifier-1. The net phase shift through the two amplifiers is zero. Now the question is that why we could not feed the output of amplifier-2 to amplifier-1 to provide regeneration needed for oscillation operation. The answer is that amplifier-1 will amplify signals over a wide range of frequencies and hence direct coupling would be result in poor frequency stability. By adding Wien bridge feedback network, the oscillator becomes sensitive to a signal of only one particular frequency. Hence good frequency stability is obtained. It is now essential that the feedback network should not introduce any phase shift between its input and output voltages.

We can vary the frequency of this oscillator by varying the two capacitors simultaneously. We can also change the range of the frequency of the oscillator by using different values of resistors R.

### Circuit Diagram

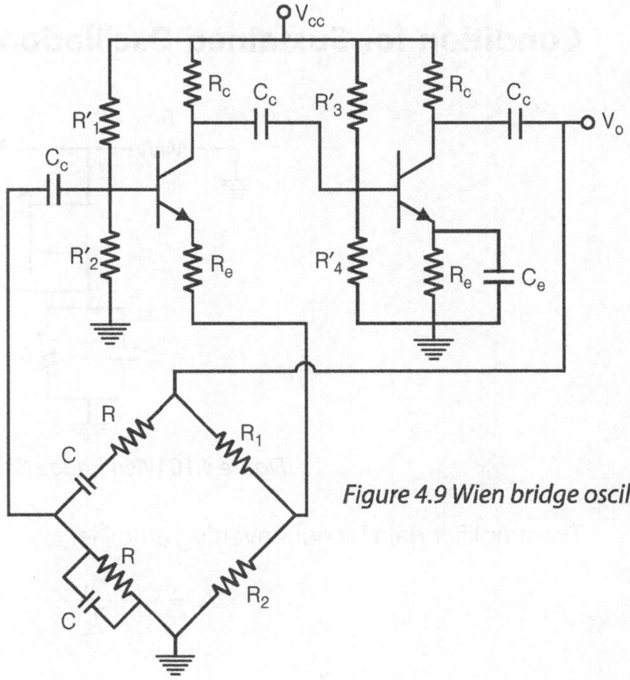


*Figure 4.9 Wien bridge oscillator*

## Frequency of The Oscillator

For bridge balance

$$\frac{R_1}{R_2} = \frac{R - jX_C}{R(-jX_C)/R - jX_C} \qquad \text{where } X_C = \frac{1}{\omega C}$$

$$\frac{R_1}{R_2} = \frac{(R - jX_C)(R - jX_C)}{R(-jX_C)} = \frac{R^2 - 2RjX_C - X_C^2}{R(-jX_C)}$$

$$R_1R(-jX_C) = R_2R^2 - 2R\,R_2jX_C - R_2X_C^2$$

Equating the real parts, we get

$$R_2R^2 = R_2X_C^2 \text{ or } X_C = R$$

or

$$\frac{1}{\omega C} = R \quad \text{or} \quad \omega = \frac{1}{RC}$$

$$\therefore\ f = \frac{1}{2\pi RC}$$

Where f is the frequency of oscillations.

## Amplitude Stabilization

The required gain of the amplifier for sustained oscillations in Wien bridge oscillator is $|A| = 3$ and $|\beta| = 1/3$.

Being a two stage amplifier, the amplifier can easily provide a much higher gain than 3. The gain of the amplifier should not be too high as it will distort the output voltage waveform of oscillator. Inorder to avoid this possibility of distortion, the amplifier gain should be controlled. This is done by introducing a negative feedback by keeping resistor $R_4$ in the emitter of $Q_1$ unbypassed. This introduces current series type negative feedback.

## Condition for Sustained Oscillations

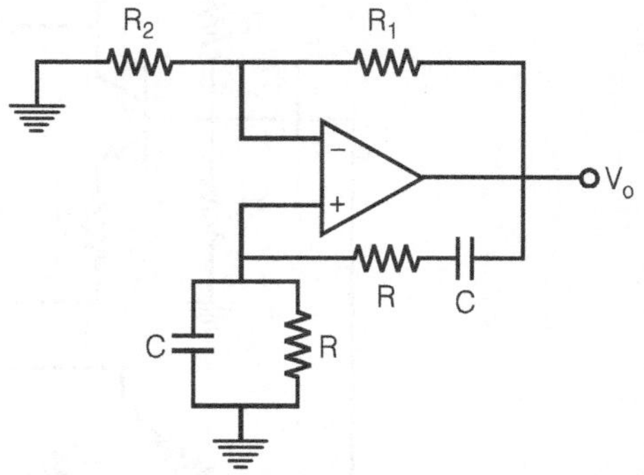


*Figure 4.10 Wien bridge Oscillator using op-amp*

The amplifier gain for non-inverting amplifier is

$$A = 1 + \frac{R_1}{R_2}$$

We know that at the oscillating frequency, the value of feedback factor is $\beta = \frac{1}{3}$ and gain

$$A_v \geq 3$$

$$\left[1+\frac{R_1}{R_2}\right] \geq 3$$

$$\frac{R_1}{R_2} \geq 2$$

$$R_1 \geq 2R_2$$

Thus $R_1$ should be greater than two times the value of $R_2$ to ensure sustained oscillations.

### Advantages

(i) This gives good frequency stability.
(ii) Overall gain is high because of two transistors.
(iii) Frequency of oscillations can be changed.
(iv) A very good quality sine wave can be obtained at the oscillator output.

### Disadvantages

(i) Poor frequency stability as the oscillator frequency is dependent only on the values of R and C components.
(ii) It requires two transistors and large number of components.

## 4.7 Comparison of RC Oscillators

| Parameter | RC phase shift oscillator | Wien bridge oscillator |
|---|---|---|
| 1. Feedback network | 3 identical RC | Wien bridge |
| 2. Phase shift introduced by feedback network | 180° | 0° |
| 3. Phase shift introduced by the amplifier | 180° | 0° |
| 4. Frequency of oscillation | $f = \frac{1}{2\pi\sqrt{6}\,RC}$ | $f = \frac{1}{2\pi RC}$ |
| 5. Value of $\beta$ | $\beta = \frac{1}{29}$ | $\beta = \frac{1}{3}$ |
| 6. Minimum value of gain | A = 29 | A = 3 |
| 7. Variable output frequency | Possible but difficult | Possible and easy |

## 4.8 LC Oscillators

- In LC oscillators the feedback network consists of inductor (L) and capacitor (C).
- These LC components determine the frequency of oscillations of LC oscillator.
- These oscillator can operate at high frequencies typically from 200 kHz to few GHz. They are not suitable for low operating frequencies because the value of L and C will be large at low frequencies.
- But as the operating frequency is increased, we need small values of inductors and capacitors which are small in size and less expensive.
- We have three types of LC oscillators:
  (i) Hartley oscillator
  (ii) Colpitts oscillator
  (iii) Clapp oscillator

### General Form of LC Oscillator

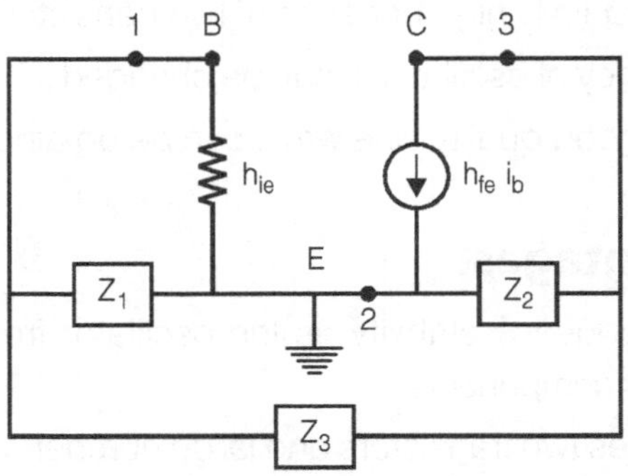


*Figure 4.11 Block diagram of LC oscillator*

The block diagram of LC oscillator is shown in Figure (4.11). The input is given between the terminals 1 and 2 and output is taken from 2 and 3. The load impedance $Z_L$ consists of $Z_2$ in parallel with the series combination of $Z_3$ and $Z_1 \parallel h_{ie}$.

The impedance across 2 and 3 = $Z_L$

$$Z_L = \left((h_{ie} \parallel Z_1) + Z_3\right) \parallel Z_2 = Z_2 \parallel \left[Z_3 + \frac{Z_1 h_{ie}}{Z_1 + h_{ie}}\right] = \frac{Z_2\left[Z_3 + \dfrac{Z_1 h_{ie}}{Z_1 + h_{ie}}\right]}{Z_2 + Z_3 + \dfrac{Z_1 h_{ie}}{Z_1 + h_{ie}}}$$

$$Z_L = \frac{Z_2[Z_3(Z_1 + h_{ie}) + Z_1 h_{ie}]}{(Z_2 + Z_3)(Z_1 + h_{ie}) + Z_1 h_{ie}} = \frac{Z_2[(Z_3 + Z_1)h_{ie} + Z_3 Z_1]}{Z_2 Z_1 + Z_3 Z_1 + h_{ie}(Z_1 + Z_2 + Z_3)}$$

The voltage feedback to the input terminals '1' and '2' is determined by potential division rule such as

$$V_f = \frac{V_o(Z_1 \parallel h_{ie})}{Z_1 \parallel h_{ie} + Z_3}$$

The feedback factor $\beta = \dfrac{V_f}{V_o}$

hence
$$\frac{V_f}{V_o} = \frac{Z_1 h_{ie}}{Z_1 + h_{ie}} \Big/ \frac{Z_1 h_{ie}}{Z_1 + h_{ie}} + Z_3$$

(or)
$$\beta = \frac{Z_1 h_{ie}}{Z_1 h_{ie} + Z_3(Z_1 + h_{ie})} = \frac{Z_1 h_{ie}}{Z_3 Z_1 + h_{ie}(Z_1 + Z_3)}$$

The voltage gain of CE amplifier is

$$A_v = \frac{-h_{fe}}{h_{ie}} Z_L$$

$$= \frac{-h_{fe}}{h_{ie}}\left[\frac{Z_2[(Z_3 + Z_1)h_{ie} + Z_3 Z_1]}{Z_3 Z_1 + Z_2 Z_1 + h_{ie}(Z_1 + Z_2 + Z_3)}\right]$$

As per Barkhausen criterion the condition for sustained oscillations is $A\beta \geq 1$, substitute the value of $A_v$ and 'β'.

$$\frac{-h_{fe}}{h_{ie}}\left[\frac{Z_2[(Z_3 + Z_1)h_{ie} + Z_3 Z_1]}{Z_3 Z_1 + Z_2 Z_1 + h_{ie}(Z_1 + Z_2 + Z_3)}\right]\frac{Z_1 h_{ie}}{Z_3 Z_1 + h_{ie}(Z_1 + Z_3)} \geq 1$$

(or)
$$\left[\frac{-h_{fe} Z_1 Z_2}{Z_3 Z_1 + Z_2 Z_1 + h_{ie}(Z_1 + Z_2 + Z_3)}\right] \geq 1$$

(or)
$$-h_{fe} Z_1 Z_2 = Z_3 Z_1 + Z_2 Z_1 + h_{ie}(Z_1 + Z_2 + Z_3)$$

$$\therefore\ h_{fe} Z_1 Z_2 + Z_3 Z_1 + Z_1 Z_2 + h_{ie}(Z_1 + Z_2 + Z_3) = 0$$

This equation is the general form of LC oscillator.
The LC oscillations produce oscillations at resonant condition.
When imganiary part is equal to zero.
Using this, frequency of oscillation is determined and, real part is equated to zero to determine the condition for oscillations.

## 4.9 Hartley Oscillator

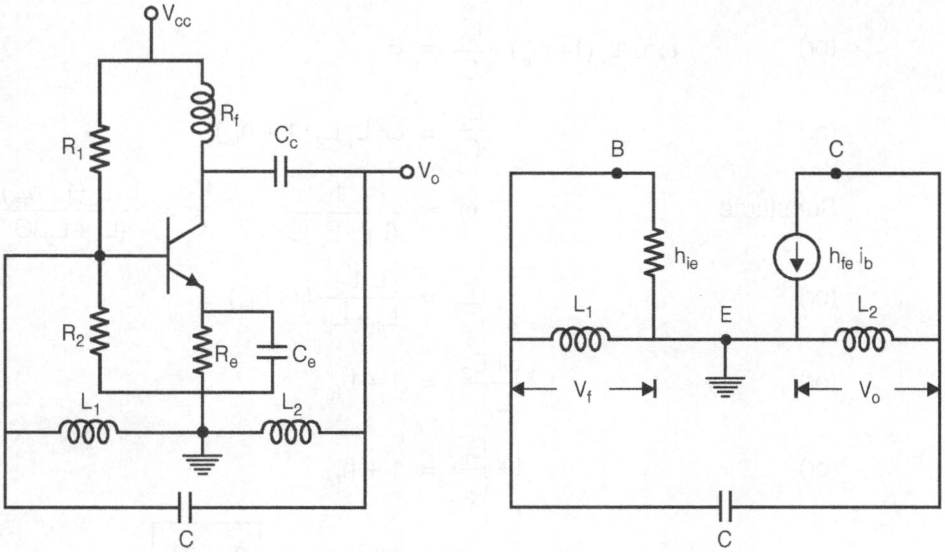


*Figure 4.12 Hartley oscillator and its equivalent circuit*

## Construction

In this case $Z_1$ and $Z_2$ are inductors and $Z_3$ is a capacitor. The biasing voltage is applied to the collector through RFC whose reactance is high compared with $Z_2$ and may be omitted in equivalent circuit. The capacitor $C_c$ has a low reactance at the frequency of oscillation, therefore it is eliminated in equivalent circuit. The significance of this capacitor is, acts as open circuit at zero frequency thus it prevents collector supply $V_{cc}$ from getting short circuited by radio frequency coil (RFC) in series with $L_2$.

The resistors $R_1$, $R_2$ and $R_e$ are used for biasing and $C_e$ to prevent degeneration. This circuit works under class 'C' mode of operation.

## Analysis

The general equation for LC oscillator is given by

$h_{fe}Z_1 Z_2 + Z_3 Z_1 + Z_1 Z_2 + h_{ie}(Z_1 + Z_2 + Z_3) = 0$

Substitute $$Z_1 = j\omega L_1 \;;\; Z_2 = j\omega L_2 \;;\; Z_3 = -\frac{j}{\omega C}$$

$$-h_{fe}\,\omega^2 L_1 L_2 + \frac{L_1}{C} - \omega^2 L_1 L_2 + h_{ie}\, j\omega\left(L_1 + L_2 - \frac{1}{\omega^2 C}\right) = 0$$

To determine the frequency of oscillation, the imaginary part is equated to zero.

$$\omega\, h_{ie}\left(L_1 + L_2 - \frac{1}{\omega^2 C}\right) = 0$$

(or) $$L_1 + L_2 = \frac{1}{\omega^2 C} \;;\; \omega^2 = \frac{1}{(L_1 + L_2)C}$$

(or) $$f = \frac{1}{2\pi\sqrt{(L_1 + L_2)C}} \text{ (or) } \frac{1}{2\pi\sqrt{L_{eq}C}}$$

Where $$L_{eq} = L_1 + L_2$$

To determine the condition for oscillation the real part is equated to zero.

$$-h_{fe}\omega^2 L_1 L_2 + \frac{L_1}{C} - \omega^2 L_1 L_2 = 0$$

(or) $$\omega^2 L_1 L_2 (1 + h_{fe}) - \frac{L_1}{C} = 0$$

(or) $$\frac{L_1}{C} = \omega^2 L_1 L_2 (1 + h_{fe})$$

Substitute $$\omega^2 = \frac{1}{(L_1 + L_2)C} \;;\; \frac{L_1}{C} = \frac{L_1 L_2 (1 + h_{fe})}{(L_1 + L_2)C}$$

(or) $$L_1 = \frac{L_1 L_2}{L_1 + L_2}(1 + h_{fe})$$

(or) $$\frac{L_1 + L_2}{L_2} = 1 + h_{fe}$$

(or) $$1 + \frac{L_1}{L_2} = 1 + h_{fe}$$

$$h_{fe} = \frac{L_1}{L_2} = \beta \qquad \boxed{\beta = \frac{L_1}{L_2}}$$

This is the condition for sustained oscillation.

## Operation

- When $V_{cc}$ is on, the collector current begins to flow and the drop in collector voltage is coupled through capacitor 'C' and $L_2$. Thus the capacitor charges to maximum voltage. This voltage acts as initial excitation for the tank circuit causing a current to flow in the LC circuit. This current induces damped harmonic oscillations across $L_1$ which drives the base of the transistor.
- The feedback voltage is in phase with input voltage, since 180° phase shift produced by transistor and another 180° being provided by the fact that the two ends of coil $L_1$ and $L_2$ are connected in opposite polarity.
- The capacitor $C_c$ blocks dc components of the collector circuit but couples ac signal. The dc component supply through RFC which also prevents ac component to flow through this path because RFC acts as short circuit for dc and open circuit for ac.
- As a result of this dc is out of tank circuit, thus energy loss due to tank circuit is reduced hence the oscillator is more stable.

# 4.10 Colpitts Oscillator

## CIRCUIT ARRANGEMENT

Figure (4.13) shows the circuit diagram of Colpitts oscillator. The circuit is the same as Hartley except that the emitter tap is connected between the capacitances $C_1$ and $C_2$.

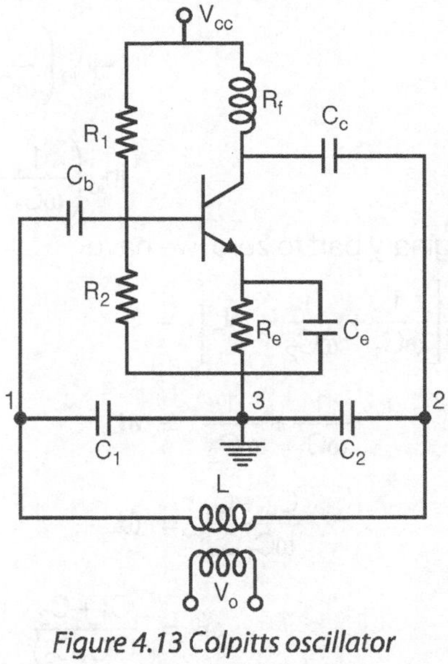


*Figure 4.13 Colpitts oscillator*

The resistors $R_1$, $R_2$ and $R_e$ provide the necessary bias condition for the circuit. The parallel combination of $R_e$ and $C_e$ in the emitter circuit is the stabilizing circuit.

The function of $C_c$ and $C_b$ is to block dc and to provide an ac path. The radio frequency choke (RFC) offers very high impedance to high frequency currents. Thus it prevents radio frequency currents from reaching the source of collector supply voltage and prevents this source from short circuiting the alternating output voltage. The frequency determining network is parallel resonant circuit consisting of capacitors $C_1$ and $C_2$ and the inductor L. The voltage developed across $C_1$ provides the regenerative feedback required for sustained oscillations.

## Circuit Operation

- When the collector supply voltage is switched on, a transient current is produced in the tank circuit. So damped harmonic oscillations are produced in the tank circuit.
- The oscillations across $C_1$ are applied to the base emitter junction and appear in the amplified form in the collector circuit.
- If terminal-1 is at positive potential with respect to terminal 3 at any instant, then terminal 2 will be at negative potential with respect to 3 at that instant because terminal 3 is grounded. Therefore, points 1 and 2 are 180° out of phase.
- A further phase shift of 180° is produced by the transistor. In other words, energy is supplied to the tank circuit in phase with the oscillations and if Aβ is greater than one, oscillations are sustained in the Circuit.

## Frequency of Oscillations

The general equation of LC oscillators is

$h_{ie}(Z_1 + Z_2 + Z_3) + Z_1 Z_2 (1 + h_{fe}) + Z_1 Z_3 = 0$

In case of Colpitts oscillator

$$Z_1 = \frac{1}{j\omega C_1} = -\frac{j}{\omega C_1} \quad ; \quad Z_2 = \frac{1}{j\omega C_2} = -\frac{j}{\omega C_2} \quad ; \quad Z_3 = j\omega L$$

Substituting these values in above equation, we get

$$h_{ie}\left(-\frac{j}{\omega C_1} - \frac{j}{\omega C_2} + j\omega L\right) + \left(-\frac{j}{\omega C_1}\right)\left(-\frac{j}{\omega C_2}\right)(1+h_{fe}) + \left(-\frac{j}{\omega C_1}\right)j\omega L = 0$$

(or)

$$-jh_{ie}\left(\frac{1}{\omega C_1} + \frac{1}{\omega C_2} - \omega L\right) - \frac{1+h_{fe}}{\omega^2 C_1 C_2} + \frac{L}{C_1} = 0$$

(or)

$$jh_{ie}\left(\frac{1}{\omega C_1} + \frac{1}{\omega C_2} - \omega L\right) + \left[\frac{1+h_{fe}}{\omega^2 C_1 C_2} - \frac{L}{C_1}\right] = 0$$

Equating imaginary part to zero, we have

$$h_{ie}\left[\frac{1}{\omega C_1} + \frac{1}{\omega C_2} - \omega L\right] = 0$$

(or)

$$\frac{1}{\omega C_1} + \frac{1}{\omega C_2} = \omega L$$

$$\frac{C_2 + C_1}{\omega C_1 C_2} = \omega L$$

(or)

$$\omega^2 = \frac{C_1 + C_2}{L(C_1 C_2)}$$

(or)

$$\omega = \sqrt{\frac{C_1 + C_2}{L C_1 C_2}}$$

$$\boxed{f = \frac{1}{2\pi}\sqrt{\frac{C_1 + C_2}{L C_1 C_2}}}$$

Equating real part to zero, we get

$$\frac{1+h_{fe}}{\omega^2 C_1 C_2} = \frac{L}{C_1}$$

(or) $$1 + h_{fe} = \frac{\omega^2 L C_1 C_2}{C_1} = \omega^2 L C_2$$

(or) $$1 + h_{fe} = \frac{C_1 + C_2}{L C_1 C_2} \times L C_2 = \frac{C_1 + C_2}{C_1} = 1 + \frac{C_2}{C_1}$$

$$h_{fe} = \frac{C_2}{C_1}$$

$$\boxed{\beta = \frac{C_2}{C_1}}$$

## 4.11 Clapp Oscillator

The clapp oscillator is a "modified form of Colpitts oscillator". The single inductor found in the Colpitts oscillator is replaced by a series LC combination.

The practical problem in Colpitts oscillator is the capacitors $C_1$ and $C_2$ are shunted to parasitic capacitances of transistor. These effect changes the capacitor values which in turn effect the frequency of oscillation.

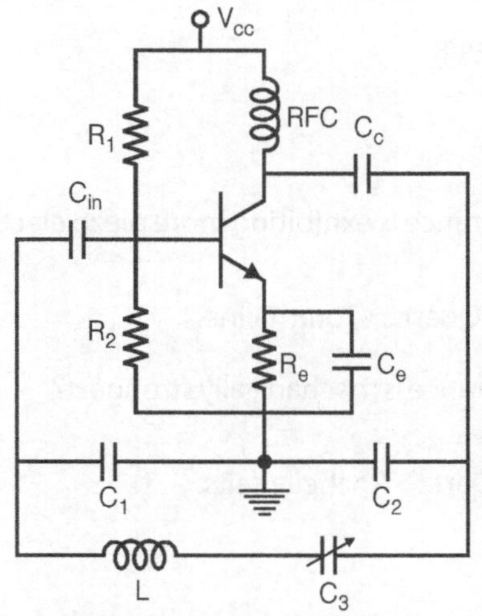


*Figure 4.14 Clapp oscillator*

To avoid such a problem a capacitor $C_3$ in series to inductor which is independent of parasitic capacitances of the transistor, is taken into account.

The design is

Assume $C_1 >> C_3$

$C_2 >> C_3$

Then $$\frac{1}{C_T} = \frac{1}{C_1} + \frac{1}{C_2} + \frac{1}{C_3}$$

$$\therefore \quad \frac{1}{C_T} = \frac{1}{C_3}$$

$$f_o = \frac{1}{2\pi\sqrt{LC_T}} = \frac{1}{2\pi\sqrt{LC_3}}$$

So in all practical applications we choose Clapp oscillator compared to Colpitts oscillator.

## 4.12 Crystal Oscillator

- A crystal oscillator is used, when accuracy and stability of the frequency of oscillations are important.
- In this oscillator, mostly quartz crystal cut in a particular direction is used. Hence one should know about the nature and cut of the quartz crystal.

### Piezo Electric Effect

"When a mechanical force or stress is applied on one face of the crystal, an alternating signal will be produced on the other side of the crystal" and vice versa.

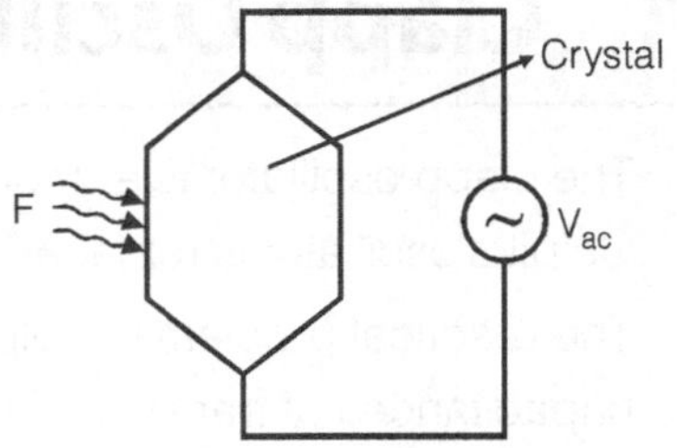


### Substances Used for The Formation of a Crystal

1. Rochelle salts
2. Quartz
3. Tourmaline

**1.** Which substance is exhibiting more piezo electric effect?
**Solution:**
Rochelle Salts > Quartz > Tourmaline.

**2.** Which substance is mechanically strongest?
**Solution:**
Tourmaline > Quartz > Rochelle salts.

### Note:

Quartz is a compromise between the piezo electric effect of Rochelle salts and the strength of Tourmaline.
Quartz is in expensive and readily available in nature, so it is widely used.

### Crystal Cuts

The natural shape of a Quartz crystal is a hexagonal prism with pyramids at the ends shown in figure.

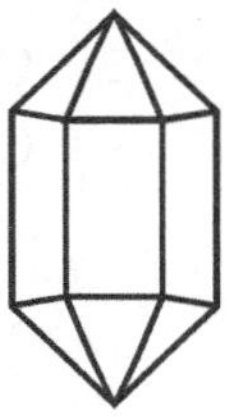

For the practical use, a rectangular slab is to be sliced out of the natural crystal shows this slab with thickness t

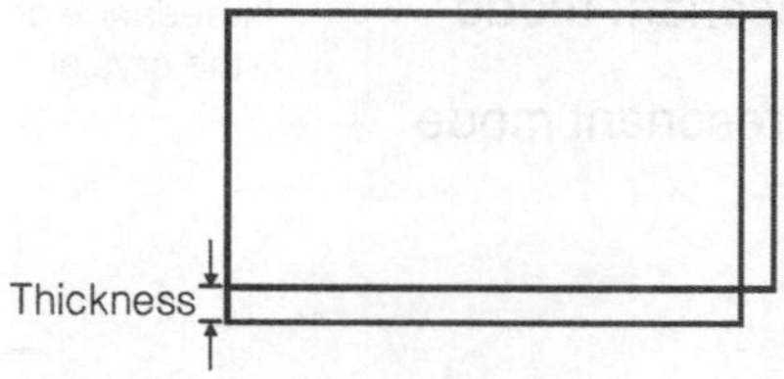


Different cuts have different piezo electric properties.

In electronic circuits, the slab is mounted between two metal plates shown in above figure.

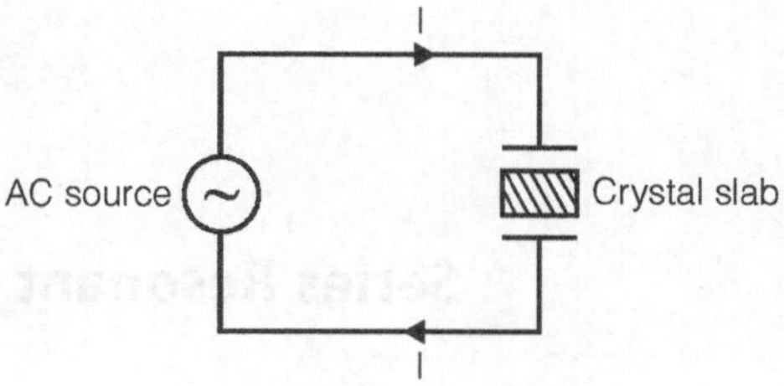


In this circuits, the amount of crystal vibration depends on the frequency of the applied voltage.

The natural frequency of vibration of the crystal is mainly depends on the thickness of the slab.

The resonant frequencies at which the crystal vibrations reach maximum can be found be changing the frequency of the applied voltage.

Since the energy for the vibrations must be supplied by the ac source, the ac current is maximum at resonant frequency.

The formula for the fundamental frequency (f) of the crystal is

$$f = \frac{k}{t}$$

Where k is a constant that depends on the cut and other factors and t is the thickness of the crystal.

It is obvious that fundamental frequency is inversely proportional to the thickness. The thinner the crystal the more fragile it becomes and the more likely it to break because of vibrations.

Because of this, Quartz crystals can work well up to 10 MHz, on the fundamental frequency.

## Equivalent Circuit of Crystal

Non vibrating mode (OFF)

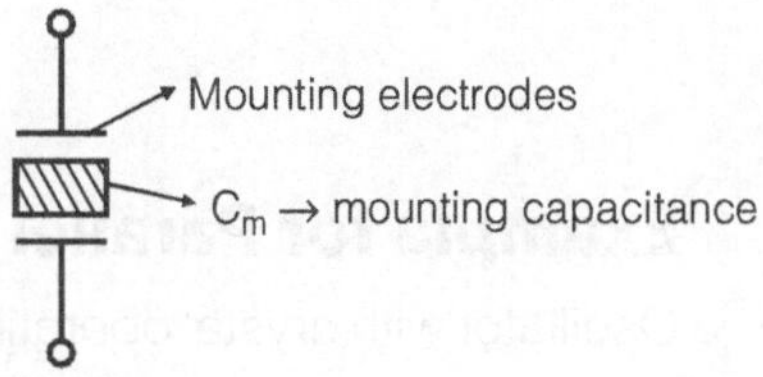


**Vibrating mode (ON)**

Electrical to mechanical conversion

R → frictional loss

L → Inertia or mass

C → Stiffness or strength of the crystal

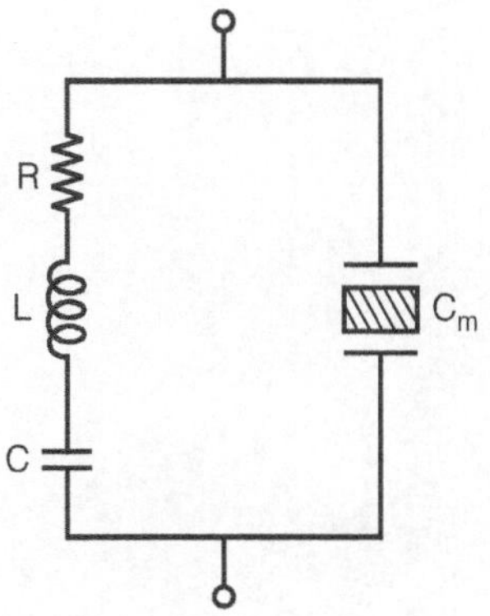

## Crystal Operating Modes

(i) Series resonant mode

(ii) Parallel resonant mode

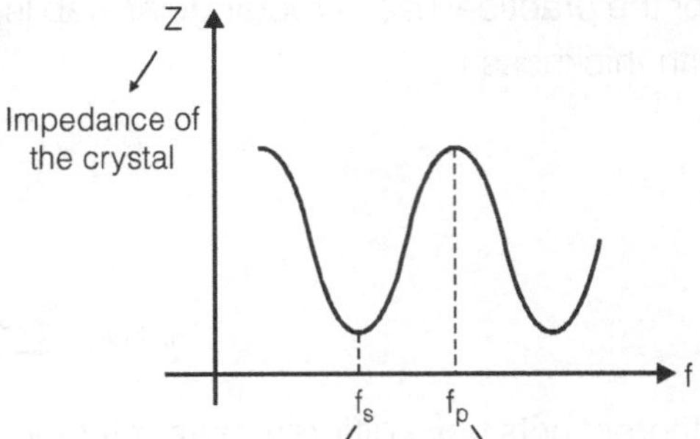


*Impedance versus frequency*

## Series Resonant Frequency (zero impedance frequency)

$$f_s = \frac{1}{2\pi\sqrt{LC}}$$

## Parallel Resonant Frequency (infinite impedance frequency)

$$f_p = \frac{1}{2\pi\sqrt{LC}}\sqrt{1+\frac{C}{C_m}}$$

As $C_m >> C$

The ratio $\frac{C}{C_m}$ is very small.

## Example for Series Resonant Mode

Oscillator with crystal operating series resonance.

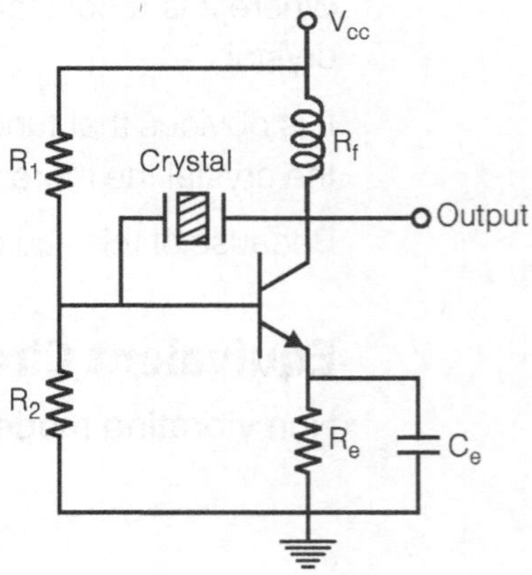


## Example for Parallel Resonant Mode

Oscillator with crystal operating in parallel resonance.

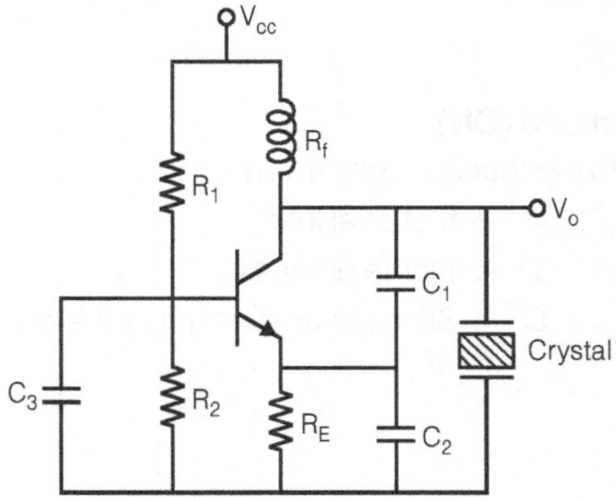

## Advantages

1. It is very simple circuit as it does not need any tank circuit other than crystal itself.
2. The Q factor, which is a measure of the quality of resonance circuit of a crystal is very high. The Q factor of a crystal may range from $10^4$ to $10^6$ where as LC circuit may have Q factor only of order of 100.
3. Good high frequency stability.

## Disadvantages

1. The crystal oscillators have a very limited tuning range. They are used for frequencies exceeding 100 kHz.
2. The crystal oscillators are fragile and therefore, can only used in low power circuits.

## SUMMARY

- Sinusoidal oscillators are electronic circuits that produce oscillations of sinusoidal nature on their own.
- Sinusoidal oscillators work on the principle of Barkhausen criterion.
- Barkhausen criterion states that oscillations are possible only when the loop gain is unity or more.
- The frequency of oscillation of an oscillator is determined on the condition that phase around the closed loop should be zero or integral multiples of $2\pi$ radians.
- RC phase shift oscillators can be built using either BJT or FET. These oscillator are useful at low frequencies. No inductors or chokes are utilised in these circuits.
- Resonant circuit oscillators operate at the resonant frequency of the tank circuit. The coupling between the output and input is via a RF transformer.
- At high frequencies, Hartley and Colpitts oscillators are used.
- Hartley oscillator has the drawback of mutual inductance and losses with bulky inductors.
- In Colpitts oscillator, frequency stability is poor due to shunting of stray capacitance with feedback network.
- Clapp oscillator eliminates the limitation of Colpitt oscillator by the use of additional capacitor in series with the inductor.
- Wien bridge oscillator employs a balance bridge in the feedback circuit. Its amplitude stability is good.
- Piezo electric is an electro mechanical phenomenon. Natural crystal like quartz exhibits this effect. Mechanical vibrations are converted to electric oscillations and vice versa.
- Quartz crystals are employed in sinusoidal oscillators. These oscillators have extreme frequency stability and are available up to large frequencies of MHz range.

## GATE AND IES QUESTIONS

**Q.1** The oscillator circuit shown in the figure is

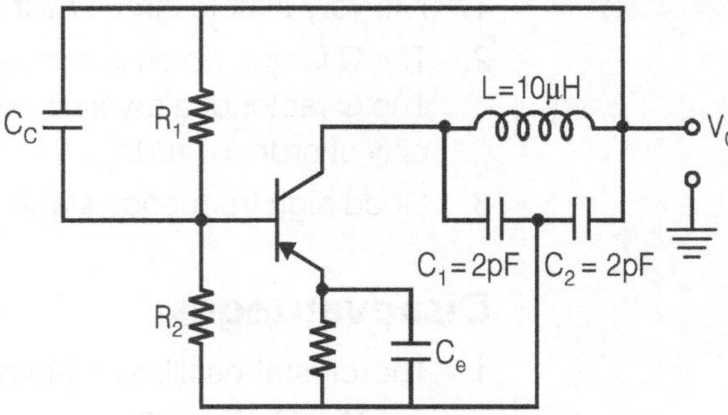


(a) Hartley oscillator with $f_{oscillation}$ = 79.6 MHz

(b) Colpitts oscillator with $f_{oscillation}$ = 50.3 MHz

(c) Hartley oscillator with $f_{oscillation}$ = 159.2 MHz

(d) Colpitts oscillator with $f_{oscillation}$ = 159.2 MHz

**Exp.** (b)

The tank circuit is having two capacitors and one inductor, so it is Colpitts oscillator and frequency is

$$f = \frac{1}{2\pi\sqrt{LC_{eq}}}$$

$$C_{eq} = \frac{C_1C_2}{C_1+C_2} = \frac{2\times 2}{4} = 1\,\text{pF}$$

$$f = \frac{1}{2\pi\sqrt{10\times10^{-6}\times10^{-12}}} = \frac{1\times10^9}{2\pi\sqrt{10}} = 50.3\text{ MHz}$$

**Q.2** The circuit in the figure employs positive feedback and is intended to generate sinusoidal oscillation. If at a frequency $f_0$, $B(f) = \frac{V_f(f)}{V_0(f)} = \frac{1}{6}\angle 0°$ then to sustain oscillation at this frequency

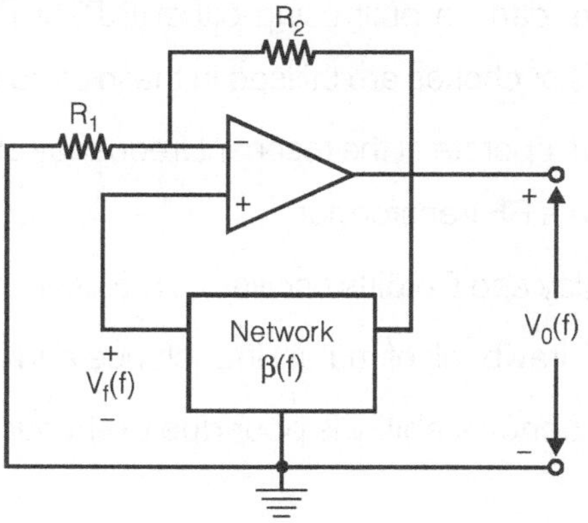


(a) $R_2 = 5R_1$

(b) $R_2 = 6R_1$

(c) $R_2 = \frac{R_1}{6}$

(d) $R_2 = \frac{R_1}{5}$

**Exp.** (a)

Barkhausen criterion for sustained oscillations $|A\beta| \geq 1$ and phase shift must be $2\pi$.

$$A = 1+\frac{R_2}{R_1}\ ;\quad \beta(f) = \frac{1}{6}\angle 0°$$

For sustained oscillation.

$$6 = 1+\frac{R_2}{R_1}$$

(or) $$R_2 = 5R_1$$

**Q.3** The oscillator circuit shown in the figure has an ideal inverting amplifier. Its frequency of oscillation (in Hz) is

(a) $\frac{1}{(2\pi\sqrt{6}RC)}$

(b) $\frac{1}{(2\pi RC)}$

(c) $\frac{1}{(\sqrt{6}RC)}$

(d) $\frac{1}{\sqrt{6}(2\pi RC)}$

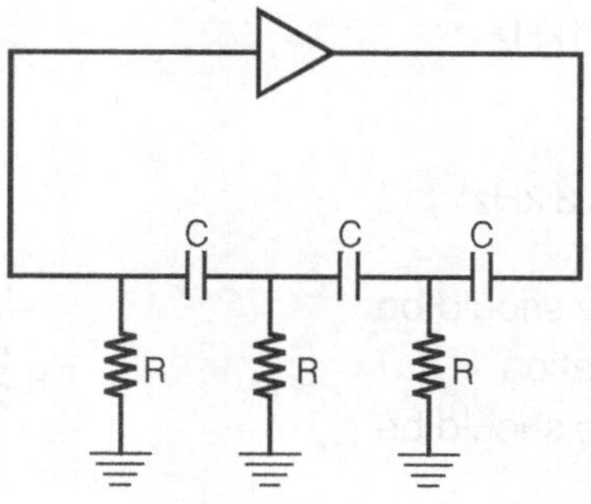


**Exp.** **(a)**

The given circuit is a RC phase shift oscillator and frequency of its oscillation is

$$f = \frac{1}{2\pi\sqrt{6}\,RC}$$

**Q.4** The value of C required for sinusoidal oscillations of frequency 1 kHz in the circuit of the figure is

(a) $\frac{1}{2\pi}\mu F$

(b) $2\pi\mu F$

(c) $\frac{1}{2\pi\sqrt{6}}\mu F$

(d) $2\pi\sqrt{6}\mu F$

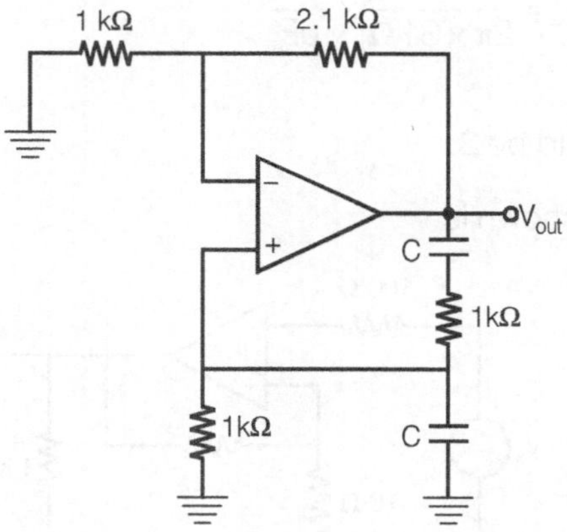


**Exp.** **(a)**

The circuit is Wien bridge oscillator.

The frequency of oscillation is

$$\omega = \frac{1}{RC}$$

$$f = \frac{1}{2\pi RC}$$

$$C = \frac{1}{2\pi Rf} = \frac{1}{2\pi \times 10^3 \times 10^3} = \frac{1}{2\pi}\mu F$$

**Q.5** A Wien bridge oscillator is shown in figure. Which of the following statements are true, if f is the frequency of oscillation.

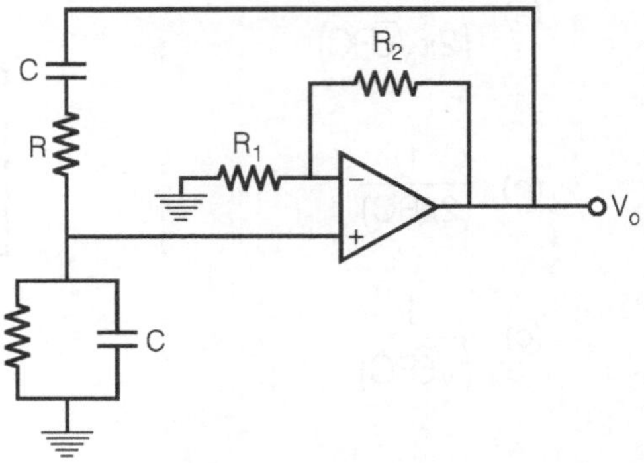


(a) For R = 1 kΩ, C = $\frac{1}{2\pi}$ μF, f = 1kHz

(b) For R = 3 kΩ, C = $\frac{1}{18\pi}$ μF, f = 3 kHz

(c) The gain of the op-amp stage should be less than two for proper operation

(d) The gain of the op-amp stage should be three for proper operation.

**Exp. (a), (b) (d)**

The circuit is Wien bridge oscillator

For $$R = 1\text{ k}\Omega, \quad C = \frac{1}{2\pi}\mu F$$

$$f = \frac{1}{2\pi RC} = \frac{1}{2\pi \times 1 k\Omega} \times \frac{2\pi}{\mu F} = 1 kHz$$

For $$R = 3\text{ k}\Omega, \quad C = \frac{1}{18\pi}\mu F$$

$$f = \frac{1}{2\pi RC} = \frac{1}{2\pi \times 3 k\Omega} \times \frac{18\pi}{\mu F}$$

$$= 3 \text{ kHz}$$

The gain of the op-amp stage should be 3.

**Q.6** The feedback factor for the circuit shown is

(a) $\frac{9}{100}$

(b) $\frac{9}{10}$

(c) $\frac{1}{9}$

(d) $\frac{1}{10}$

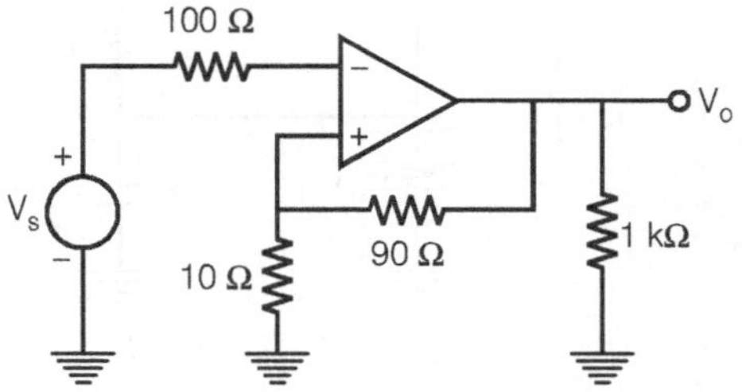


**Exp. (d)**

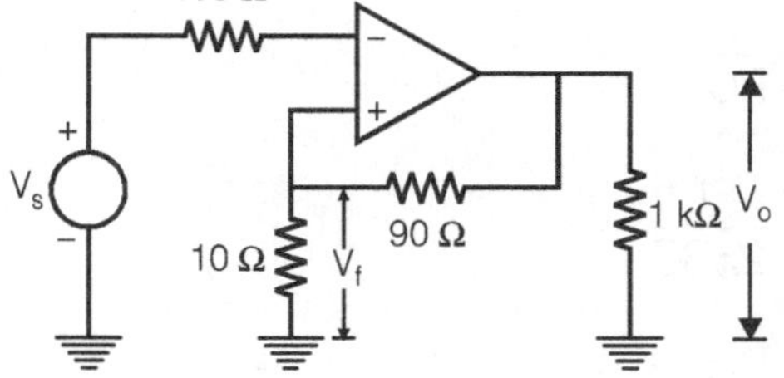


$$\beta = \frac{V_f}{V_o} = \frac{10}{10+90} = \frac{1}{10}$$

**Q.7** For the oscillator circuit shown in figure, the expression for the time period of oscillation can be given by (where $\tau = RC$)

(a) $\tau \ln 3$

(b) $2\tau \ln 3$

(c) $\tau \ln 2$

(d) $2\tau \ln 2$

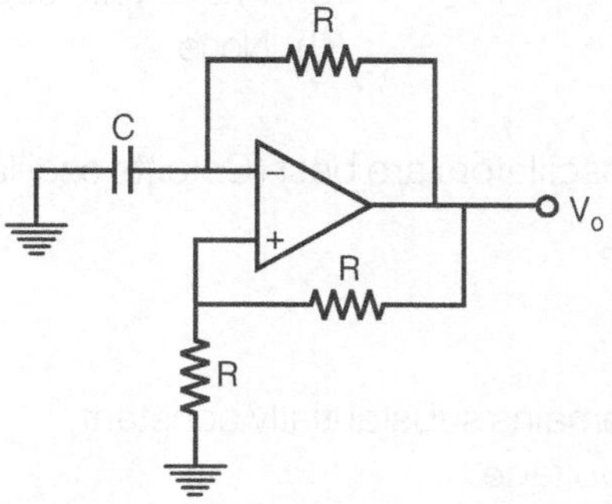


**Exp. (b)**

The circuit is astable multivibrator.

The time period for the oscillation is given by

$$T = 2RC\,ln\left(\frac{1+\beta}{1-\beta}\right) \qquad \left(\text{where } \beta = \frac{R}{R+R} = \frac{1}{2}\right)$$

$$= 2RC\,ln\left(\frac{1+\frac{1}{2}}{1-\frac{1}{2}}\right) = 2RC\,ln3$$

**Q.8** A higher-Q quartz crystal exhibits series resonance at the frequency $\omega_s$ and parallel resonance at the frequency $\omega_p$. Then

(a) $\omega_s$ is very close to, but less than $\omega_p$
(b) $\omega_s$ is very close to, but greater than $\omega_p$
(c) $\omega_s << \omega_p$
(d) $\omega_s >> \omega_p$

**Exp. (a)**

Series resonance at the frequency $\omega_s$ is very close to, but less than parallel resonance at the frequency $\omega_p$.

**Q.9** If the Q of a single stage tuned amplifier is doubled, then its bandwidth

(a) Remains same
(b) Becomes half
(c) Becomes double
(d) Becomes fourtimes

**Exp. (b)**

$$Q \propto \frac{1}{BW}$$

**Q.10** Positive feedback, in an amplifier circuit will _______ the gain of amplifier.

(a) increase
(b) decrease
(c) not alter
(d) None

**Exp. (a)**

By using positive feedback, an amplifier circuit will increase the gain of amplifier.

**Q.11** For generating a 1 kHz signal, the most suitable circuit is

(a) Hartley oscillator
(b) Colpitts oscillator
(c) Tuned-collector oscillator
(d) Wien bridge oscillator

**Exp. (d)**

For audio range of frequencies RC oscillators are best (Wien bridge oscillator).

**Q.12** To generate a 1 MHz signal, the most suitable circuit is
(a) Wien bridge oscillator
(b) Phase shift oscillator
(c) Colpitts oscillator
(d) None

**Exp. (c)**
For radio range of frequencies LC oscillators are best (Colpitts oscillator).

**Q.13** We use a crystal oscillator because
(a) It gives high output voltage
(b) It works at high efficiency
(c) The frequency of oscillations remains substantially constant
(d) It requires very low dc supply voltage

**Exp. (c)**
The use of a crystal oscillator is the frequency of oscillations remains substantially constant.

**Q.14** Oscillator is a
(a) ac to dc converter
(b) dc to ac converter
(c) none

**Exp. (b)**
Oscillator is a dc to ac converter.

**Q.15** The condition which is required to get sustained oscillation is called
(a) Barkhausen criteria
(b) $(A\beta) = 1$
(c) Phase should be zero
(d) All of above

**Exp. (d)**
The condition which is required to get sustained oscillation is called Barkhausen criteria.

$$|A\beta| = 1 \quad \text{(or)} \quad \text{phase should be zero.}$$

**Q.16** In transistor tuned oscillator, frequency of oscillations is determined by
(a) RC network
(b) Tank circuit
(c) RC parallel circuit
(d) None

**Exp. (b)**
In tuned oscillators, frequency of oscillations is determined by tank circuit.

**Q.17** The minimum voltage gain required by transistor, when it is used in RC phase shift oscillator is
(a) 44.5
(b) 29
(c) 3
(d) none

**Exp. (b)**
The minimum voltage gain required by RC phase shift oscillator is 29.

**Q.18** The minimum gain required by the amplifier is _______, when it is used as a Wien bridge oscillator
(a) 3
(b) $\sqrt{29}$
(c) 44.5
(d) none

**Exp. (a)**
The minimum gain required by Wien bridge is 3.

**Q.19** The phase shift introduced by Wien bridge is
(a) 90°
(b) 180°
(c) 0°
(d) Zero

**Exp. (c)**
The phase shift introduced by Wien bridge is 0°.

**Q.20** Crystals are used to convert mechanical energy into electrical energy and vice versa. This phenomena is known as

(a) piezo conductive effect
(b) piezo magnetic
(c) piezo electric effect
(d) none

**Exp. (c)**

Crystals are used to convert mechanical energy into electrical energy and vice versa. This phenomena is called as piezo electric effect.

**Q.21** Piezo electric effect is used in following crystals

(a) Rochelle salt
(b) Quartz
(c) Both (a) and (b)
(d) none

**Exp. (c)**

Rochelle salt and Quartz will have piezo electric effect.

**Q.22** In crystal oscillator, frequency determining parameters are

(a) thickness of crystal
(b) angle of cut
(c) physical size of crystal
(d) all of the above

**Exp. (d)**

In crystal, frequency depends on

- Thickness of crystal.
- Angle of cut.
- Physical size of crystal.

**Q.23** The Q of a crystal is of the order of

(a) 100
(b) 1000
(c) 50
(d) more than 10,000

**Exp. (b)**

The Q of a crystal is of the order of 1000.

**Q.24** In the Clapp oscillator circuit shown below, the expression for $f_o$ is

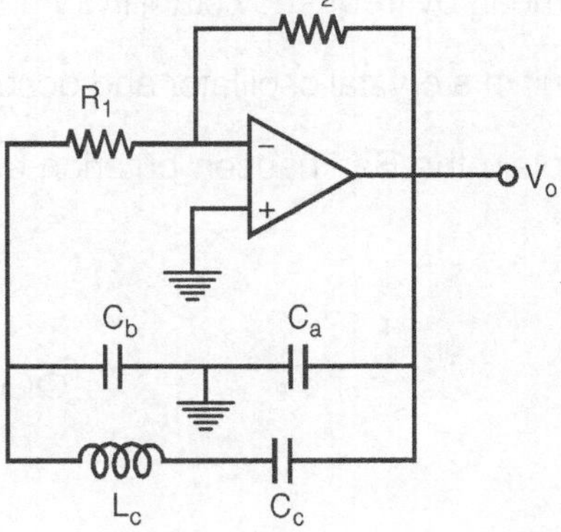


(a) $f_o = \frac{1}{2\pi\sqrt{L_C C_C}}$

(b) $f_o = \frac{1}{2\pi L_C R_C}$

(c) $f_o = \frac{1}{2\pi\sqrt{6}\sqrt{L_C R_C}}$

(d) none

**Exp. (a)**

The $f_o$ for Clapp oscillator is

$$f_o = \frac{1}{2\pi\sqrt{L_C C_C}}$$

**Q.25** In BJT RC phase shift oscillator, condition for sustained oscillations is

(a) $h_{fe} \geq 4K + 23 + \frac{29}{K}$ (b) $h_{fe} \geq \frac{L_b}{L_a}$

(c) $h_{fe} \geq \frac{L_b}{L_a} + \frac{h_{ie}}{R_c}\frac{L_a}{L_b}$ (d) none

**Exp. (a)**

In RC phase shift oscillator the condition for sustained oscillations is

$$-h_{fe}\frac{R_C}{R} \geq -29 - 23K - 4K^2 \qquad \text{where } K = \frac{R_C}{R}$$

$$h_{fe} \geq 4K + 23 + \frac{29}{K}$$

## REVIEW QUESTIONS

1. Draw the circuit of a tuned collector oscillation and explain.
2. With a neat circuit explain the working of a Clapp oscillation.
3. Draw the circuit of a Wien's bridge oscillator using Op-amp and explain it with necessary expressions for the frequency of oscillation.
4. Explain the circuit of RC phase shift oscillator and derive the conditions for oscillations and frequency of oscillation.
5. Draw the circuit of Colpitts oscillator and derive the conditions for oscillation and frequency of oscillation.
6. Given the circuit of a Hartley oscillator and explain how the oscillations are maintained and derive the condition for oscillation and frequency of oscillation.
7. What do you mean by frequency stability of an oscillator and explain how will you achieve it.
8. Draw the circuit of a crystal oscillator and describe its working.
9. State and explain the Barkhausen criterion and also derive an expression for the condition for oscillation.

OOOO

# CHAPTER 5

# POWER AMPLIFIERS

## About This Chapter

- Different amplifier classes.
- Harmonic distortion in power amplifiers.
- Analysis of different types of class-A amplifiers
- Class-B amplifier analysis with minimum harmonic distortion.
- Cross over distortion.
- Class-AB amplifier configuration.
- Class-C amplifier configuration.
- Class-D amplifier configuration.
- Class-S amplifier configuration

# 5 POWER AMPLIFIERS

## 5.1 Introduction

- An amplifying system consists of several stages in cascade. The input and intermediate stages amplify small signal excitations to a value large enough to drive the final device.
- The output stage feeds a transducer such as a cathode-ray oscilloscope, a loud speaker, or a servomotor.
- Thus, the final amplifier stage must be capable of delivering a large voltage or current for appreciable amount of power. This requires an amplifier which is referred to as power amplifier.
- Power amplifiers are those amplifiers, that are the final stage of an amplifier. Power amplifiers find extensive use in audio frequencies.
- For small-signal amplifiers, we had been using the small-signal techniques (such as h-parameter) which are based on linear model.
- However, for large-signal amplifiers, we cannot use linear models and analysis has to be done graphically using the output characteristic graphs of transistors. This non-linearity arises because the output characteristic curves are not equidistant.

## 5.2 Difference Between Voltage Amplifier and Power Amplifier

The comparison between voltage amplifier and power amplifier is shown below in the form of table:

| Characteristic | Voltage amplifier | Power Amplifier |
|---|---|---|
| $\beta$ | high | low |
| $R_C$ | high | low |
| Input voltage | low (few mV) | high (few V) |
| Power output | low | high |
| Collector current | low | high |
| Output impedance | high | low |
| Coupling | RC coupling | transformer or tuned circuit |

**Why voltage amplifier can not be used as power amplifiers?**

**Ans:** Consider the simple circuit of CE voltage amplifier shown in Figure (5.1).

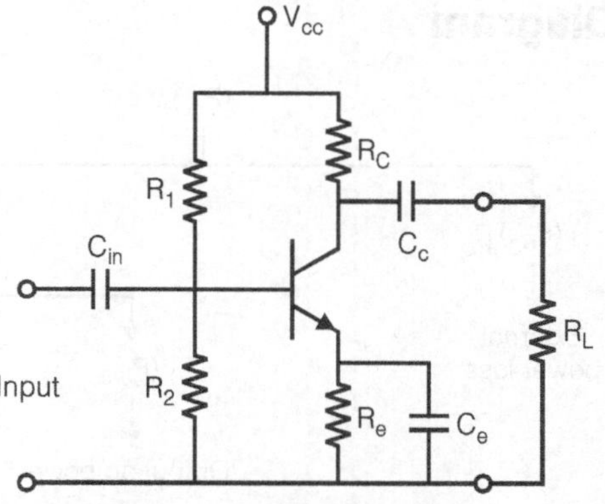


*Fig. 5.1 CE amplifier circuit*

Let $(I_c)_Q$ be the quiescent collector current. If $(V_{CE})_Q$ be the quiescent voltage between collector and emitter, then

$$V_{cc} = R_c(I_C)_Q + (V_{CE})_Q + R_e(I_C)_Q$$

Where $R_c$ is the collector load and $R_e$ is the emitter resistor.

$$V_{CC} - (V_{CE})_Q = (I_C)_Q\,(R_C + R_e)$$

Power taken from supply = $V_{CC}\,(I_C)_Q$

dc power that goes to transistor = $(V_{CE})_Q\,(I_C)_Q$

When ac signal is applied, only a part of this power is converted by the amplifier in to useful ac.

$$\begin{aligned}\text{Uselss power} &= \text{Power taken from supply} - \text{power goes to transistor}\\ &= V_{CC}(I_C)_Q - (V_{CE})_Q\,(I_C)_Q\\ &= (I_C)_Q\left[V_{CC} - (V_{CE})_Q\right]\\ &= (I_C)_Q\left[(I_C)_Q(R_C + R_e)\right]\\ &= (I_C)_Q^2\,(R_C + R_e)\end{aligned}$$

The above equation gives the power that goes waste in unnecessarily heating the resistors. Due to this fact, a voltage amplifier can not use as power amplifier.

## Conclusion

To reduce the wastage, we can short circuit $R_e$ and $R_C$. But we can not short circuit resistor $R_e$ because this is a part of biasing network. Secondly, we can not short circuit $R_c$ because $R_L$ will be short circuited in ac equivalent circuit i.e. no power will be delivered to load.

Now the best possible way is to replace $R_C$ by a choke (its dc resistance is almost zero and power loss in choke is almost zero). Hence by replacing $R_C$ by choke, power loss may be avoided. This is the reason that why we use transformer coupling in power amplifiers rather than RC coupling.

# 5.3 Power Amplifiers

## Power Diagram

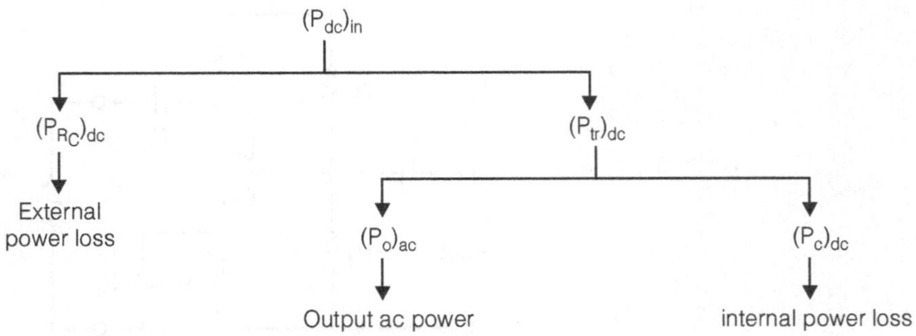


Where $(P_{dc})_{in}$ → Input dc power $V_{CC}(I_C)_Q$

$(P_{R_C})_{dc}$ → Power loss across $R_C$

$(P_{tr})_{dc}$ → dc power delivered to the transistor

$(P_C)_{dc}$ → Power loss across collector junction of the transistor

$(P_o)_{ac}$ → ac output power

The general expression for power amplifiers.

DC power input = AC power output + losses

"The power amplifiers are large signal amplifiers which raise the power level of the signals".

(or)

"The power amplifiers may also be defined as a device which converts dc power to ac power and whose action is controlled by the input signal".

# 5.4 Power Amplifier Classes

## Class-A

"The output current $I_c$ conducts for 360° of input signal is called as class-A operation".

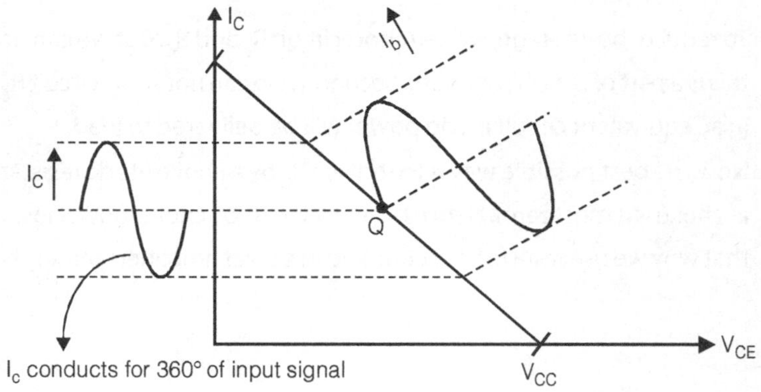

## Class-B

"The output current $I_c$ conducts only 180° of the input signal is called class-B operation".

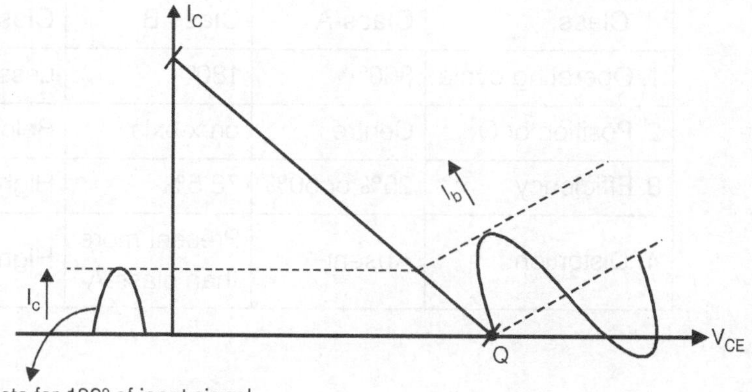


$I_c$ conducts for 180° of input signal

## Class-C

"The output current $I_C$ conducts less than 180° of the input signal is called class-C operation"

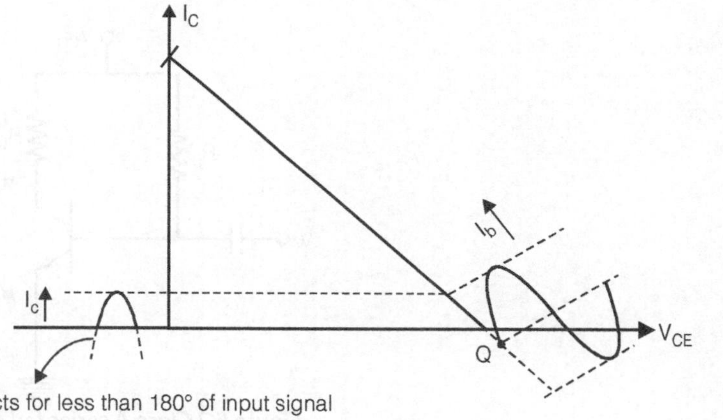


$I_c$ conducts for less than 180° of input signal

## Class-AB

"The output current $I_c$ conducts more than 180° but less than 360° of input signal is called class-AB operation".

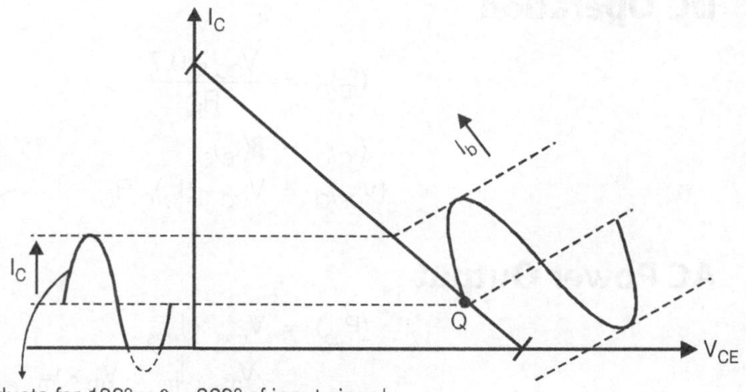


$I_c$ conducts for $180° < \theta < 360°$ of input signal

## 5.5 Comparison of Amplifier Classes

| Class | Class-A | Class-B | Class-C | Class-AB |
|---|---|---|---|---|
| 1. Operating cycle | 360° | 180° | Less than 180° | 180° to 360° |
| 2. Position of Q | Centre | on x-axis | Below x-axis | Above x-axis |
| 3. Efficiency | 25% or 50% | 78.5% | High (almost 100%) | 50% to 78.5% |
| 4. Distortion | Absent | Present more than class-A | Highest | Present less than Class-B |

## 5.6 Class-A Power Amplifier

### 1. CLASS A SERIES FED POWER AMPLIFIER

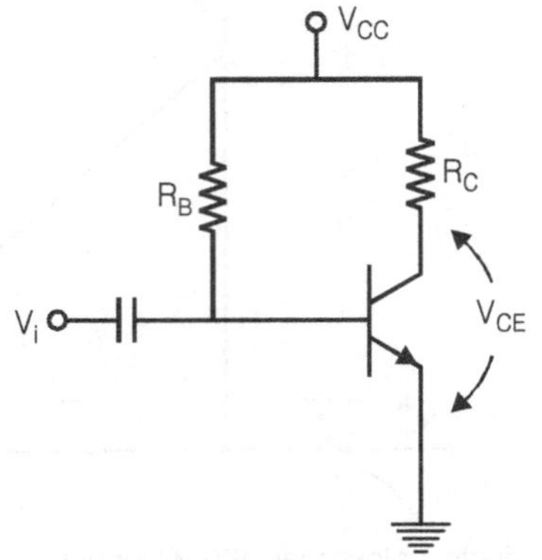


*Figure 5.2 Class-A series fed amplifier*

#### DC Power Input

$$P_{dc} = V_{CC}(I_C)_Q$$

#### DC Operation

$$(I_B)_Q = \frac{V_{CC} - 0.7}{R_B}$$

$$(I_C)_Q = \beta(I_B)_Q$$

$$(V_{CE})_Q = V_{CC} - (I_C)_Q R_C$$

#### AC Power Output

$$(P_{ac}) = V_{rms} \times I_{rms}$$

$$= \frac{V_m}{\sqrt{2}} \times \frac{I_m}{\sqrt{2}} = \frac{V_m \times I_m}{2}$$

## AC Operation

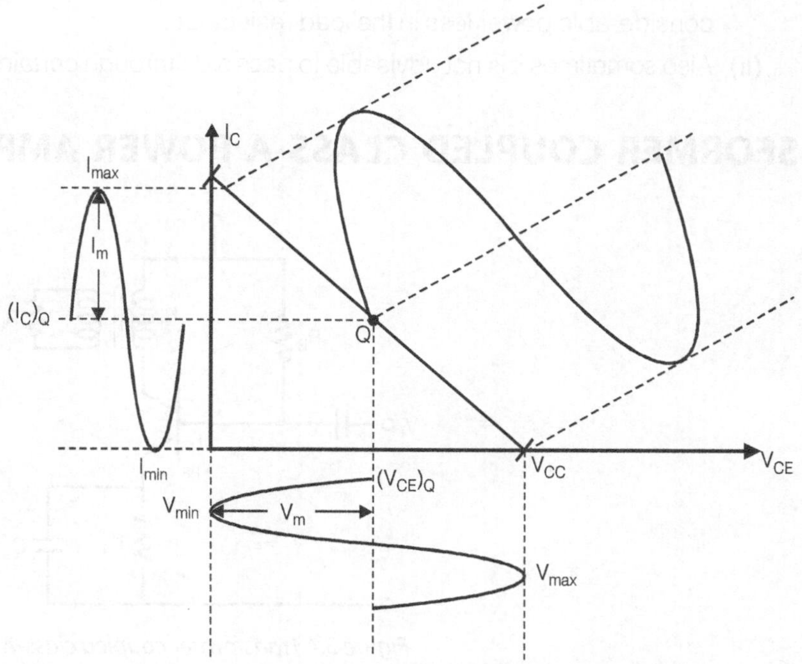


*Figure 5.3 AC output for maximum input signal*

We have,
$$V_m = \frac{V_{max} - V_{min}}{2}$$
$$I_m = \frac{I_{max} - I_{min}}{2}$$
$$(P_{ac}) = \frac{1}{2} \cdot \frac{(V_{max} - V_{min})}{2} \frac{(I_{max} - I_{min})}{2}$$
$$= \frac{1}{8}(V_{max} - V_{min})(I_{max} - I_{min}) \quad \text{...(i)}$$

## Maximum Efficiency

To calculate maximum efficiency of series fed class-A

We have
$$V_{min} = 0$$
$$V_{max} = V_{CC}$$
$$I_{min} = 0$$
$$I_{max} = 2(I_C)_Q$$

Substitute above conditions in equation (i)
$$\%\eta = \frac{(P_{ac})}{(P_{dc})} = \frac{\dfrac{(V_{CC} - 0)\left[2(I_C)_Q - 0\right]}{8}}{V_{CC}(I_C)_Q}$$
$$= \frac{1}{4} = 0.25 = 25\%$$

∴ So the maximum efficiency in a class-A series fed power amplifier is 25%.

## Disadvantages

(i) Since there is a quiescent current $(I_C)_Q$ passing through the load resistance, there will be considerable power loss in the load resistance.

(ii) Also sometimes it is not advisable to pass a dc through certain loads such as load speakers.

# 2. TRANSFORMER COUPLED CLASS-A POWER AMPLIFIER

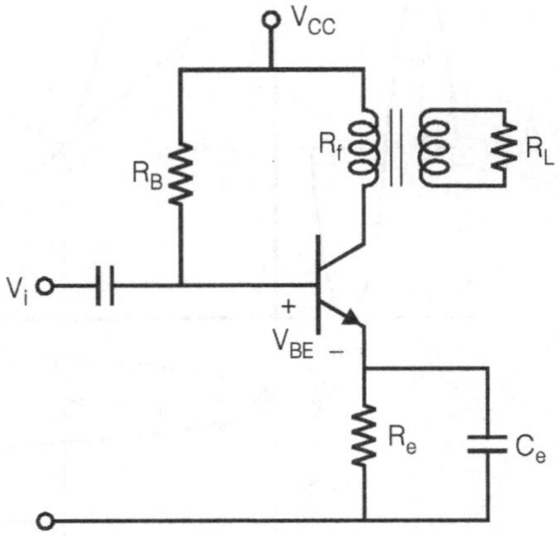


*Figure 5.4 Transformer coupled class-A*

The above mentioned disadvantages of series fed connection can be overcome by employing a transformer, known as the output transformer.

For coupling the load $R_L$ to the collector circuit of the transistor. Figure (5.4) shows a transformer-coupled amplifier. The primary of the transformer, having negligible dc resistance, is connected in the collector circuit. The secondary of transformer is connected to the load $R_L$.

## Impedance Matching

Let us consider only the transformer and the load $R_L$.

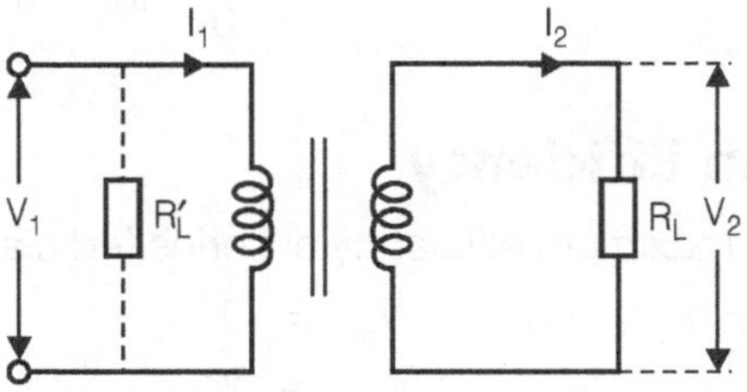


*Figure 5.5 Transformer*

Let the turn-ratio between the primary and the secondary of the transformer be N. Therefore the voltage ratio and current ratio can be written as

$$\frac{V_1}{V_2} = N \quad \text{and} \quad \frac{I_2}{I_1} = N$$

Therefore $$V_1 = NV_2$$

and $$I_1 = \frac{I_2}{N}$$

From the above expression, we get

$$\frac{V_1}{I_1} = \frac{NV_2}{I_2/N} = N^2 \cdot \frac{V_2}{I_2}$$

Since the secondary current $I_2$ is equal to $\left(\frac{V_2}{R_L}\right)$, therefore $\left(\frac{V_2}{I_2}\right) = R_L$.

Similarly $\left(\frac{V_1}{I_1}\right)$ represents the effective load resistance, on the primary side of the transformer which is denoted by $R_L'$.

We can write

$$\boxed{R_L' = N^2 R_L}$$

This means that, the effective load resistance $R_L'$ appearing on the collector circuit of the transistor is $N^2$ times the load resistance connected to the secondary.

## Analysis

DC power input

$$P_{DC} = V_{CC}(I_C)_Q$$

DC operation

$$(I_B)_Q = \frac{V_{CC} - 0.7}{R_B}$$

$$(I_C)_Q = \beta(I_B)_Q$$

$$(V_{CE})_Q = V_{CC} - (I_C)_Q R_f$$

Since the primary of the transformer has negligible dc resistance, the voltage drop under quiescent condition is also negligible. Therefore $(V_{CE})_Q$ is almost equal to $V_{CC}$.

$$\therefore \quad (V_{CE})_Q = V_{CC}$$

## Maximum Efficiency

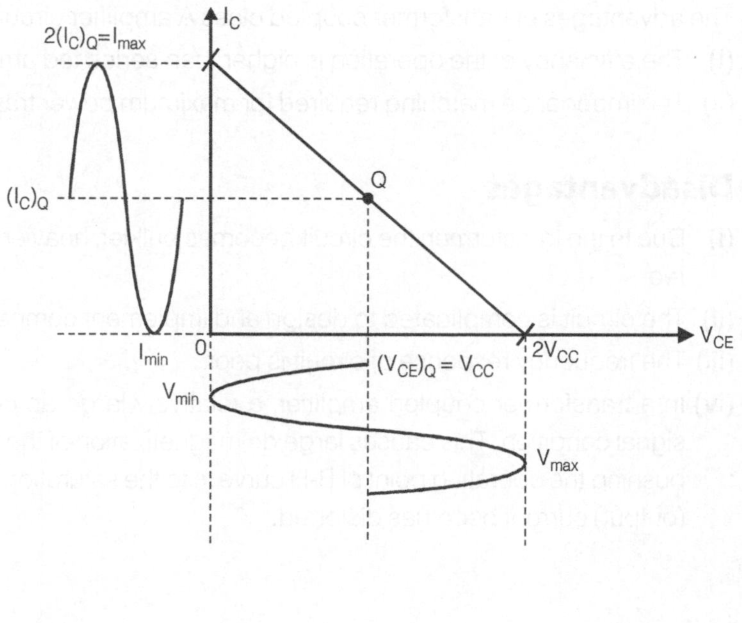

$$(P_o)_{ac} = \frac{(V_{max} - V_{min})(I_{max} - I_{min})}{8} \quad \text{...(i)}$$

We have from figure

$$V_{min} = 0$$
$$V_{max} = 2V_{CC}$$
$$I_{min} = 0$$
$$I_{max} = 2(I_C)_Q$$

Substitute in equation (1), we get

$$(P_o)_{ac} = \frac{2V_{CC} \times 2(I_C)_Q}{8}$$
$$= \frac{1}{2}V_{CC}(I_C)_Q$$

Maximum efficiency

$$\%\eta = \frac{(P_o)_{ac}}{(P_{in})_{dc}}$$
$$= \frac{1/2\, V_{cc}(I_C)_Q}{V_{CC}(I_C)_Q} = \frac{1}{2} = 50\%$$

Hence in class-A transformer coupled power amplifier the efficiency become 50%.

## Power Dissipation

Power dissipation $\boxed{P_d = P_{dc} - P_{ac}}$

At zero signal conditions (i.e. $P_{ac} = 0$)
The above equation becomes

$$(P_d)_{max} = P_{dc}$$

i.e. $(P_d)_{max} = V_{CC}(I_C)_Q$

That means,
maximum power dissipation happens in class-A at zero signal conditions.

## Advantages

The advantages of transformer coupled class A amplifier circuit are

**(i)** The efficiency of the operation is higher than series fed amplifier.

**(ii)** The impedance matching required for maximum power transfer is possible.

## Disadvantages

**(i)** Due to the transformer, the circuit becomes bulkier, heavier and costlier compared to series fed.

**(ii)** The circuit is complicated to design and implement compared to series fed.

**(iii)** The frequency response of circuit is poor.

**(iv)** In a transformer coupled amplifier, a relatively large dc collector current flows under no-signal condition. This causes large dc magnetization of the transformer core, which leads to pushing the operating point of B-H curve in to the saturation region. Therefore the secondary (output) current becomes distorted.

# 5.7 Distortion in Amplifiers

## HARMONIC DISTORTION

"The harmonic distortion means the presence of the frequency components in the output waveform, which are not present in the input signal".

The component with frequency same as the input signal is called fundamental frequency component. The additional frequency components present in the output signal are having frequency components which are integer multiples of fundamental frequency components or harmonics. The fundamental frequency component is not considered as a harmonic.

%$n^{th}$ harmonic distortion

$$\% D_n = \frac{|B_n|}{|B_1|} \times 100$$

$$\% D_2 = \frac{|B_2|}{|B_1|} \times 100$$

$$\% D_3 = \frac{|B_3|}{|B_1|} \times 100$$

Where fundamental frequency component has an amplitude $B_1$, and $n^{th}$ harmonic component has an amplitude of $B_n$.

Total harmonic distortion

$$\%D = \sqrt{D_2^2 + D_3^2 + D_4^2 + \cdots} \times 100\%$$

Where D is total harmonic distortion.

### Power Output Due to Distortion

$$P_{ac} = \frac{I_m^2 R_L}{2}$$

Where $I_m = \dfrac{I_{max} - I_{min}}{2}$

$$B_1 = \frac{I_{max} - I_{min}}{2}$$

Here $B_1$ = Fundamental harmonic component magnitude

Hence $I_m = B_1$

$$P_{ac} = \frac{1}{2} B_1^2 R_L$$

With distortion, the power delivered to the load increases proportional to amplitude of harmonic components.

$(P_{ac})_D$ = AC power output with harmonic distortion

$$= \frac{1}{2} B_1^2 R_L + \frac{1}{2} B_2^2 R_L + \frac{1}{2} B_3^2 R_L + \cdots$$

$$(P_{ac})_D = \frac{1}{2} B_1^2 R_L \left[1 + \frac{B_2^2}{B_1^2} + \frac{B_3^2}{B_1^2} + \cdots\right]$$

$$(P_{ac})_D = \frac{1}{2} B_1^2 R_L \left[1 + D_2^2 + D_3^2 \cdots\right]$$

$$\boxed{(P_{ac})_D = P_{ac}\left[1 + D^2\right]}$$

Where $D^2 = D_2^2 + D_3^2 + \cdots$

## 5.8 Class-B Power Amplifier

- When the collector current flows only during positive half cycle of the input signal, the amplifier is called as class-B amplifier.
- In class-B operation, the transistor is biased in such a way that at zero signal, the collector current is zero. Hence no biasing system is required in class-B power amplifier.
- Figure (5.6) shows the output waveform of class-B amplifier.

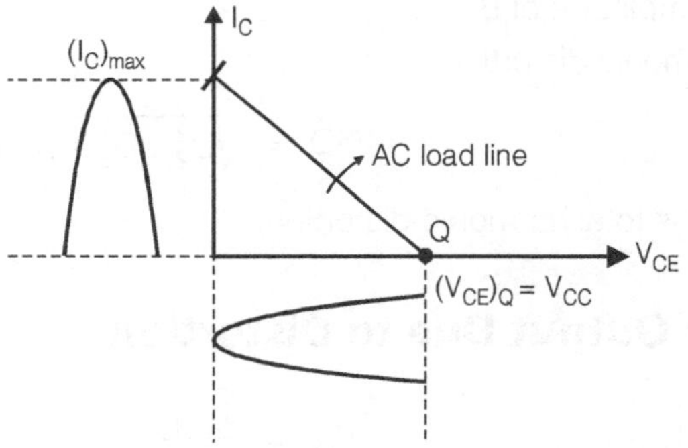


*Figure 5.6 Class-B operation*

As shown in Figure (5.6), during the positive half cycle of input signal, the circuit is forward biased. Now the collector current flows.

On the other hand, during negative half cycle of the input, the circuit is reverse biased. In this case, no collector current flows.

The following points are to be noted:

(i) As the negative half-cycle is totally absent from the output, the signal distortion is high.

(ii) Zero signal input represents the best condition for class-B amplifier because collector current is zero.

(iii) When signal increases, the transistor dissipate more power.

(iv) As average current is less than class-A amplifier the power dissipation is less i.e. the overall efficiency is increased.

## Overall Efficiency of Class-B Amplifier

The power input is given by

$$(P_{in})_{dc} = V_{CC} \times I_{dc}$$

Where $I_{dc}$ is the average or dc current drawn from the supply.

The value of $I_{dc}$ from rectifiers is $\frac{I_m}{\pi}$ (HWR).

$$\therefore \quad (P_{in})_{dc} = \frac{V_{CC} \times I_m}{\pi}$$

$$\text{RMS value of collector current} = \frac{I_m}{\sqrt{2}}$$

$$\text{RMS value of output voltage} = \frac{V_m}{\sqrt{2}}$$

Where $V_m = V_{CC}$

Hence output power during half cycle.

$$(P_o)_{ac} = \frac{1}{2} \times \frac{I_m}{\sqrt{2}} \times \frac{V_{CC}}{\sqrt{2}}$$

$$= \frac{I_m \times V_{CC}}{4}$$

The overall efficiency is given by

$$\eta_{overall} = \frac{(P_o)_{ac}}{(P_{in})_{dc}}$$

$$= \frac{I_m \times V_{CC}}{4} \times \frac{\pi}{V_{CC} \times I_m}$$

$$= \frac{\pi}{4} = 0.785 = 78.5\%$$

Let us explain that why class-B operation is more efficient than class-A operation the reason as follow:

We have seen that in class-A operation, under no signal condition, the transistor has to dissipate maximum power and least power when input signal is maximum. On the other hand, in class-B operation, under no signal condition, the transistor dissipates zero power and it increases with increased input signal.

The average current in class-B operation is $I_m/\pi$. This current is less than that in class-A operation. Hence the power dissipated by the transistor under class-B operation is less than under class-A operation.

So operation B is more efficient than A operation. The theoretical efficiency in class-B operation is 78.5%, while it is only 50% in class-A operation.

# 5.9 Class-B Push Pull Amplifier

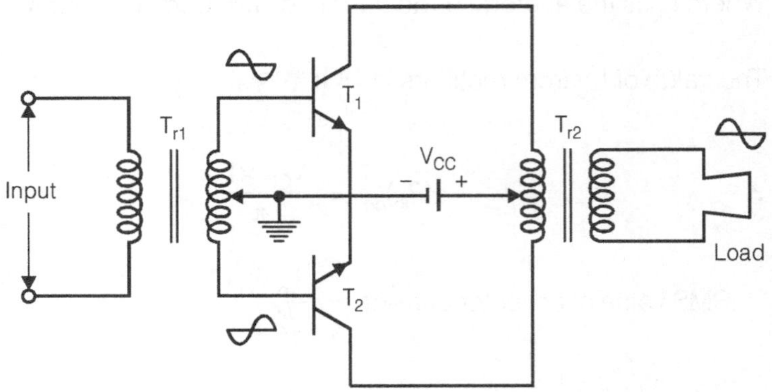


*Figure 5.7 Class B push pull amplifier*

- We have seen that class-A reduces the harmonic distortion but the conversion efficiency is only 50%. To increase the efficiency and output power for a given transistor dissipation capability, class-B push pull amplifier is used.
- The circuit arrangement of class-B push-pull amplifier is that the transistor are biased at cut-off.
- This is done by providing zero bias on the base of each transistor by connecting the base and emitter together.
- As shown in the figure, the bases of the two CE connected identical transistors have been connected to the opposite ends of the secondary of the input transformer $T_{r1}$ and collector ends to the primary of the output transformer $T_{r2}$.
- In order to get the balanced circuit, the emitters of the transistors are connected to the centre tap of the secondary of transformer $T_{r1}$ and $V_{CC}$ is connected to the centre tap on the primary of transformer $T_{r2}$. The two bases have been earthed.

## Circuit Operation

- When no signal is applied, both the transistors are cut-off. Hence no current is drawn from supply source $V_{CC}$. So there is no power wastage on standby.
- When the input signal applied, the phase splitter transformer $T_{r1}$ produces two signals which are 180° out of phase with each other. The transistors $T_1$ and $T_2$ are driven by these two signals.
- During positive half cycle of the signal, transistor $T_1$ conducts, because its base is driven positive. Now collector current $i_{c1}$ flows. Mean while, transistor $T_2$ does not conduct because its base has negative voltage. Thus $i_{c2}$ is zero. In this way one positive half cycle of output signal appears across secondary.
- During negative half of the input signal, transistor $T_2$ becomes forward-biased and allows a current $i_{c2}$ to flow, while transistor $T_1$ becomes non conducting. Thus only one transistor conducts at a time.

## Equivalent Model of Class-B Push Pull

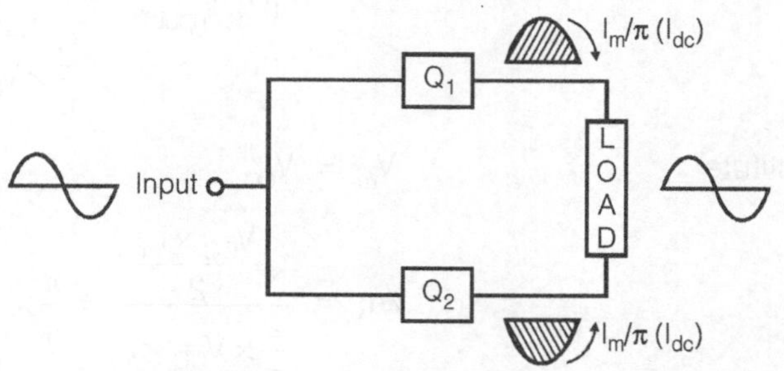


*Figure 5.8 Block diagram of class-B push pull*

# ANALYSIS

DC power input $P_{DC} = V_{CC} \times I_{dc} = \frac{2}{\pi} \times V_{CC} \times I_m$

Where $I_{dc} = \frac{2I_m}{\pi}$

$$(P_{ac})_{output} = V_{rms} \times I_{rms} = \frac{V_m}{\sqrt{2}} \times \frac{I_m}{\sqrt{2}} = \frac{V_m \times I_m}{2}$$

## Graphical Analysis

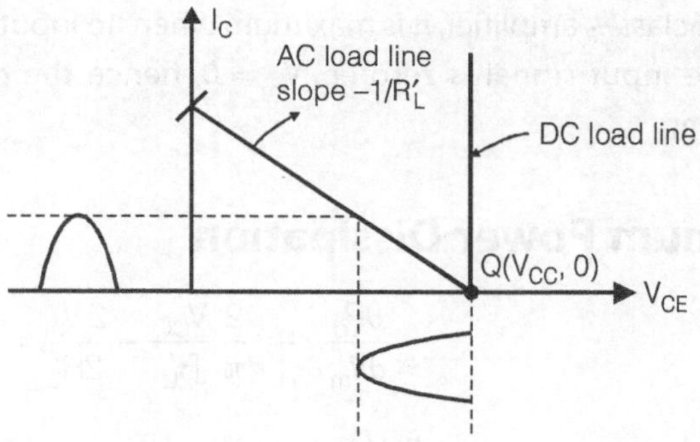


## Maximum Efficiency

For maximum efficiency calculations:

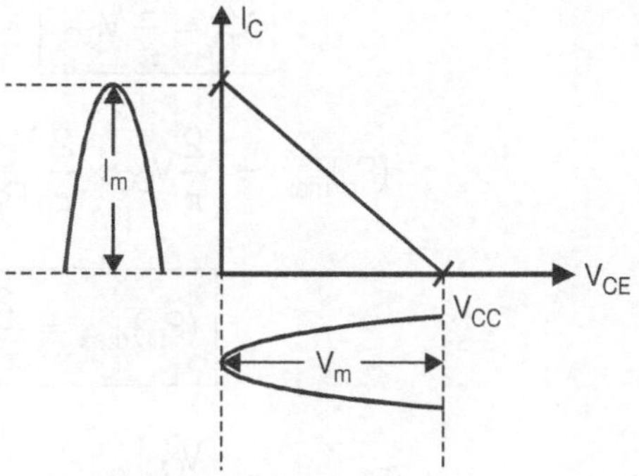


Here $V_m = V_{CC}$

$$\%\eta = \frac{(P_{ac})_{output}}{(P_{dc})_{input}} = \frac{\frac{V_m \times I_m}{2}}{\frac{2}{\pi} \times V_{CC} \times I_m}$$

Substitute $$V_m = V_{CC}$$

$$\%\eta = \frac{\frac{V_{CC} \times I_m}{2}}{\frac{2}{\pi} \times V_{CC} \times I_m} = \frac{\pi}{4} \times 100\% = 78.5\%$$

The slope of ac load line can be represented in terms of $V_m$ and $I_m$ as

$$\frac{1}{R_L'} = \frac{I_m}{V_m}$$

$$\boxed{R_L' = \frac{V_m}{I_m}}$$

## Power Dissipation

$$P_d = P_{dc} - P_{ac} = \frac{2}{\pi} V_{CC} I_m - \frac{V_m I_m}{2}$$

$$P_d = \frac{2}{\pi} V_{CC} \frac{V_m}{R_L'} - \frac{V_m^2}{2R_L'}$$

In case of class-A amplifier, it is maximum when no input signal is there, but in class-B operation, when the input signal is zero i.e. $V_m = 0$, hence the power dissipation is zero and not the maximum.

## Maximum Power Dissipation

$$\frac{\partial P_d}{\partial V_m} = \frac{2}{\pi} \frac{V_{CC}}{R_L'} - \frac{2V_m}{2R_L'} = 0$$

$$\frac{2}{\pi} \frac{V_{CC}}{R_L'} = \frac{V_m}{R_L'}$$

$$\boxed{V_m = \frac{2}{\pi} V_{CC}}$$ For maximum power dissipation.

$$(P_d)_{max} = \frac{2}{\pi} V_{CC} \times \frac{2}{\pi} \frac{V_{CC}}{R_L'} - \frac{4}{\pi^2} \frac{V_{CC}^2}{2R_L'} = \frac{4}{\pi^2} \frac{V_{CC}^2}{R_L'} - \frac{2}{\pi^2} \frac{V_{CC}^2}{R_L'}$$

$$\boxed{(P_D)_{max} = \frac{2}{\pi^2} \frac{V_{CC}^2}{R_L'}}$$

$$(P_{ac})_{max} = \frac{V_{CC}^2}{2R_L'}$$

$$(P_d)_{max} = \frac{4}{\pi^2} \times (P_{ac})_{max}$$

The maximum power dissipation per transistor is $\frac{(P_d)_{max}}{2}$.

$$(P_d)_{max} \text{ per transistor} = \frac{2}{\pi^2}(P_{ac})_{max}$$

## HARMONIC DISTORTION IN CLASS-B PUSH PULL

Let the base currents are sinusoidal in nature.

$$i_{b1} = i_{bm} \cos \omega t \quad \text{and} \quad i_{b2} = -i_{bm} \cos \omega t$$

$$i_{c1} = (I_C)_Q + B_0 + B_1 \cos \omega t + B_2 \cos \omega t + \cdots$$

$$i_{b2} = -i_{bm} \cos \omega t = i_{bm} \cos(\omega t + \pi)$$

$$i_{c2} = I_{CQ} + B_0 + B_1 \cos(\omega t + \pi) + B_2 \cos 2(\omega t + \pi) + \cdots$$

$$= I_{CQ} + B_0 - B_1 \cos \omega t + B_2 \cos 2\omega t - B_3 \cos 3\omega t + \cdots$$

$$i_L = i_{c1} - i_{c2}$$

$$= I_{CQ} + B_0 + B_1 \cos \omega t + B_2 \cos 2\omega t + B_3 \cos 3\omega t + \cdots - (I_{CQ} + B_0 - B_1 \cos \omega t + B_2 \cos 2\omega t - B_3 \cos 3\omega t + \cdots)$$

$$i_L = 2B_1 \cos \omega t + 2B_3 \cos 3\omega t + \cdots$$

The even harmonics are eliminated, similarly, the dc component also gets eliminated.

$$\%D_3 = \frac{|B_3|}{|B_1|} \times 100, \quad \%D_5 = \frac{|B_5|}{|B_1|} \times 100$$

Hence total harmonic distortion is

$$\%D = \sqrt{D_3^2 + D_5^2 + D_7^2 + \cdots} \times 100$$

### Advantages of Class-B Push Pull

(i) The efficiency is much higher than class-A

(ii) When there is no input signal, the power dissipation is zero.

(iii) The even harmonics get cancelled. This reduces the harmonic distortion.

(vi) There is no possibility of dc saturation of core.

(v) Due to transformer, impedance matching is possible

### Disadvantages

(i) Two centre tap transformers are necessary.

(ii) The transformers, make the circuit bulky and hence costlier.

(iii) Frequency response in poor.

# 5.10 Complementary Symmetry Class-B Amplifier

- This amplifier, shown in Figure (5.9) does away with the input transformer and output transformer.
- The input transformer has the function of providing two input signals 180° out of phase, which make only one transistor (either $Q_1$ or $Q_2$) to conduct at a time.
- However in this circuit, the same signal voltage is applied simultaneously to the two transistors. One of them ($Q_1$) a npn and another ($Q_2$) a pnp.

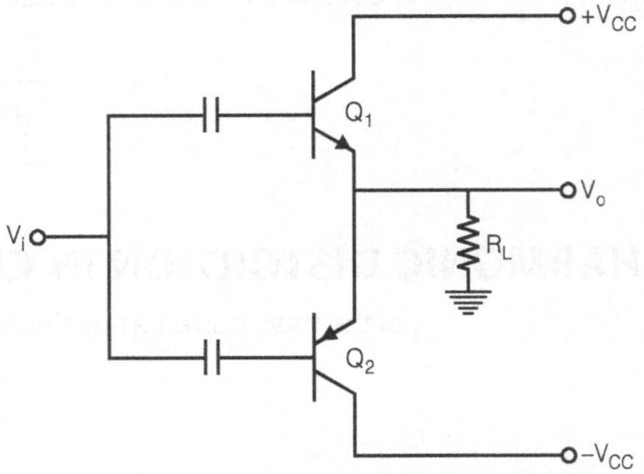


*Figure 5.9 Complementary symmetry class-B*

- On the positive half cycle of $V_i$, $Q_1$ is forward biased and therefore conducts, but $Q_2$ being pnp is reverse biased and therefore does not conduct.
- Just the opposite takes place when the signal $V_i$ is the negative half cycle.
- Thus the current $i_1$ and $i_2$ flow in one half cycle each but both flow through $R_L$. Therefore the total current flowing through $R_L$ is in both half cycles.
- The theoretical efficiency of this circuit is also 78.5%.

# 5.11 Comparison Between Push Pull and Complementary Symmetry

| | Push Pull | Complementary Symmetry |
|---|---|---|
| 1. | Both the transistor are similar either pnp or npn | Transistor are npn and pnp |
| 2. | The transformer is used to connect the load as well as input | The circuit is transformerless |
| 3. | The impedance matching is possible due to output transformer | Impedance matching is possible due to common collector circuit |
| 4. | Frequency response is poor | Frequency response is improved |
| 5. | Dual power supply is not required | Dual power supply is required |
| 6. | Efficiency is higher than class-A | Efficiency is higher than push pull |

## 5.12 Cross Over Distortion

- In the above discussions, we have assumed that the transistor $Q_1$ start conducting as soon as the input signal becomes even slightly positive.
- However, we know that the base-emitter diode does not go into conduction till it reaches "cut-in" voltage (which about 0.2 V germanium and 0.5 V for silicon transistors).
- Therefore the transistor does not have a perfectly sinusoidal collector current. This is shown in Figure (5.10), where we find that when the total collector curve crosses the x-axis there is non sinusoidal behaviour, which is known as cross over distortion.

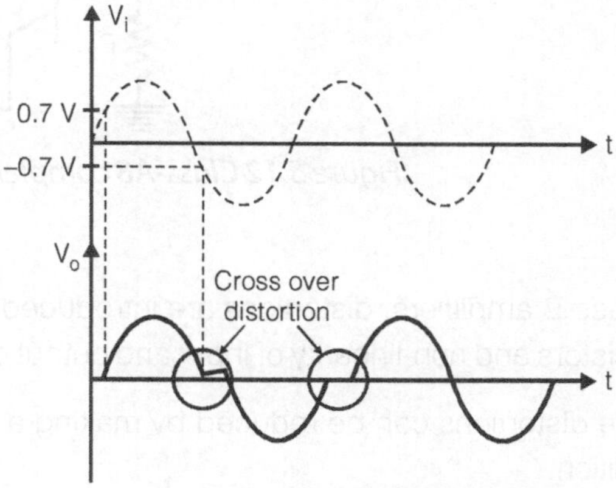


*Figure 5.10 Cross over distortion*

## 5.13 Class-AB Amplifier

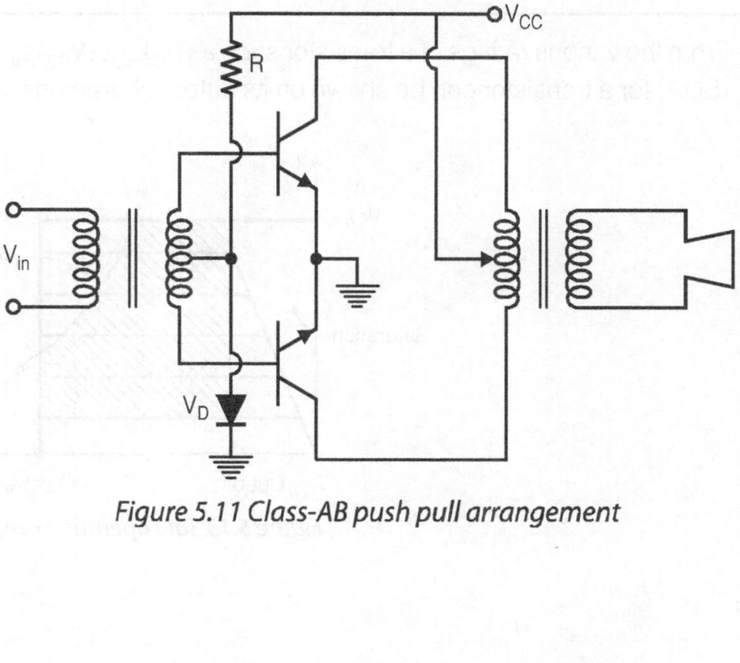


*Figure 5.11 Class-AB push pull arrangement*

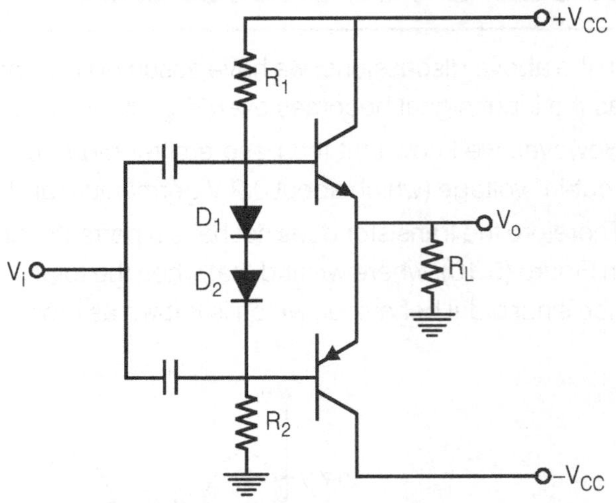


*Figure 5.12 Class-AB complementary symmetry*

- In class-B amplifiers, distortions are introduced due to cross over distortion, mismatch of transistors and non-linearity of input and output characteristics.
- These distortions can be reduced by making a small stand by current flow of zero-signal condition.
- In class-AB mode, the base bias during quiescent condition is such that current flows for more than 180° but less than 360° and hence the name class-AB.
- Class-AB gives less distortion than class-B but efficiency is less than class-B.

## 5.14 Safe Operating Area (SOA) For a Transistor

From the various ratings of a transistor such as $(I_C)_{max}$, $(V_{CE})_{max}$, $(P_d)_{max}$, the safe operating area (SOA) for a transistor can be shown on its output characteristics.

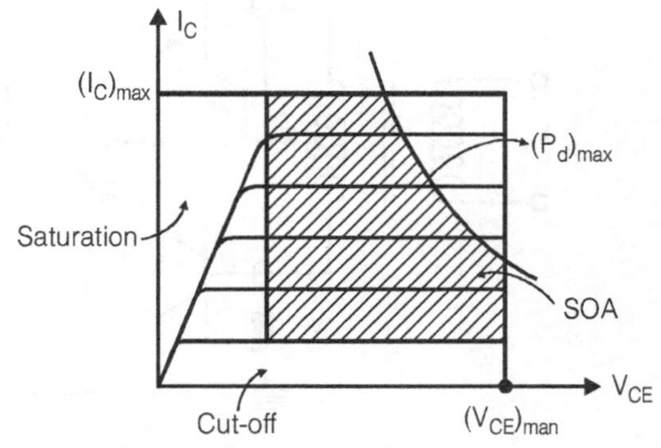


*Figure 5.13 Safe operating area*

### Key point

When transistor is operated in a safe operating area, it is said thermal stable.

### Heat Sink for Power Transistors

$$P_d = V_{CE} I_C$$

Let $T_j$ be the junction temperature which due to power dissipation.
The lower limit of a semiconductor is taken as

$$-65°C \le T_j \le (T_j)_{max}$$

Where $(T_j)_{max}$ is specified by the manufactures.
Thus practically it is necessary to keep the junction temperature less than $(T_j)_{max}$ specified for the power transistor used in power amplifiers.

### Derating of Power Transistors

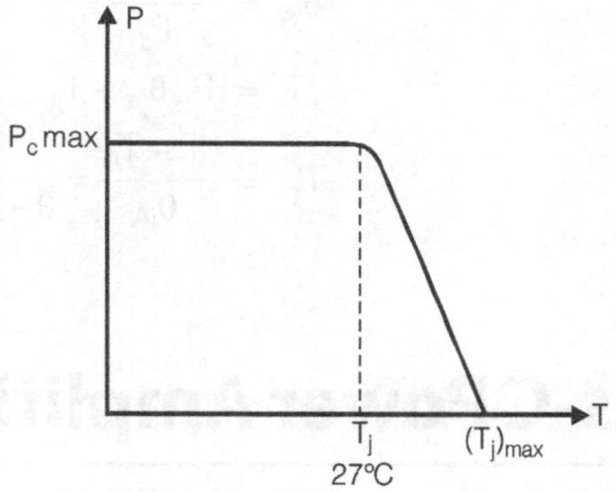


$$\Delta P = D \cdot f\,(T_2 - T_1) = \frac{mW}{°C} \times °C = mW$$

Where D · f means derating factor.

## 5.15 Thermal Analogy of Power Transistors

$$\theta = \frac{T_2 - T_1}{P_d} \frac{°C}{mW}$$

$$P_d = \frac{T_2 - T_1}{\theta} mW$$

Now to develop the thermal-electric analogy. Let us define some parameters are:

$T_j \rightarrow$ Junction temperature
$T_C \rightarrow$ Case temperature
$T_A \rightarrow$ Ambient temperature
$\theta_{jA}$ = Total thermal resistance (junction to ambient)
$\theta_{jC}$ = Transistor thermal resistance (junction to case)
$\theta_{CS}$ = Insulator thermal resistance (case to heat sink)
$\theta_{SA}$ = Heat sink thermal resistance (heat sink to ambient)

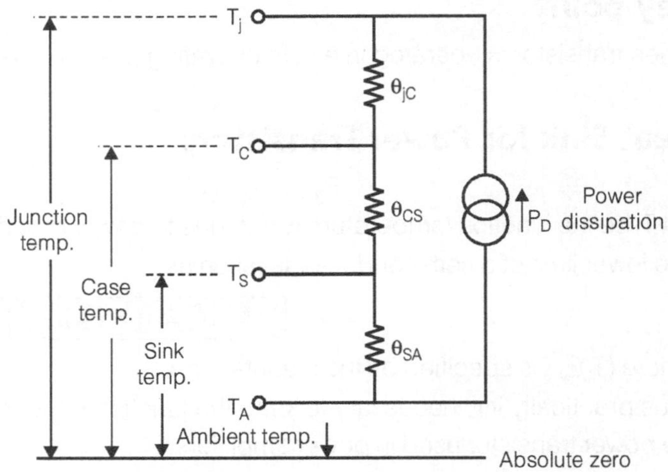


$$\theta_{jA} = \theta_{jC} + \theta_{CS} + \theta_{SA}$$

But

$$\theta_{jA} = \frac{T_j - T_A}{P_d}$$

$$T_j = P_d\,\theta_{jA} + T_A$$

$$P_D = \frac{T_j - T_A}{\theta_{jA}} = \frac{T_j - T_A}{\theta_{jC} + \theta_{CS} + \theta_{SA}}$$

## 5.16 Class-C Power Amplifier

- Class-C amplifiers can not be used as audio amplifier.
- However we can use the class-C amplifier in tuned circuits used in communication.
- The LC tuned circuit as load for transistor Q.
- The dc source-$V_{BB}$ biases the transistor below cut off.
- The ac input voltage $V_i$ is applied to the base of the transistor. The voltage is super imposed on $-V_{BB}$ as shown.

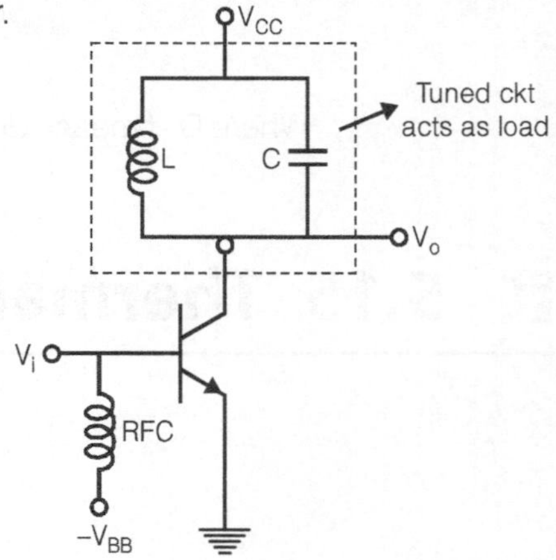


*Figure 5.14 Class-C power amplifier*

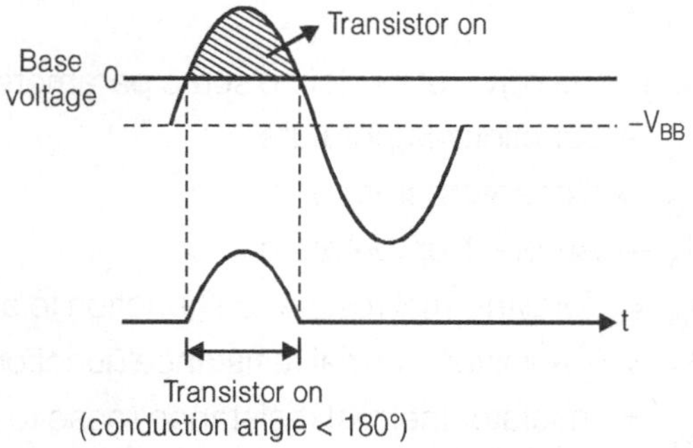


The transistor will be forward biased and will conduct for an angle which is less than 180°.

The transistor collector current flows in the form of pulses of duration less than 180°.

When these current pulses are applied to the LC resonant circuit a full cycle of output signal at the resonant frequency of the tuned circuit is obtained.

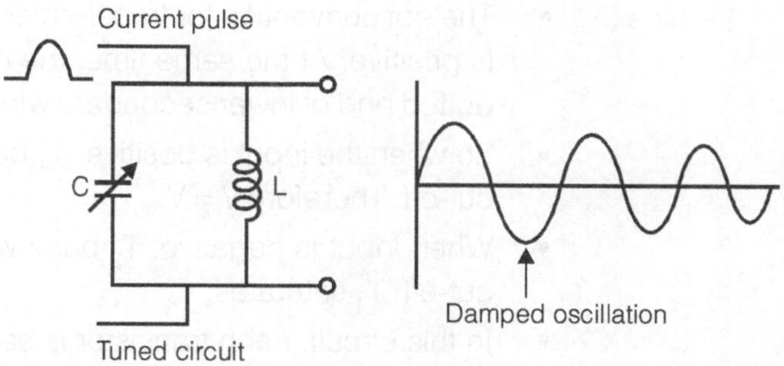


### Property of Tuned Circuit

- To understand the class-C amplifier, we must know an important property of a tuned circuit.
- **The property states that if we apply a current pulse to a tuned circuit then it generates damped voltage oscillations at its output.**
- The amplitude of oscillations is proportional to the size of current pulse and the decay rate is proportional to time constant.
- If we apply a train of current pulses at frequency $f_r$ which is equal to resonant frequency of tuned circuit, then we will obtain a sinusoidal signal at frequency $f_r$ at the output.

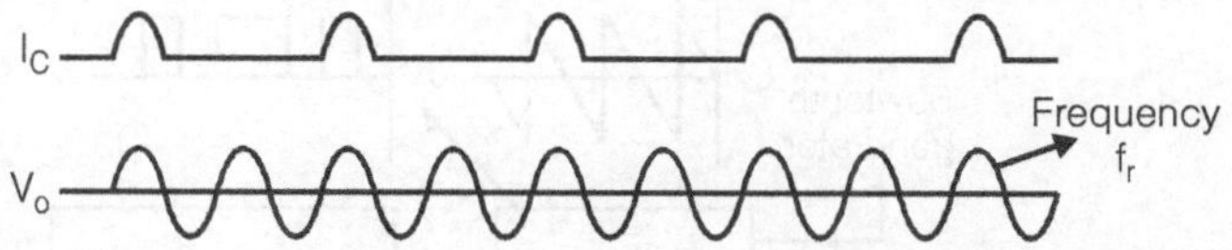


This shows class-C operation is limited to use at only one frequency i.e. resonant frequency $f_r$.

## 5.17 Class-D Amplifiers

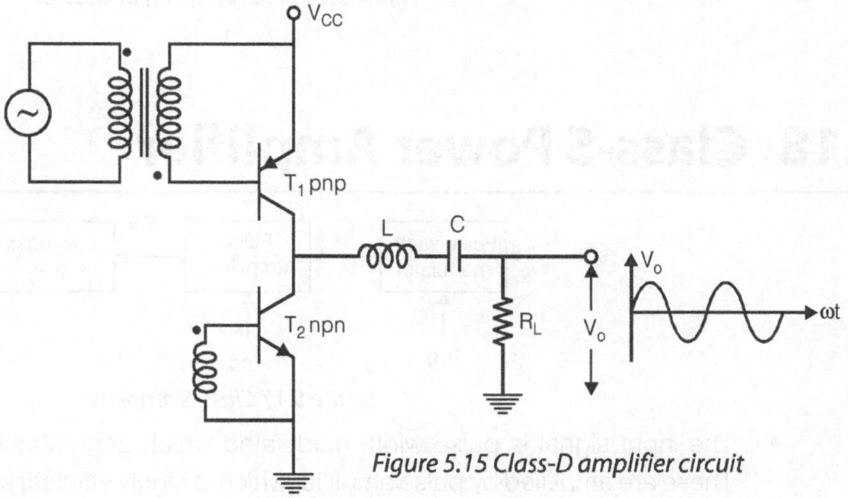


*Figure 5.15 Class-D amplifier circuit*

- A push pull connection of two transistors in common emitter configuration of complementary transistors (one pnp the other npn) is employed.
- When the input is positive, $T_1$ is cut-off and $T_2$ saturates.
- During the negative half cycle of the input, $T_1$ saturates and $T_2$ is cut-off. Therefore the output voltage is a square wave with voltage changing between 0 and $V_{CC}$.

- The dot convention for transformer is, when input is positive. The dotted end of the primary is positive. At the same time, the dotted end of upper secondary winding is positive and dotted end of lower secondary winding is positive.
- So when the input is positive, $T_1$ base which is n-type (pnp) gets positive voltage. So $T_1$ is cut-off. Therefore $V = V_{CC}$.
- When input is negative, $T_2$ base which is p-type (npn) will get negative voltage. So $T_2$ is cut-off, $T_1$ saturates.
- In this circuit, each transistor is saturated for almost 180° of the cycle. So each transistor acts like a switch rather than like a current source.
- When the transistor saturates, the power dissipation
$$P_D = V_{CE\,(sat)}\, I_{C\,(sat)}$$
It is very small, since $V_{CE\,(sat)}$ is nearly zero. When the transistor is cut-off, $P_D = 0$. Therefore average power dissipation over the cycle is very small. Therefore $\eta = 100\%$.
- The output of the collectors of transistors, is a square wave with 0-$V_{CC}$ voltages. This is given to a series resonant circuit. So the output will be a sine wave (oscillator circuit).
∴ These are used in transmitters because their efficiency ($\eta$) is high ≈ 100% and pulse and digital circuits.

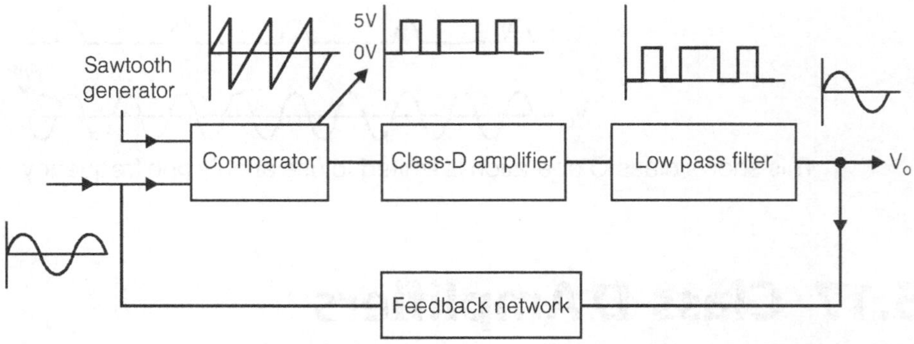


*Figure 5.16 Block diagram of class-D*

## 5.18 Class-S Power Amplifier

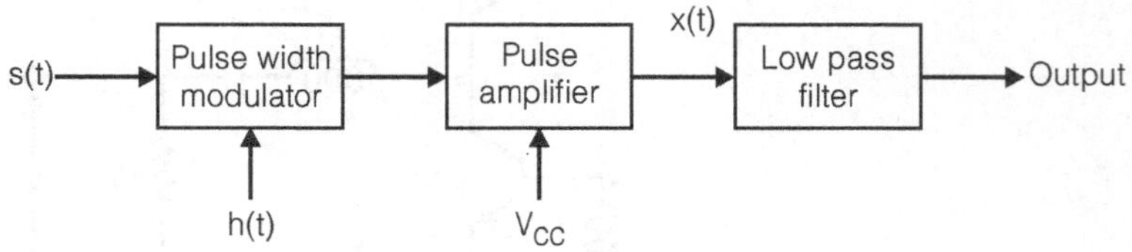


*Figure 5.17 Class-S amplifier*

- The input signal is pulse width modulated which generates constant amplitude pulses. These are amplified by pulse amplifier which is highly efficient switching amplifier like class-D amplifier.
- The modulating frequency, as much higher than the signal frequency, is completely removed using the low pass filter at last stage
- The basic difference between class-D and class-S amplifiers is that the output stage of class-D uses tuned circuit which is tuned to fundamental frequency of the input, while the output stage of class-S uses the low pass filter which recovers the input signal.

## SUMMARY

- Power amplifiers are those that involve large signal amplification.
- Linear approximation of the active devices is not possible and so graphical analysis is employed and the analysis depends on the selection of operating point.
- The conversion efficiency is the parameter that gives the amplifier the ability to convert the dc power to ac signal power.
- When the current flows for 100% time of the full cycle at the output of the amplifier, the amplifier is class A type. In this, the conversion efficiency of the amplifier is very small and distortion due to harmonics is minimum.
- To improve the efficiency of the class A amplifier, transformer coupling is used. To reduce the harmonic distortion, two transistors in push-pull configuration are employed. This eliminates the even harmonics totally.
- Class *B* amplifier has current only for 50% of the cycle and thus conversion efficiency is improved to a large extent. The non-linearity in the input of the transistor causes crossover distortion.
- Crossover distortion can be eliminated by the use of class *AB* amplifier where a small current is allowed to be passed above 50% of the cycle, which eliminates the threshold voltage effect of the input of the transistor.
- Efficiency of the amplifier tends to a very large value when the output current of the amplifier is in form of short bursts. This is achieved in the class C amplifier where the harmonic distortion is at its peak.
- Tuned parallel resonant circuits are used to generate the required voltage signals in the class C amplifier. These circuits are referred to as tuned amplifiers.
- Single-tuned amplifiers have the drawback of small bandwidth and high *Q* of the tank circuit makes the reproducibility of the signal poor.
- Double-tuned amplifier utilizes two tuned circuits that are coupled properly.
- Staggered tuned amplifier increases the bandwidth of the tuned amplifier by use of two tuned circuits, which are slightly tuned away from resonant input frequency.

## GATE AND IES QUESTIONS

**Q.1** In a class-B push pull amplifier $V_{CC}$ = 25 V, peak voltage output amplitude $V_m$ = 16 V ; $R_L = 8\ \Omega$. The dc power $P_{dc}$ supplied is given by

(a) 10 W (b) 16 W
(c) 20 W (d) 32 W

**Exp.** (d)

$$P_{dc} = V_{CC} \times I_{dc} = 2\,V_{CC} \times \frac{I_m}{\pi} = 32 \text{ Watts}$$

**Q.2** In a class-C power amplifier has an input signal with frequency of 200 kHz and the width of collector current pulses of 0.1 μS. Then the duty cycle of the amplifier will be

(a) 1% (b) 2%
(c) 10% (d) 20%

**Exp.** (b)

$$T = \frac{1}{f} = \frac{1}{200 \times 10^3} = 5 \times 10^{-6} = 5\,\mu S$$
$$\tau = 0.1\,\mu S$$
$$\text{Duty cycle} = \frac{\tau}{T} \times 100 = \frac{0.1}{5\,\mu S} \times 100 = 2\%$$

**Q.3** A class-B push pull type amplifier with transformer coupled load uses two transistors rated 10 W each. What is the maximum power output one can obtain at the load from this circuit?

(a) 40 W
(b) 50 W
(c) 60 W
(d) 70 W

**Exp.** (b)

$$(P_D)_{max} = \frac{4}{\pi^2}(P_{ac})_{max}$$

Per transistor Q

$$(P_D)_{max} = \frac{2}{\pi^2}(P_{ac})_{max}$$

$$(P_{ac})_{max} = \frac{\pi^2}{2} \times (P_D)_{max} = \frac{(3.14)^2}{2} \times 10 = 50\text{ W}$$

**Q.4** To avoid thermal run-away in the design of an analog circuit, the operating point of the BJT should be such that it satisfies the condition

(a) $V_{CE} = \frac{1}{2} V_{CC}$
(b) $V_{CE} \le \frac{1}{2} V_{CC}$
(c) $V_{CE} > \frac{1}{2} V_{CC}$
(d) $V_{CE} \le 0.78\, V_{CC}$

**Exp.** (b)

To avoid thermal run-away

$$V_{CE} \le \frac{1}{2} V_{CC}$$

**Q.5** An amplifier has a d.c. power supply of 15V and draws a current of 10mA. It produces an output of 5V peak to peak across a load resistance of 600 Ω for a signal frequency of 1 kHz. What will be its a.c. power output?

(a) 260 mW
(b) 20.8 mW
(c) 520 mW
(d) 40.6 mW

**Exp.** (b)

$$(P_o)_{ac} = \frac{V_o^2}{2R_L}(P-P) = \frac{5^2}{2 \times 600} = 20.8\text{ mW}$$

**Q.6** A dc to dc converter has an efficiency of 80% and is supplying a load of 24 W at 240 V. What is the current drawn from the battery if the converter is working from a battery of 12 V?

(a) 0.1 A
(b) 2.0 A
(c) 2.5 A
(d) 10 A

**Exp. (c)**

$$\eta = \frac{P_o}{P_i} \quad ; \quad P_i = \frac{P_o}{\eta} = \frac{24}{0.8}$$

$$P_i = 30\text{ W} \quad ; \quad I_i = \frac{P_i}{V_i} = \frac{30}{12} = 2.5\text{ A}$$

**Q.7** Which one of the following statements is correct?
The efficiency of class-B push-pull amplifiers is much higher than that of class-A amplifiers primary because
(a) The distortion is kept with in acceptable limits.
(b) One half of the input signal is amplified using one transistor and the other half is phase-inverted and fed to the other transistor.
(c) Matched pair transistors are used in the class-B push pull operation.
(d) The quiescent dc current is avoided.

**Exp. (d)**
The efficiency of class-B push pull is higher than class-A because the quiescent dc current is avoided.

**Q.8** Match **List-I** (class of amplifiers) with **List-II** (angle of conduction) and select the correct answer using the codes given below the lists:

| List-I | List-II |
|---|---|
| A. Class-A | 1. Less than 180° |
| B. Class-B | 2. 180° |
| C. Class-AB | 3. Greater than 180° |
| D. Class-C | 4. 360° |
| | 5. Usually 50° to 60° |

**Codes:**

| | A | B | C | D |
|---|---|---|---|---|
| (a) | 2 | 3 | 4 | 5 |
| (b) | 4 | 5 | 2 | 3 |
| (c) | 4 | 2 | 3 | 1 |
| (d) | 1 | 2 | 3 | 4 |

**Exp. (c)**

Class-A – 360°
Class-B – 180°
Class-AB – Greater than 180°
Class-C – Less than 180°

**Q.9** A push-pull amplifier suppresses
(a) only the second harmonic component
(b) only the odd harmonic components
(c) only the even harmonic components
(d) both odd and even harmonic components

**Exp. (c)**
Class-B push pull amplifier suppresses the even harmonic components.

**Q.10** Thermal run-away will take place if the quiescent point is such that
(a) $V_{CE} > \frac{1}{2}V_{CC}$ (b) $V_{CE} < V_{CC}$
(c) $V_{CE} < 2\,V_{CC}$ (d) $V_{CE} < \frac{1}{2}V_{CC}$

**Exp. (a)**

Thermal run-away will take place if $V_{CE} > \frac{1}{2} V_{CC}$

**Q.11** The condition to be satisfied to prevent thermal run-away in a transistor amplifier where ($P_c$ = power dissipated at collector, $T_j$ = junction temperature, $T_A$ = ambient temperature, $\theta$ = thermal resistance) is

(a) $\frac{\partial P_c}{\partial T_j} > \frac{1}{\theta}$
(b) $\frac{\partial P_c}{\partial T_A} < \frac{1}{\theta}$
(c) $\frac{\partial P_c}{\partial T_j} < \frac{1}{\theta}$
(d) $\frac{\partial P_c}{\partial T_A} > \frac{1}{\theta}$

**Exp. (c)**

$\frac{\partial P_c}{\partial T_j} < \frac{1}{\theta}$ (condition for thermal stability).

**Q.12** Using transistors,

1. Class-A power amplifier has a minimum efficiency of 50%.
2. Class-B push pull power amplifier gives rise to cross-over distortion.
3. Class-AB push pull power amplifier has higher efficiency than class-B push pull amplifier.
4. Class-C power amplifier is generally used with tuned load for RF applications.

(a) 1, 2, 3, and 4
(b) 2 and 4 only
(c) 3 and 4 only
(d) 1 and 2 only

**Exp. (b)**

- Class-B push pull power amplifier gives rise to cross-over distortion.
- Class-C power amplifier is generally used with tuned load for RF applications.

**Q.13** A class-A transformer coupled transistor power amplifier is required to deliver a power output of 10 Watts. The maximum power rating of the transistor should not be less than

(a) 3 W
(b) 10 W
(c) 20 W
(d) 40 W

**Exp. (c)**

The efficiency of class-A transformer coupled transistor power amplifier is 50%.

**Q.14** A power amplifier delivers 50 W output at 50% efficiency. The ambient temperature is 25°C. If the maximum allowable junction temperature is 150°C. Then the maximum thermal resistance $\theta_{jc}$ that can be tolerated is

(a) 2.5°C/W
(b) 5°C/W
(c) 10°C/W
(d) none

**Exp. (a)**

$$\theta_{jc} = \frac{T_j - T_A}{P_D} = \frac{150° - 25°}{50\text{ W}} = 2.5°\text{C/W}$$

**Q.15** Cross-over distortion behaviour is characteristic of

(a) Class-A
(b) Class-B
(c) Class-AB
(d) Common base

**Exp. (b)**

Cross-over distortion occurs in class-B

**Q.16** The output power of a power amplifier is several times less than its input power. This is possible because
(a) The power amplifier introduces negative resistance.
(b) The power amplifier converts a part of the input dc power into ac output power
(c) Positive feedback exists in the circuit
(d) Step up transformer is used in the circuit

**Exp.** **(b)**
Power amplifier converts a part of the input dc power into ac output power.

**Q.17** As compared to class-B power amplifier a class-C power amplifier gives ________ of the current signal waveform
(a) Less distortion
(b) Same distortion
(c) More distortion
(d) None

**Exp.** **(c)**
Class-C power amplifier gives more distortion than class-B.

**Q.18** In the case of a class-C power amplifier, the load must be
(a) pure resistance
(b) a parallel tuned circuit
(c) a series tuned circuit
(d) none

**Exp.** **(b)**
Parallel tuned circuit is taken as load for class-C power amplifier.

**Q.19** Class-A power amplifier systems are
(a) Linear
(b) Some what non-linear
(c) non-linear
(d) none

**Exp.** **(c)**
Transistor is a non linear device.

**Q.20** A radio frequency signal contains following three frequencies: 800 kHz, 825 kHz and 880 kHz. This signals need to be amplified. The amplifier used should be
(a) audio frequency amplifier
(b) wideband amplifier
(c) tuned voltage amplifier
(d) video amplifier

**Exp.** **(c)**
The frequencies are very close which is called tuned voltage amplifier.

**Q.21** Tuned voltage amplifiers are not used
(a) in public address systems
(b) in radio receivers
(c) where a band of frequencies is to be selected and amplified
(d) in television receivers

**Exp.** **(a)**
Tuned voltage amplifiers are not used in public address systems.

**Q.22** Class-D power amplifiers are used in
(a) public address system
(b) pulse and digital circuits
(c) tuned amplifiers
(d) none

**Exp.** **(b)**
Class-D power amplifier output will be high or low i.e. pulses digital circuits.

**Q.23** The difference between class-D and Class-S power amplifier are

(a) low pass circuit (b) tuned circuit

(c) amplifier (d) none

**Exp.** (a)

The only difference block in low pass circuit.

**Q.24** Transformer and emitter follower can both be used for impedance matching at the output of an audio amplifier. The basic relationship between the input power $P_{in}$ and output power $P_{out}$ in both the cases is

(a) $P_{in} = P_{out}$ for both transformer and emitter follower.

(b) $P_{in} > P_{out}$ for both transformer and emitter follower.

(c) $P_{in} < P_{out}$ for transformer and $P_{in} = P_{out}$ for emitter follower.

(d) $P_{in} = P_{out}$ for transformer and $P_{in} < P_{out}$ for emitter follower

**Exp.** (d)

Input and output power of a transformer is same

$$P_{in} = P_{out}$$

For emitter follower,

$$\text{Voltage gain } (A_V) = 1$$

$$\text{Current gain } (A_i) > 1$$

$$\text{Power } (P_{out}) = A_V A_i P_{in}$$

Since emitter follower has a high current gain

So $$P_{out} > P_{in}$$

## REVIEW QUESTIONS

1. Sketch the circuit of a transformer-coupled class-A amplifier. Briefly explain the operation of the circuit?
2. Distinguish between small signal and large signal amplifiers. How are the power amplifiers classified?
3. Derive the expressions for maximum theoretical efficiency for series fed class-A amplifier?
4. What are the advantages and disadvantages of transformer coupling?
5. Derive the expression for maximum theoretical efficiency in the case of class-B push pull amplifier. Why it is named so? What are its advantages and disadvantages?
6. Explain class-B complementary symmetry with a neat sketch?
7. Explain about class-C power amplifier?
8. Explain about class-D and class-S power amplifiers mention their salient features and applications.
9. How are the tuned amplifiers classified? Explain the salient features of each of them.
10. What is cross-over distortion? How it is reduced in class-AB

OOOO

# CHAPTER 6

# MULTISTAGE AMPLIFIERS

## About This Chapter

In this part of the chapter, we study about

- The current gain or voltage gain obtained from a single transistor is generally less. To acheive the desired amplification, we use multistage.
- Cascading of single stage amplifiers.
- Lower cut-off frequency, upper cut-off frequency, when n-stages are cascaded are given.
- Methods of coupling in multistage.
- Relation between rise time and bandwidth.
- Study of cascode and Darlington pair circuits.
- Boot strap emitter follower and Millers theorem proofs are discussed.
- Effect of cascading on bandwidth.

# 6 MULTISTAGE AMPLIFIERS

## 6.1 Introduction

- One single-stage transistor amplifier can not provide all the necessary performance required from an amplifier. Hence two or more single stages of amplification are frequently used to achieve the desired amplification.
- The object of coupling is
  (i) To transfer ac output of one stage to the input of the next stage and
  (ii) To isolate the dc conditions from one stage to the next stage.
- Requirements of a multistage amplifier
  (i) Gain should be sufficiently large.
  (ii) Input impedance should match with source impedance.
  (iii) Output impedance should match with load impedance.
  (iv) Bandwidth should be adequately large.

## 6.2 Characteristics of Cascaded Amplifier

The overall performances of the cascaded amplifier are:

### Overall Voltage Gain

Overall voltage gain $A_V$ of a multistage amplifier is equal to the product of the voltage gains of each stage.

$$A_V = \frac{V_o}{V_i}$$

Suppose it is a three stage cascaded amplifier, then

$$A_V = \frac{V_o}{V_{o_2}} \times \frac{V_{o2}}{V_{o_1}} \times \frac{V_{o_1}}{V_i}$$

Where $$\frac{V_o}{V_{o_2}} = A_{V_3}$$

$$\frac{V_{O_2}}{V_{O_1}} = A_{V_2}$$

$$\frac{V_{O_1}}{V_i} = A_{V_1}$$

$$\therefore \quad A_V = A_{V_3} \times A_{V_2} \times A_{V_1}$$

### Overall Current Gain

Overall current gain $A_I$ of a multistage amplifier is equal to the product of the current gains of each stage.

$$A_I = A_{I_3} \times A_{I_2} \times A_{I_1}$$

### Overall Input Resistance

The overall input resistance of a cascaded configuration is equal to the input resistance of the first stage.

### Overall Output Resistance

The overall output resistance of a cascaded amplifier is equal to output resistance of last stage. For analysis, one has to draw the ac equivalent circuit and then start with the last stage and step by step, proceed towards first stage. Also, it is convenient to calculate for each stage the following quantities in the given sequence:

**(a)** Current gain
**(b)** Input resistance
**(c)** Voltage gain
**(d)** Output resistance

After calculating the above quantities for each stage, the overall quantities can be calculated.

## 6.3 Gain in Decibels

The gain expressed as a ratio of output and input voltage is called as linear gain.
But sometimes it is more convenient to express the gain on the logarithmic scale rather than linear scale. Therefore gain is defined in decibels.

(i) Power gain dB = $10\log_{10}{}^{(P_o/P_i)}$

(ii) Overall voltage gain in dB = $20\log_{10}{}^{(V_o/V_i)}$

**Why to express gain in dB?**

**(a)** On the logarithmic scale, we can conveniently express very small as well as very large values of gain.

**Explanation:** A voltage gain of 0.000001 can be expressed as –120 dB and a voltage gain of 1000000 can be expressed as +120 dB.

**(b)** Human ears have a logarithmic response.

## 6.4 Selection of an Amplifier Configuration For Cascade Connection

### Selection of Configuration For Input Stage

| S.N. | Configuration | Input resistance | Select |
|---|---|---|---|
| 1. | CB | Very low | Yes |
| 2. | CE | Low | Yes |
| 3. | CC | High | No |

If the source resistance is low, then CB configuration should be selected for input stage and if source resistance is medium CE is selected.

### Selection of Configuration for Output Stage

| S.N. | Configuration | Input resistance | Select |
|---|---|---|---|
| 1. | CB | Very High | No |
| 2. | CE | High | No |
| 3. | CC | Low | Yes |

Therefore for loads such as load speakers which have a very low resistance, a common collector configuration should be used as output stage.

### Selection of Configuration for Middle Stages

| S.N. | Configuration | Input resistance | Select |
|---|---|---|---|
| 1. | CB | Medium | No |
| 2. | CE | High | Yes |
| 3. | CC | Low | No |

Therefore for a high voltage gain CE configuration is preferred for middle stages of cascaded amplifiers.

## 6.5 Methods of Coupling in Multistage Amplifiers

The coupling techniques are:

(i) RC coupling

(ii) Transformer coupling

(iii) Direct coupling

In RC coupling, a resistor and a capacitor are used as a coupling device. The capacitor connects the output of one stage to the input of next stage to pass ac signal and to block the dc bias voltages. The amplifier using R-C coupling is called the R-C coupled amplifier.

In transformer coupling, transformer is used as the coupling device. The amplifier using transformer coupling is called the transformer coupled amplifier. In this case there is no need of using a coupling capacitor because the secondary of the coupling transformer conveys the ac component directly to the base of the second stage. More over, the secondary winding also provides a base return path and so base resistance is not required.

In direct coupling or dc coupling, the individual amplifier stage bias conditions are so designed that the two stages may be directly connected without the necessity of dc isolation. This coupling is used where it is desirable to connect the load directly in series with the output terminal of the active circuit element such as in case of head phones, loudspeakers etc. The amplifier using direct coupling is called the direct-coupled amplifier.

**RC coupling is the most commonly used coupling between the two stages of a cascaded or multistage amplifier because it is cheaper in cost and is very compact circuit and provides excellent frequency response.**

## 6.6 RC Coupled Amplifiers

- A two-stage RC coupled amplifier using npn transistors in CE configuration is shown in Figure (6.1).
- The two npn transistors are identical and have a common power supply $V_{CC}$.

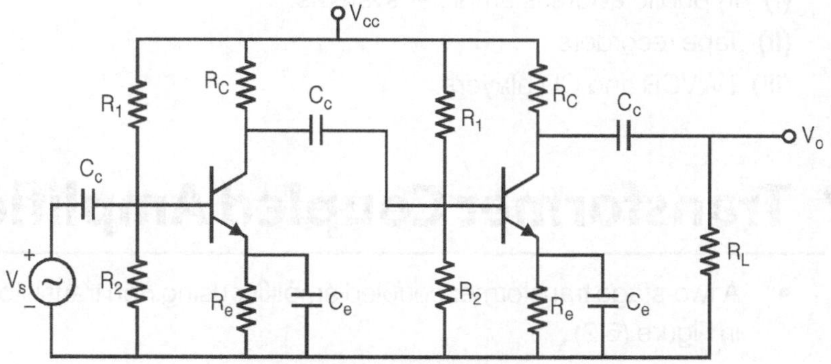


*Figure 6.1 RC coupled Amplifier*

- $R_1$, $R_2$ and $R_e$ are the biasing resistors used separately for the two stages. Voltage divider biasing is being used.
- Due to the use of coupling capacitors the dc voltages will not be coupled from one stage to the other. Therefore the quiescent point of the next stage will not be affected due to coupling. Thus due to RC coupling the dc operating point in any stage remain unaffected.
- R-C network gives a wide band frequency response without introduction of peaks at any frequencies. Therefore R-C coupling can be used for audio frequency amplifiers.
- The frequency response drops off at low frequencies due to the coupling capacitors and at high frequencies due to shunting effects of internal capacitors of the transistor.
- Due to reduction in gain at low frequencies. The R-C coupled amplifiers are not suitable for amplification of low frequency signals.

Typical frequency response is:

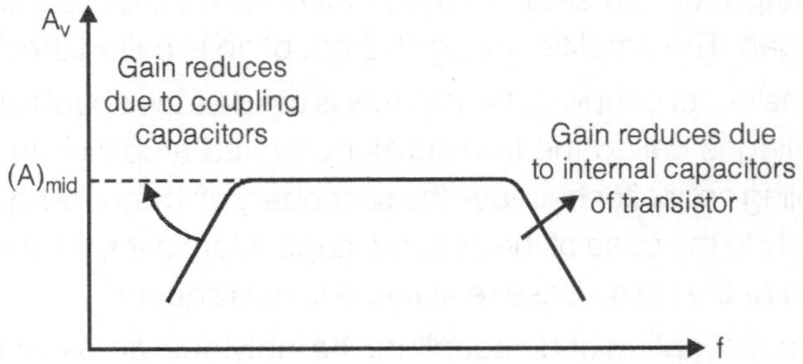


## Advantages

(i) Wide frequency response (large bandwidth).
(ii) It is the most convenient coupling.
(iii) Due to the coupling capacitor $C_c$, Q point is unchanged.
(iv) The distortion in the output is low.
(v) No core distortion.

## Disadvantages

(i) Gain reduces at low frequencies due to coupling capacitors.
(ii) Ageing can make these amplifiers noisy.
(iii) No impedance matching.
(iv) Overall voltage gain is less.

## Applications

(i) In public address amplifier systems.
(ii) Tape recorders.
(iii) TV, VCR and CD players.

# 6.7 Transformer Coupled Amplifier

- A two-stage transformer coupled amplifier using npn transistors in CE configuration is shown in Figure (6.2).

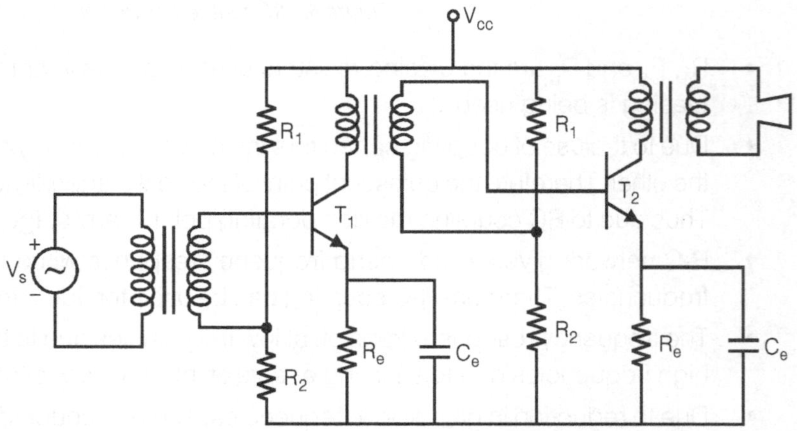


*Figure 6.2 Transformer coupled amplifier*

- The collector resistors are not being used.
- The coupling from one stage to the other takes place through impedance matching transformer.
- The transformer only couples ac signals. Therefore dc voltages are not coupled through the transformers from one stage to the other. Therefore Q point of the next stage is not affected.
- The transformer coupling is specially useful in the audio power amplifiers.
- The frequency response of transformer coupled amplifiers is poor as compared to that of an RC coupled amplifiers.

Different frequency components at the input are not amplified equally due to the presence of the leakage inductance and inter winding capacitance of transformers.

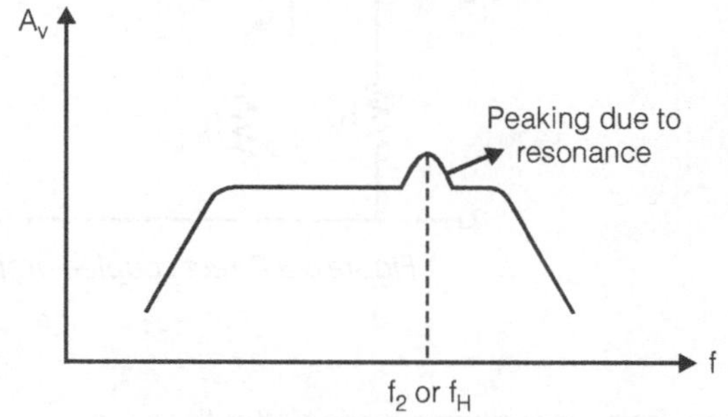


- It is possible to convert the transformer coupled amplifier into a tuned voltage amplifier by connecting capacitors across the transformer windings.
- Due to low winding resistance of the transformer windings, and absence of collector resistors, there is no power loss.

## Advantages

(i) Higher voltage gain than R-C coupled amplifier.
(ii) Impedance matching between the stages is possible.
(iii) DC biasing of individual stages will remain unchanged.

## Disadvantages

(i) Coupling transformer is expensive and bulky.
(ii) Frequency response is poor.
(iii) There is a possibility of core saturation.
(iv) It produces hum in the output.
(v) Low efficiency due to losses in transformer.

## Applications

(i) For amplification of radio frequency signals.
(ii) In power amplifiers as class-C.
(iii) For impedance matching.
(iv) For transferring power to a lower impedance load such as loud speaker.

## 6.8 Direct Coupled Amplifiers

- Due to direct coupling, both dc and ac voltages are coupled from first stage to the second stage. Due to this Q point changes.

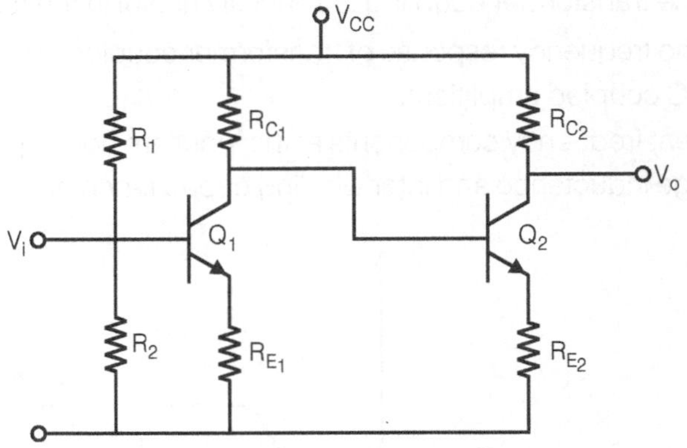


*Figure 6.3 Direct coupled amplifier*

- The low frequency response is better than that of RC coupled amplifier. This is due to the absence of coupling capacitors in the circuit.

- Due to dc amplification which is coupled to the next stage causes "Drift" problem.

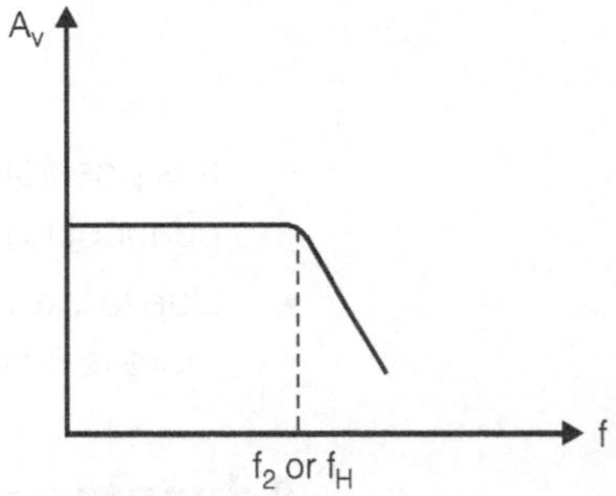


### Advantages

(i) Due to absence of coupling capacitors, the gain does not reduce on low frequency side.
(ii) This amplifier can amplify even dc signals.
(iii) Wide frequency response.
(iv) Reduced cost and complexity due to absence of coupling capacitors.

### Disadvantages

(i) The Q point changes in direct coupled.
(ii) The output waveform has dc shift.
(iii) Poor frequency response at high frequencies.
(iv) Poor temperature stability.

### Applications

(i) In operational amplifiers.
(ii) In analog computation.
(iii) In linear power supplies (voltage regulators).

## 6.9 Comparison of Different Coupling Techniques

| Parameter | R-C Coupling | Transformer Coupling | Direct Coupling |
|---|---|---|---|
| 1. Coupling element | R and C | Transformer | — |
| 2. Type of coupling signal | Only ac | Only ac | dc and ac |
| 3. Ability to block dc | dc is blocked | dc is blocked | dc can not blocked |
| 4. Impedance matching | Not provided | Provided | Not provided |
| 5. Frequency response | Drops off at low frequencies | Not uniform | Improved at low frequency side |
| 6. DC amplification | Not possible | Not possible | Possible |
| 7. Drift | Absent | Absent | Present |
| 8. Application | AF amplifiers | Power amplifiers | Operational amplifiers |

## 6.10 Rise Time-bandwidth Relation

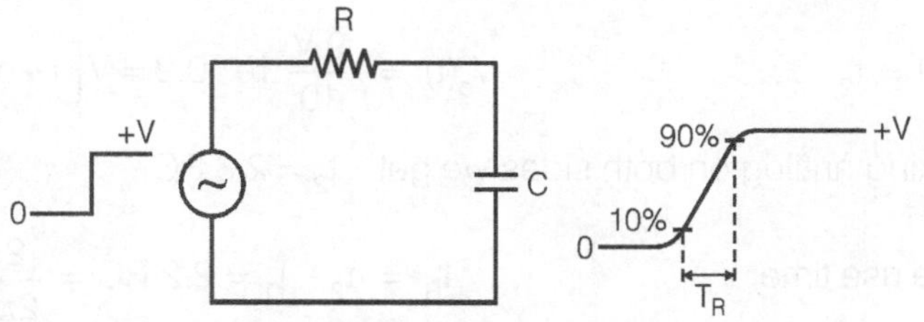


*Figure 6.11 Relation between 3dB frequency and rise time*

$$V_i = Ri(t) + \frac{1}{C}\int i\,dt$$

Taking Laplace transformer on both sides.

$$\frac{V_i}{s} = RI(s) + \frac{1}{C}\left[\frac{I(s) - I(0)}{S}\right] \qquad \text{[Assume I(0) = 0]}$$

$$= \left[R + \frac{1}{CS}\right] I(s)$$

or
$$I(s) = \frac{V_i}{S\left[R+\frac{1}{CS}\right]} = \frac{V_i CS}{S[RCS+1]} = \frac{V_i}{R\left[s+\frac{1}{RC}\right]}$$

Taking inverse Laplace on both sides

$$i(t) = \frac{V_i}{R}e^{-t/RC} = \frac{V_i}{R}e^{-t/\tau}$$

The voltage across the capacitor

$$V_c(t) = \frac{1}{C}\int i\,dt$$

$$\therefore \quad V_c(t) = \frac{1}{C}\int_0^t \frac{V_i}{R}e^{-t/RC} = \frac{V_i}{RC}\left[\frac{e^{-t/RC}}{-1/RC}\right]_0^t$$

$$V_c(t) = V\left(1-e^{-t/RC}\right] \quad \text{...(i)}$$

Where
$V_c$ = Voltage across capacitor
$V$ = Total change in input voltage
$t$ = Time after input transition
$RC$ = Time constant

At the 10% point, $V_{out} = 0.1$ V. At 90% point, $V_{out} = 0.9$ V, substitute these voltage in equation (i).

At $t = t_1$
$$V_o(t) = \frac{V}{10} \text{ i.e. } 0.1V = V\left[1-e^{-t_1/RC}\right]$$

Taking antilog on both sides we get $t_1 = 0.1$ RC

At $t = t_2$,
$$V_o(t) = \frac{9\,V}{10} \text{ i.e. } 0.9 = V\left[1-e^{-t_2/RC}\right]$$

Taking antilog on both sides we get $t_2 = 2.3$ RC

The rise time
$$t_R = t_2 - t_1 = 2.2\text{ RC} = \frac{2.2}{2\pi f_2}$$

Where
$$f_2 = \frac{1}{2\pi RC} = \text{upper 3dB frequency}$$

$$\text{Bandwidth} = f_2 - f_1 \approx f_2 \text{ thus}$$

$$t_R = \frac{2.2}{2\pi BW} = \frac{0.35}{BW}$$

$$\therefore \; t_R = \frac{0.35}{BW}$$

## 6.11 Cascode Amplifier (CE-CB)

- It is one approach to solve the low impedance problem of a common base circuit.
- The cascode amplifier gives the high input impedance of a common emitter amplifier as well as good voltage gain.
- Another important feature of cascode is providing high frequency performance of a common base circuit.
- This circuit is particularly useful in the tuned RF amplifiers in TV circuits.
- It is also used as wide band amplifier.

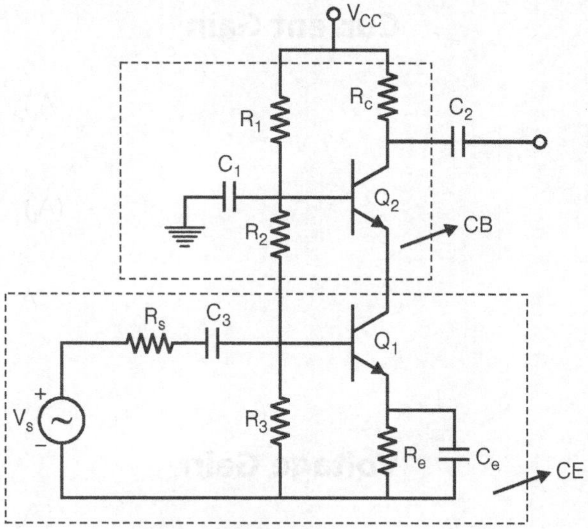


*Figure 6.12 Cascode amplifier*

## 6.12 Darlington Pair (CC-CC)

The Darlington emitter follower is also called as high input resistance circuit. The input resistance is in the order of kΩ in emitter follower circuit. To achieve larger input resistance the circuit known as Darlington connection is used. It consists of two cascaded emitter followers connected in such a way that the input resistance of the second transistor forms the emitter load for the first. The Darlington connection is not useful with three or more transistors because the leakage current of the first transistor is amplified by the second transistor which results in overall all very high leakage current.

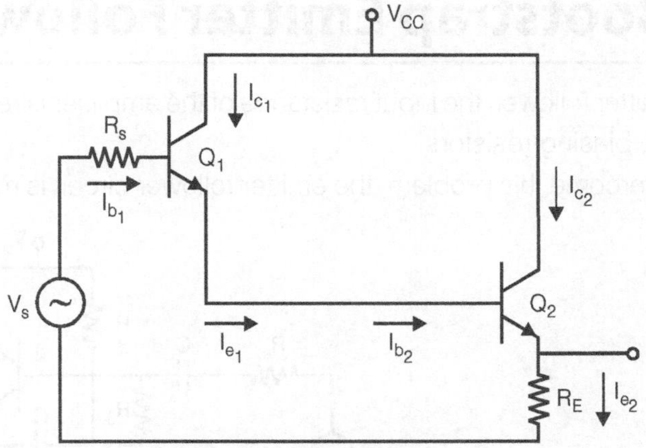


*Fig. 6.13 Darlington Pair*

Darlington pair can also be called as high input impedance circuit. It improves current gain and input impedance of emitter follower.

## ANALYSIS

### Current Gain

$$(A_I)_2 = \frac{I_{e_2}}{I_{b_2}} = \frac{I_{b_2} + I_{c_2}}{I_{b_2}} = \frac{(1+h_{fe})\, I_{b_2}}{I_{b_2}} = (1+h_{fe})$$

$$(A_I)_1 = \frac{I_{e_1}}{I_{b_1}} = \frac{I_{b_1} + I_{c_1}}{I_{b_1}} = \frac{I_{b_1}(1+h_{fe})}{I_{b_1}} = (1+h_{fe})$$

$$A_I = A_{I_1} \cdot A_{I_2} = (1+h_{fe})^2$$

$$\boxed{A_I = (1+h_{fe})^2}$$

### Voltage Gain

$$A_V = (A_V)_1 \times (A_V)_2 \approx 1$$

### Input Resistance

$$A_V = \frac{V_o}{V_s} = \frac{I_{e_2}}{I_{b_1}} \times \frac{R_L}{R_{in}} = A_I \times \frac{R_L}{R_{in}}$$

$$R_{in} = \frac{A_I \times R_L}{A_V}$$

$$\boxed{R_{in} = (1+h_{fe})^2\, R_L}$$

## 6.13 Bootstrap Emitter Follower

In emitter follower, the input resistance of the amplifier is reduced because of the shunting effect of the biasing resistors.

To overcome this problem, the emitter follower circuit is modified as shown in Figure (6.14).

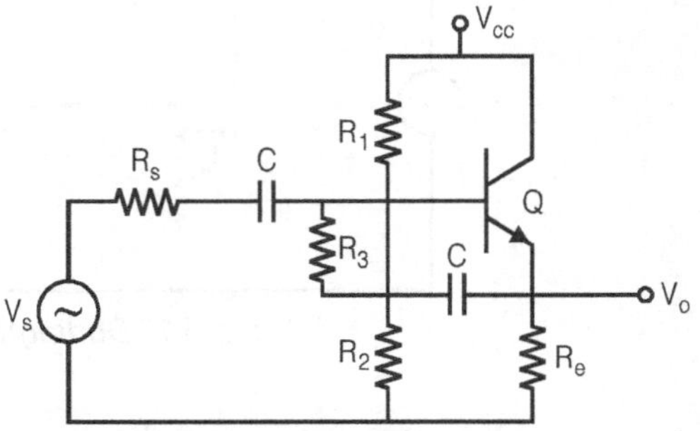


*Figure 6.14: Boot strap emitter follower*

Here, two additional components are used, resistance $R_3$ and capacitor C.

For dc signal, capacitor C acts as a open circuit and therefore resistance $R_1$, $R_2$ and $R_3$ provides necessary biasing to keep transistor in the active region.

For ac signal, the capacitor acts as a short circuit. In other words, $R_3$ is connected between input and output. For such connection effective input resistance is given by Millers theorem.
If millers theorem is applied to the bootstrap circuit

$$R_3' = \frac{R_3}{1-k}$$

Where $k = A_v$

If $R_3 = 100$ k and $A_v = 0.99$

Then $R_3' = \frac{100\text{ k}}{1-0.99} = 10000\text{ k}$

So $R_3'$ is extremely large.

# 6.14 Millers Theorem

## STATEMENT

Consider a network as shown below with impedance Z connected between input and output terminal can be resolved into two components $Z_1$ and $Z_2$.

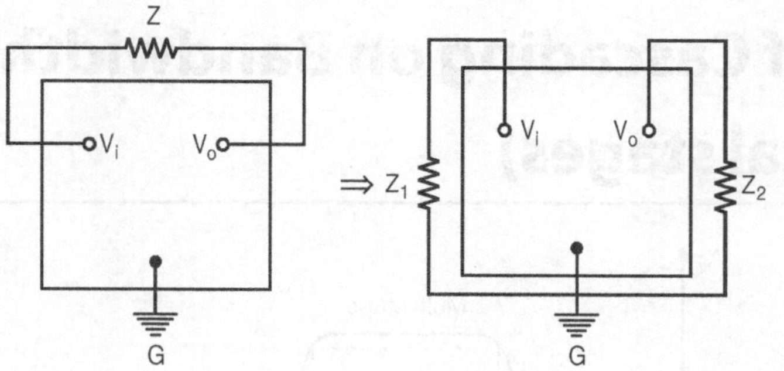


Where $Z_1 = \frac{Z}{1-k}$ ; $Z_2 = \frac{Zk}{k-1}$

## PROOF

### Case-1

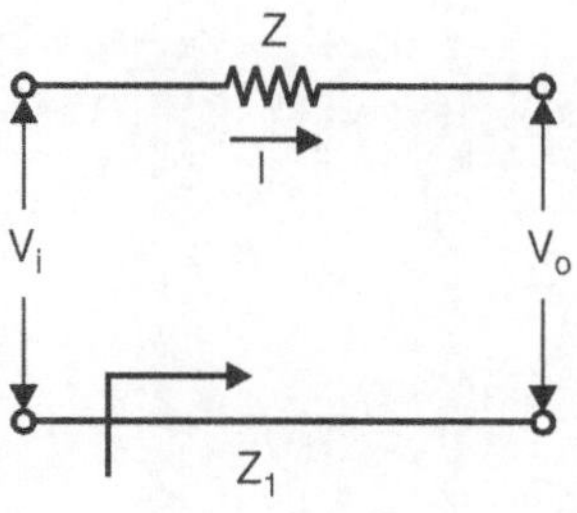


$$Z_1 = \frac{V_i}{I}$$

$$I = \frac{V_i - V_o}{Z} = \frac{V_i\left(1-\frac{V_o}{V_i}\right)}{Z}$$

$$= \frac{V_i[1-A_v]}{Z}$$

$$Z_1 = \frac{V_i}{I} = \frac{Z}{1-A_v} = \frac{Z}{1-k}$$

Where $\frac{V_o}{V_i} = A_v = k$

### Case-2

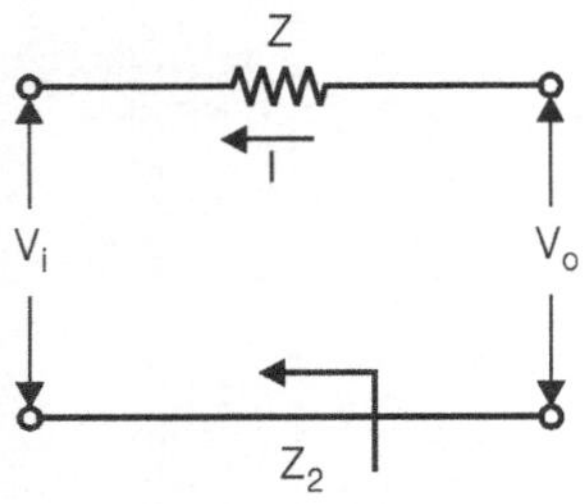


$$Z_2 = \frac{V_o}{I}$$

$$I = \frac{V_o - V_i}{Z} = \frac{V_o\left[1 - \frac{V_i}{V_o}\right]}{Z} = \frac{V_o\left[\frac{A_v - 1}{A_v}\right]}{Z}$$

$$Z_2 = \frac{V_o}{I} = \frac{Z}{\frac{A_v - 1}{A_v}} = \frac{Z A_v}{A_v - 1}$$

$$\boxed{Z_2 = \frac{Zk}{k-1}}$$

## 6.15 Effect of Cascading on Bandwidth (identical stages)

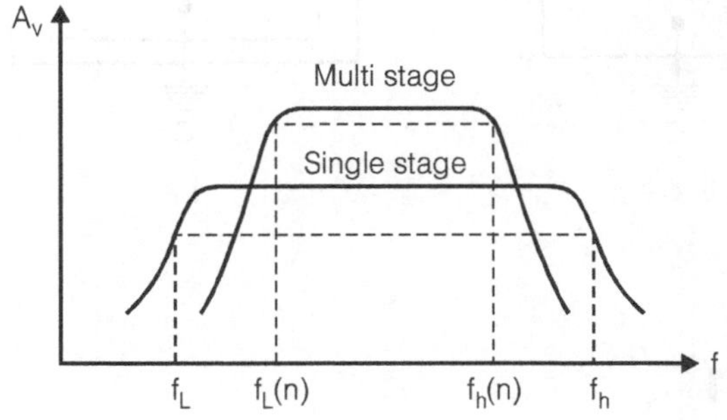


## CASE-1

### Higher Cut-off Frequency $f_H(n)$

For single stage, $$(A_v)_H = \frac{(A_v)_m}{1 + j\left(\frac{f}{f_H}\right)}$$

$$\left|\frac{(A_v)_H}{(A_v)_m}\right| = \frac{1}{\sqrt{1 + (f/f_H)^2}}$$

For n stages, $$\left|\frac{(A_v)_H}{(A_v)_m}\right|_{overall} = \left[\frac{1}{\sqrt{1+(f/f_H)^2}}\right]^n$$

When $f = f_H(n)$

The upper cut-off frequency of overall amplifier

$$\left|\frac{(A_v)_H}{(A_v)_m}\right|_{overall} = \frac{1}{\sqrt{2}}$$

$$\left[\frac{1}{\sqrt{1+(f_H(n)/f_H)^2}}\right]^n = \left(\frac{1}{2}\right)^{1/2}$$

$$1+\left(\frac{f_H(n)}{f_H}\right)^2 = 2^{1/n}$$

$$\boxed{f_H(n) = f_H\sqrt{2^{1/n}-1}}$$

Where n is number of stages.

## CASE-2

### Lower Cut-off Frequency $f_L(n)$

For single stage, $$(A_v)_L = \frac{(A_v)_m}{1-j(f_L/f)}$$

$$\left|\frac{(A_v)_L}{(A_v)_m}\right| = \frac{1}{\sqrt{1+(f_L/f)^2}}$$

For n stages,

$$\left|\frac{(A_v)_L}{(A_v)_m}\right|_{overall} = \left(\frac{1}{1+(f_L/f)^2}\right)^n$$

When $f = f_L(n)$, the lower cut-off frequency of overall amplifier.

$$\frac{1}{\sqrt{1+(f_L/f_L(n))^2}} = \frac{1}{\sqrt{2}}$$

$$1+(f_L/f_L(n))^2 = 2^{1/n}$$

$$\frac{f_L}{f_L(n)} = \sqrt{2^{1/n}-1}$$

$$\boxed{f_L(n) = \frac{f_L}{\sqrt{2^{1/n}-1}}}$$

# 6.16 Effect of Cascading on Bandwidth (non identical stages)

Let $A_1, A_2, \cdots, A_n$ are gains of 1st, 2nd ⋯ nth stages.

Let $f_{L_1}, f_{L_2}, \cdots, f_{L_n}$ are lower cut-off frequencies of 1st, 2nd ⋯ nth stages.

Let $f_{h_1}, f_{h_2}, \cdots, f_{h_n}$ are higher cut-off frequencies of 1st, 2nd ⋯ nth stages.

## Overall Lower Cut-off Frequency

Consider

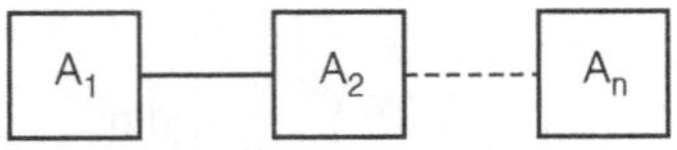


$$A^n = A_1 \cdot A_2 \cdots A_n$$

$$A^n_{Lf} = A_{1Lf} \cdot A_{2Lf} \cdots A_{nLf}$$

$$\frac{A^n}{\sqrt{1+\left(\frac{f^n_L}{f}\right)^2}} = \frac{A_1}{\sqrt{1+\left(\frac{f_{L1}}{f}\right)^2}} \cdot \frac{A_2}{\sqrt{1+\left(\frac{f_{L2}}{f}\right)^2}} \cdots \frac{A_n}{\sqrt{1+\left(\frac{f_{Ln}}{f}\right)^2}}$$

$$\sqrt{1+\left(\frac{f^n_L}{f}\right)^2} = \sqrt{1+\left(\frac{f_{L1}}{f}\right)^2+\left(\frac{f_{L2}}{f}\right)^2+\cdots+\left(\frac{f_{Ln}}{f}\right)^2+\left(\frac{f_{L1}\,f_{L2}}{f^2}\right)^2+\left(\frac{f_{L1}\cdot f_{L3}}{f^2}\right)^2+\left(\frac{f_{L1}\,f_{L2}\cdots f_{Ln}}{f^2}\right)^2}$$

$$1+\left(\frac{f^n_L}{f}\right)^2 = 1+\left(\frac{f_{L1}}{f}\right)^2+\left(\frac{f_{L2}}{f}\right)^2+\cdots \quad \text{(Neglecting higher order terms)}$$

$$\boxed{f^n_L = \sqrt{f^2_{L1}+f^2_{L2}+\cdots}}$$

Similarly

$$f^n_h = \frac{1}{\sqrt{\frac{1}{f^2_{h1}}+\frac{1}{f^2_{h2}}+\cdots+\frac{1}{f^2_{hn}}}}$$

$$\boxed{\frac{1}{(f^n_h)^2} = \frac{1}{f^2_{h1}}+\frac{1}{f^2_{h2}}+\cdots+\frac{1}{f^2_{hn}}}$$

## SUMMARY

- To achieve the desired amplification, we use multi stage.
- In a cascade connection at the input stage CB is preferred at the output stage CC is preferred at the middle stage CE is preferred.
- There are three coupling techniques:
  - (i) RC coupling
  - (ii) Transformer coupling
  - (iii) Direct coupling.
- RC coupling is the most commonly used coupling between two stages of a cascaded amplifier.
- The relation between rise time and bandwidth is $t_r = 0.35/BW$.
- Cascode (CE-CB) is a special type which is used to imrpove $Z_i$ and $A_v$ of CB. To improve bandwidth of CE.
- Darlington pair (CC-CC) is called as high input impedance circuit.
- Effect of cascading on bandwidth:

| Identical Stages | Non identical Stages: |
|---|---|
| (i) $f_H(n) = f_H\sqrt{2^{1/n}-1}$ | (i) $f_L^n = \sqrt{f_{L1}^2 + f_{L2}^2 + f_{L3}^2 + \cdots}$ |
| (ii) $f_L(n) = f_L\big/\sqrt{2^{1/n}-1}$ | (ii) $\frac{1}{(f_h^n)^2} = \frac{1}{f_{h1}^2} + \frac{1}{f_{h2}^2} + \frac{1}{f_{h3}^2} + \cdots$ |

## GATE AND IES QUESTIONS

**Q.1** An amplifier using BJT has two identical stages each having a lower cut-off (3 dB) frequency of 64Hz due to coupling capacitor. The emitter bypass capacitor also provides a lower cut-off (3 dB) frequency due to emitter degeneration alone of 64 Hz. The lower (3 dB) frequency of the overall amplifier is nearly

(a) 100 Hz (b) 128 Hz
(c) 156 Hz (d) 244 Hz

**Exp.** (a)

$$f_L' = \frac{f_L}{\sqrt{2^{1/n}-1}} = \frac{64}{\sqrt{2^{1/2}-1}} = 100\text{ Hz.}$$

**Q.2** An RC amplifier stage has a bandwidth of 500 kHz. What will be the rise time of this amplifier stage?

(a) 0.35 μs (b) 0.7 μs
(c) 1.0 μs (d) 2.0 μs

**Exp.** (b)

$$t_r = \frac{0.35}{BW} = 0.7\ \mu s$$

**Q.3** An amplifier has two identical cascaded stages. Each stage has a bandwidth of 20 kHz. The overall bandwidth shall approximately be equal to
(a) 10 kHz (b) 12.9 kHz
(c) 20 kHz (d) 28.3 kHz

**Exp. (b)**

$$BW' = BW\sqrt{2^{1/n}-1} = 12.9\,\text{kHz}$$

**Q.4** A transistor RC coupled amplifier is designed for a voltage and band gain of 20. But a measurement at a particular frequency shows the gain to be only 14. What is the likely phase shift at this frequency?
(a) 180° (b) 135°
(c) 90° (d) 45°

**Exp. (b)**

$$\frac{(A_v)_h}{(A_v)_m} = \frac{1}{1+j\frac{f}{f_h}}$$

$$A_v = 14 = \frac{20}{\sqrt{2}} \quad (-3\text{dB})$$

$$\theta_h = 180° - \tan^{-1}\frac{f}{f_h} = 180° - \tan^{-1}1$$

$$\theta_h = 135°$$

**Q.5** The two stages of a cascade amplifier have individual upper cut-off frequencies $f_1$=5MHz and $f_2$=3.33 MHz. What is the best approximation for the upper cut-off frequency of the cascade combination?
(a) 4.16 MHz (b) 3.33 MHz
(c) 2 MHz (d) 5.00 MHz

**Exp. (c)**

$$f_h^n = \frac{1}{\sqrt{\frac{1}{f_{h1}^2}+\frac{1}{f_{h2}^2}+\cdots}} \quad \text{(non identical stages)} = 2\text{ MHz}$$

**Q.6** Consider the following statements:
In frequency response of n-stage amplifiers,

1. upper cut-off frequency is $\left(2^{1/n}-1\right)^{1/2}$ times that of a single stage
2. lower cut-off frequency is $\left(2^{1/n}-1\right)^{1/2}$ times that of a single stage
3. lower cut-off frequency is $\left(2^{1/n}-1\right)^{-1/2}$ times that of a single stage
4. upper cut-off frequency is $\left(2^{1/n}-1\right)^{-1/2}$ times that of a single stage

Which of the statements given above are correct?
(a) 1 and 2 (b) 1 and 3
(c) 2 and 4 (d) 3 and 4

**Exp. (b)**

$$f'_h = f_h\sqrt{2^{1/n}-1} \quad ; \quad f'_l = \frac{f_l}{\sqrt{2^{1/n}-1}}$$

**Q.7** What is the purpose of impedance matching between the output of previous stage and input of next stage in a cascaded amplifier?
(a) To achieve high efficiency (b) To achieve maximum power transfer
(c) To achieve reduced distortion (d) To achieve reduced noise

**Exp. (b)**
To achieve maximum power transfer.

**Q.8** If a common emitter amplifier with an emitter resistance $R_e$ has an overall transconductance gain of –1 mA/V, a voltage gain of –4 and desensitivity of 50, then the value of the emitter resistance $R_e$ would be
(a) 50 Ω (b) 0.98 Ω
(c) 50 kΩ (d) 0.98 kΩ

**Exp. (c)**

$$\frac{I_c}{V_i} = g_m = -1 \text{ mA/V}$$

$$\frac{V_o}{V_i} = A_v = -4$$

$$1 + A\beta = 50$$

$$1 - 4\beta = 50$$

$$\frac{V_f}{V_o} = \beta = \frac{49}{4} = 12.25$$

$$\frac{V_o}{I_c} = \frac{4}{1\text{mA}}$$

$$\frac{I_c R_e}{I_c \times \frac{4}{1\text{mA}}} = 12.25 = \beta$$

$$R_e = 12.25 \times \frac{4}{1\text{mA}} = 50 \text{ k}\Omega$$

**Q.9** As compared to the impedance of the signal source, the input impedance of a voltage amplifier should be
(a) small (b) large
(c) equal (d) none

**Exp. (b)**

$Z_i = \infty$ → Voltage amplifier → $Z_o = 0$

**Q.10** A radio frequency signal contains following three frequencies:
800 kHz, 875 kHz and 880 kHz. This signals need to be amplified. The amplifier used should be
(a) audio frequency amplifier (b) wide band amplifier
(c) tuned voltage amplifier (d) video amplifier

**Exp. (c)**
As the frequencies are close to each other, the amplifier used should be tuned voltage amplifier.

**Q.11** A narrow band amplifier is one that has a passband
(a) limited to 2000 Hz only
(b) limited to audio frequency range only
(c) approx 10% of its centre frequency
(d) none

**Exp. (c)**
A narrow band amplifier will have very small pass band approximate 10% of centre frequency.

**Q.12** Tuned voltage amplifiers are not used
(a) public address systems
(b) in radio receivers
(c) where a band output frequencies is to be selected and amplified
(d) television receivers

**Exp. (a)**
Tuned voltage amplifiers are not used in public address system.

**Q.13** In a series resonant circuit, the impedance at resonance is
(a) minimum
(b) maximum
(c) equal
(d) none

**Exp. (a)**
Always for series resonant circuit impedance will be less.

**Q.14** An ac circuit resonates at 1000 kHz. If has quality factor 50. The bandwidth (BW) and half power points are
(a) BW = 10 kHz; $f_1$ = 1000 kHz; $f_2$ = 1010 kHz
(b) BW = 20 kHz; $f_1$ = 990 kHz; $f_2$ = 1010 kHz
(c) BW = 20 kHz; $f_1$ = 1000 kHz; $f_2$ = 1020 kHz
(d) BW = 20 kHz; $f_1$ = 995 kHz; $f_2$ = 1005 kHz

**Exp. (c)**

$$Q = \frac{f_c}{BW}$$

**Q.15** In a multistage RC coupled amplifier the coupling capacitor
(a) limits the low frequency response
(b) limits the high frequency response
(c) does not affect the frequency response
(d) none

**Exp. (a)**
The coupling capacitor limits the low frequency response.

**Q.16** A cascode amplifier stage is equivalent is
(a) CE – CB
(b) CE – CE
(c) CE – CC
(d) CC – CC

**Exp. (a)**
Cascode amplifier (CE-CB) is a special type which improves $Z_i$ and $A_v$.

**Q.17** Which of the following amplifier has higher voltage gain and impedance matching?
(a) RC coupling
(b) Inductive coupling
(c) Transformer coupling
(d) Direct coupling

**Exp. (c)**
Transformer coupling will have higher voltage gain and impedance matching.

**Q.18** Which of the following amplifier is having minimum possible non linear distortion
(a) RC coupling (b) inductive coupling
(c) transformer coupling (d) direct coupling

**Exp. (a)**
Non linear distortion is less in RC coupling.

**Q.19** Effect of cascading several amplifier stages is to
(a) reduce the overall gain
(b) reduce the overall frequency response
(c) increases overall gain and reduce the frequency response
(d) double tuned circuits

**Exp. (c)**
Cascading of several amplifiers result in increase in voltage gain and decrease in bandwidth.

**Q.20** When a multistage transistor amplifier is required to amplifier dc signals also, we must use
(a) RC coupling (b) direct coupling
(c) transformer coupling (d) double tuned circuits

**Exp. (b)**
To amplify dc signals also along with ac signals direct coupling is best.

**Q.21** Transformer coupling provides high efficiency because the
(a) collector voltage is stepped up (b) dc resistance in the collected circuit is low
(c) collector voltage is stepped down (d) flux linkages are in complete

**Exp. (b)**
The collector resistance $R_c$ is replaced by inductor where the dc resistance is low.

**Q.22** The lower 3 dB frequency of n stage amplifier equals to

(a) $\frac{f_L}{\sqrt{2^{1/n}-1}}$ (b) $f_L\sqrt{2^{1/n}-1}$

(c) $\frac{f_L}{n}$ (d) $nF_L$

**Exp. (a)**

$$f_L^n = \frac{f_L}{\sqrt{2^{1/n}-1}}$$

**Q.23** The higher 3 db frequency of n stage amplifier equals

(a) $f_h\sqrt{2^{1/n}-1}$ (b) $\frac{f_h}{\sqrt{2^{1/n}-1}}$

(c) $nf_H$ (d) $\frac{f_H}{n}$

**Exp. (a)**

$$f_h^n = f_h\sqrt{2^{1/n}-1}$$

**Q.24** The rise time $t_r$ of an amplifier in reproducing a square wave is related to the higher 3 db frequency $f_H$ by

(a) $t_r = \dfrac{0.35}{\sqrt{f_h}}$ (b) $t_r = \dfrac{0.35}{f_h}$

(c) $t_r = 0.35\, f_H$ (d) $t_r = 0.35\sqrt{f_h}$

**Exp. (b)**

$$t_r = \frac{0.35}{f_h}$$

**Q.25** In transformer coupled amplifiers
(a) dc load line ius nearly vertical
(b) ac load line is nearly vertical
(c) both dc as well as ac load line is nearly vertical
(d) both load linear are nearly horizontal

**Exp. (a)**
In transformer coupling, dc load line is vertical because $(I_C)_Q = 0$.

**Q.26** Non linear distortion in an amplifier is caused by
(a) non linear dynamic characteristic of active device
(b) reactive elements in the circuit
(c) ripple components in circuit
(d) high temperature of operation

**Exp. (a)**
Non linear distortion in an amplifier is caused by non linear characteristics of active device.

## REVIEW QUESTIONS

1. Obtain the expression for voltage gain $A_v$ in LF, MF and HF ranges in case of single stage BJT amplifier.
2. Explain above the classification of amplifiers based on type of coupling and bandwidth.
3. Draw the circuit for cascode amplifier and give the advantages of it.
4. Draw the circuit of Darlington pair and calculate $A_I$, $A_v$ and $Z_i$.
5. Explain the drawback of Darlington pair and give a modified design to over come the drawback.
6. Explain the effect of cascading on bandwidth.

OOOO

# CHAPTER 7

# LINEAR AND NON LINEAR WAVE SHAPING

## About This Chapter

- This process by which the waveform of a non sinusoidal signal is altered by transmitting the signal through a linear network is termed as linear wave shaping.
- Study of step, pulse square, ramp and exponential input when transmitted through a high pass circuit.
- Study of step, pulse square, ramp and exponential input when transmitted through a low pass circuit.

A wave shaping circuit which controls the shape of waveform by removing or clipping a portion of the applied wave is known as clipping circuit. For a clipping circuit atleast two components a diode and a resistor are required. Sometimes a dc battery is also used to fix the clipping level. Clipper are used in radar, digital and other electronic devices.

The important diode clippers are:

- Positive clipper
- Negative clipper
- Combination clipper

Clamping is a process of introducing a dc level in to an ac signal. A clamping circuit is a device that places either the positive or negative peak of a signal at a desired level.

# 7 NON LINEAR WAVE SHAPING

## LINEAR WAVE SHAPING

## 7.1 Introduction

- A linear network is a network made up of linear elements only. A linear network can be described by linear differential equations.
- Resistor, capacitor and inductor are linear circuit elements.
- The principle of superposition and the principle of homogeneity hold good for linear networks.
- In pulse circuitry, there are a number of waveforms, which appear very frequently. The most important of these are sinusoidal, step, pulse, square wave, ramp and exponential waveforms.
- When a sinusoidal voltage is applied to a linear circuit, the output is also sinusoidal in nature.
- There will only be a change in the amplitude of the signal and there may be a phase shift between the input and output waveforms.
- No other periodic waveform preserves its shape precisely when transmitted through a linear network.

**"The process by which the waveform of a non sinusoidal signal is altered by transmitting the signal through a linear network is termed as linear wave shaping".**

## 7.2 High Pass RC Circuit

Let $v_i$ = Input voltage

$v_o$ = Output voltage

Since the reactance of the capacitor decreases as the frequency increases $\left(X_C = \frac{1}{2\pi fc}\right)$,

The capacitor offers very small impedance to harmonics of higher orders.

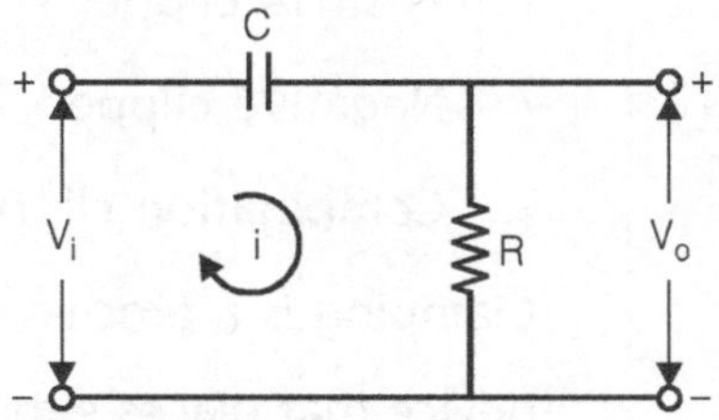


*Figure. 7.1 High pass RC circuit*

At the high frequency components of the input voltage, it pass through the network with very little attenuation. For this reason, the circuit is termed as "high pass RC circuit".

Applying KVL to the circuit

$$V_C + V_R = V_i$$

Where

$$V_C = \text{Voltage across capacitor}$$
$$V_R = \text{Voltage across resistor}$$

but,

$$V_C = \frac{1}{C}\int i\,dt, \quad V_R = V_o$$

$$\frac{1}{C}\int i\,dt + V_o = V_i$$

$$\text{Current } i = \frac{V_R}{R} = \frac{V_o}{R}$$

$$\therefore \quad \frac{1}{C}\int \frac{V_o}{R}dt + V_o = V_i$$

Differentiating w.r.t. t the above equation transforms into

$$\frac{1}{C}\left[\frac{V_o}{R}\right] + \frac{dV_o}{dt} = \frac{dV_i}{dt}$$

$$\text{(or)} \quad \boxed{\left(\frac{1}{CR}\right)V_o + \frac{dV_o}{dt} = \frac{dV_i}{dt}}$$

The total response of the circuit to the applied forcing function consists of transient response and steady state response.

We have transient response $= \frac{dV_o}{dt}$

Steady state response $= \left(\frac{1}{RC}\right)V_o$

## 7.3 RC Differentiator

For a high pass RC circuit

$$\frac{dV_o}{dt} = \frac{dV_i}{dt} - \left(\frac{1}{RC}\right)V_o$$

If the time constant $\tau = RC$ is for less than T.

The transient response becomes insignificant, and the steady state response becomes predominant.

Ignoring the transient response

$$\left(\frac{1}{RC}\right)V_o = \frac{dV_i}{dt} \quad \text{(or)} \quad V_o = RC\frac{dV_i}{dt}$$

Since R and C are of fixed magnitude

$$\boxed{V_o \propto \frac{d}{dt}V_i}$$

That is, the output is proportional to the time derivative of the input. Hence the circuit is called RC differentiator.

# 7.4 Sinusoidal Input

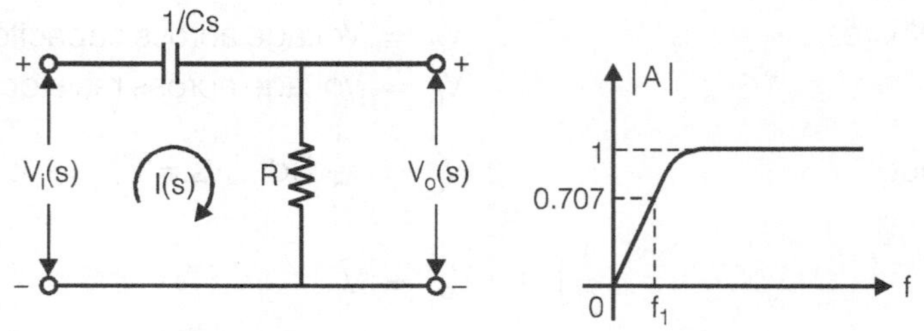


*Figure 7.2 High pass RC circuit and frequency response*

The figure shows the Laplace transformed high pass RC circuit. The gain versus frequency curve of a high pass circuit excited by a sinusoidal input is shown in Figure (7.2).

## Expression For the Lower Cut-off Frequency

The magnitude of the steady-state gain A, and the angle θ by which the output leads the input are given by

$$A = \frac{V_o(s)}{V_i(s)} = \frac{R}{R+\frac{1}{Cs}} = \frac{1}{1+\frac{1}{RCs}}$$

$$\text{Putting} \quad s = j\omega, \quad A = \frac{1}{1-j\frac{1}{\omega RC}} = \frac{1}{1-j\frac{1}{2\pi fRC}}$$

$$|A| = \frac{1}{\sqrt{1+\left(\frac{1}{2\pi fRC}\right)^2}} \quad \text{and} \quad \theta = -\tan^{-1}\frac{1}{2\pi fRC}$$

At the lower cut-off frequency $f_1$, $|A| = \frac{1}{\sqrt{2}}$

$$\frac{1}{\sqrt{1+\left(\frac{1}{2\pi f_1 RC}\right)^2}} = \frac{1}{\sqrt{2}}$$

$$\frac{1}{2\pi f_1 RC} = 1 \text{ i.e. } f_1 = \frac{1}{2\pi RC}$$

This is the expression for the lower cut-off frequency of a high pass circuit.

For a sinusoidal input $V_i$, the output signal $V_o$ increases in amplitude with increasing frequency. The frequency at which the gain is $1/\sqrt{2}$ of its maximum value is called the lower cut-off or lower 3-dB frequency. For a high pass circuit, there is no upper cut-off frequency because all high frequency signals are transmitted with zero attenuation. Therefore $f_2 = \infty$. Hence bandwidth $= f_2 - f_1 = \infty$.

## 7.5 Response of Step, Pulse, Square, Ramp, Exponential for High Pass Circuit

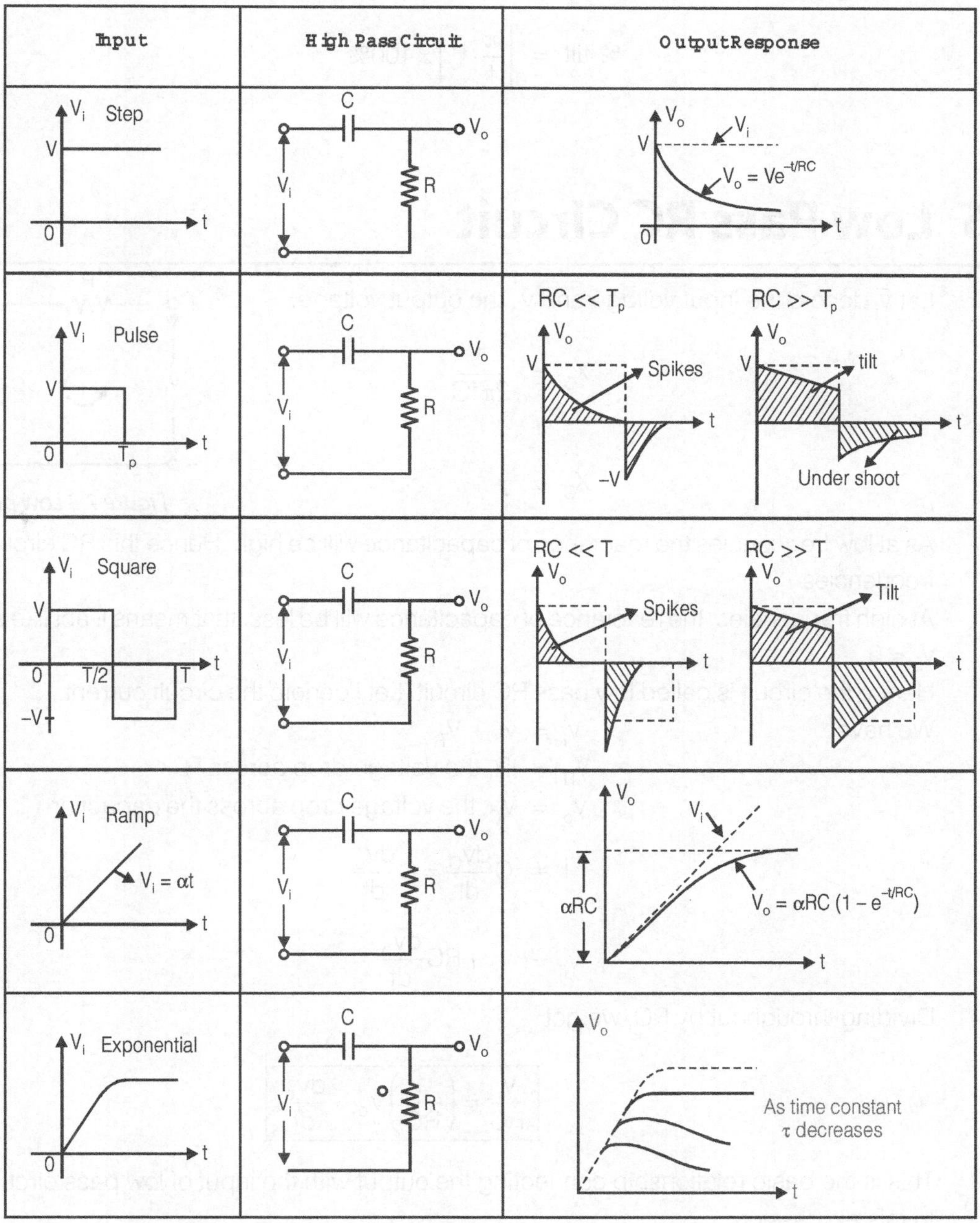


### Important Relations

(i) Expression for percentage tilt

$$\% \text{ tilt} = \frac{1-e^{-T/2RC}}{1+e^{-T/2RC}} \times 200 \quad \text{(or)} \quad \% \text{ tilt} = \frac{T}{2RC} \times 100\%$$

**(ii)** Relation between lower cut-off frequency ($f_L$) and time constant ($\tau$)

$$f_L = \frac{1}{2\pi RC} = \frac{1}{2\pi\tau}$$

**(iii)** Relation between $f_L$ and percentage tilt

$$\%\ \text{tilt} = \left[\frac{\pi}{f} \cdot f_L\right] \times 100\%$$

## 7.6 Low Pass RC Circuit

Let $v_i$ denote the input voltage and $v_o$ the output voltage.

$$X_C = \frac{1}{2\pi fC}$$

$$X_C \propto \frac{1}{f}$$

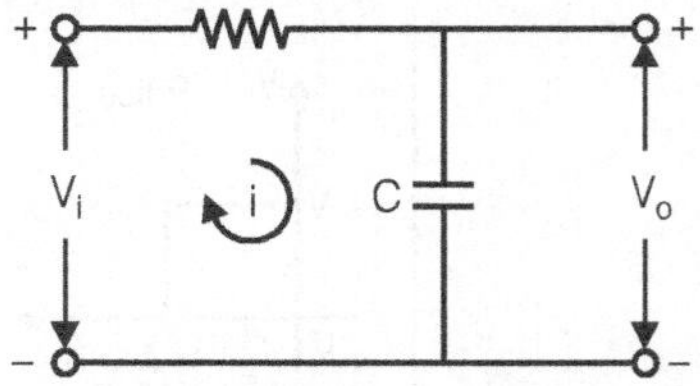


*Figure 7.3 Low pass circuit*

As at low frequencies the reactance of capacitance will be high. Hence this RC circuit allows low frequencies.

At high frequencies, the reactance of capacitance will be less, that means it act like short circuit; $v_o = 0$.

Hence the circuit is called low pass RC circuit. Let i denote the circuit current.

We have

$$v_i = v_o + V_R$$

$V_R$ = iR, the voltage drop across R

$v_o$ = $V_C$, the voltage drop across the capacitor

$$i = C\frac{dv_C}{dt} = C\frac{dv_o}{dt}$$

$$v_i = v_o + RC\frac{dv_o}{dt}$$

Dividing throughout by RC, we get

$$\boxed{\frac{v_i}{RC} = \left(\frac{1}{RC}\right)v_o + \frac{dv_o}{dt}}$$

This is the basic relationship connecting the output with the input of low pass circuit.

In this expression,

$\left(\frac{1}{RC}\right)v_o$ represents the forced or steady state response.

$\frac{dv_o}{dt}$ represents transient or natural response.

# 7.7 Low Pass RC Circuit as Integrator

For a low pass RC circuit

$$\left(\frac{1}{RC}\right)v_o + \frac{dv_o}{dt} = \frac{v_i}{RC}$$

RC is the time constant of the circuit.

Let it be assumed that RC >> T, for a large time constant, the output takes a fairly long time to come to steady state and as such the transient response is very predominant and the steady state response becomes insignificant and hence can be ignored.

Putting $$\left(\frac{1}{RC}\right)v_o = 0$$

$$\frac{dv_o}{dt} = \left(\frac{1}{RC}\right)v_i$$

Integrating both sides with respect to t, we get

$$v_o = \int \frac{1}{RC} \cdot v_i \, dt$$

or $$\boxed{v_o = \frac{1}{RC}\int v_i \, dt}$$

Since R and C are fixed in magnitude, $v_o \propto \int v_i \, dt$.

That is, the output is proportional to the time integral of the input. For this reason, the low pass RC circuit with RC >> T is called RC integrator.

# 7.8 Sinusoidal Input

The Laplace transformed low pass RC circuit is shown in Figure (7.4). The gain versus frequency curve of a low pass circuit excited by a sinusoidal input is shown in Figure (7.4).

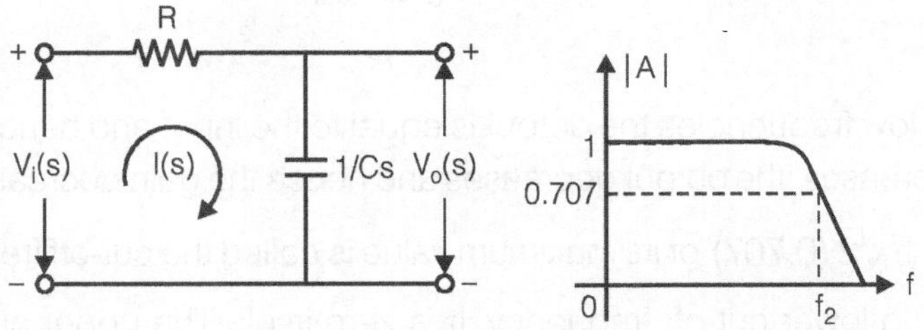


*Figure 7.4 Low pass RC circuit and frequency response*

The magnitude of the steady state gain A is given by

$$A = \frac{V_o(s)}{V_i(s)} = \frac{\frac{1}{Cs}}{R+\frac{1}{Cs}}$$

$$= \frac{1}{1+RCs} = \frac{1}{1+j\omega RC}$$

$$= \frac{1}{1+j2\pi fRC}$$

$$|A| = \frac{1}{\sqrt{1+(2\pi fRC)^2}}$$

At the upper cut-off frequency $f_2$, $|A| = \frac{1}{\sqrt{2}}$

$$\frac{1}{\sqrt{2}} = \frac{1}{\sqrt{1+(2\pi f_2 RC)^2}}$$

$$2 = 1+(2\pi f_2 RC)^2$$

The upper cut-off frequency, $f_2 = \frac{1}{2\pi RC}$

So $$A = \frac{1}{1+j\frac{f}{f_2}}$$

and $$|A| = \frac{1}{\sqrt{1+\left(\frac{f}{f_2}\right)^2}}$$

The angle θ by which the output leads the input is given by

$$\theta = \tan^{-1}\frac{f}{f_2}$$

At low frequencies the output is equal to the input and hence the gain is unity. As the frequency increases, the output decreases and hence the gain decreases. The frequency at which the gain is $1/\sqrt{2}$ (0.707) of its maximum value is called the cut-off frequency. For a low pass circuit, there is no lower cut-off frequency. It is zero itself. The upper cut-off frequency is the frequency at which the gain is $1/\sqrt{2}$ i.e. 70.7% of its maximum value. The bandwidth of the low pass circuit is equal to the upper cut-off frequency $f_2$ itself.

# 7.9 Response of Step, Pulse, Square, Ramp and Exponential for Low Pass Circuit

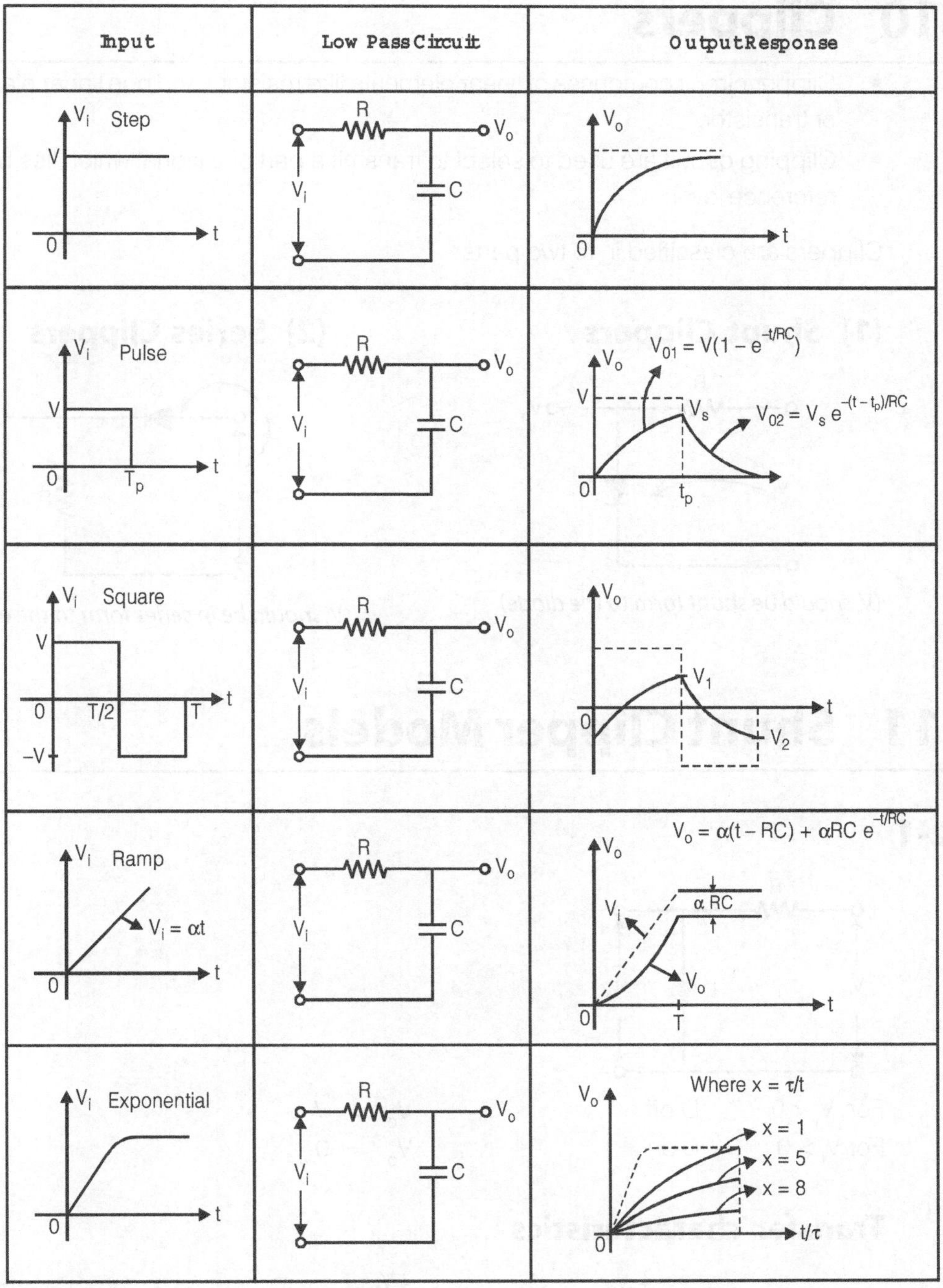


## Important Relations

1. Relation between rise time and time constant $t_r = 2.2\,RC = 2.2\tau$

2. Relation between rise time and bandwidth $t_r = \dfrac{0.35}{BW}$

# NON LINEAR WAVE SHAPING

## 7.10 Clippers

- Clipper circuit comprises of linear elements like resistors and non linear elements like diode or transistor.
- Clipping circuit are used to select to transmit a part of a signal which lies below or above a reference level.

Clippers are classified in to two parts:

### (1) Shunt Clippers

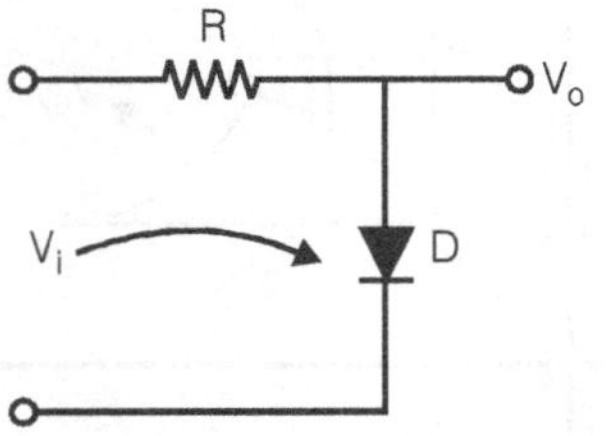


*($V_i$ should be shunt form to the diode)*

### (2) Series Clippers

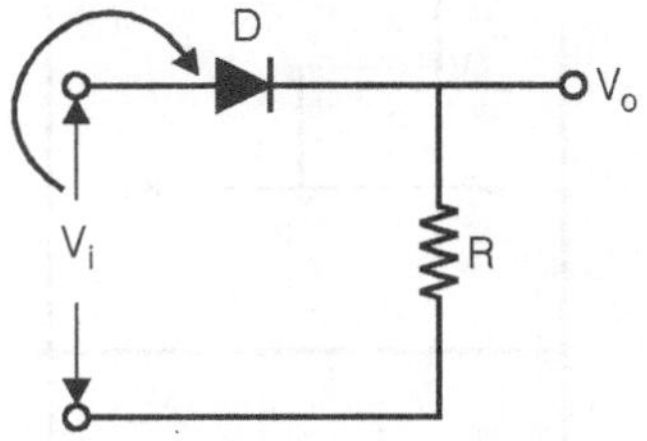


*($V_i$ should be in series form to the diode)*

## 7.11 Shunt Clipper Models

### MODEL-1

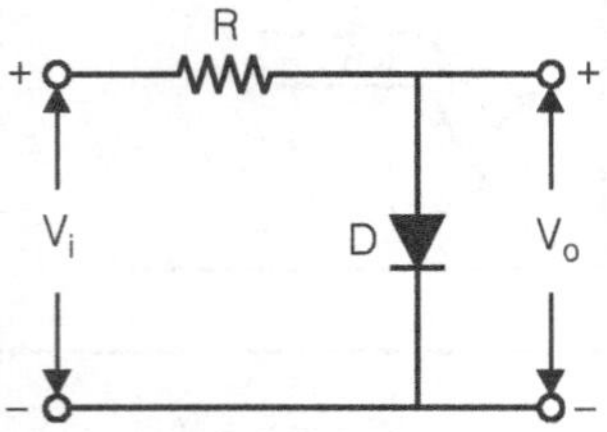


For $V_i < 0$ D off $\therefore$ $V_o = V_i$

For $V_i > 0$ D on $\therefore$ $V_o = 0$

### Transfer characteristics

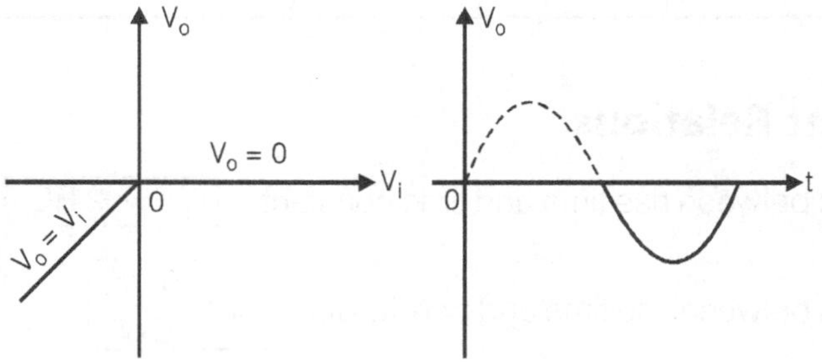

## MODEL-2

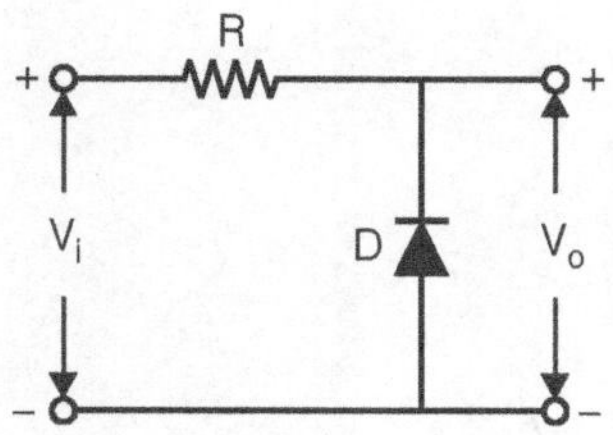


For $V_i < 0$ D on $\therefore$ $V_o = 0$

For $V_i > 0$ D off $\therefore$ $V_o = V_i$

### Transfer characteristics

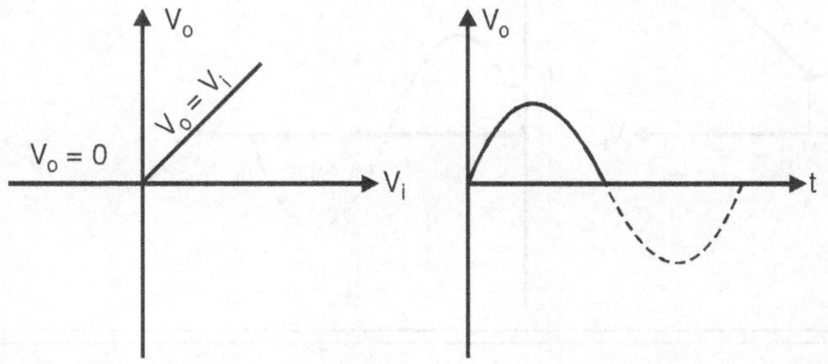


## MODEL-3

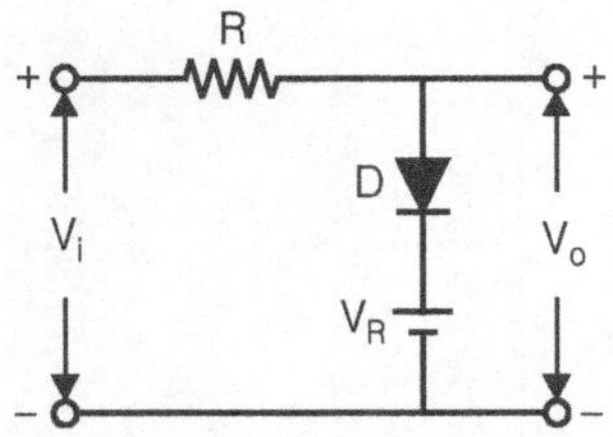


For $V_i < V_R$ D off $\therefore$ $V_o = V_i$

For $V_i > V_R$ D on $\therefore$ $V_o = V_R$

### Transfer characteristics

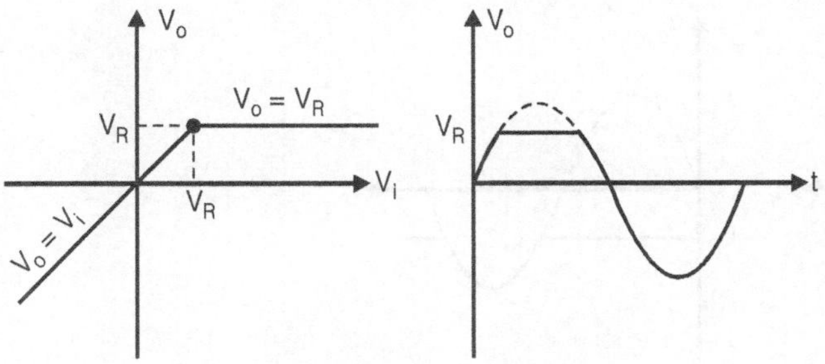

## MODEL-4

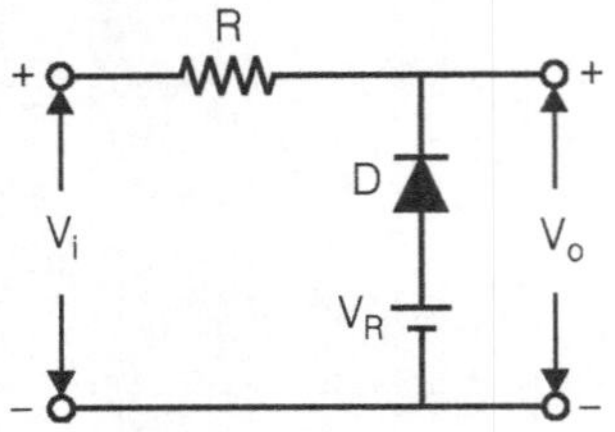


For $V_i < V_R$ D on $\therefore$ $V_o = V_R$
For $V_i > V_R$ D off $\therefore$ $V_o = V_i$

### Transfer characteristics

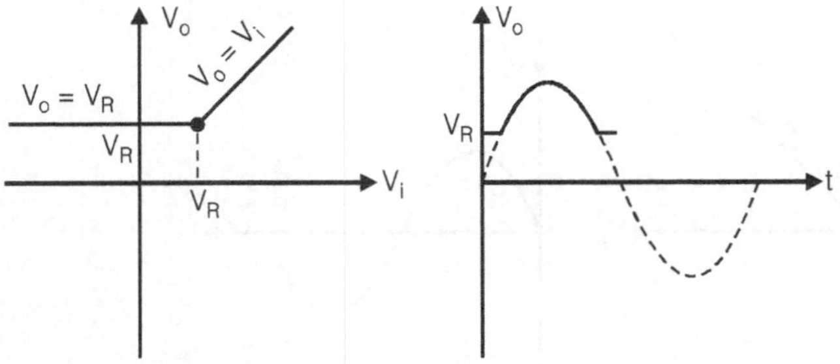


## MODEL-5

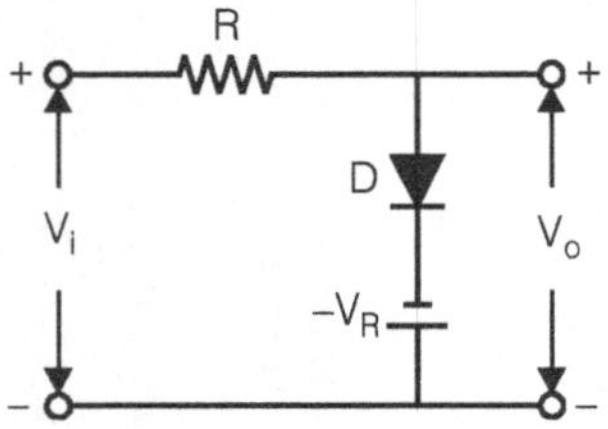


For $V_i < -V_R$ D off $\therefore$ $V_o = V_i$
For $V_i > -V_R$ D on $\therefore$ $V_o = -V_R$

### Transfer characteristics

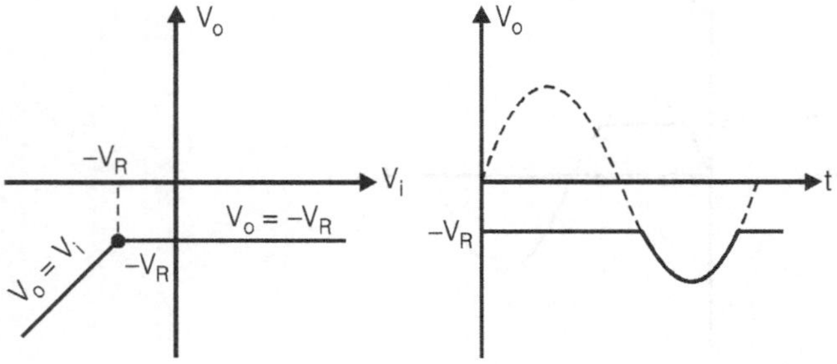

## MODEL-6

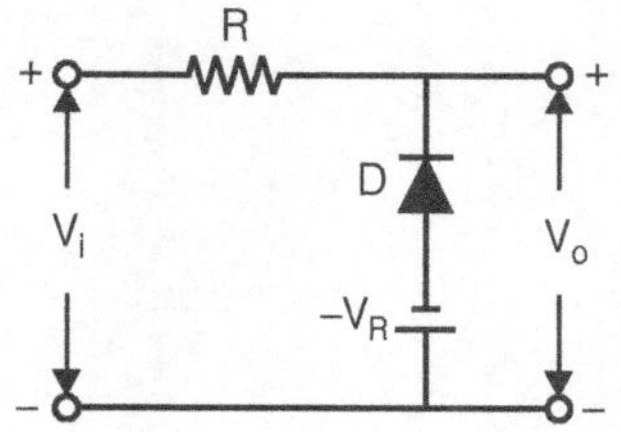


For $V_i < -V_R$ D on $\therefore$ $V_o = -V_R$

For $V_i > -V_R$ D off $\therefore$ $V_o = V_i$

### Transfer characteristics

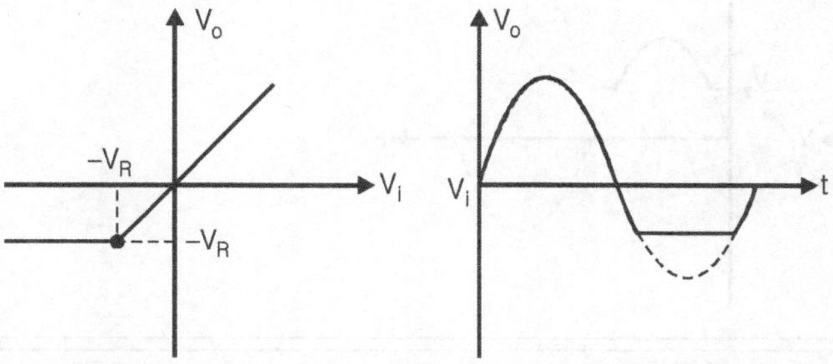


## MODEL-7

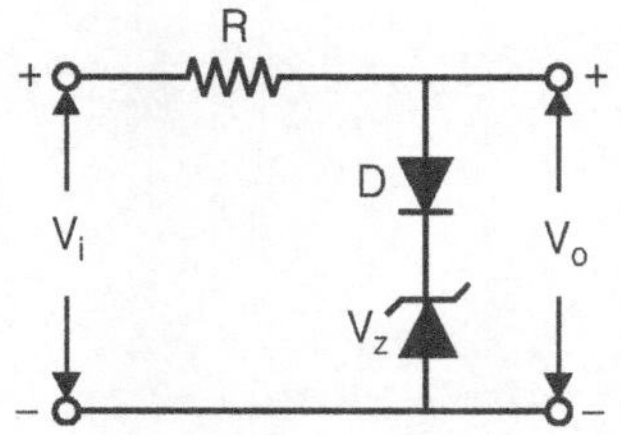


For $V_i < V_z$ D off $\therefore$ $V_o = V_i$

For $V_i > V_z$ D on $\therefore$ $V_o = V_z$

### Transfer characteristics

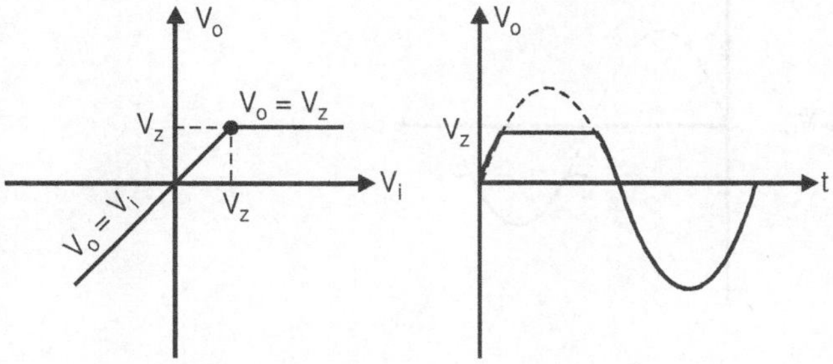

## MODEL-8

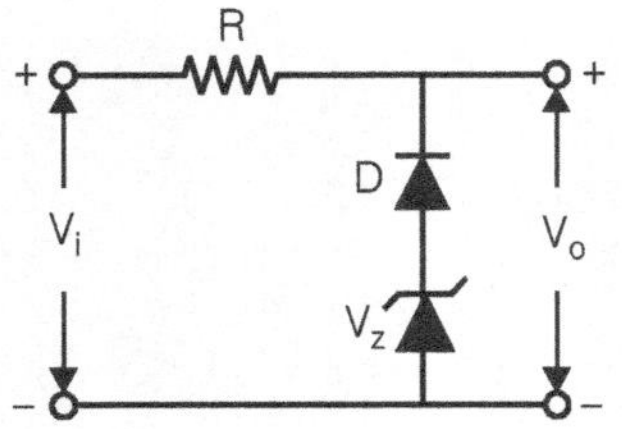


| | | | |
|---|---|---|---|
| For $V_i < V_z$ | D on | ∴ | $V_o = V_z$ |
| For $V_i > V_z$ | D off | ∴ | $V_o = V_i$ |

### Transfer characteristics

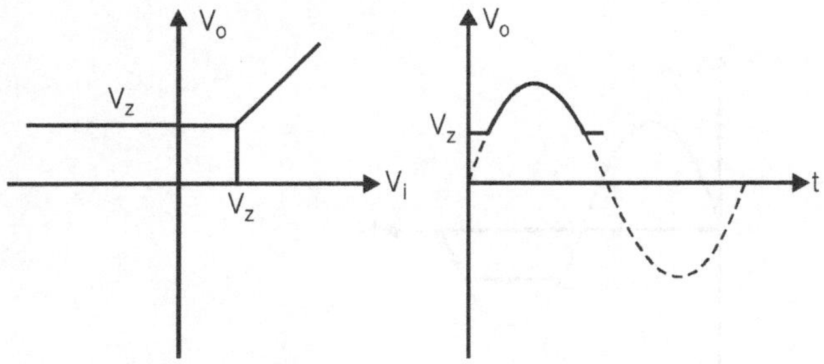


## MODEL-9

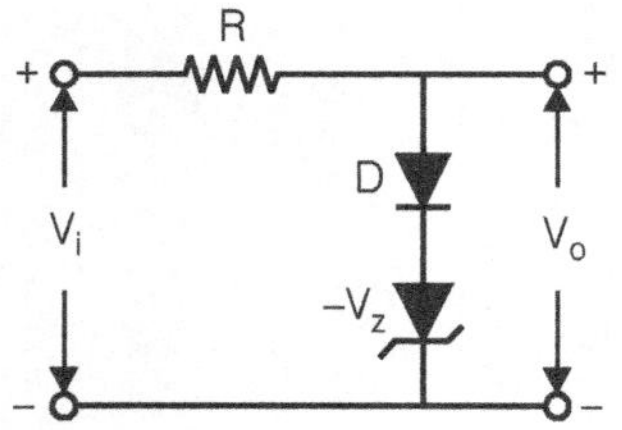


| | | | |
|---|---|---|---|
| For $V_i < -V_z$ | D off | ∴ | $V_o = V_i$ |
| For $V_i > -V_z$ | D on | ∴ | $V_o = -V_z$ |

### Transfer Characteristics

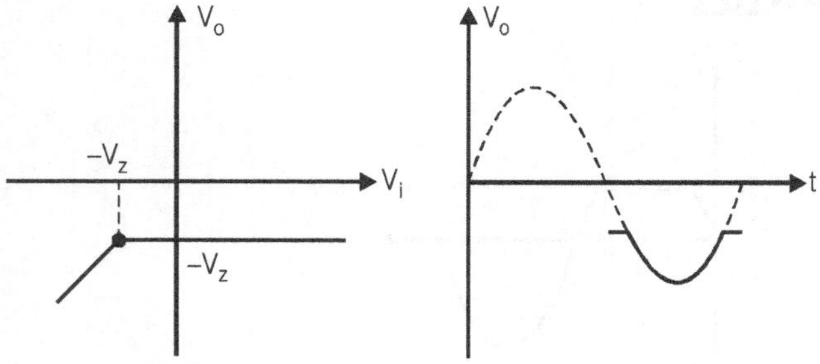

## MODEL-10

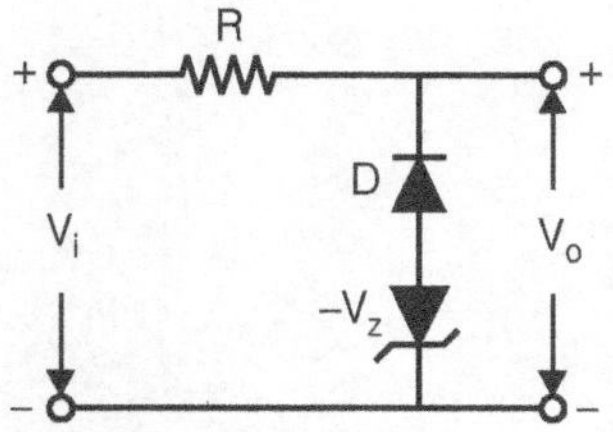


For $V_i < -V_z$ D on $\therefore$ $V_o = -V_z$

For $V_i > -V_z$ D off $\therefore$ $V_o = V_i$

### Transfer Characteristics

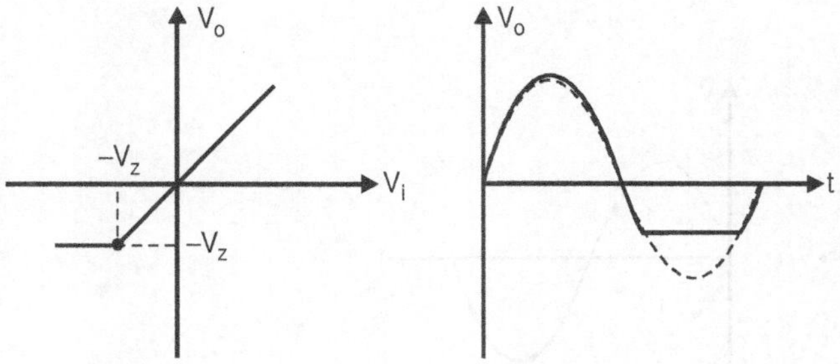


# 7.12 Series Clipper Models

## MODEL-1

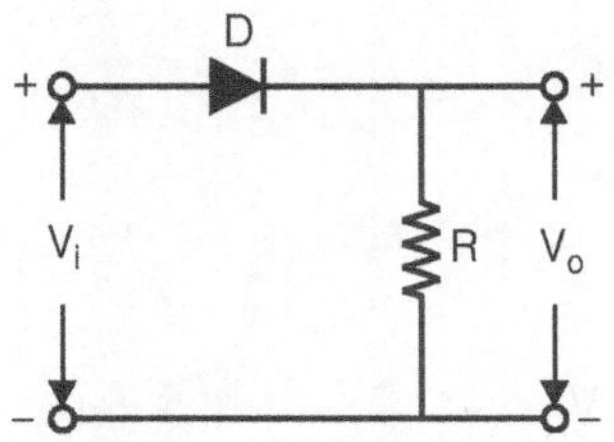


For $V_i < 0$ D off $\therefore$ $V_o = 0$

For $V_i > 0$ D on $\therefore$ $V_o = V_i$

### Transfer Characteristics

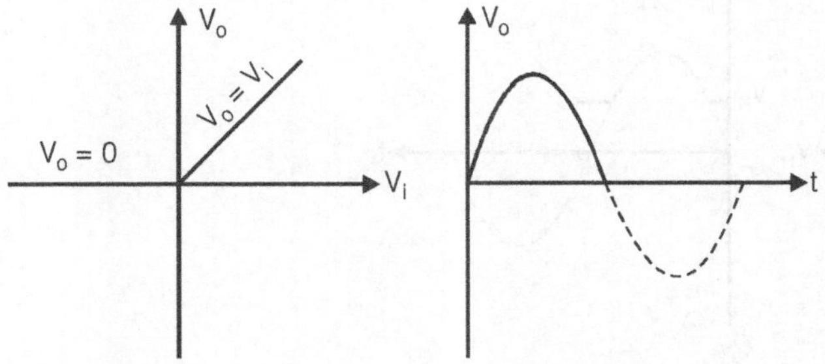

## MODEL-2

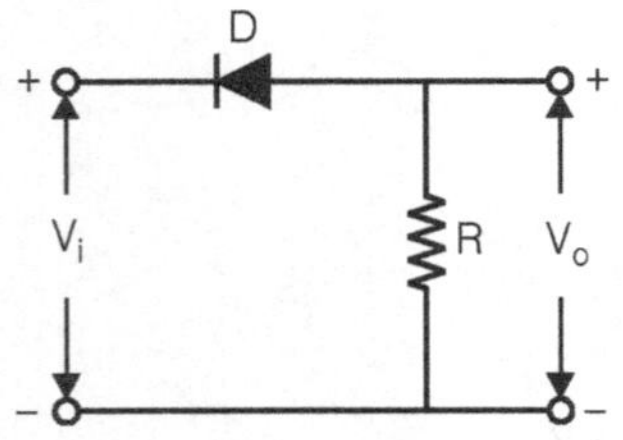


For $V_i < 0$ D on $\therefore$ $V_o = V_i$
For $V_i > 0$ D off $\therefore$ $V_o = 0$

### Transfer Characteristics

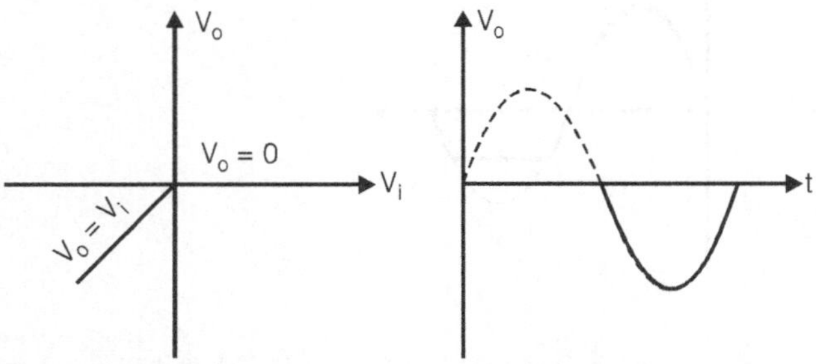


## MODEL-3

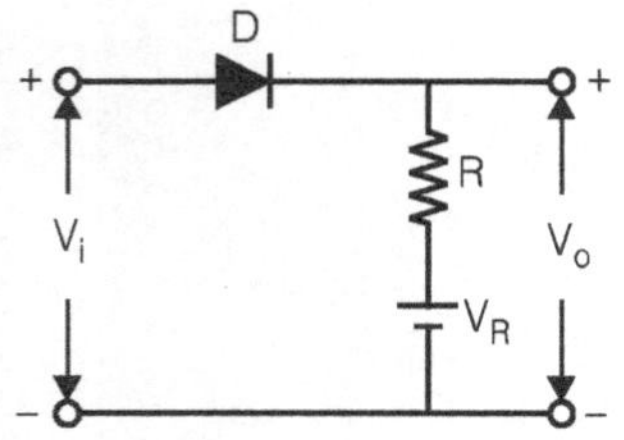


For $V_i < V_R$ D off $\therefore$ $V_o = V_R$
For $V_i > V_R$ D on $\therefore$ $V_o = V_i$

### Transfer Characteristics

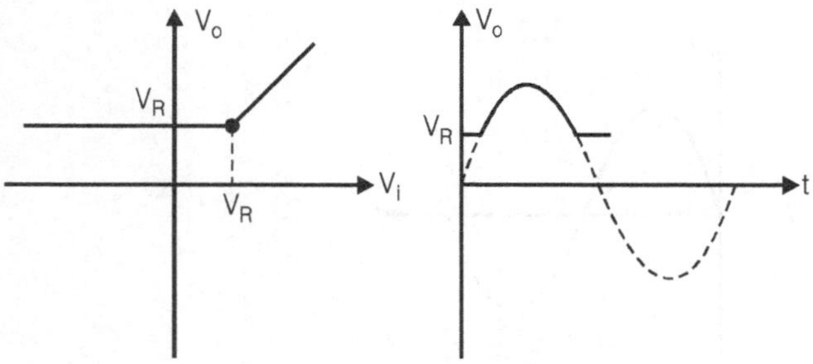

## MODEL-4

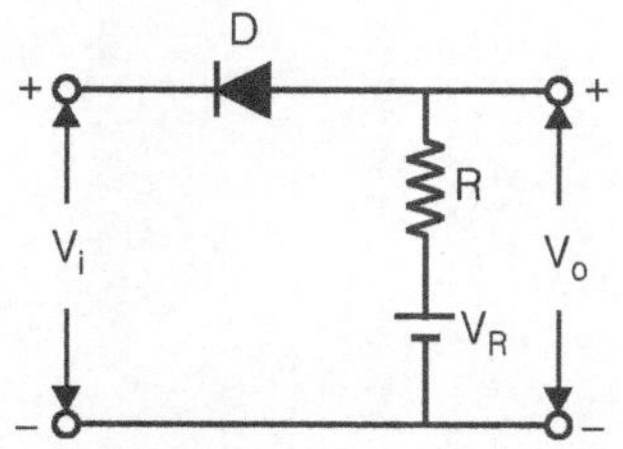


For $V_i < V_R$ D on $\therefore$ $V_o = V_i$

For $V_i > V_R$ D off $\therefore$ $V_o = V_R$

### Transfer Characteristics

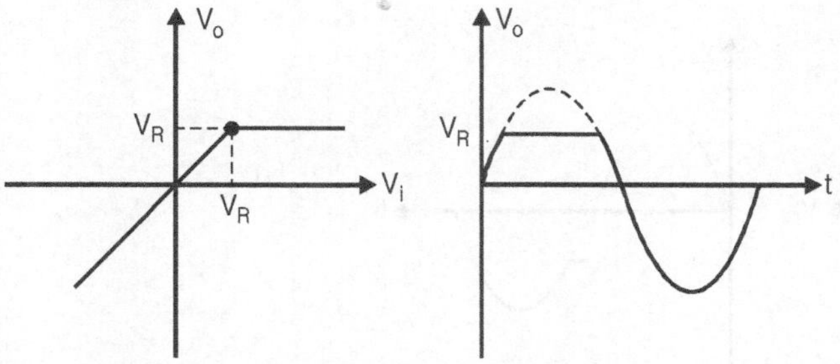


## MODEL-5

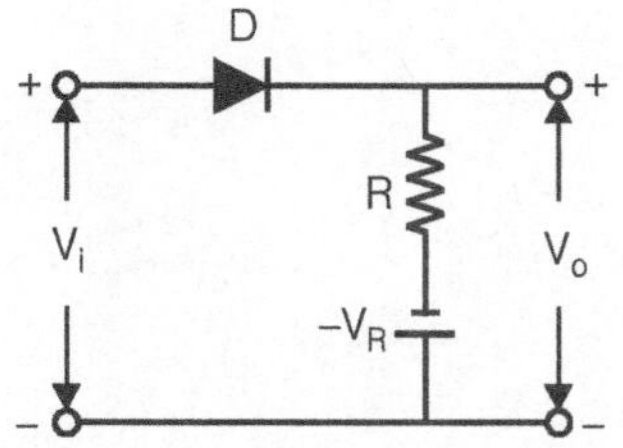


For $V_i < -V_R$ D off $\therefore$ $V_o = -V_R$

For $V_i > -V_R$ D on $\therefore$ $V_o = V_i$

### Transfer Characteristics

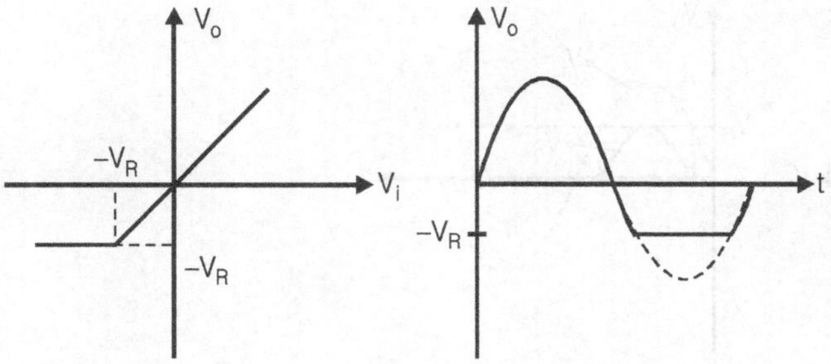

## MODEL-6

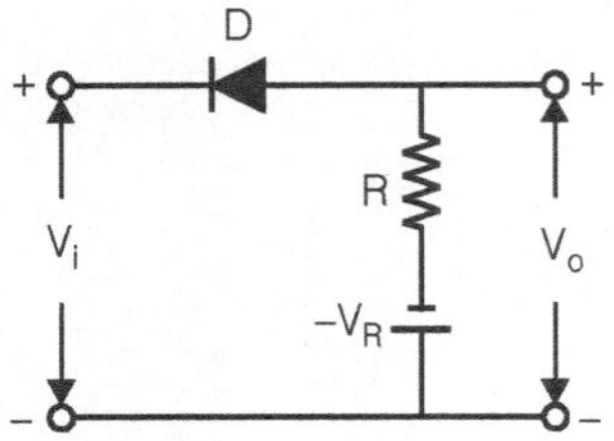


| | | | |
|---|---|---|---|
| For $V_i < -V_R$ | D on | $\therefore$ | $V_o = V_i$ |
| For $V_i > -V_R$ | D off | $\therefore$ | $V_o = -V_R$ |

### Transfer Characteristics

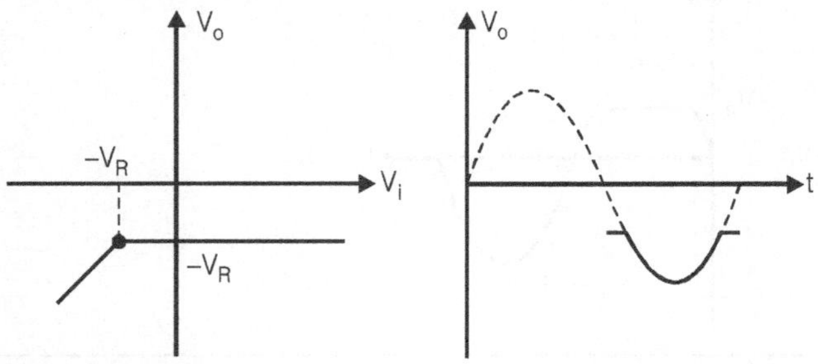


## MODEL-7

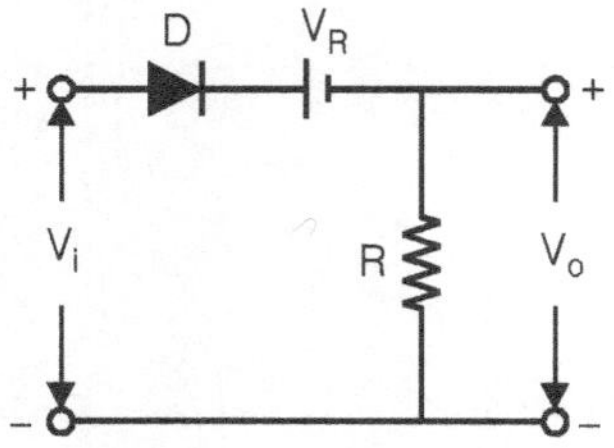


| | | | |
|---|---|---|---|
| For $V_i < V_R$ | D off | $\therefore$ | $V_o = 0$ |
| For $V_i > V_R$ | D on | $\therefore$ | $V_o = V_i - V_R$ |

### Transfer Characteristics

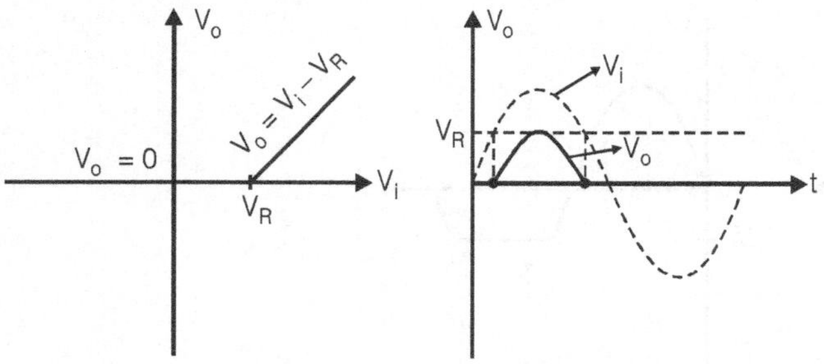

## MODEL-8

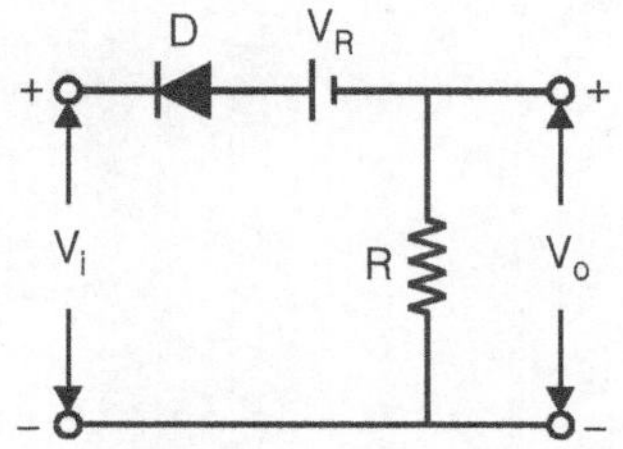


For $V_i < V_R$ D on $\therefore$ $V_o = V_i - V_R$
For $V_i > V_R$ D off $\therefore$ $V_o = 0$

### Transfer Characteristics

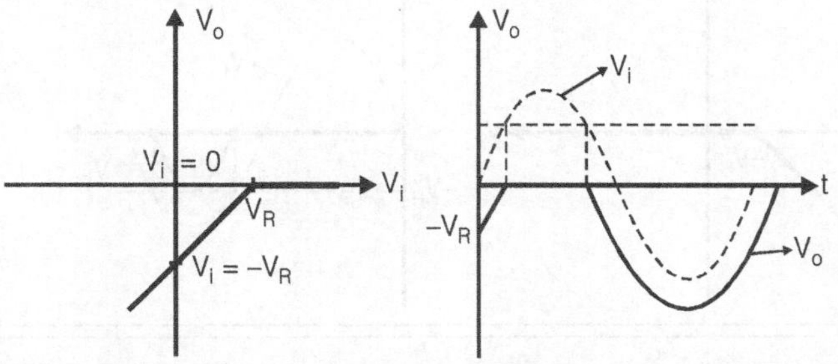


## MODEL-9

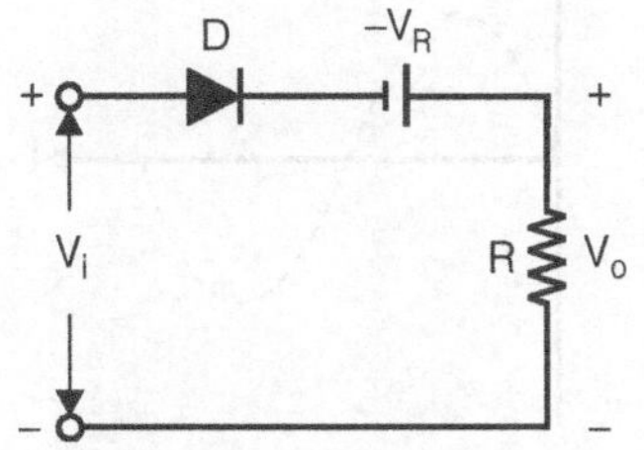


For $V_i < -V_R$ D off $\therefore$ $V_o = 0$
For $V_i > -V_R$ D on $\therefore$ $V_o = V_i + V_R$

### Transfer Characteristics

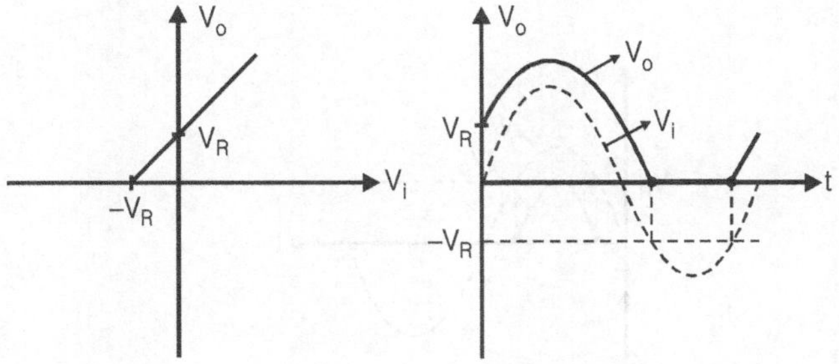

## MODEL-10

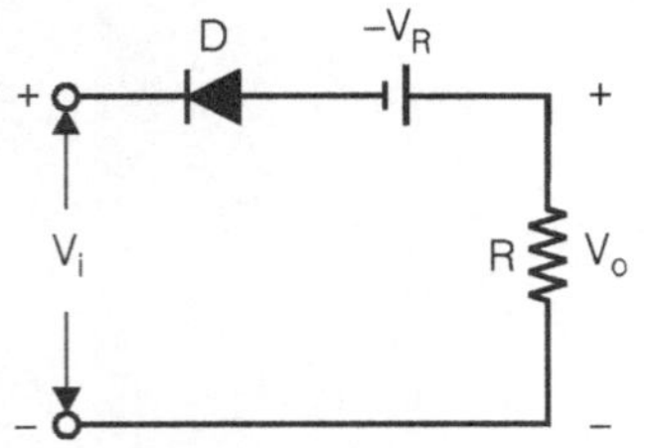


For $V_i < -V_R$ D on $\therefore$ $V_o = V_i + V_R$

For $V_i > -V_R$ D off $\therefore$ $V_o = 0$

### Transfer Characteristics

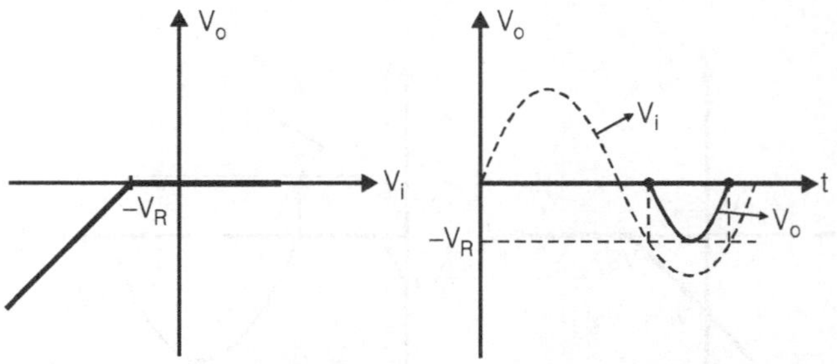


## MODEL-11

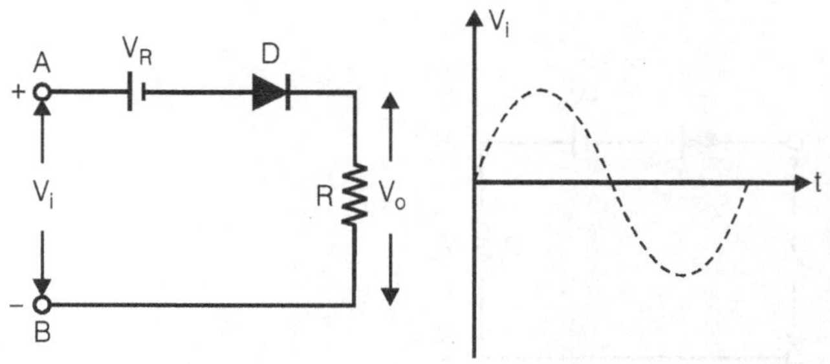


- In positive half cycle of the signal is applied between terminals A and B. Due to this signal diode is forward bias but no current will flow till the signal exceeds the opposing battery voltage of $V_R$. Hence, the upper part of positive signal will pass through the circuit and appears $V_o$ across R.
- The negative half cycle of the signal is clipped off as diode is reverse bias.

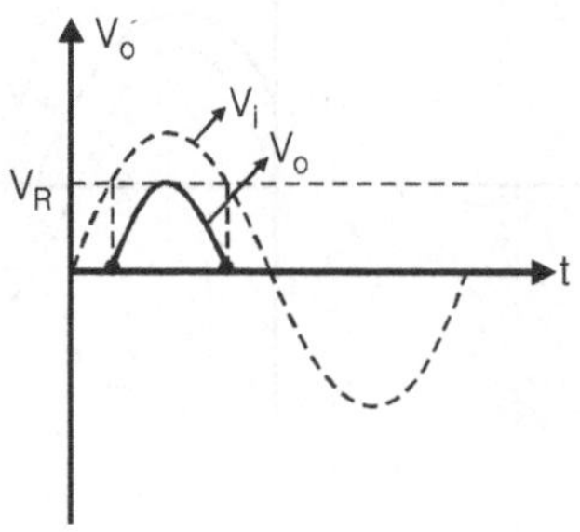

## MODEL-12

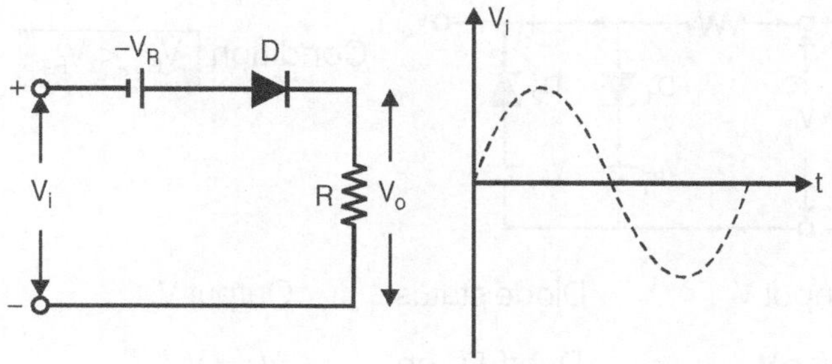


- During positive half cycle, when diode D is forward bias, the voltage across the load R will be sum of signal voltage and battery voltage. The signal voltage is lifted by $V_R$.
- During negative half cycle the diode is reverse bias, but current will flow upto $-V_R$, below $-V_R$ is clipped off.

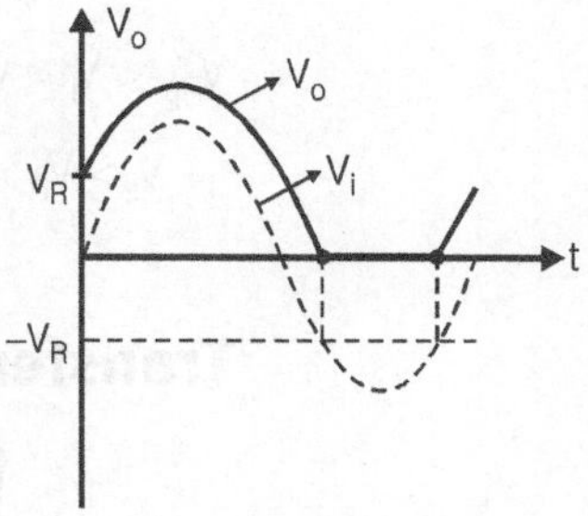


# 7.13 Clipping at Two Independent Levels

## MODEL-1

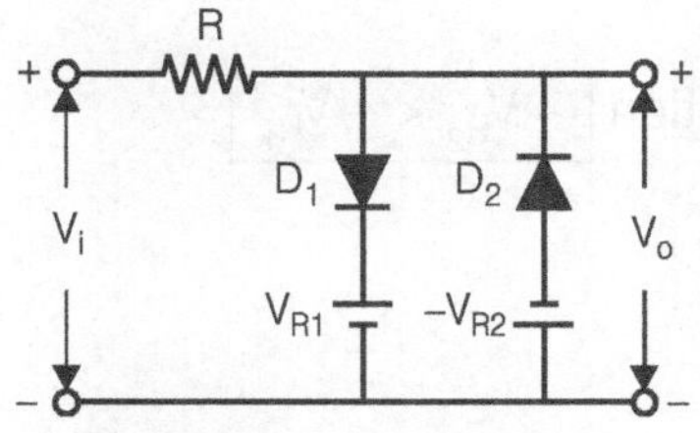


| Input $V_i$ | Output $V_o$ | Diode status |
|---|---|---|
| $V_i < -V_{R_2}$ | $V_o = -V_{R_2}$ | $D_1$ off, $D_2$ on |
| $-V_{R_2} < V_i < V_{R_1}$ | $V_o = V_i$ | $D_1$ off, $D_2$ off |
| $V_i > V_{R_1}$ | $V_o = V_{R_1}$ | $D_1$ on, $D_2$ off |

### Transfer Characteristics

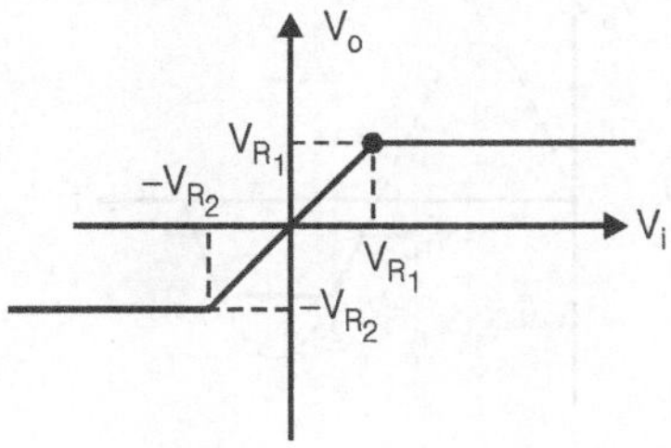


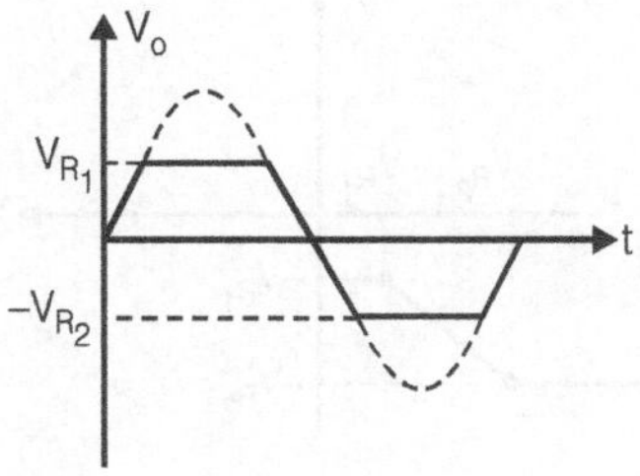

## MODEL-2

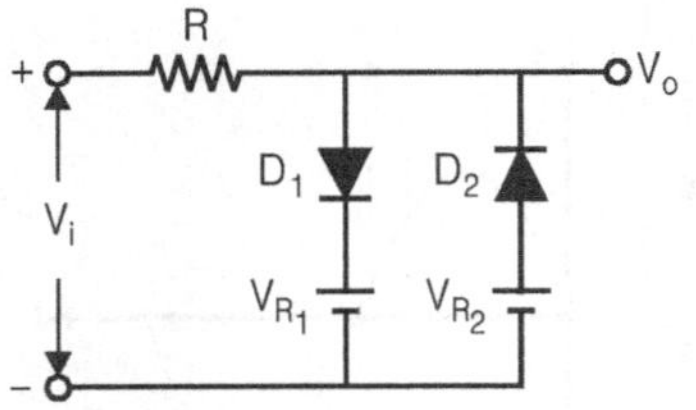


Condition $V_{R_2} < V_{R_1}$

| Input $V_i$ | Diode status | Output $V_o$ |
|---|---|---|
| $V_i < V_{R_2}$ | $D_1$ off, $D_2$ on | $V_o = V_{R_2}$ |
| $V_{R_2} < V_i < V_{R_1}$ | $D_1$ off, $D_2$ off | $V_o = V_i$ |
| $V_i > V_{R_1}$ | $D_1$ on, $D_2$ off | $V_o = V_{R_1}$ |

### Transfer Characteristics

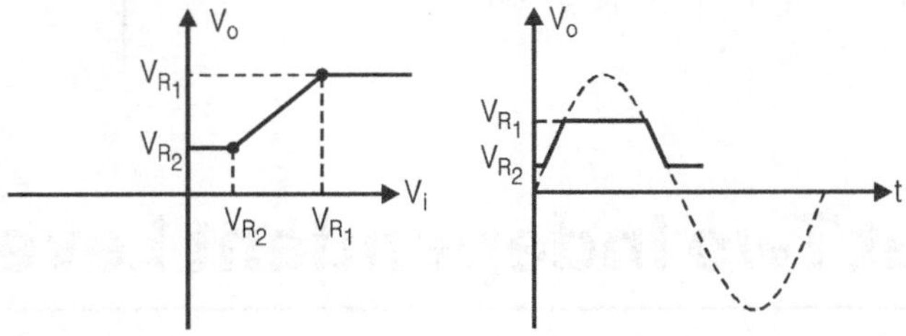


## MODEL-3

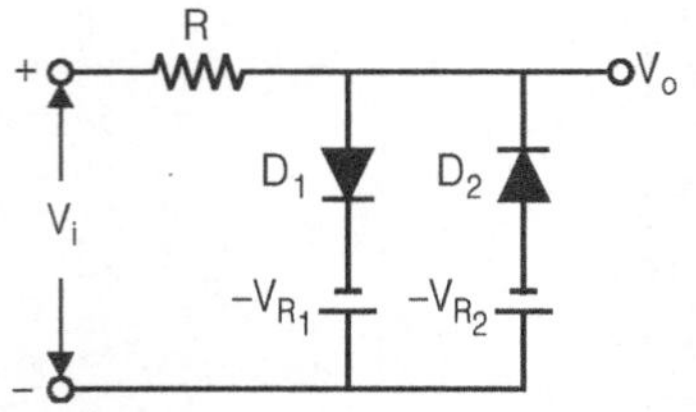


Condition $-V_{R_2} < -V_{R_1}$

| Input $V_i$ | Diode status | Output $V_o$ |
|---|---|---|
| $V_i < -V_{R_2}$ | $D_1$ off, $D_2$ on | $V_o = -V_{R_2}$ |
| $-V_{R_2} < V_i < -V_{R_1}$ | $D_1$ off, $D_2$ off | $V_o = V_i$ |
| $V_i > -V_{R_1}$ | $D_1$ on, $D_2$ off | $V_o = -V_{R_1}$ |

### Transfer Characteristics

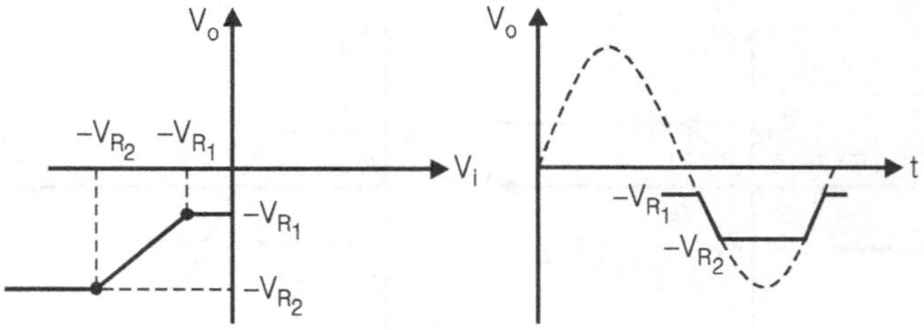

## MODEL-4

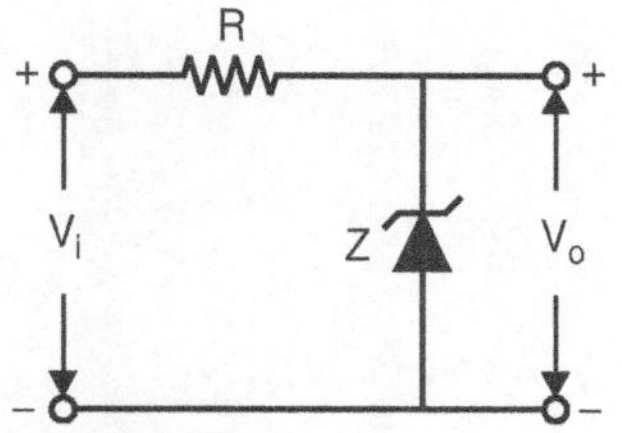


For $V_i < -V_\gamma$ Z is on ∴ $V_o = V_\gamma$

For $-V_\gamma < V_i < V_z$ Z is off ∴ $V_o = V_i$

For $V_i > V_z$ Z breaks ∴ $V_o = V_z$

### Transfer Characteristics

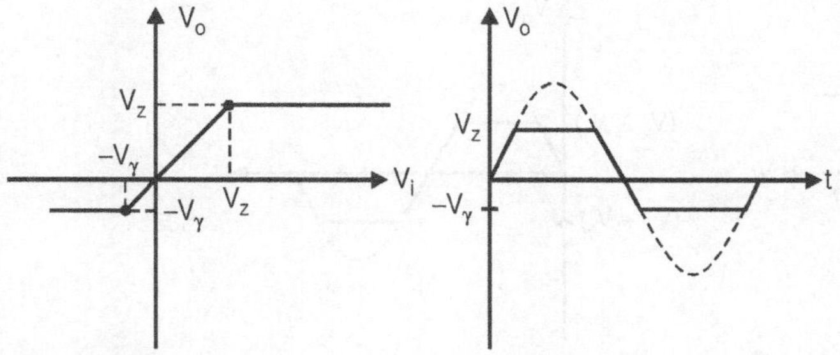


## MODEL-5

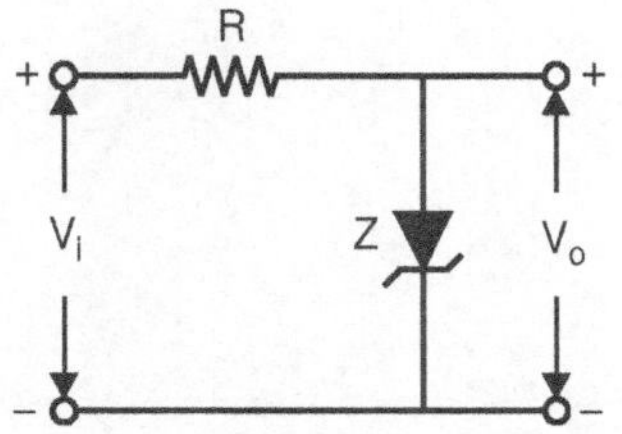


For $V_i < -V_z$ Z breaks ∴ $V_o = -V_z$

For $-V_z < V_i < V_\gamma$ Z off ∴ $V_o = V_i$

For $V_i > V_\gamma$ Z on ∴ $V_o = V_\gamma$

### Transfer Characteristics

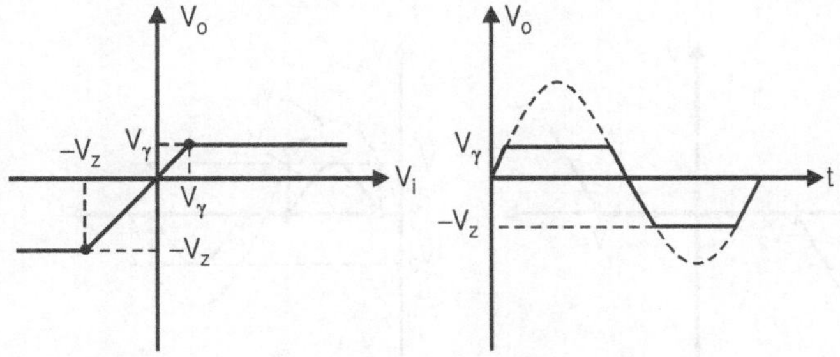

## MODEL-6

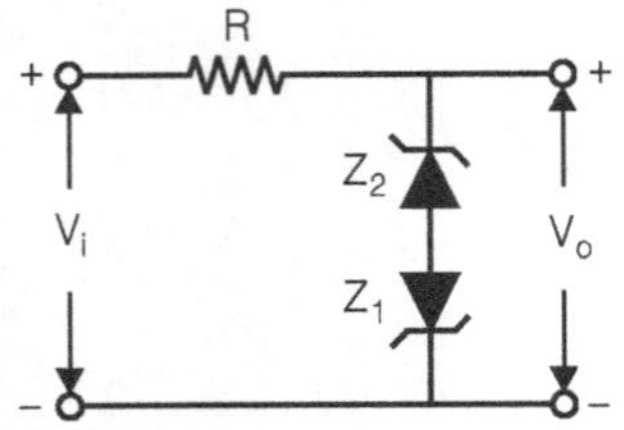


| | | | |
|---|---|---|---|
| For $V_i < -(V_z + V_\gamma)$ | $Z_1$ breaks, $Z_2$ on | ∴ | $V_o = -(V_z + V_\gamma)$ |
| For $-(V_z + V_\gamma) < V_i < (V_z + V_\gamma)$ | $Z_1$ off, $Z_2$ off | ∴ | $V_o = V_i$ |
| For $V_i > (V_z + V_\gamma)$ | $Z_1$ on, $Z_2$ breaks | ∴ | $V_o = V_z + V_\gamma$ |

### Transfer Characteristics

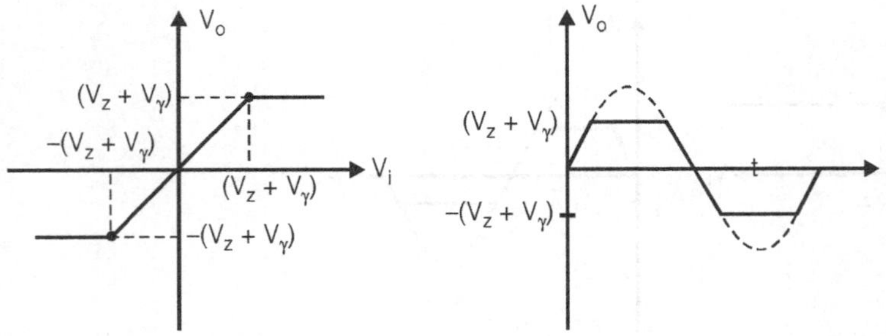


## MODEL-7

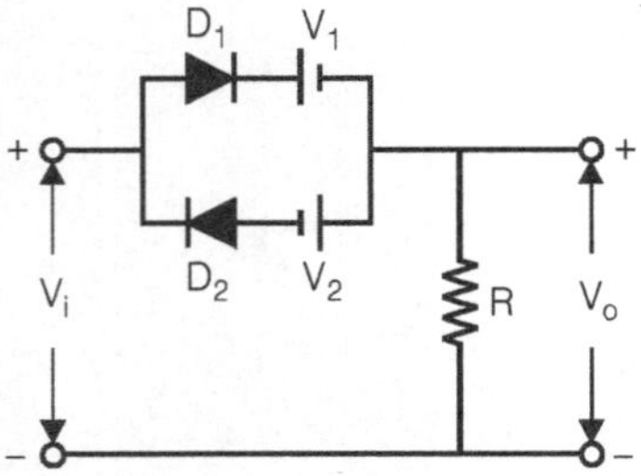


| | | |
|---|---|---|
| For $V_i < -V_2$ | $D_1$ off, $D_2$ on | $V_o = V_i + V_2$ |
| For $-V_2 < V_i < V_1$ | $D_1$ off, $D_2$ off | $V_o = 0$ |
| For $V_i > V_1$ | $D_1$ on, $D_2$ off | $V_o = V_i - V_1$ |

### Transfer Characteristics

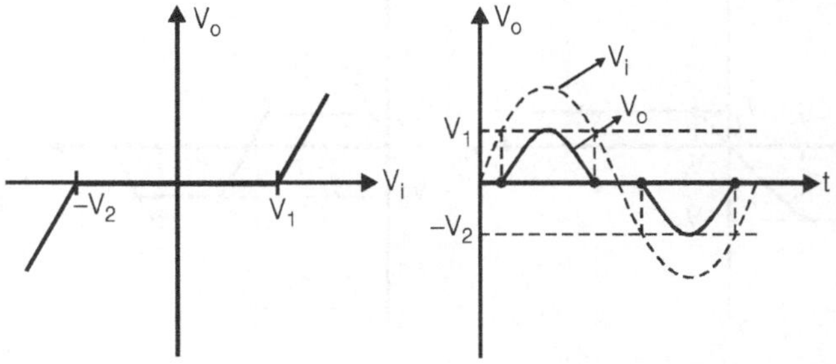

# 7.14 Clamping Circuits

## INTRODUCTION

- Clamping is a process of introducing a dc level into an ac signal. A clamping circuit finds use when the dc value of the signal is to be shifted from one level to another level.
- The basic requirements of a clamping circuit are:
  - Diode
  - Capacitor
  - Resistor
- The capacitor blocks the passage of direct current. Therefore when a non sinusoidal signal is transmitted through a capacitive coupling circuit, it loses its dc component.
- To restore the dc component at a later stage, the signal should pass through a clamping circuit.
- In a clamping circuit, the waveforms of input and output remain the same, just there is a vertical shift either upward or downward.

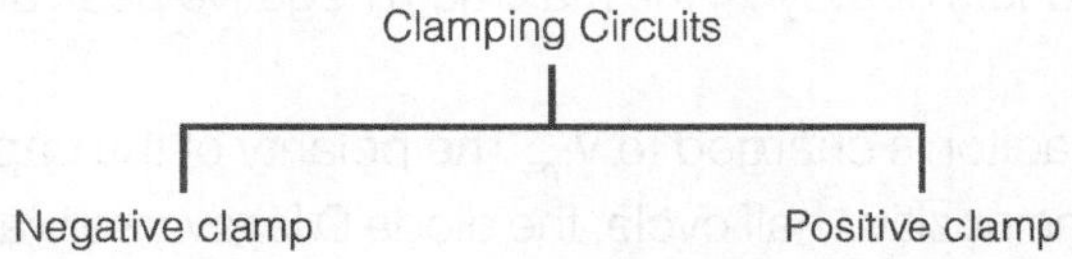


## NEGATIVE CLAMPER

When a clamper shifts the original signal in vertical downward direction, it is known as negative clamper. This type of clamper introduces a negative dc value.

Let the input be a sinusoidal signal given as

$$V_i = V_m \sin \omega t$$

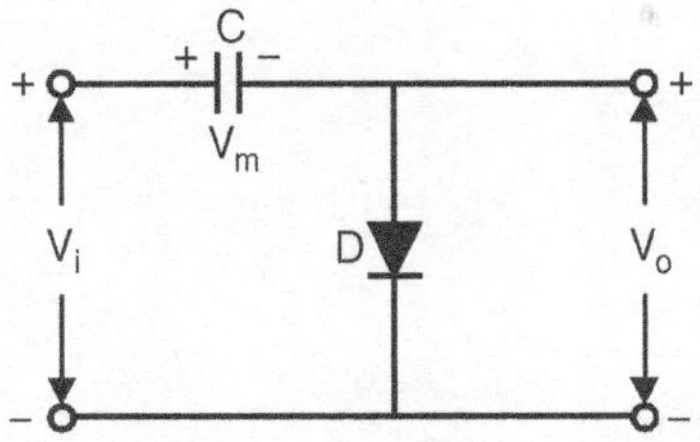


During the positive half cycle, diode D is forward biased and conducts heavily. At the maximum positive peak of the applied signal, the capacitor is charged to $V_m$. The polarity across the capacitor is shown in the figure.

Therefore the voltage across the capacitor opposes the input voltage $V_i$, so the output voltage is given by

$$V_o = \text{output voltage} = V_i - V_m$$

During the negative half cycle, the diode gets reverse biased and does not conduct. During the subsequent positive half cycles, the voltage across the capacitor remains fixed at $V_m$, and when steady state is reached, we have

$$\boxed{V_o = V_i - V_m}$$

The charged capacitor acts as a voltage source $V_m$. The equivalent circuit is as shown:

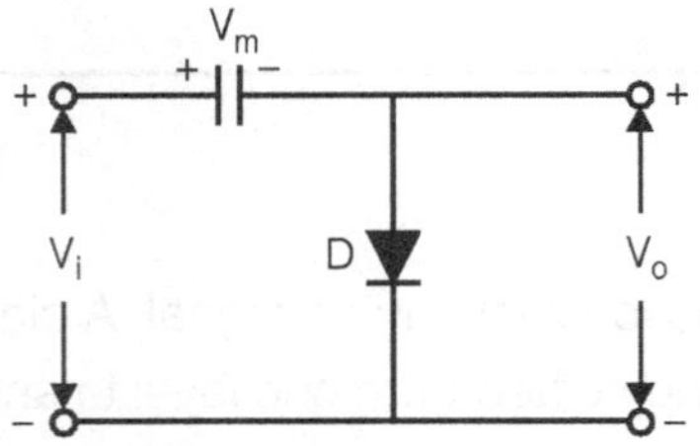


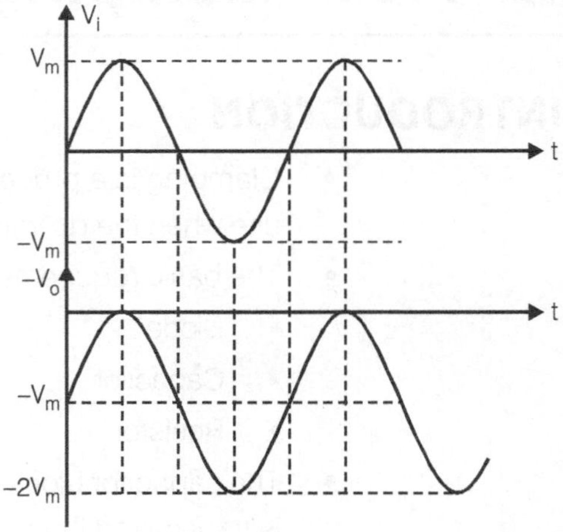


For $V_i = 0$, $V_o = -V_m$

For $V_i = V_m$, $V_o = 0$

For $V_i = -V_m$, $V_o = -2V_m$

## POSITIVE CLAMPER

When a clamper shifts the original signal in vertical upward direction, it is known as positive clamper. This type of clamper introduces a positive dc value.

Let the input be a sinusoidal signal given as

$$V_i = V_m \sin \omega t$$

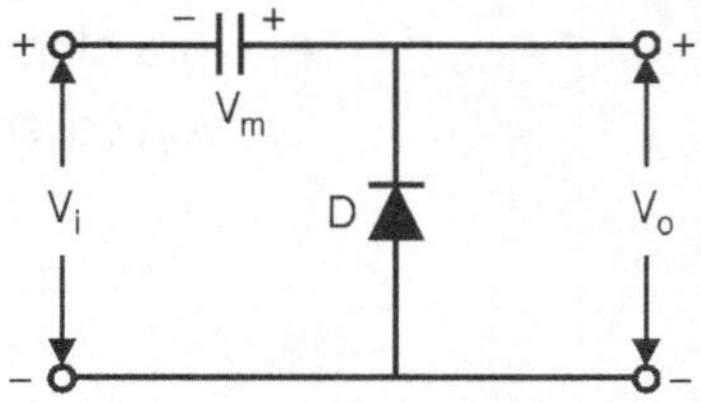


During the negative half cycle, the diode D is forward biased and conducts heavily. At the maximum negative peak of applied signal.

The capacitor is charged to $V_m$. The polarity of the capacitor is shown in the figure.

During the positive half cycle, the diode D is reverse biased and does not conduct. The capacitor charged to $V_m$ behaves as a battery and the output $V_o$ is given by

$$\boxed{V_o = V_i + V_m}$$

Based on the above relationship between $V_o$ and $V_i$, the output voltage waveform is plotted. It is seen that the negative peaks of the input signal are clamped to zero level.

Peak to peak amplitude of output voltage = $2V_m$, which is the same as that of the input signal.

$V_i = V_m$ $\quad V_o = 2V_m$

$V_i = 0$ $\quad V_o = V_m$

$V_i = -V_m$ $\quad V_o = 0$

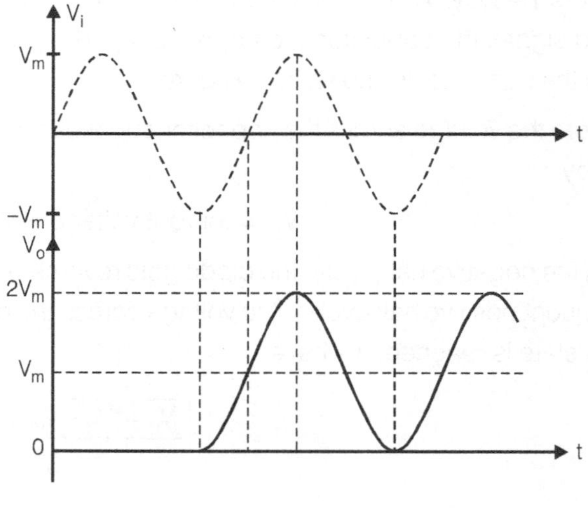

## SUMMARY

- The process by which a non sinusoidal signal is altered by transmission through a linear network is called linear wave shaping.
- A high pass circuit attenuates all low frequency signals and transmits only signals of high frequency.
- The lower cut-off frequency of a high pass circuit is $f_L = 1/2\pi RC$ and upper cut-off frequency is $\infty$.
- A low pass circuit passes low frequency signals and attenuates high frequency signals.
- The upper cut-off frequency of a low pass circuit is equal to $f_h = 1/2\pi RC$.
- Rise time is defined as time taken by the output waveform to rise from 10% to 90% of its final steady state value.
- Rise time is directly proportional to the time constant and inversely proportional to bandwidth.
- Clippers are used to remove a part of input waveform.
- Clipping circuits do not require energy storage elements.
- Clipping circuits are available in three types:
  - Series clippers
  - Shunt clippers
  - Two independent level clippers.
- A clipping circuit may be used to convert a sine wave into a square wave.
- The clamping circuit is also referred as dc restorer.
- Clamping circuit only changes the dc level of the signal but does not affect its shape.
- Clamping circuits may be
  - Positive clampers
  - Negative clampers

## GATE AND IES QUESTIONS

**Q.1** The wave shape of $V_o$ in figure is

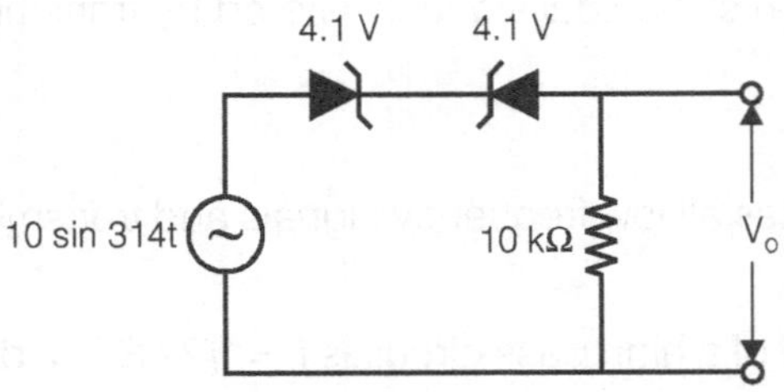


(a)

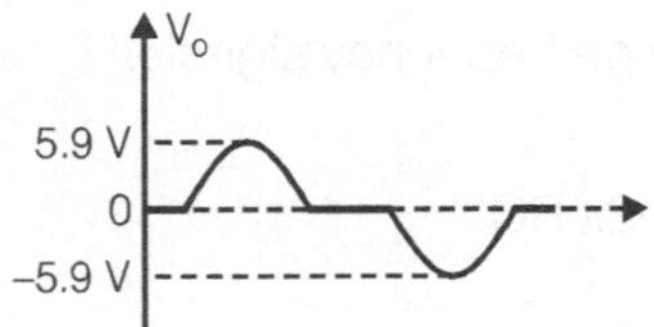


(b)

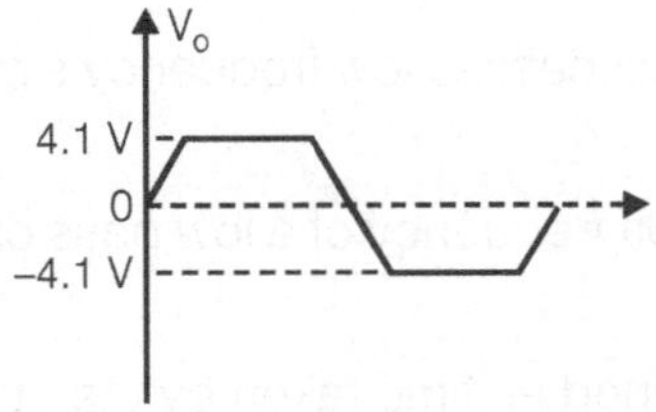


(c)

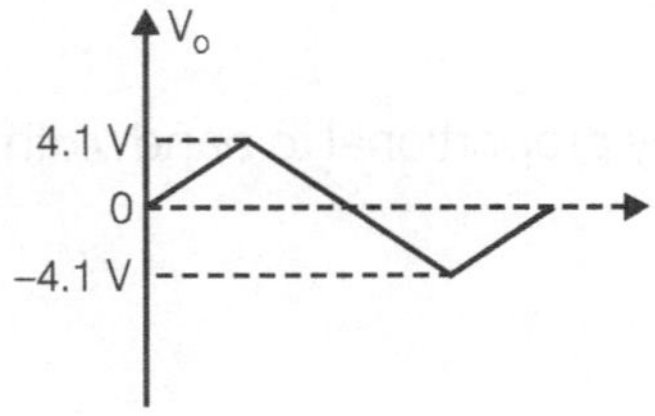


(d)

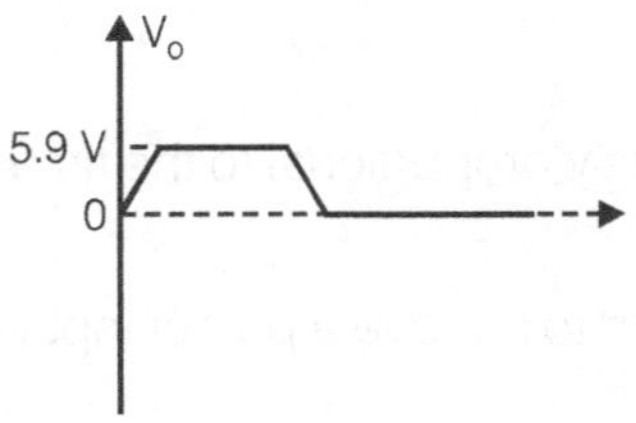


**Exp.** **(a)**

- During positive half cycle $D_2$ acts as a voltage regulator, $D_1$ acts as a short circuit, when input applied is greater than 4.1 V, $D_2$ starts conducting.
- Similarly during negative half cycle. $D_1$ acts as a voltage regulator, $D_2$ is forward bias.

**Q.2** For the circuit shown below, assume that the zener diode is ideal with a break down voltage of 6 V. The waveform observed across R is

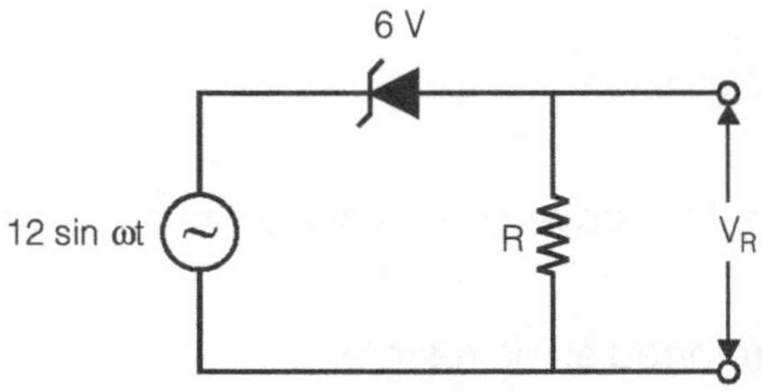


(a)

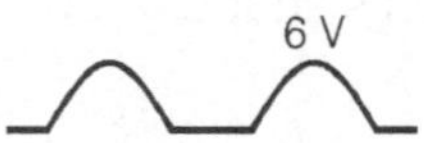


(b)

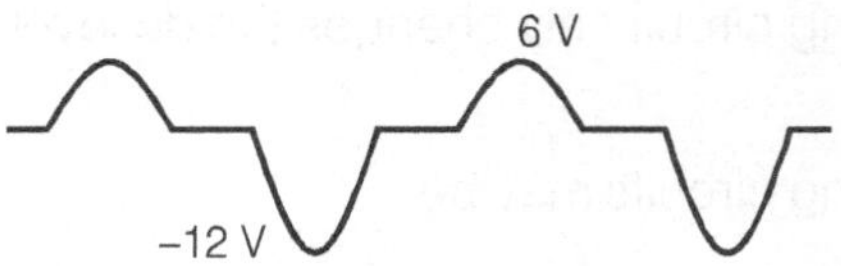


(c)

(d)

**Exp.** (b)

In forward bias Zener diode works as normal diode. Thus for negative half cycle Zener is forward bias

$$V_o = V_{in}$$

For positive half cycle

When $0 < V_{in} < 6$ V

D is OFF $V_R = 0$

When $V_{in} > 6$ V, diode conducts

$$V_R = V_i - 6$$

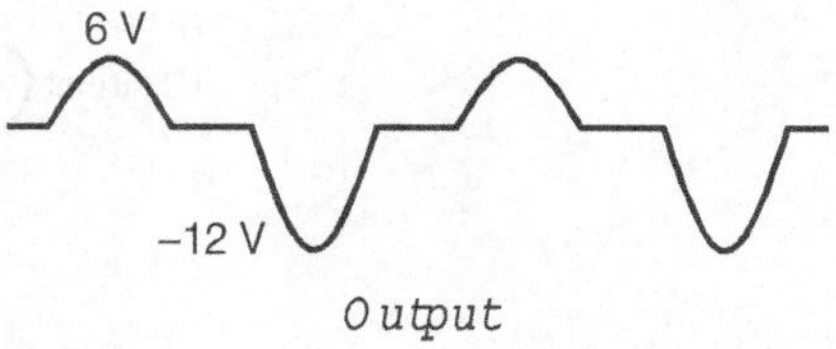


**Q.3** Assuming that the diodes in the given circuit are ideal, the voltage $V_o$ is

(a) 4 V

(b) 5 V

(c) 7.5 V

(d) 12.12 V

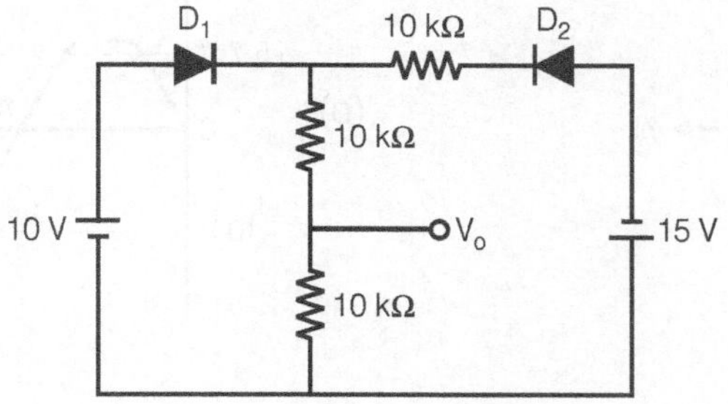


**Exp.** (b)

Given circuit is

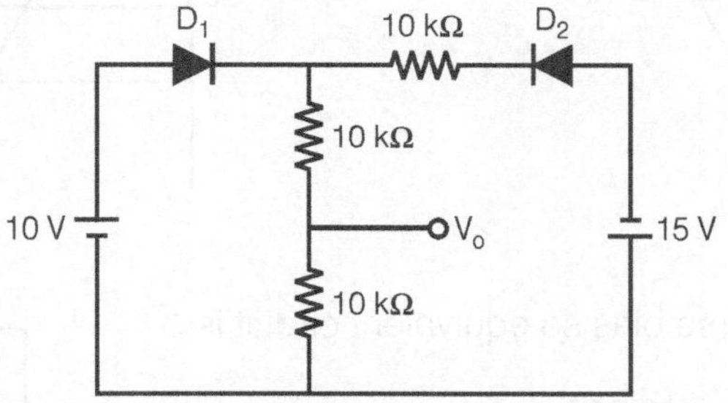


We can observe that diode $D_2$ is always OFF so equivalent circuit is

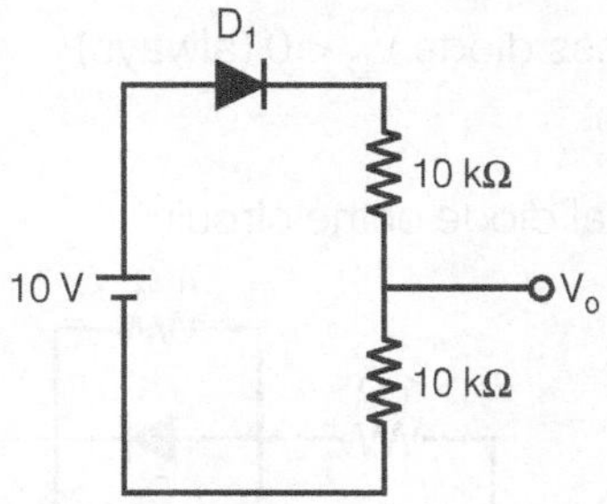


$D_1$ is ON in this condition

$$V_o = \frac{10}{10+10} \times 10 = 5 \text{ V}$$

**Q.4** The equivalent circuits of a diode, during forward biased and reverse biased conditions, are shown in the figure

(I) 

(II) 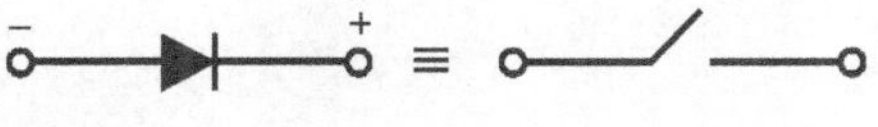

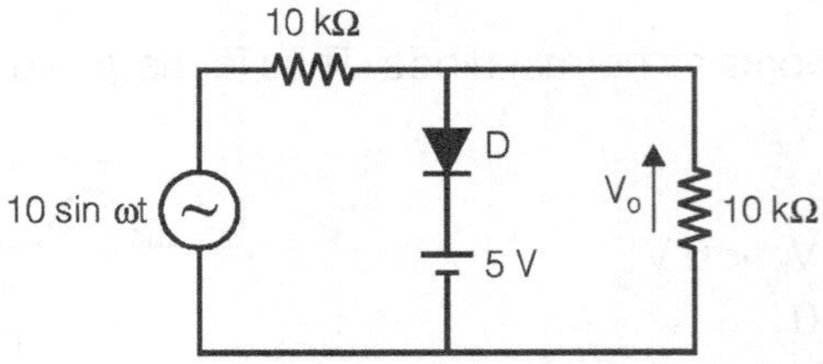


If such a diode is used in clipper circuit of figure given above, output voltage $V_o$ of the circuit will be

(a)

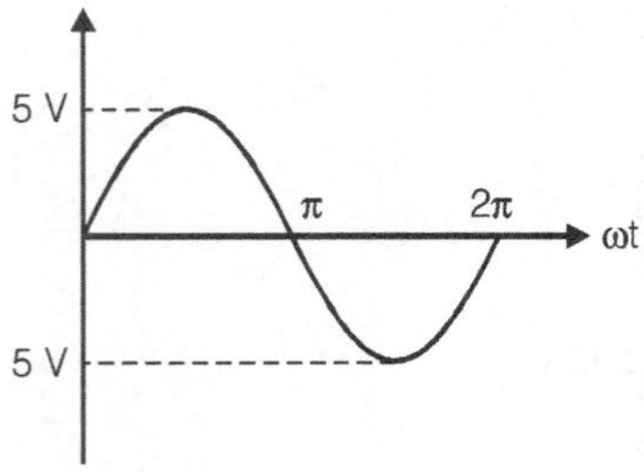


(b)

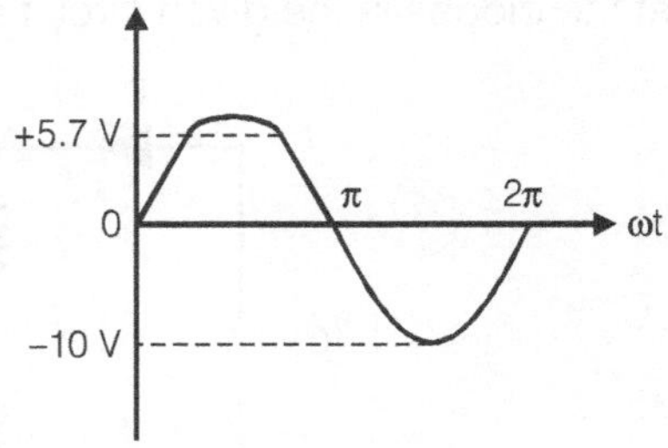


(c)

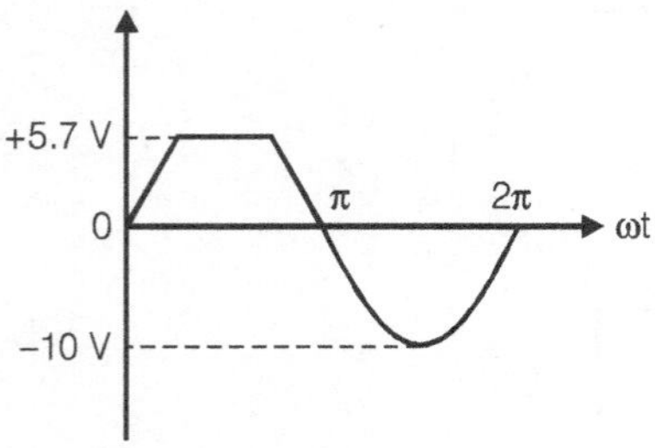


(d)

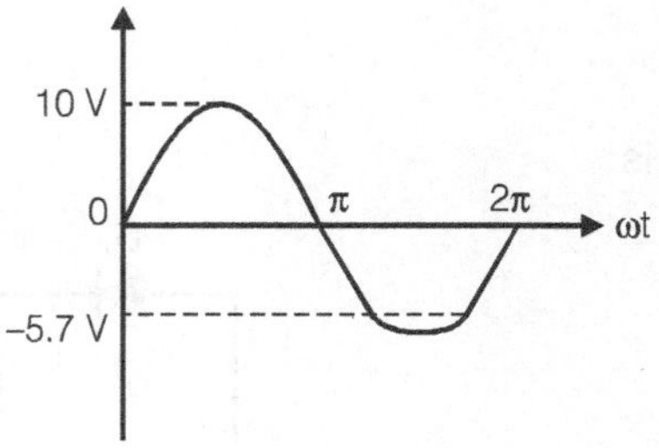


**Exp.** (a)

Assume the diode is in reverse bias so equivalent circuit is

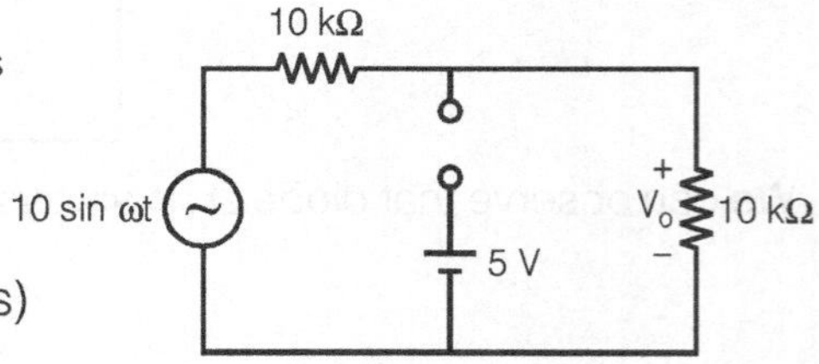


Output voltage $V_o = \dfrac{10 \sin\omega t}{10+10} \times 10 = 5 \sin \omega t$

Due to resistor divider, voltage across diode $V_D < 0$ (always)

So $V_o = 5 \sin \omega t$

**Q.5** What are the states of the three ideal diode of the circuit

(a) $D_1$ ON, $D_2$ OFF, $D_3$ OFF

(b) $D_1$ OFF, $D_2$ ON, $D_3$ OFF

(c) $D_1$ ON, $D_2$ OFF, $D_3$ ON

(d) $D_1$ OFF, $D_2$ ON, $D_3$ ON

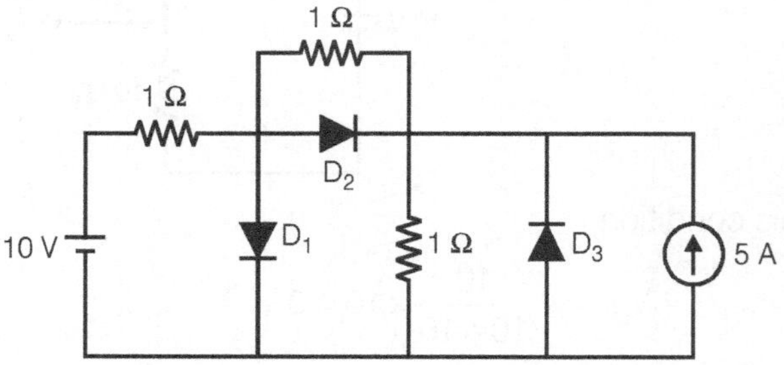


**Exp.** (a)

According to 10 V battery $D_1$ is ON, then $D_2$ will be OFF.

$D_3$ is OFF because 5 A current flow is opposite to the diode

Hence $D_1$ ON, $D_2$ OFF, $D_3$ OFF.

**Q.6** Assuming the diodes $D_1$ and $D_2$ of the circuit shown in figure are ideal ones, the transfer characteristics of the circuit will be

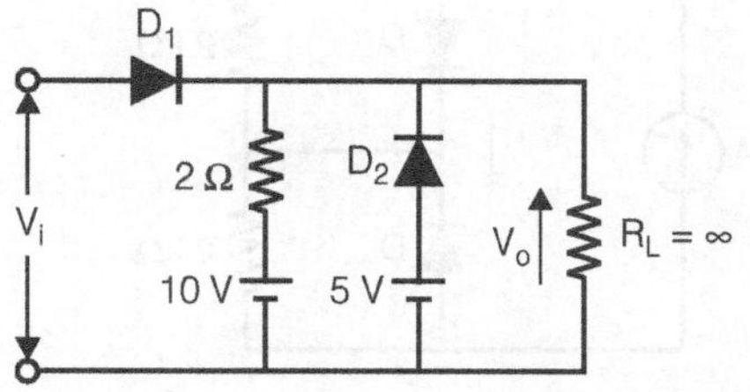


(a)
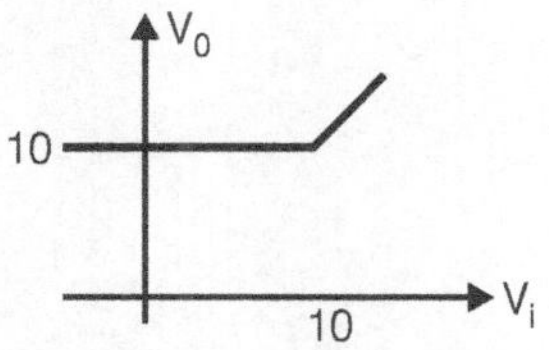


(b)
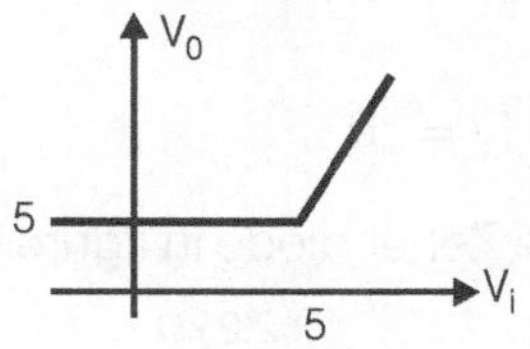


(c)
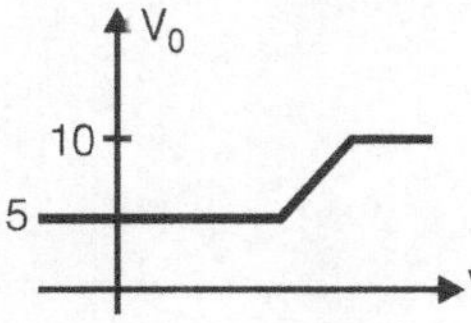


(d)
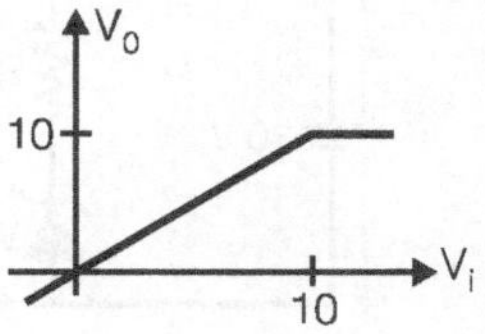


**Exp.** (a)

In the circuit when $V_i < 10$ V, both $D_1$ and $D_2$ are OFF

So equivalent circuit is

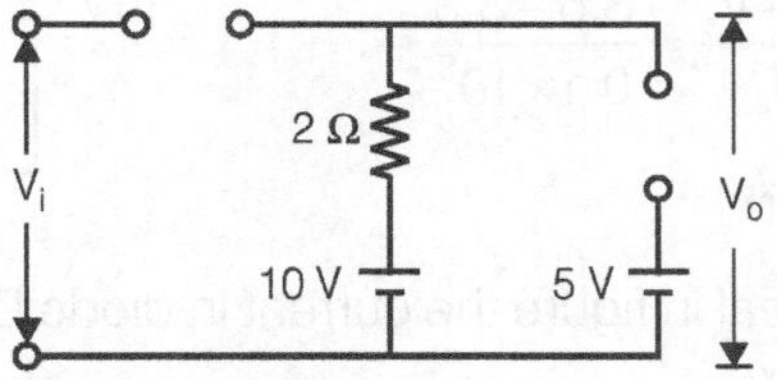


Output $V_o = 10$ V

When $V_i > 10$ V ($D_1$ is forward bias and D2 is OFF), so equivalent circuit is

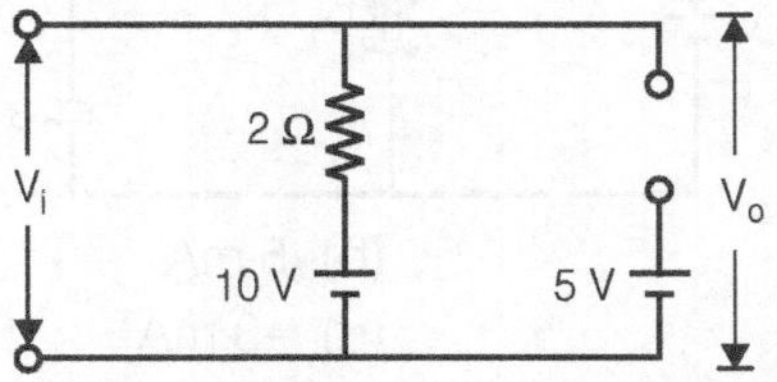


Output $V_o = V_i$

Transfer characteristics of the circuit

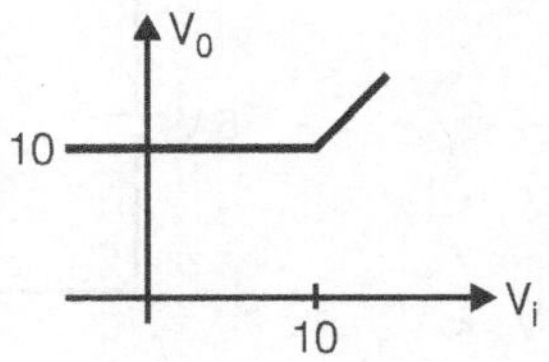

**Q.7** Assume that $D_1$ and $D_2$ in figure are ideal diodes. The value of the current I is

(a) 0 mA

(b) 0.5 mA

(c) 1 mA

(d) 2 mA

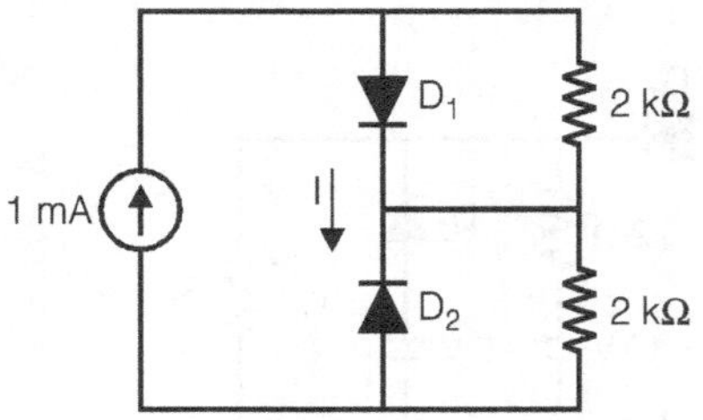


**Exp. (a)**

Diode $D_2$ is RB

$\therefore \qquad I = 0$

**Q.8** The current through the Zener diode in figure is

(a) 33 mA

(b) 3.3 mA

(c) 2 mA

(d) 0 mA

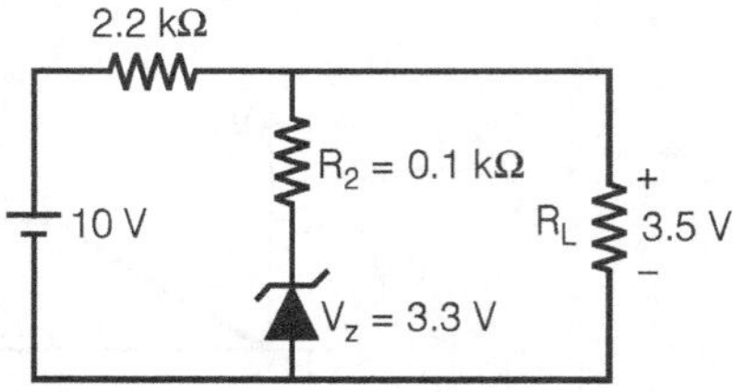


**Exp. (c)**

Given circuit

In the circuit $\quad V_1 = 3.5$ V

Current through Zener

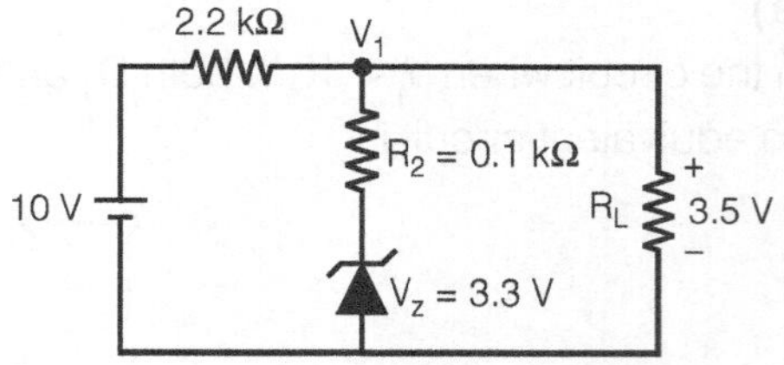


$$I_z = \frac{V_1 - V_z}{R_z} = \frac{3.5 - 3.3}{0.1 \times 10^3}$$

$$\therefore \qquad I_z = 2 \text{ mA}$$

**Q.9** Assuming that the diodes are ideal in figure the current in diode $D_1$ is

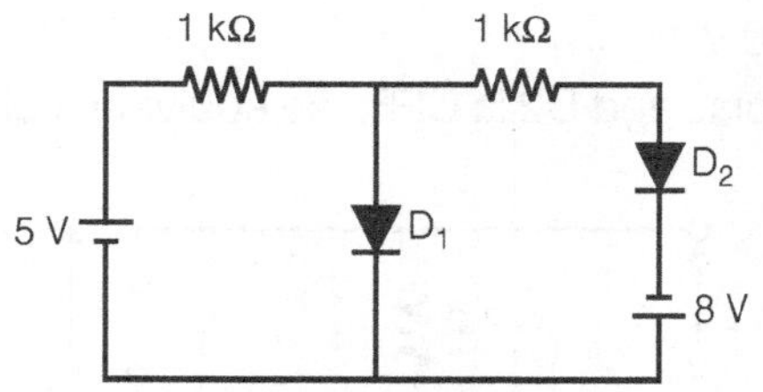


(a) 9 mA

(b) 5 mA

(c) 0 mA

(d) –3 mA

**Exp. (c)**

Assume $D_1$ is RB, $D_2$ is FB

Apply node equation

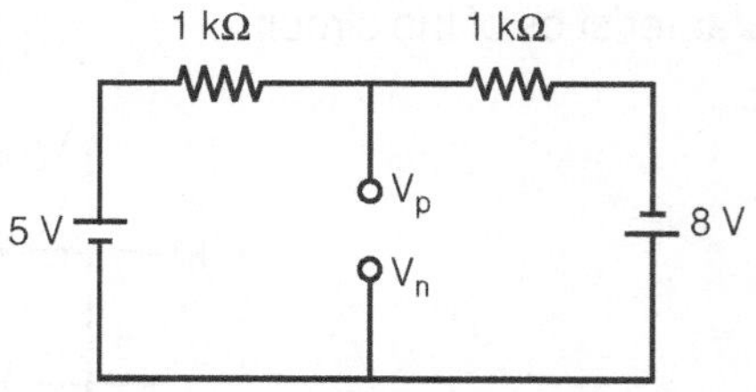


$$\frac{V_p - 5}{1} + \frac{V_p + 8}{1} = 0$$

$$2V_p = -3$$

$$V_p = -1.5$$

$$V_n = 0 \quad ; \quad V_p < V_n$$

$$I_{D_1} = 0 \text{ mA}$$

**Q.10** A voltage signal 10 sin ωt is applied to the circuit with ideal diodes as shown in figure. The maximum and minimum values of the waveform $V_{out}$ of the circuit are respectively.

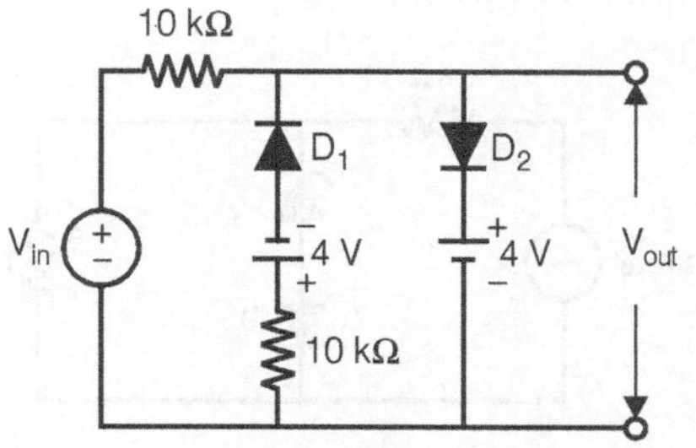


(a) +10 V and –10 V (b) +4 V and –4 V
(c) +7 V and – 4 V (d) +4 V and –7 V

Exp. (d)

In positive half cycle when ($V_{in}$ > 4 V) diode $D_2$ conducts and $D_1$ OFF so the equivalent circuit is

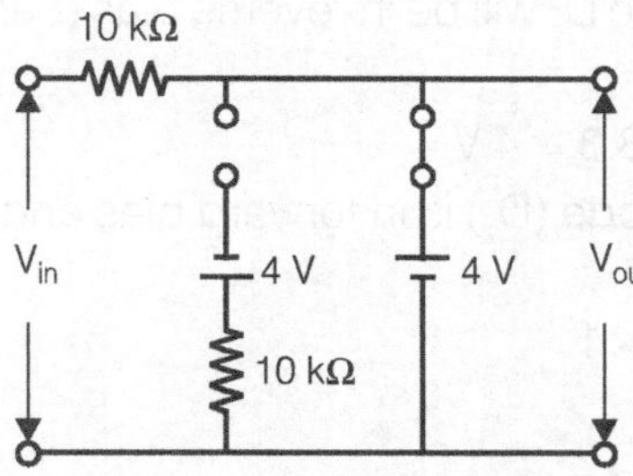


∴ $V_o = +4\text{ V}$

In negative half cycle diode $D_1$ conducts and $D_2$ will be OFF so that

Apply kVL

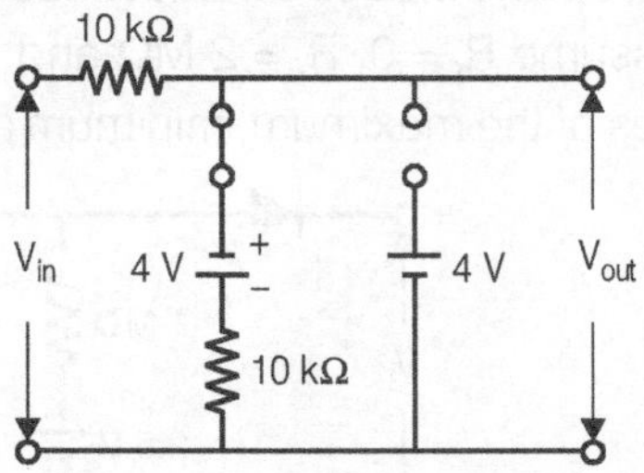


$$V_{in} - 10I + 4 - 10I = 0$$

$$\frac{V_{in} + 4}{20} = I$$

$$V_{in} = -10\text{ V}$$

(Maximum value in negative half cycle)

$$I = \frac{-10+4}{20}$$

$$= \frac{-3}{10}\text{ mA}$$

$$\frac{V_{in} - V_{out}}{10} = I$$

$$\frac{-10 - V_{out}}{10} = \frac{-3}{10}$$

$$V_{out} = -(10 - 3) = -7\text{ V}$$

**Q.11** The cut in voltage of both Zener diode $D_z$ and diode D shown in figure is 0.7 V, while breakdown voltage of $D_z$ is 3.3 V and reverse break down voltage of D is 50 V. The other parameters can be assumed to be the same as those of an ideal diode. The values of the peak output voltage ($V_o$) are

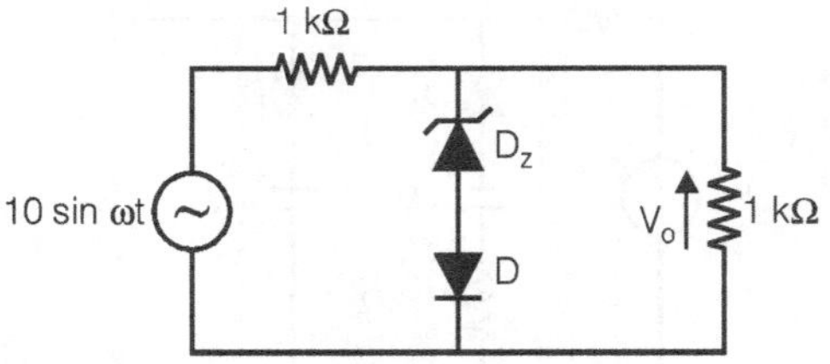


(a) 3.3 V in the positive half cycle and 1.4 V in the negative half cycle
(b) 4 V in the positive half cycle and 5 V in negative half cycle
(c) 3.3 V in both positive and negative half cycles
(d) 4 V in both positive and negative half cycles

**Exp.** (b)

In the positive half cycle Zener diode $D_z$ will be in reverse bias (behaves or a constant voltage source) and diode D is forward bias.

Output $V_o = 0.7 + 3.3 = 4$ V

In the negative half cycle, Zener diode ($D_z$) is in forward bias and diode (D) is in reverse bias mode so

$$V_o = \frac{10}{(1+1)} \times 1$$

$$V_o = 5 \text{ V}$$

∴ 4 V in the positive half cycle and 5 V in the negative half cycle.

**Q.12** A symmetrical 5 kHz square wave whose output varies between +10 V and –10 V is impressed upon the clipping circuit shown. Assume $R_f = 0$, $R_r = 2$ MΩ and $V_\gamma = 0$. Sketch the steady state output waveform indicating numerical values of the maximum, minimum and constant portions

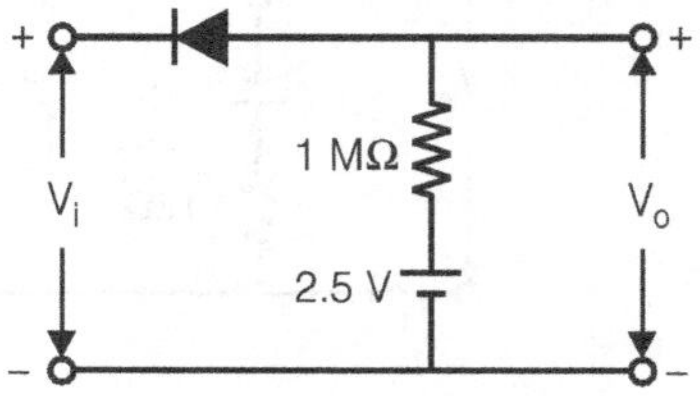


**Solution:**

Here $R_f = 0, \quad R_r = 2$ MΩ

and $V_\gamma = 0$

The diode will conduct only when

$V_i < 2.5$ V

and when $V_i > 2.5$ V diode is open

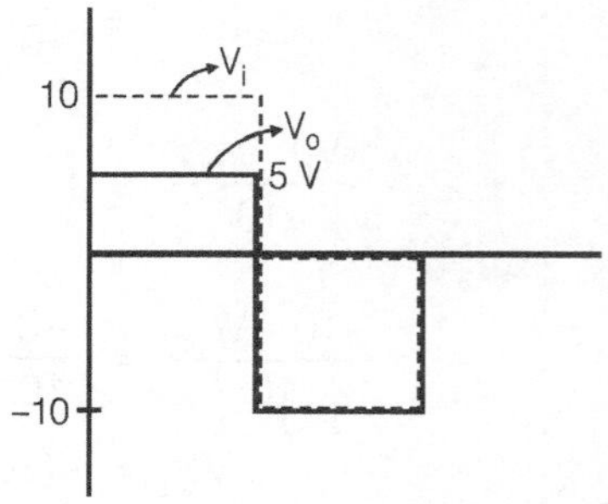


$$V_o = 2.5 + \left(\frac{V_i - 2.5}{3\text{ M}\Omega}\right) \times 1\text{M}\Omega$$

When $V_i = 10$ V

$$V_o = 2.5 + \frac{10 - 2.5}{3\text{M}\Omega} = 2.5 + 2.5 = 5\text{ V}$$

**Q.13** In the following limiter circuit, an input voltage $V_i = 10 \sin 100\pi t$ is applied. Assume that the diode drop is 0.7 V. When it is forward biased. The Zener breakdown voltage is 6.8 V. The maximum and minimum values of the output voltage respectively are

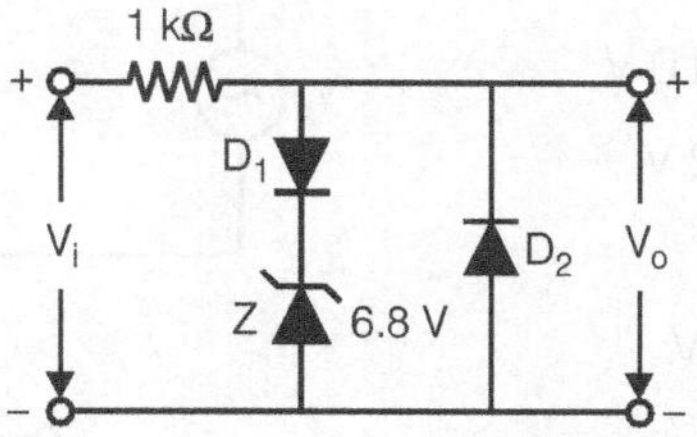


(a) 6.1 V, –0.7 V
(b) 0.7 V, –7.5 V
(c) 7.5 V, –0.7 V
(d) 7.5 V, –7.5 V

**Exp. (c)**

For the positive half of $V_i$, the diode $D_1$ is forward bias, $D_2$ is reverse bias and the Zener diode is in breakdown stage $V_i > 6.8$ V.

The output voltage is

$$V_o = 0.7 + 6.8 = 7.5 \text{ V}$$

For the negative half of $V_i$, $D_2$ is forward bias thus

$$V_o = -0.7 \text{ V}$$

**Q.14** Select the correct output ($V_o$) wave-shape for a given input $V_i$ in the clamping network given below:

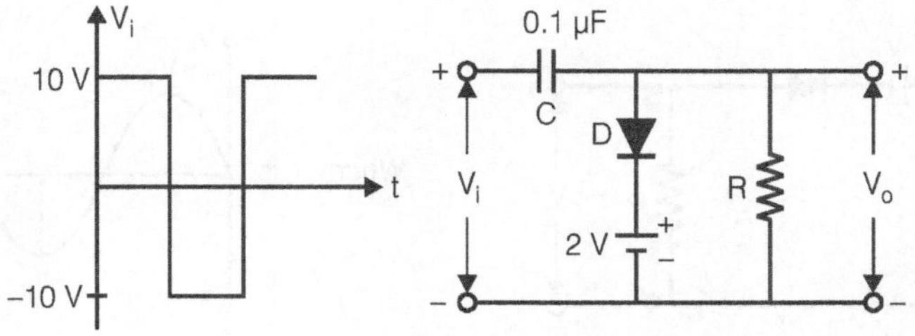


(a)
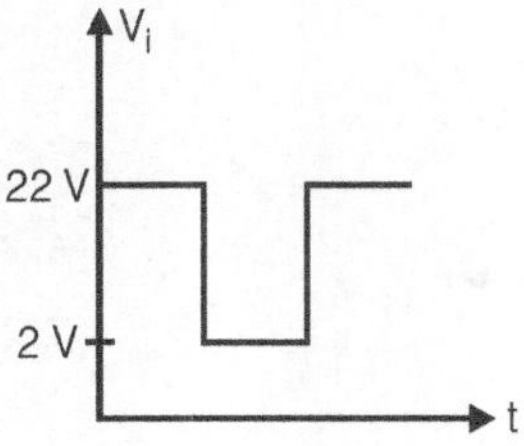


(b)
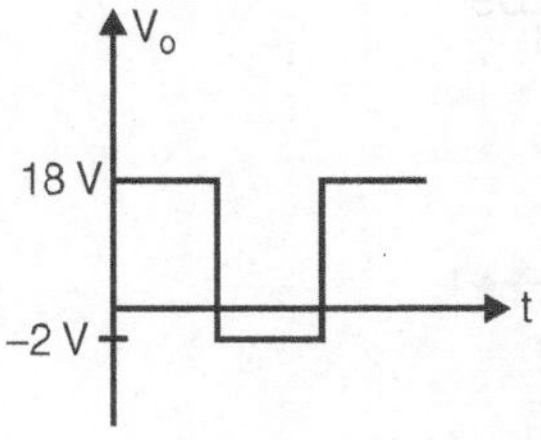


(c)
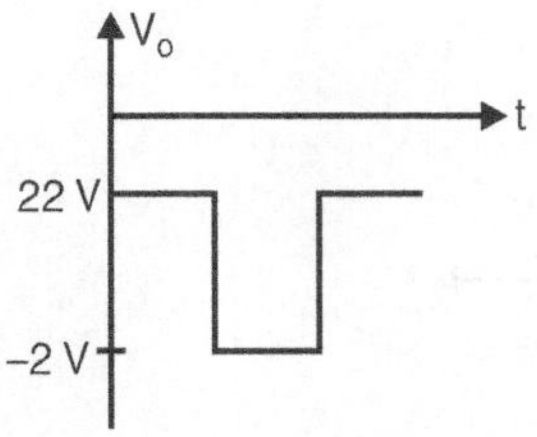


(d)
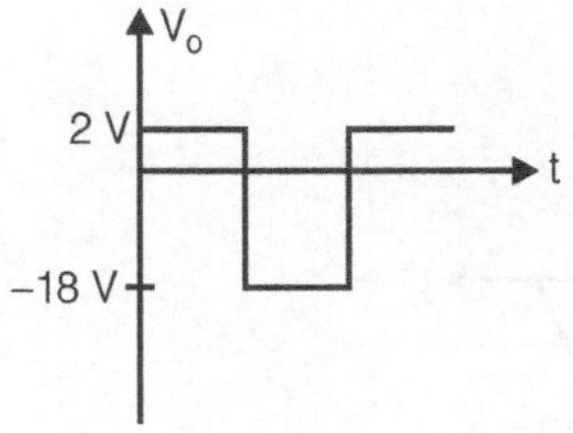


**Exp. (d)**

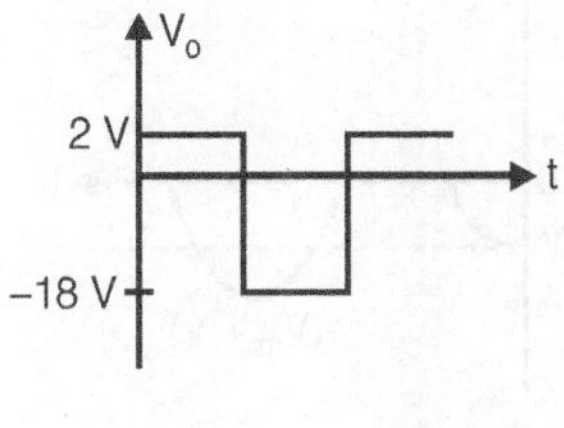

**Q.15** For an input of $V_s = 5 \sin \omega t$, (assuming ideal diode), circuit shown in the figure will behave as a

(a) clipper, sinewave clipped at –2 V

(b) clamper, sinewave clamped at –2 V

(c) clamper, sinewave clamped at 0 V

(d) clipper, sine wave clipped at 2 V

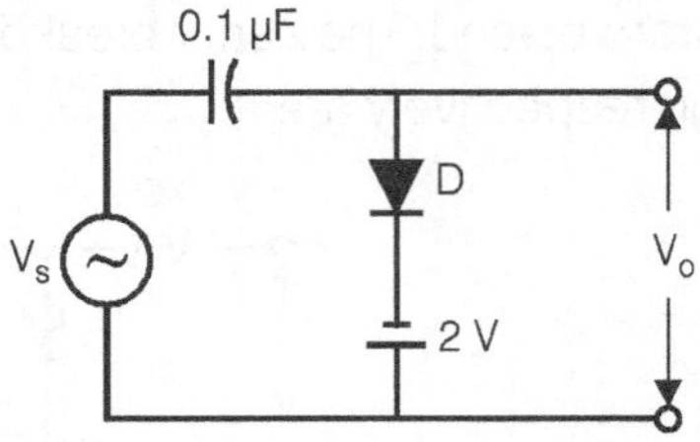


**Exp.** **(b)**

clamper, sinewave clamped at –2 V.

**16.** Shown below is the circuit consisting of ideal diodes at time t = 0 sec, $\frac{1}{200}$ sec, $\frac{1}{100}$ sec, $\frac{1}{50}$ sec, $V_i$ is 0, +5 V, 0, –5 V respectively, $V_o$ at the same timings shall be

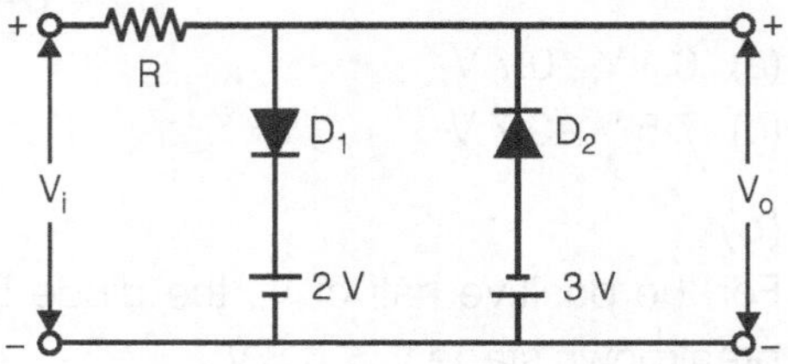


(a) 0, +5 V, 0, –5 V
(b) +2 V, +2 V, –3 V, –3 V
(c) +2 V, –3 V, +2 V, –3 V
(d) 0, +2 V, 0, –3 V

**Exp.** **(d)**

It is parallel clipper which clips the voltage at +2 V, –3 V respectively.

**17.** Consider the following ideal diode in the circuit

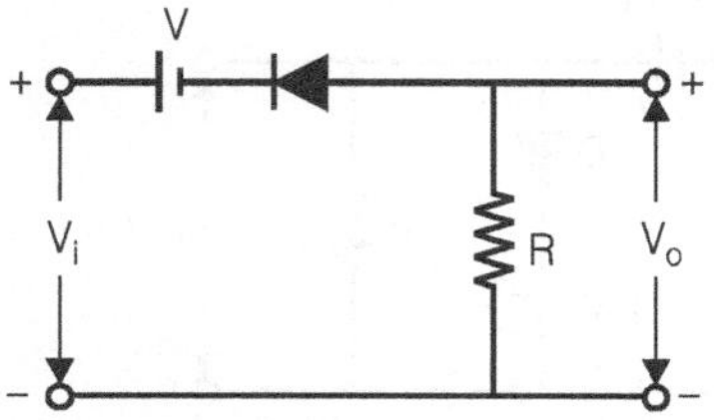


Where $V_i$ is

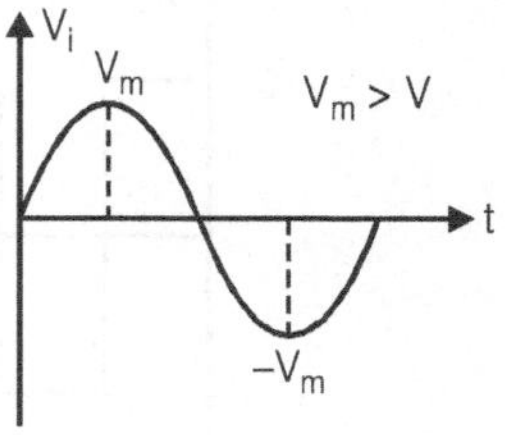


Corresponding $V_o$ can be

(a) 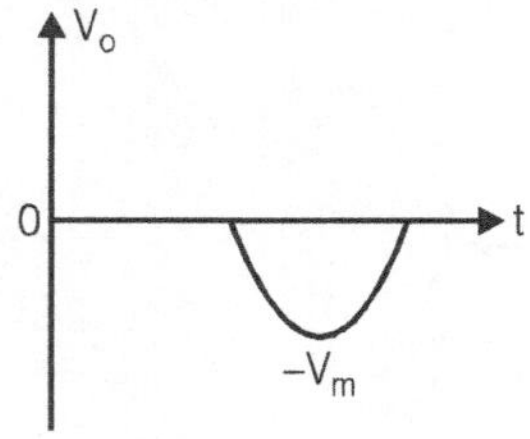


(b) 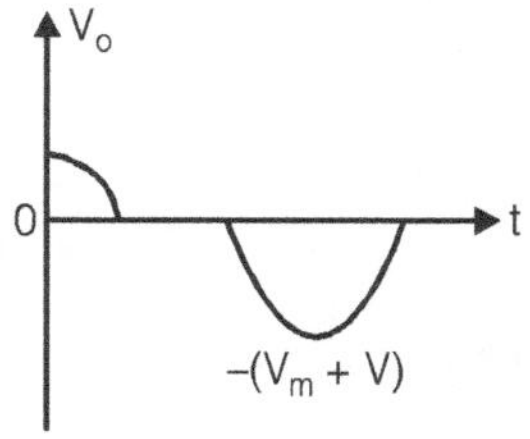


(c) 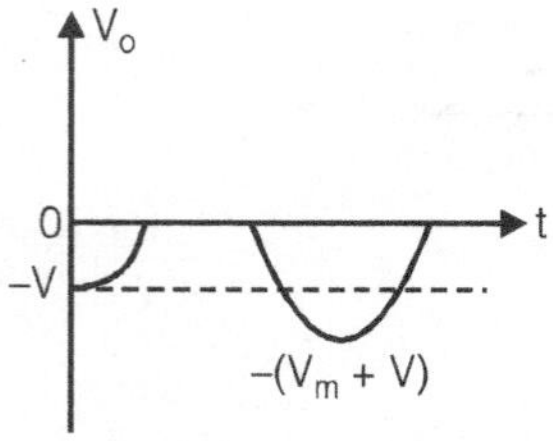


(d) 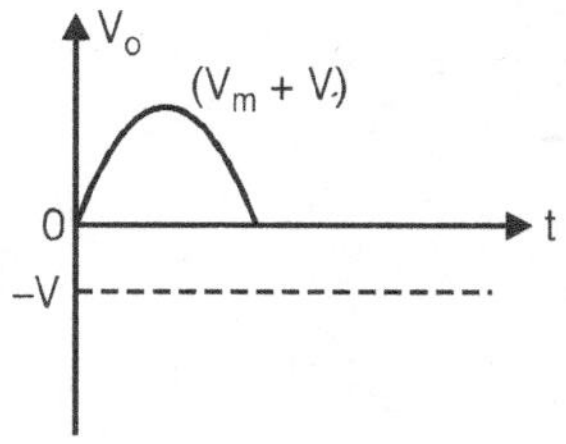


**Exp.** **(c)**

In positive half cycle D is RB.

when $V_i = 0, \quad V_o = 0 - V = -V$

when $V_i = -V_m, \quad V_o = -V_m - V = -(V_m + V)$

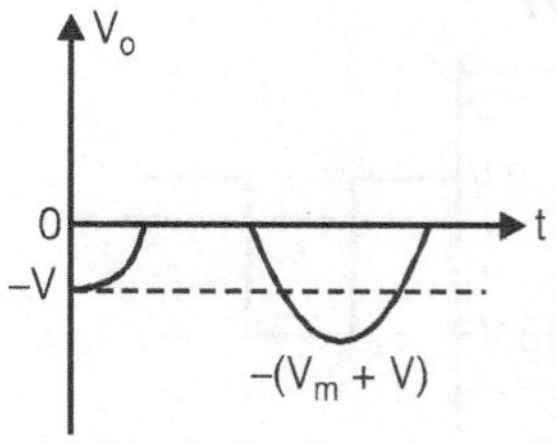

18. Consider the following limiter circuit an input voltage $V_i$ = 10.5 sin 100πt is applied. Assume that the diode drop is 0.7 V forward bias. The zener breakdown voltage is 6.3 V

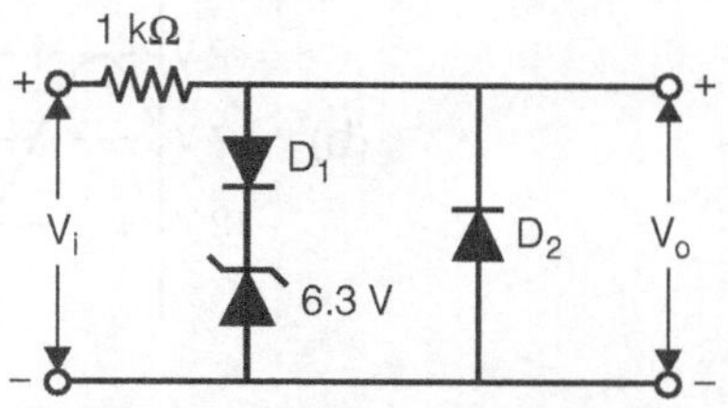


The maximum and minimum values of the output voltage respectively are

(a) 3 V, –0.7 V
(b) 0.7 V, –3 V
(c) 7 V, 3 V
(d) 7 V, –0.7 V

Exp. (d)

For positive value of $V_i$, $D_1$ is FB and Zener diode regulates the voltage at 6.3 V across it.

$$V_o = 6.3 + 0.7 = 7 \text{ V}$$

and for negative half of $V_i$,

$D_2$ is FB $\quad V_o = -0.7$ V

19. Shown below is the circuit with ideal diode

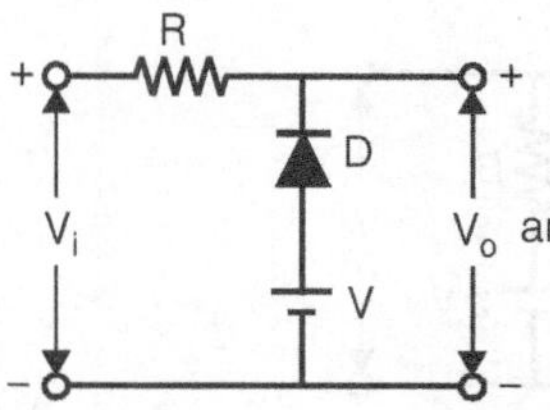


and $V_i$ is

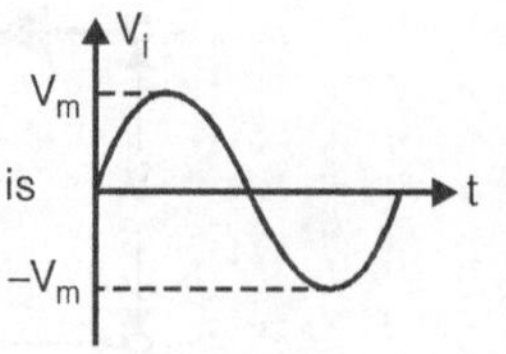


The output voltage $V_o$ is

(a)

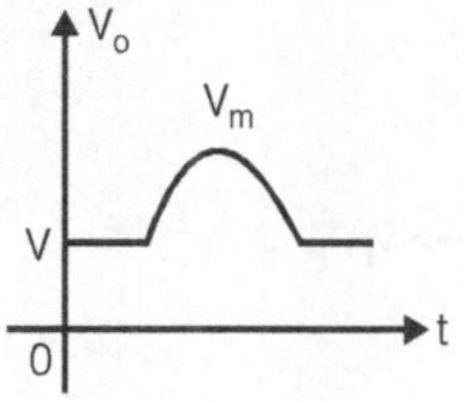


(b)

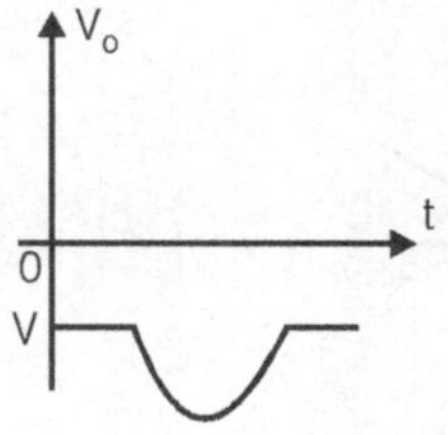


(c)

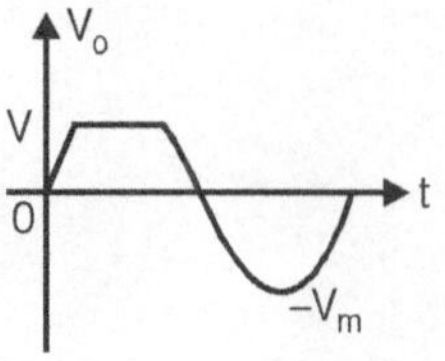


(d)

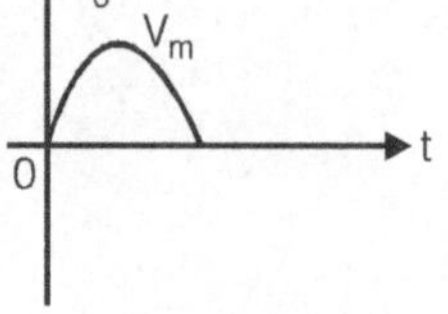


Exp. (a)

This a clipper circuit when diode is in upward direction its clips above the reference voltage.

20. Suppose an input sinusoidal waveform with ±5 V maximum and minimum position is applied to the circuit below:

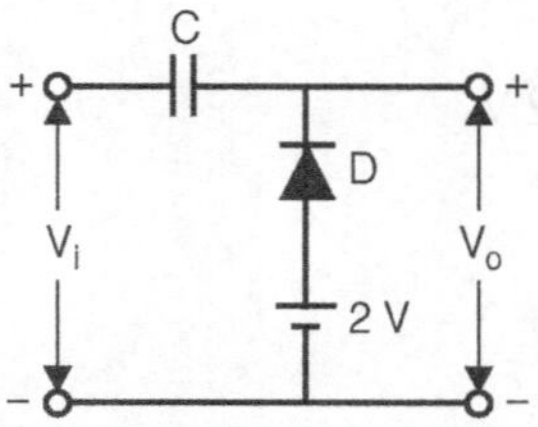

The output voltage waveform is

(a) 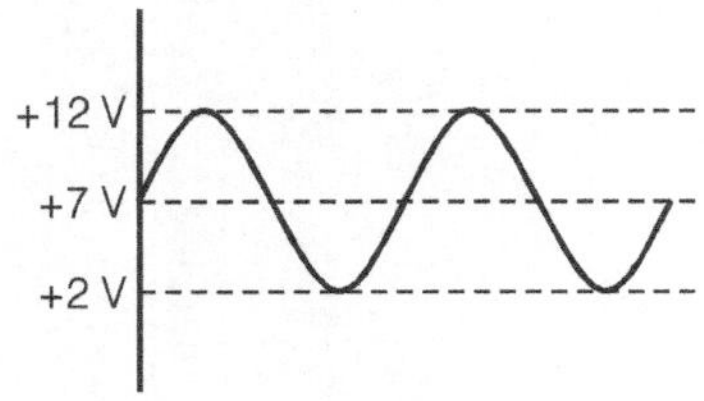


(b) 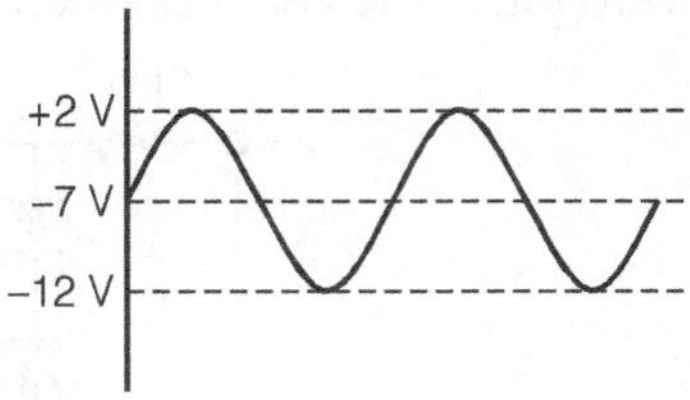


(c) 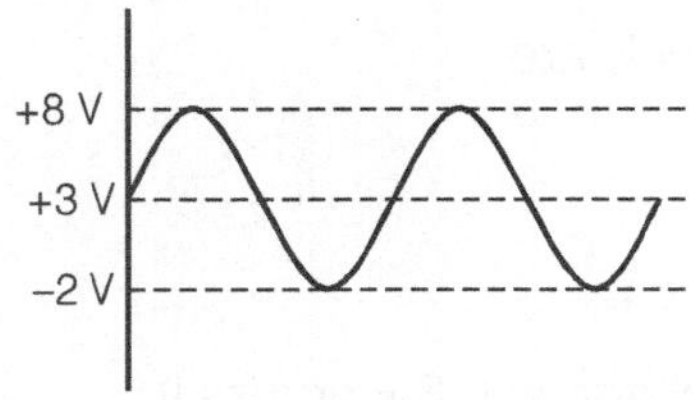


(d) 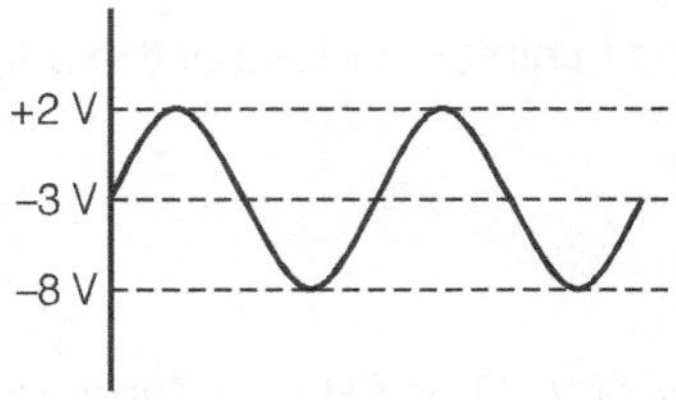


**Exp.** **(a)**

It is a positive clamper, it clamps the signal above 2 V.

**21.** Consider the circuit below (ideal diode)

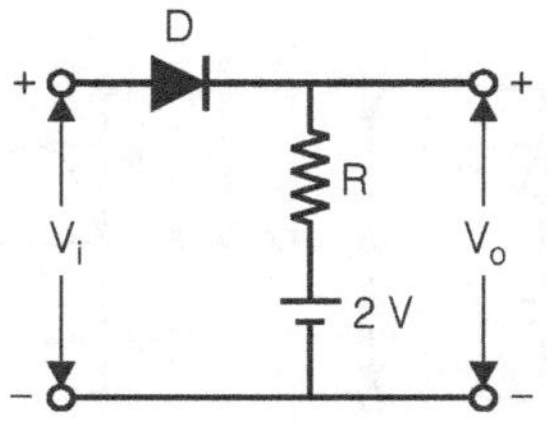


(a) 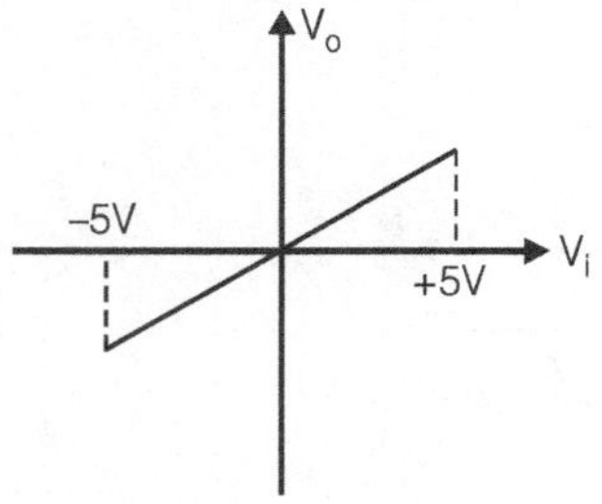


(b) 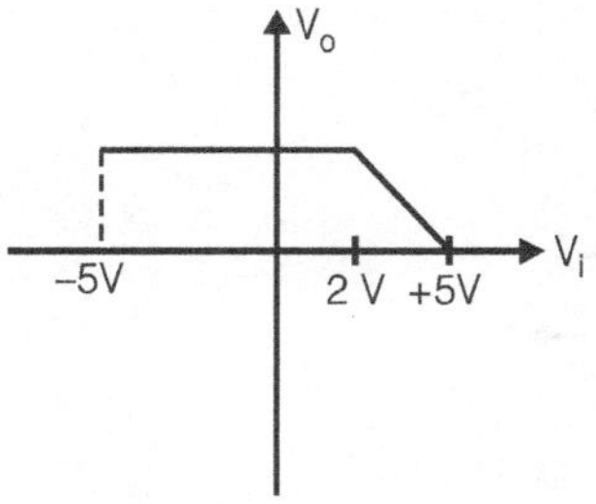


(c) 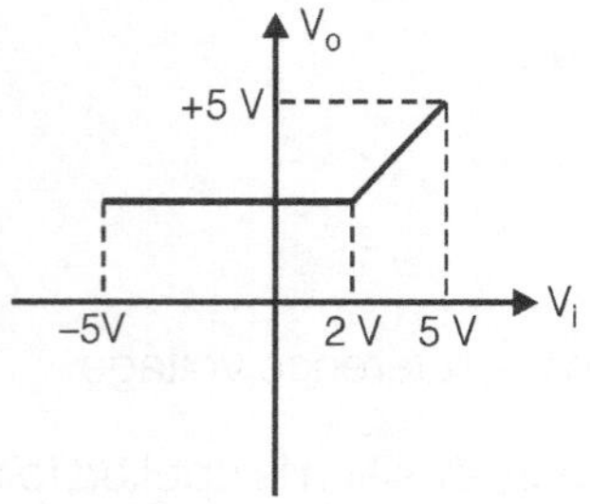


(d) 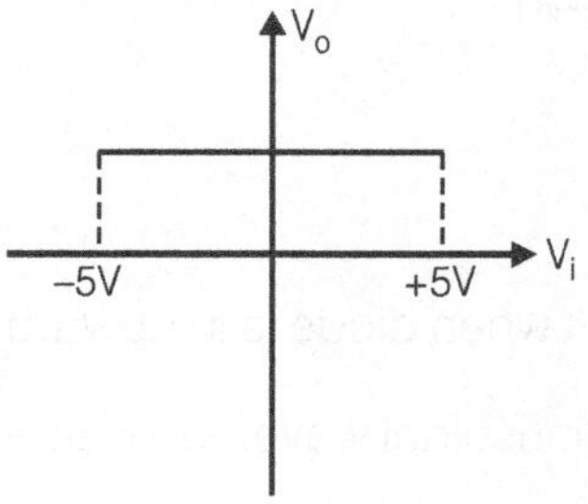


**Exp.** **(c)**

Till $V_i$ = 2 V, diode is RB, Hence $V_o$ = 2 V.

After 2 V to 5 V, diode is FB

$V_o$ follows $V_i$.

**22.** Consider the following circuit:

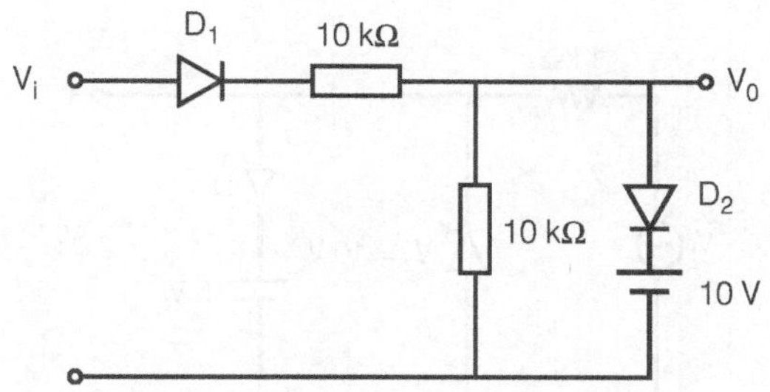


For the circuit shown above, which one of the following is correct statement?

(a) $D_2$ does not conduct for any value of $V_i$
(b) $V_0$ = 10 V for all values of $V_i$ > 10 V
(c) $V_0$ = 0V for all values of $V_i$ < 0 V
(d) $V_0$ = 10 V for all values of $V_i$ > 0V

**Exp. (c)**

For $V_i < 0$, diode $D_1$ is RB

$$V_o = 0 \quad \text{For } V_i < 0$$

For $0 < V_i < 10$, $D_1$ is FB, $D_2$ is RB

$$V_o = \frac{10}{10+10}V_i = \frac{V_i}{2}$$

$V_i > 20$ V, both diodes $D_1$ and $D_2$ ON

$$V_o = 10\text{ V}$$

**23.** In the voltage regulator shown in the figure, the load current can vary from 100 mA to 500 mA. Assuming that the zener diode is ideal (i.e. the zener knee current is negligible small and zener resistance is zero in the breakdown region). The value of R is

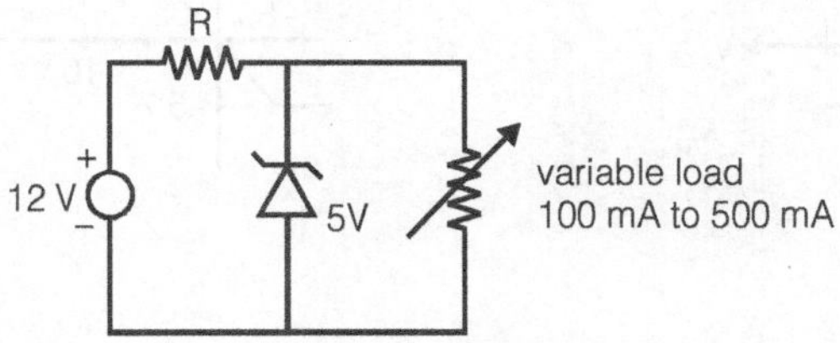


(a) 7 Ω
(b) 70 Ω
(c) 70/3 Ω
(d) 14 Ω

**Exp. (d)**

$$\frac{V_{in} - V_z}{R} \geq I_L + I_Z$$

| $I_L = 100$ mA | $I_L = 500$ mA |
|---|---|
| $\frac{12-5}{R} \geq I_L$ | $\frac{12-5}{R} \geq I_L$ |
| $\frac{12-5}{R} \geq 100$ mA | $\frac{12-5}{R} \geq 500$ mA |
| $R \leq 70\Omega$ | $R \leq 14\ \Omega$ |

The best suitable value for $R \leq 14\ \Omega$.

**24.** A clipper circuit is shown below:

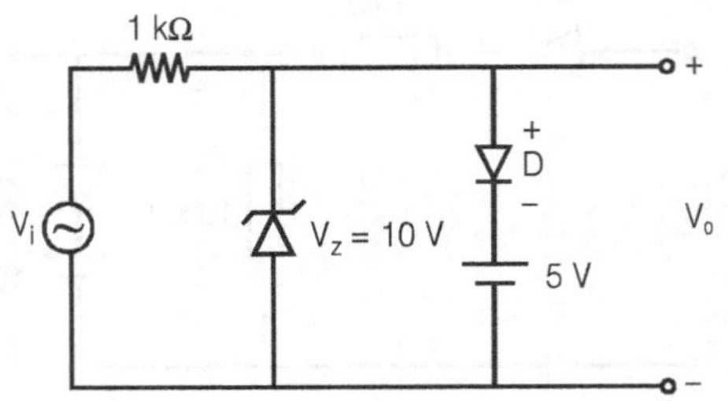


Assuming forward voltage drops of the diodes to be 0.7 V, the input-output transfer characteristics of the circuit is

(a)

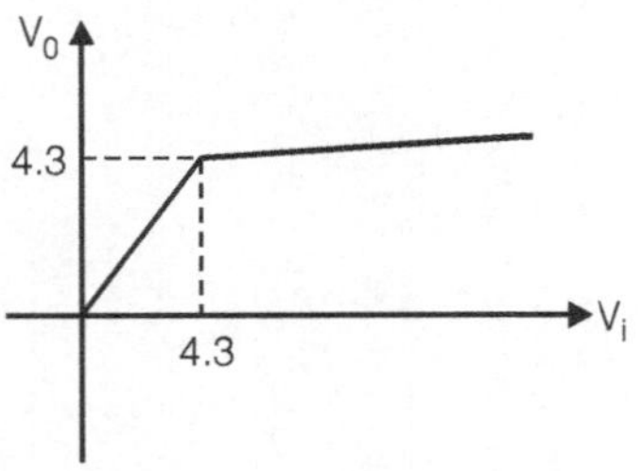


(b)

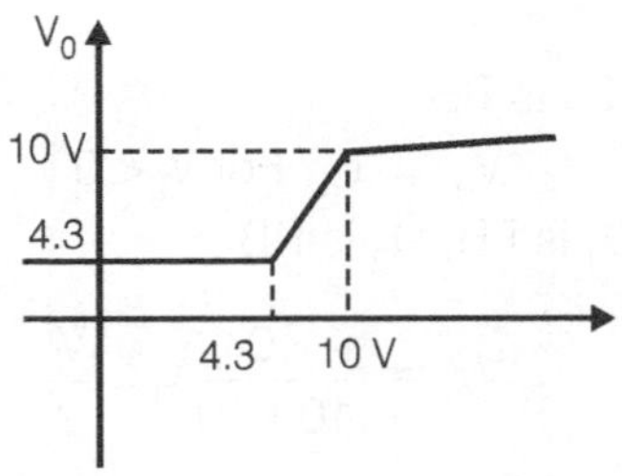


(c)

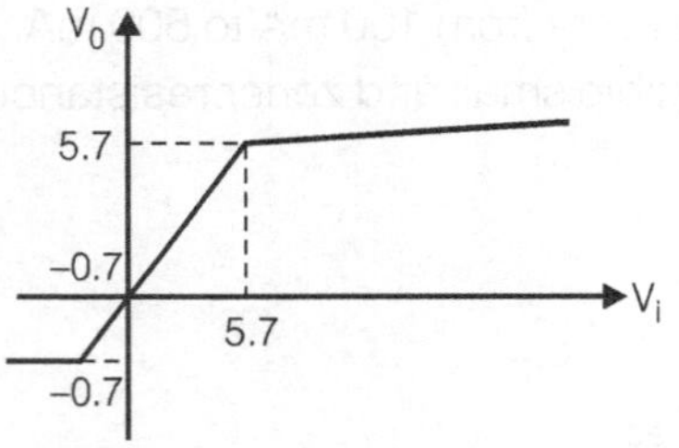


(d)

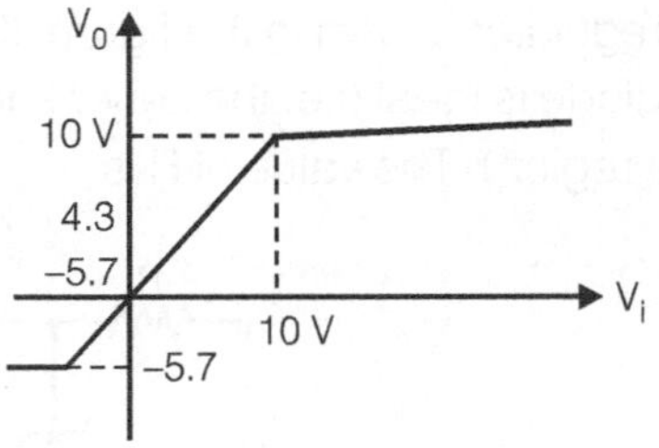


**Exp. (c)**

For positive voltage the waveform clips at +5 +0.7 = 5.7 V.

For negative voltage, at –0.7 V, because Zener conducts.

OOOO

# CHAPTER 8

# RECTIFIERS, FILTERS AND VOLTAGE REGULATORS

## About This Chapter

For the operation of most of the electronic devices and circuits, a d.c. source is required. It is advantageous to convert a.c. supply into d.c. voltage. The process of converting a.c. voltage into d.c. voltage is called as rectification. This is achieved with rectifiers and filters. Simple rectifier circuits do not provide a perfect d.c. Actually, they give unidirectional current which does not have a constant value rather it is pulsating in nature. In some applications such as unidirectional (but not perfect d.c.) voltage may be acceptable. Where more perfect d.c. is needed, filters are used after the rectifier. Thus a transformer, a rectifier and a suitable filter can jointly act to give the required d.c. voltage from a.c. supply lines. However it does not ensure a dc voltage of constant magnitude. So, to maintain constant dc voltage at the output voltage regulators are used.

# 8 RECTIFIERS, FILTERS AND VOLTAGE REGULATORS

## 8.1 Introduction

For most of the electronic circuits, the required dc voltage are in the range of a few volts to fe tens of volts. Hence it becomes necessary to first step down the ac voltage from the 230 V mains supply.

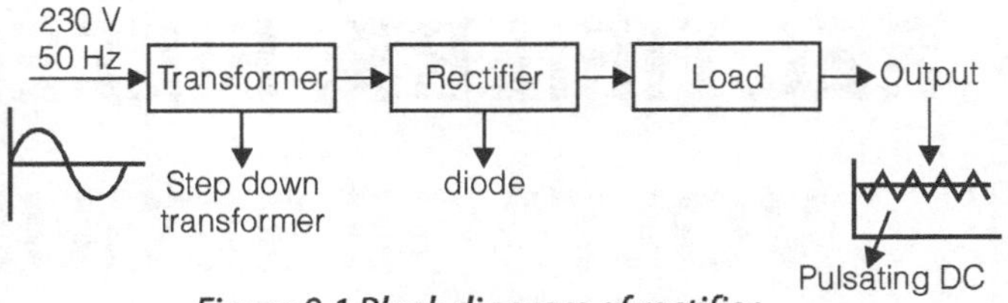


*Figure 8.1 Block diagram of rectifier*

In almost all the circuits, we shall include a transformer (which could be either a step-down transformer for low voltage dc). A block diagram Figure. (8.1) shows generalised arrangement consisting of

- Transformer
- Rectifier
- Load

"Rectifier is a device which converts the sinusoidal ac voltage into pulsating DC".

## 8.2 Rectifier Parameters

1. Average current $I_{dc}$
2. DC output voltage $V_{dc}$
3. RMS load current $I_{rms}$
4. Ripple factor
5. Voltage regulation
6. Rectification efficiency (η)
7. Transformer utilization factor (TUF.)
8. Peak inverse voltage
9. Form factor
10. Peak factor

**Practical Rectifier Circuits are:**

(i) Half wave rectifier (ii) Full wave rectifier with centre tap (iii) Bridge rectifier

# 8.3 Half Wave Rectifier

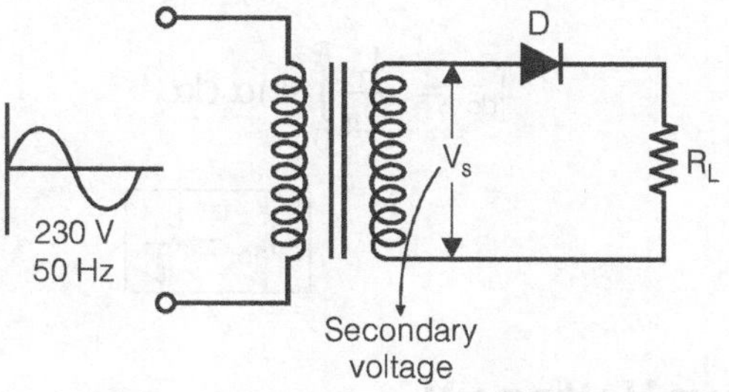


*Figure 8.2 Half-wave rectifier*

$V_s = V_m \sin\omega t \quad 0 < \omega t < 2\pi$

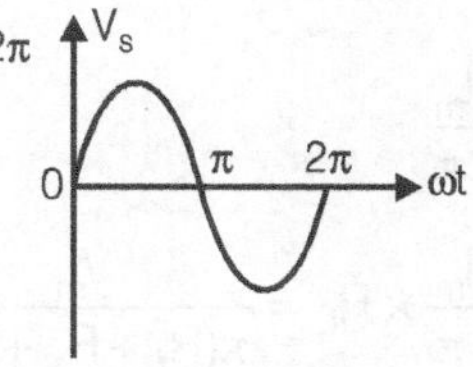


$$I_L = I_m \sin\omega t \qquad 0 < \omega t < \pi$$
$$= 0 \qquad \pi < \omega t < 2\pi$$

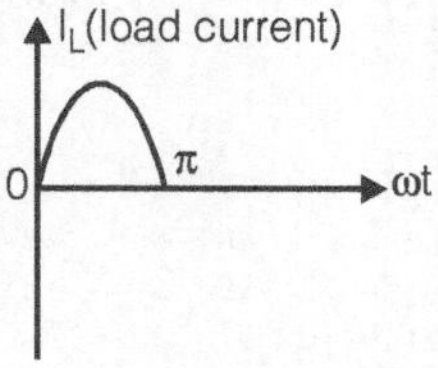


## CIRCUIT

The half wave rectifier circuit using a semiconductor diode with a load resistance $R_L$ but no smoothing filter. The diode is connected in series with the secondary of the transformer and the load resistance $R_L$, the primary of the transformer is being connected to the ac supply mains.

## WORKING

- The ac voltage across the secondary winding changes polarity after every half cycle.
- During the positive half cycles of the input ac voltage i.e. when upper end of the secondary winding is positive w.r.t. its lower end. The diode is forward biased and therefore, conducts current. If the forward resistance of the diode is assumed to be zero (in practice however, a small resistance exists).
  The input voltage during the positive half cycles is directly applied to the load resistance $R_L$, making its upper end positive w.r.t. its lower end. The waveforms of the output current and output voltage are of the same shape as that of the input ac voltage.
- During the negative half cycles of the input ac voltage i.e. when the lower end of the secondary winding is positive w.r.t. its upper end, the diode is reverse biased and so does not conduct. Thus during the negative half cycles of the input ac voltage the current through and voltage across the load remains zero. Thus for negative half cycles no power is delivered to the load.

### (i) Average Current ($I_{dc}$)

$$I_{dc} = \frac{\text{area under the curve}}{2\pi} = \frac{1}{2\pi}\int_0^{2\pi} I\,d\alpha \qquad (\therefore I = I_m \sin\alpha)$$

$$= \frac{I_m}{2\pi}\int_0^{2\pi} \sin\alpha\, d\alpha$$

For half-wave rectifier

$$I = I_m \sin\alpha \qquad 0 < \omega t < \pi \qquad (\therefore \omega t = \alpha)$$
$$= 0 \qquad \pi < \omega t < 2\pi$$

$$I_{dc} = \frac{I_m}{2\pi}\int_0^{\pi} \sin\alpha \, d\alpha$$

$$\boxed{I_{dc} = \frac{I_m}{\pi}}$$

## (ii) DC Output Voltage $V_{dc}$

$$V_{dc} = I_{dc} \times R_L$$

From HWR

$$I_{dc} = \frac{I_m}{\pi}$$

$$V_{dc} = \frac{I_m}{\pi} \times R_L = \frac{V_m}{\pi(R_f + R_s + R_L)} \times R_L$$

Where

$$I_m = \frac{V_m}{R_f + R_s + R_L}$$

$R_f \rightarrow$ Forward resistance of diode

$R_s \rightarrow$ Secondary resistance

$R_L \rightarrow$ Load resistance

$$V_{dc} = \frac{V_m}{\pi} \cdot \frac{R_L}{R_f + R_s + R_L}$$

$$V_{dc} = \frac{V_m}{\pi} \cdot \frac{1}{\frac{R_f + R_s}{R_L} + 1}$$

$$\boxed{(V_{dc})_{NL} \text{ (No load)} = \frac{V_m}{\pi}} \quad \text{Where } R_L \rightarrow \infty$$

## (iii) RMS Load Current $I_{rms}$

$$I_{rms} = \left[\frac{1}{2\pi}\int_0^{2\pi} I^2 \, d\alpha\right]^{1/2} = \left[\frac{I_m^2}{2\pi}\int_0^{2\pi} \sin^2\alpha \, d\alpha\right]^{1/2}$$

In HWR

$$I_{rms} = \left[\frac{I_m^2}{2\pi}\int_0^{\pi} \sin^2\alpha \, d\alpha\right]^{1/2}$$

$$\boxed{I_{rms} = \frac{I_m}{2}}$$

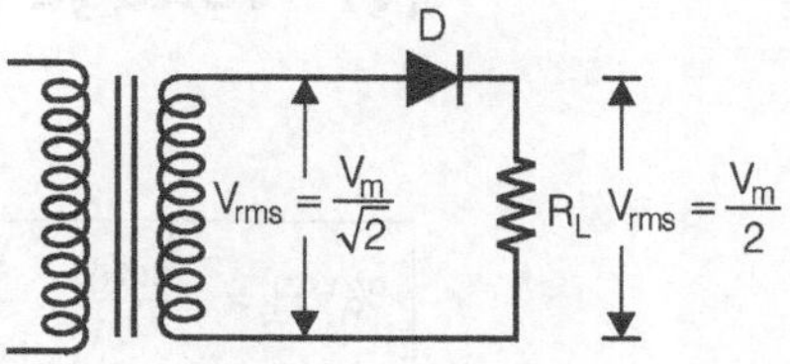


$$V_{rms} = \frac{V_m}{\sqrt{2}}$$

(RMS voltage calculated across secondary of transformer)

$$V_{rms} = \frac{V_m}{2}$$

(rms voltage across the load)

## (iv) Ripple Factor

$$\text{Ripple factor} = \frac{\text{rms value of alternating component}}{\text{Average value}} = \frac{I'_{rms}}{I_{dc}}$$

Assume

$I' \rightarrow$ alternating current

$I_{dc} \rightarrow$ DC current

$I \rightarrow$ Total current

$I'_{rms} \rightarrow$ rms value of alternating current

$\therefore$

$$\text{ripple factor} = \frac{I'_{rms}}{I_{dc}}$$

**Proof:**

$$I'_{rms} = \left[\frac{1}{2\pi}\int_0^{2\pi} I'^2 \, d\alpha\right]^{1/2} = \left[\frac{1}{2\pi}\int_0^{2\pi} (I - I_{dc})^2 \, d\alpha\right]^{1/2}$$

Where

$$I = I_{dc} + I'$$

$$I'^2_{rms} = \frac{1}{2\pi}\int_0^{2\pi} I^2 \, d\alpha - 2I_{dc}\frac{1}{2\pi}\int_0^{2\pi} I \, d\alpha + \frac{I_{dc}^2}{2\pi}\int_0^{2\pi} d\alpha$$

$$I'^2_{rms} = I^2_{rms} - 2I^2_{dc} + I^2_{dc}$$

$$I'_{rms} = \sqrt{I^2_{rms} - I^2_{dc}}$$

$$\boxed{r.f = \sqrt{\left(\frac{I_{rms}}{I_{dc}}\right)^2 - 1}}$$

For half-wave rectifier

$$I_{rms} = \frac{I_m}{2}$$

$$I_{dc} = \frac{I_m}{\pi}$$

$$r.f = \sqrt{\left(\frac{I_m/2}{I_m/\pi}\right)^2 - 1} = \sqrt{\left(\frac{\pi}{2}\right)^2 - 1} = 1.21$$

It is disadvantage i.e 121% of ac component present in dc value.

## (v) Voltage Regulation

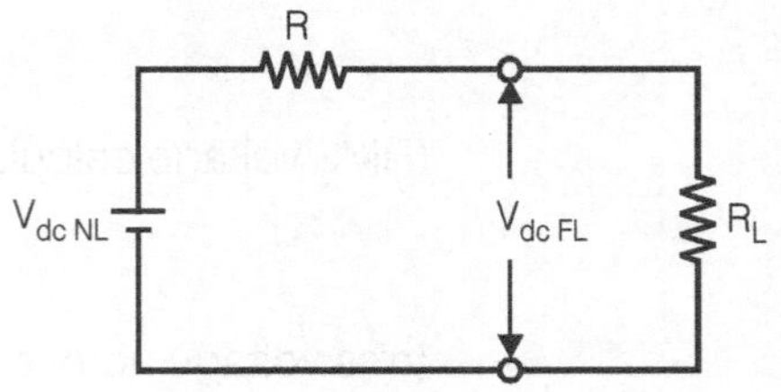


$$\%VR = \frac{V_{dcNL} - V_{dcFL}}{V_{dcFL}} \times 100\%$$

Voltage regulation should be zero in ideal case practically it should be less in value.

$$I_{dc} = \frac{I_m}{\pi} = \frac{V_m}{(R_f + R_s + R_L)\pi}$$

$$I_{dc}(R_f + R_s) + \underbrace{I_{dc} R_L}_{V_{dc\,FL}} = \underbrace{V_m / \pi}_{V_{dcNL}}$$

$$\%VR = \frac{I_{dc}(R_f + R_s)}{\frac{V_m}{\pi} - I_{dc}(R_f + R_s)}$$

## (vi) Rectification Efficiency

Rectification efficiency is represented by $\eta$

$$\eta = \frac{\text{DC power delivered to load}}{\text{AC input power}} = \frac{I_{dc}^2 R_L}{I_{rms}^2 (R_f + R_s + R_L)}$$

From HWR

$$\eta = \frac{\left(\frac{I_m}{\pi}\right)^2 R_L}{\left(\frac{I_m}{2}\right)^2 (R_f + R_s + R_L)}$$

$$\eta = \frac{4}{\pi^2} \frac{1}{\frac{R_f + R_s}{R_L} + 1}$$

If $R_L \to \infty$

$$\eta = \frac{4}{\pi^2} \times 100\% = 40.6\%$$

## (vii) Transformer Utilization Factor (TUF)

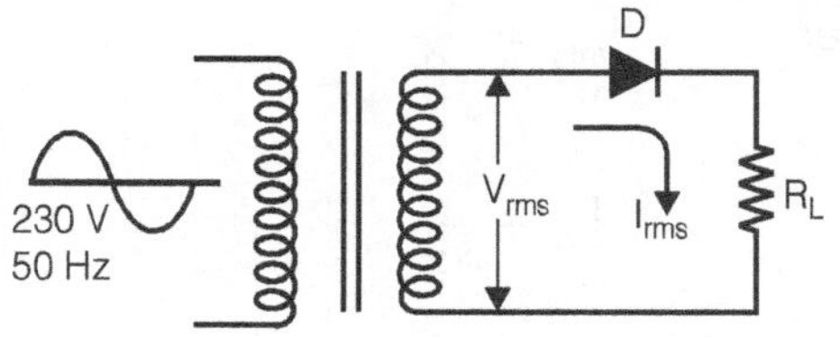


$$TUF = \frac{\text{DC power delivered to load}}{P_{ac}\text{ rated secondary}} = \frac{P_{DC}}{P_{ac}\text{ rated secondary}}$$

$$= \frac{I_{dc}^2 R_L}{V_{rms} \times I_{rms}} = \frac{I_{dc}^2 R_L}{\frac{V_m}{\sqrt{2}} \cdot \frac{I_m}{2}}$$

From HWR
$$TUF = \frac{\left(\frac{I_m}{\pi}\right)^2 R_L}{\frac{I_m^2}{2\sqrt{2}}(R_f + R_s + R_L)} = 0.28 \qquad (R_L \to \infty)$$

Transformer utilization factor for half wave rectifier is 28%.

## (viii) Peak Inverse Voltage

PIV: It is the maximum reverse bias voltage of a diode which can withstand in a circuit.

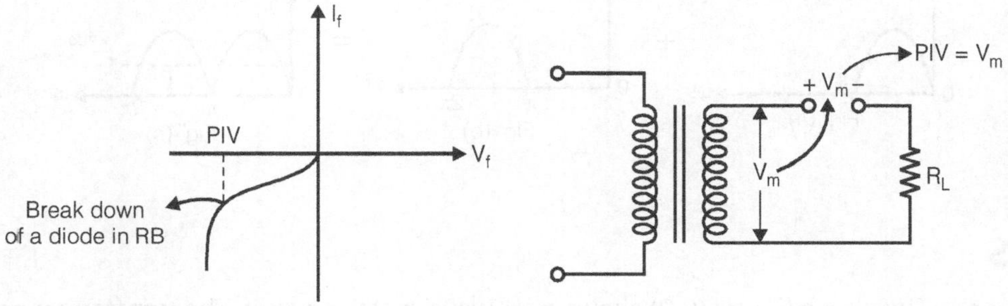


In half wave rectifier **PIV = $V_m$**

## (ix) Form Factor

$$\text{Form factor} = \frac{\text{RMS value}}{\text{Average value}}$$

$$= \frac{I_{rms}}{I_{dc}}$$

From HWR
$$\text{Form factor} = \frac{I_m/2}{I_m/\pi} = \frac{\pi}{2} = 1.57$$

## (x) Peak factor

$$\text{Peak factor} = \frac{\text{Peak value}}{\text{rms value}}$$

$$= \frac{I_m}{I_{rms}}$$

From HWR
$$\text{Peak factor} = \frac{I_m}{I_m/2} = 2$$

**Note:** Ripple frequency = f (HWR)

# 8.4 Full Wave Rectifier

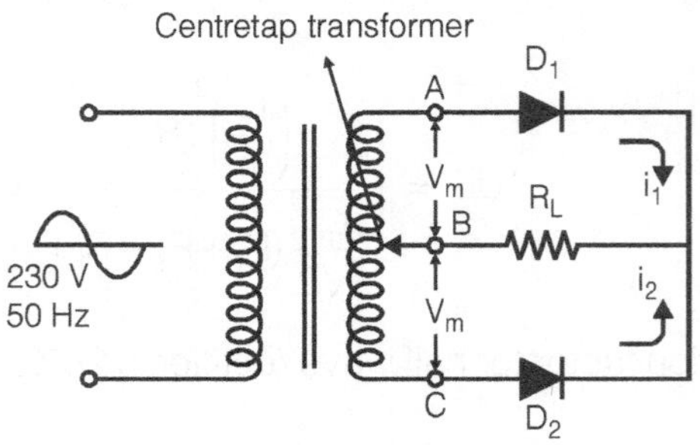


*Figure 8.3 Full-wave rectifier*

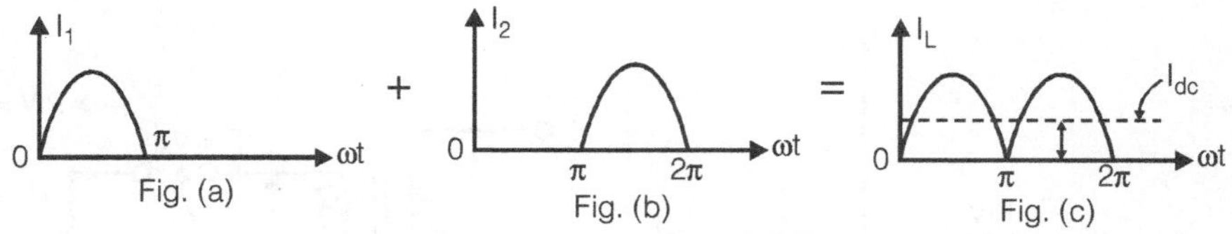


## ANALYSIS

- The above Figure (8.3) shows a full wave rectifier circuit. The transformer secondary has a centre tap and each half cycle give a peak voltage of $V_m$. In each half there is one diode i.e. $D_1$ and $D_2$. The load resistance $R_L$ is common to both halves.
- On the positive half cycle, when the point A is positive w.r.t. B, the diode $D_1$, conducts and current $I_1$ flows through $R_L$. During this half cycle, the point C is negative w.r.t. point B and hence the diode $D_2$ does not conduct. Therefore $I_2 = 0$.
- On the negative half cycle, the point C is positive w.r.t. point B. Hence the diode $D_2$ conducts and current $I_2$ flows through $R_L$. During this cycle, the point A is negative w.r.t. point B and hence the diode $D_1$ does not conduct. Therefore $I_1 = 0$.

  Fig. (a) and (b) shows the waveforms of currents $I_1$ and $I_2$. Since both $I_1$ and $I_2$ flow through the load $R_L$. The total current I through $R_L$ is $I = I_1 + I_2$ which is obtained by adding the two waveform and is shown in fig. (c).

  We shall obtain the various parameters for full wave rectifier, in a manner similar to the half wave circuit.

### (i) Average Current ($I_{dc}$)

$$I_{dc} = \frac{1}{2\pi}\int_0^{2\pi} I\,d\alpha = \frac{I_m}{2\pi}\left[\int_0^{\pi}\sin\alpha\,d\alpha + \int_{\pi}^{2\pi}\sin\alpha\,d\alpha\right]$$

In fullwave rectifier, the waveform of I in the period $\pi \le \alpha \le 2\pi$ is the same as the period $0 \le \alpha \le \pi$. Therefore the area under the curve will be twice that of the halfwave case.

$$\boxed{I_{dc} = \frac{2I_m}{\pi}}$$

## (ii) DC Output Voltage ($V_{dc}$)

$$V_{dc} = I_{dc} R_L = \frac{2I_m}{\pi} R_L = \frac{2V_m}{\pi(R_f + R_s + R_L)} R_L = \frac{\frac{2V_m}{\pi}}{\frac{R_f + R_s}{R_L} + 1}$$

If $\frac{R_f + R_s}{R_L} << 1$

$$(V_{dc})_{NL} = \frac{2V_m}{\pi}$$

It is verified that $V_{dc}$ of a full wave rectifier is double that of half wave.

## (iii) RMS Load Current ($I_{rms}$)

$$I_{rms} = \left[\frac{1}{2\pi}\int_0^{2\pi} I^2 \, d\alpha\right]^{1/2} \qquad \text{(Where } I = I_m \sin \alpha\text{)}$$

$$I_{rms} = \left[\frac{1}{2\pi}\int_0^{2\pi} I_m^2 \sin^2 \alpha \, d\alpha\right]^{1/2} = \frac{I_m}{\sqrt{2}}$$

$$I_{rms} = \frac{I_m}{\sqrt{2}}$$

Similarly

$$V_{rms} = \frac{V_m}{\sqrt{2}}$$

## (iv) Ripple Factor

$$\text{ripple factor} = \sqrt{\left(\frac{I_{rms}}{I_{dc}}\right)^2 - 1}$$

For full-wave rectifier, we have seen that

$$I_{rms} = \frac{I_m}{\sqrt{2}}$$

and

$$I_{dc} = \frac{2I_m}{\pi}$$

$$\text{ripple factor} = \sqrt{\left(\frac{\frac{I_m}{\sqrt{2}}}{\frac{2I_m}{\pi}}\right)^2 - 1} = \sqrt{\left(\frac{\pi}{2\sqrt{2}}\right)^2 - 1} = 0.482$$

The ripple content is much lower compared to the halfwave rectifier. Therefore the output of a full wave rectifier is a better d.c. than that of half wave.

## (v) Voltage Regulation

$$\%\text{Voltage regulation} = \frac{V_{dcNL} - V_{dcFL}}{V_{dcFL}} \times 100\%$$

For a full wave rectifier,

$$I_{dc} = \frac{2I_m}{\pi} = \frac{\frac{2V_m}{\pi}}{R_f + R_s + R_L}$$

$$I_{dc}(R_f + R_s) + I_{dc} R_L = \frac{2V_m}{\pi}$$

$$I_{dc}(R_f + R_s) + V_{dcFL} = V_{dc\,NL}$$

$$V_{dc\,FL} = \frac{2V_m}{\pi} - I_{dc}(R_f + R_s)$$

and

$$V_{dc\,NL} - V_{dc\,FL} = I_{dc}(R_f + R_s)$$

$$\%\text{voltage regulation} = \frac{I_{dc}(R_f + R_s)}{\frac{2V_m}{\pi} - I_{dc}(R_f + R_s)}$$

## (vi) Rectification Efficiency

$$\eta = \frac{P_{dc}}{P_{ac}} \times 100\%$$

For a full wave rectifier circuit

$$P_{dc} = (I_{dc})^2 R_L = \left(\frac{2I_m}{\pi}\right)^2 R_L$$

$$P_{ac} = I_{rms}^2 (R_f + R_s + R_L) = \left(\frac{I_m}{\sqrt{2}}\right)^2 (R_f + R_s + R_L)$$

∴ Rectification efficiency

$$\eta = \frac{\left(\frac{2I_m}{\pi}\right)^2 R_L}{\left(\frac{I_m}{\sqrt{2}}\right)^2 (R_f + R_s + R_L)} \times 100\%$$

$$= \frac{4}{\pi^2}\frac{I_m^2}{I_m^2} \cdot 2 \cdot \frac{R_L}{R_f + R_s + R_L} \times 100$$

$$= \frac{8}{\pi^2} \times \frac{1}{\frac{R_f + R_s}{R_L} + 1} \times 100\%$$

If $(R_f + R_s) << R_L$ then

$$\eta = \frac{8}{\pi^2} \times 100\% = 0.812 \times 100\% = 81.2\%$$

## (vii) Transformer Utilization Factor (TUF)

In case of fullwave rectifier, the TUF is calculated as primary and secondary TUF separately and take the average of two values.

**(a) TUF (secondary)**

$$P_{ac}(\text{rated secondary}) = P_{ac}(\text{rated) one half} + P_{ac}(\text{rated) other half}$$

$$= \frac{V_m}{\sqrt{2}}\cdot\frac{I_m}{2}+\frac{V_m}{\sqrt{2}}\cdot\frac{I_m}{2} = \frac{2\cdot V_m\cdot I_m}{2\sqrt{2}}=\frac{V_m\cdot I_m}{\sqrt{2}}$$

$$(\text{TUF})\text{ secondary} = \frac{\left(\frac{2I_m}{\pi}\right)^2 R_L}{\frac{V_m I_m}{\sqrt{2}}}$$

Also $V_m = I_m(R_f + R_s + R_L)$

$$\text{TUF (secondary)} = \frac{\left(\frac{2I_m}{\pi}\right)^2 R_L}{I_m(R_f+R_s+R_L)\cdot\frac{I_m}{\sqrt{2}}} = \frac{4\sqrt{2}}{\pi^2}\cdot\frac{R_L}{R_f+R_s+R_L} = \frac{4\sqrt{2}}{\pi^2}\cdot\frac{1}{\frac{R_f+R_s}{R_L}+1}$$

If $(R_f + R_s) << R_L$, then

$$(\text{TUF})\text{ secondary} = \frac{4\sqrt{2}}{\pi^2} = 0.574$$

**(b) TUF (Primary)**

Since the primary winding carries a full wave (sinusoidal) current, its rated capacity is utilized to the same extent as the rectification efficiency.

$\therefore$ TUF (primary) = 0.812

$$\text{Therefore average TUF} = \frac{\text{TUF (sec)} + (\text{TUF})_{\text{primary}}}{2} = \frac{0.574+0.812}{2} = 0.693$$

## (viii) Peak Inverse Voltage (PIV)

In positive half cycle, the equivalent circuit of FWR is

We note that PIV for a diode in half wave circuit is $V_m$, where the PIV for diodes in full wave circuit is $2V_m$.

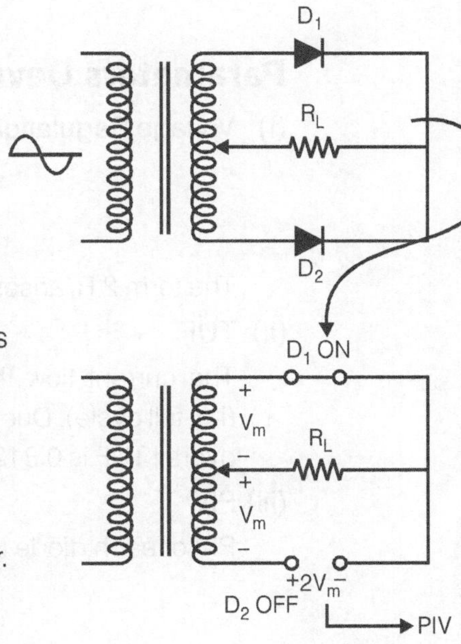


## Advantage and Disadvantage of Full Wave Rectifier

1. Amount of ripple is much lower (r.f = 0.482) as compared to half wave (r.f = 1.21).
2. Rectification efficiency is high ($\eta$ = 0.812).
3. TUF is better (0.693) then that of half wave (0.287).
4. No problem of core saturation
5. Requires centre-tapped secondary of the transformer.

# 8.5 Bridge Rectifier

To reduce the drawbacks of FWR with centre tap, bridge rectifier is used. The Figure (8.4) shows the bridge rectifier circuit. There are four diodes $D_1$, $D_2$, $D_3$ and $D_4$ which form the four arms of the bridge. The ac from the transformer secondary, in fed to the two corners and the load resistance $R_L$ is connected to the other two corners.

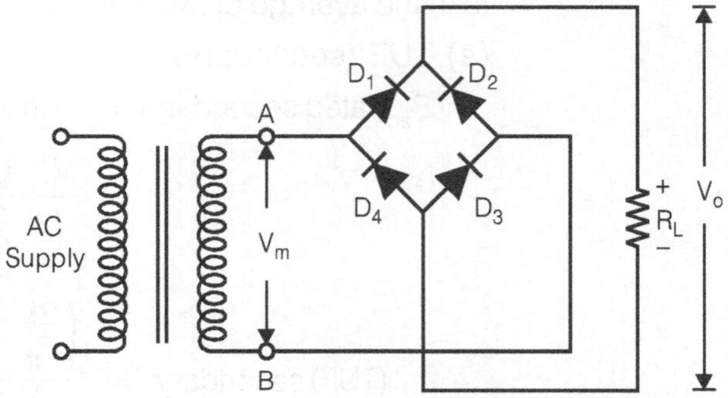

*Figure 8.4 Bridge rectifier*

- During positive half cycle of input the point A is positive, diodes $D_1$, $D_3$ forward biased and diodes $D_2$, $D_4$ reverse biased then, the current flows through diode $D_1$, the load $R_L$ and through diode $D_3$ back to the negative polarity of the transformer secondary.
- During negative half cycle of input 'B' is more positive than A, thus $D_2$, $D_4$ forward biased, and diodes $D_1$, $D_3$ reverse biased, then the direction of current flow will be through the diode $D_2$, load $R_L$ and diode $D_4$.
- In both the case the current flowing through resistor $R_L$ is in same direction. Thus it is called unidirectional current. Thus we obtain full-wave rectification.
- The following expressions are same as FWR with centre tap:

(i) Average current $I_{dc} = \frac{2I_m}{\pi}$ (ii) DC output voltage $(V_{dc})_{NL} = \frac{2V_m}{\pi}$

(iii) RMS load current $I_{rms} = \frac{I_m}{\sqrt{2}}$ (iv) Ripple-factor = 0.48

(v) Rectification efficiency $(\eta) = 0.812$

## Parameters Deviating From FWR

**(i) Voltage Regulation**

$$V_{dc\,FL} = \frac{2V_m}{\pi} - I_{dc}(R_s + 2R_f)$$

The term 2 $R_f$ arises because two diodes are conducting in series at any given time.

**(ii) TUF**

The current flow through both the primary and the secondary windings are sinusoidal (i.e. full cycle). Due to this, the TUF of the both the primary and the secondary are 0.812 and overall TUF is 0.812.

**(iii) PIV**

PIV of each diode in bridge rectifier is $V_m$.

### Advantage and Disadvantage

(i) A centre tap is not required in the transformer secondary.

(ii) Bridge circuit requires four diodes. At any given time two of them conduct in series, causing voltage drop due to two diodes. This additional voltage drop can not be ignored in low voltage dc power supplies.

Comparison of rectifier circuits

| S.N. | Parameter | Halfwave | Fullwave | Bridge |
|---|---|---|---|---|
| 1 | Number of diodes | 1 | 2 | 4 |
| 2 | PIV of diodes | $V_m$ | $2V_m$ | $V_m$ |
| 3 | $I_{rms}$ | $\frac{I_m}{2}$ | $\frac{I_m}{\sqrt{2}}$ | $\frac{I_m}{\sqrt{2}}$ |
| 4 | $I_{dc}$ | $\frac{I_m}{\pi}$ | $\frac{2I_m}{\pi}$ | $\frac{2I_m}{\pi}$ |
| 5 | DC output voltage | $\frac{V_m}{\pi}$ | $\frac{2V_m}{\pi}$ | $\frac{2V_m}{\pi}$ |
| 6 | Ripple factor | 1.21 | 0.482 | 0.482 |
| 7 | Ripple frequency | f | 2f | 2f |
| 8 | Rectification efficiency $\eta$ | 40.6% | 81.2% | 81.2% |
| 9 | TUF | 0.287 | 0.693 | 0.812 |

# FILTERS

## 8.6 Introduction

- The output of the rectifying circuit consists of a dc component as well as an ac component of the voltage. This is due to the pulsating nature of the output of any rectifier circuit.
- We have the ripple content in the half wave rectifier is 121% or that of full wave or bridge rectifier is 48%. But, for a dc power supply, one requires that the output of the supply should not have any ripple, that is, the ac component should be zero.
- This can be achieved by employing filters in cascade with rectifying circuits. The basic components of filters are the inductors and capacitors.
- Some of the important filters are the following:
  - Capacitor filter
  - Inductor filter
  - L section filter
  - $\pi$ section filter

# 8.7 Capacitor Filter

A capacitor, in parallel to the load, provides a easier by pass for the ripples voltage through it due to low impedance at ripple frequency and leave the dc to appear across the load.

## Half-wave Rectifier with Capacitor Filter

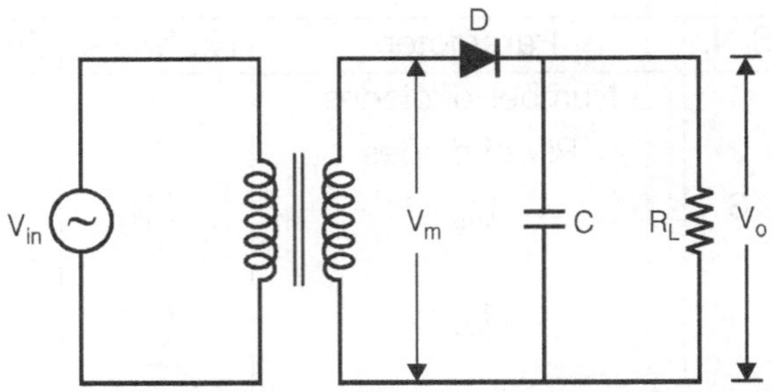


*Figure 8.5 Half wave rectifier with capacitor filter*

## Operation

- Let us assume that $R_L$ has infinite resistance. In the positive half cycle, the diode D is forward bias (assume no voltage drop in diode D), the capacitor will get charged to $V_m$ volts.
- In negative half cycle, the diode does not conduct. Since the diode resistance is infinite in the reverse bias, the capacitor can not discharge. Hence capacitor will remain charged at $V_m$ volts.
- Assume practical situation where $R_L$ is not infinite.
- In positive half cycle, the capacitor charges to $V_m$ volts, and in negative half cycle, capacitor discharges through $R_L$.

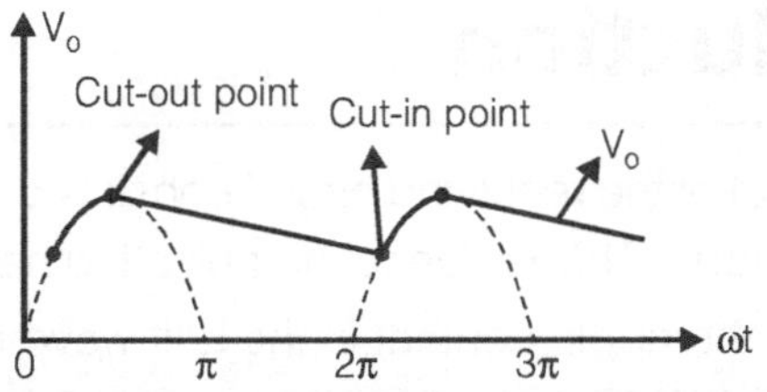


*Figure 8.6 Voltage waveform of HWR with capacitor filter*

- From the Figure (8.6), charging time is quite less than discharging time ($C \times R_L$).
- The angle at which the diode starts conducting is called cut-in point and the angle at which the diode stops conducting is called cut-out point.
- The average (dc) voltage output is higher with filter. At heavy loads ($R_L$ is very high) $V_{dc}$ will approach to peak value $V_m$.
- The value of ripple voltage is much lower. The ripple depends on the value of C and $R_L$. For a given load $R_L$, larger capacitor will give lower ripple.

## Expression for Ripple Factor

Let $T_1$ is the charging time.
$T_2$ is the discharging time.
The amount of charge lost by the capacitor at $T_2$ is

$$Q_{discharge} = I_{dc} \cdot T_2$$

The charge gained by the capacitor in the short interval of time $T_1$ is

$$Q_{charge} = C \cdot V_r \quad \text{(where } V_r = \text{ripple voltage)}$$

We know $$Q_{charge} = Q_{discharge}$$

Then, $$I_{dc} T_2 = C \times V_r$$

$$V_r = \frac{I_{dc} T_2}{C}$$

From the figure

$$\text{Total time } T = T_1 + T_2$$

$$T_2 >> T_1$$

$$T \approx T_2$$

Hence $$T \simeq T_2 = \frac{1}{f}$$

$$V_r = \frac{I_{dc}}{Cf}$$

The waveform is assumed as triangular waveform, hence its rms value is

$$V_{rms} = \frac{V_r}{2\sqrt{3}} = \frac{I_{dc}}{2\sqrt{3}\, fC}$$

We know $$V_{dc} = I_{dc} R_L$$

ripple factor $$r = \frac{V_{rms}}{V_{dc}} = \frac{\frac{I_{dc}}{2\sqrt{3}\, fC}}{I_{dc} R_L}$$

$$r = \frac{1}{2\sqrt{3}\, fCR_L}$$

### Conclusions

**(i)** If $R_L$ is small, ripple is high. If $R_L$ is high, ripple is small.
**(ii)** A larger capacitor will have lower ripple.

## Full-wave Rectifier with Capacitor Filter

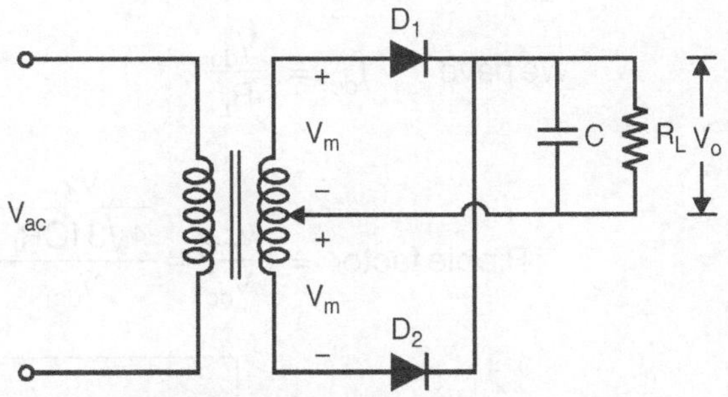


*Figure 8.7 Full wave rectifier with capacitor filter*

- During positive half cycle the diode $D_1$ is forward bias, $D_2$ is reverse bias. The capacitor charges upto the peak value of the input voltage $V_m$.
- When the input voltage falls below $V_m$, then both diodes become reverse bias, the capacitor starts discharging through the load resistor, it will continue until the input voltage exceeds the capacitor voltage.
- In the negative half cycle of input $D_1$ is reverse bias and diode $D_2$ is forward bias. When input voltage exceeds the capacitor voltage, capacitor starts charging.

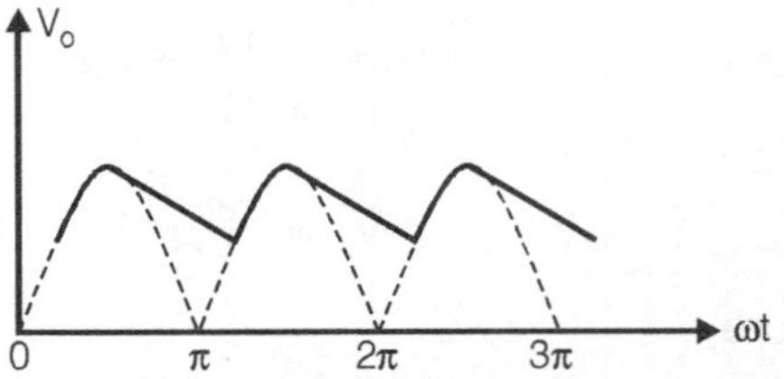


*Figure 8.8 Voltage waveform of FWR with capacitor filter*

Let $T_1$ = Charging time

$T_2$ = Discharging time

The amount of charge lost by the capacitor during the time interval $T_2$ is

$$Q_{discharge} = I_{dc} \times T_2$$

The amount of charge stored by the capacitor during the time interval $T_1$ is

$$Q_{charge} = C \times V_r$$

We know $Q_{charge} = Q_{discharge}$

$$C \times V_r = I_{dc} \times T_2$$

$$V_r = \frac{I_{dc}}{C} \times T_2$$

Here $T_1 << T_2$

$$T_2 = \frac{T}{2} = \frac{1}{2f}$$

then $V_r = \dfrac{I_{dc}}{2fC}$

As the ripple waveform is assumed as triangular wave

$$V_{rms} = \frac{V_r}{2\sqrt{3}} = \frac{I_{dc}}{4\sqrt{3}\,fC}$$

We have $I_{dc} = \dfrac{V_{dc}}{R_L}$

$$\text{Ripple factor} = \frac{V_{rms}}{V_{dc}} = \frac{\frac{V_{dc}}{4\sqrt{3}\,fCR_L}}{V_{dc}}$$

$$r = \frac{1}{4\sqrt{3}\,fCR_L}$$

Thus ripple can be reduced by increasing C or $R_L$ with a resulting increase in dc output voltage.

## 8.8 Inductor Filter

An inductor, in series with the load, prevents the passage of the ripple current, while allowing the dc. The characteristic of an inductor to oppose any change in current. That means inductor filter will try to smoothen the ripples of the current.

### Inductor Filter with Half Wave Rectifier

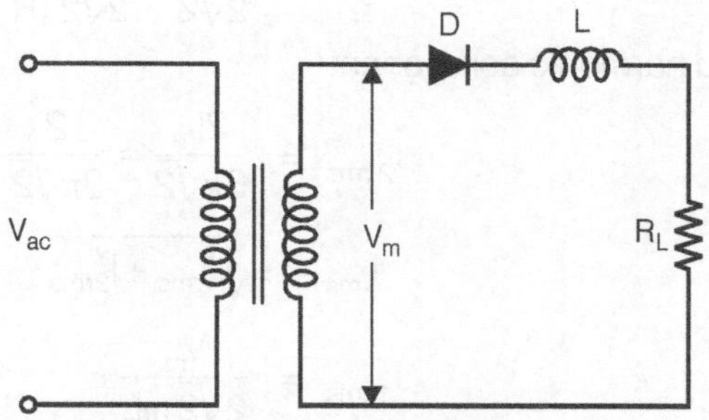


*Figure 8.9 HWR with inductor filter*

### Operation

- When an ac voltage is applied to the primary winding of transformer, it will be transferred to secondary winding of the transformer.
- During the positive half cycle of input, diode D starts conducting and permits the current to pass through it. But the inductor does not permit any variation in current passing through it.
- During the negative half cycle of input, the diode D is reverse bias, thus it act as open circuit and does not permit any current to pass through it.

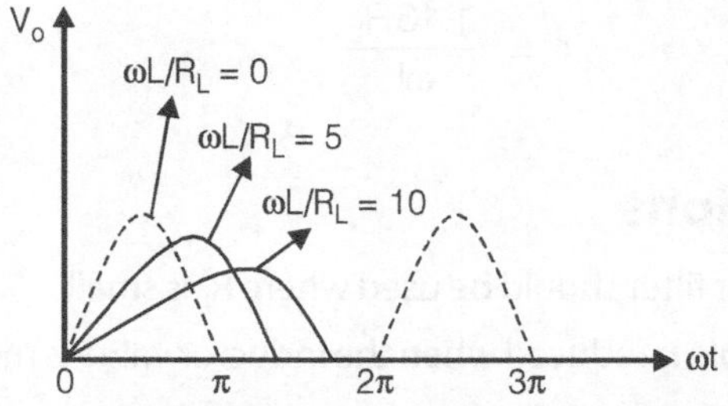


*Figure 8.10 Voltage waveform of HWR with inductor filter*

The Fourier series expression of half-wave rectifier is given by

$$I(t) = \frac{I_m}{\pi} + \frac{I_m}{2}\sin\omega t - \frac{2I_m}{3\pi}\cos 2\omega t + \cdots$$

First term represents dc or average value

$$I_{dc} = \frac{I_m}{\pi} = \frac{V_m}{\pi R_L}$$

The second and third terms represents first and second order ripple components respectively (neglecting higher order terms).

$$I_{ripple} = \frac{I_m}{2}\sin\omega t - \frac{2I_m}{3\pi}\cos 2\omega t$$

$I_{rms}$ for ripple is $\frac{I_{max}}{\sqrt{2}}$

In first harmonic component

$$I_{max} = \frac{I_m}{2}$$

and second harmonic $$I_{max} = \frac{2I_m}{3\pi}$$

$$I_{1\,rms} = \frac{I_m}{2\sqrt{2}} = \frac{V_m}{2\sqrt{2}\,(R + j\omega L)}$$

For second harmonic component

$$I_{2\,rms} = \frac{2I_m}{3\pi\sqrt{2}} = \frac{2}{3\pi\sqrt{2}}\frac{V_m}{(R + j2\omega L)}$$

$$I_{rms} = \sqrt{I_{1rms}^2 + I_{2rms}^2} \quad \text{As } \omega L >> R_L$$

$$I_{1\,rms} = \frac{V_m}{2\sqrt{2}\,\omega L}$$

$$I_{2\,rms} = \frac{V_m}{3\sqrt{2}\,\pi\,\omega L}$$

Ripple factor $$r = \frac{V_{ac}}{V_{dc}} = \frac{I_{ac}R_L}{I_{dc}R_L} = \frac{I_{ac}}{I_{dc}} = \frac{\sqrt{\left(\frac{V_m}{2\sqrt{2}\,\omega L}\right)^2 + \left(\frac{V_m}{3\sqrt{2}\pi\,\omega L}\right)^2}}{\frac{V_m}{\pi R_L}} = \frac{\frac{V_m}{\omega L}\sqrt{\left(\frac{1}{8} + \frac{1}{18\pi^2}\right)}}{\frac{V_m}{\pi R_L}}$$

$$r = \frac{\pi R_L}{\omega L}\sqrt{\left(\frac{1}{8} + \frac{1}{18\pi^2}\right)}$$

$$r = \frac{1.13R_L}{\omega L}$$

## Conclusions

**(i)** Inductor filter should be used where $R_L$ is small.

**(ii)** The ripple is reduced when the inductor value is more.

## Full-wave Rectifier with Inductor Filter

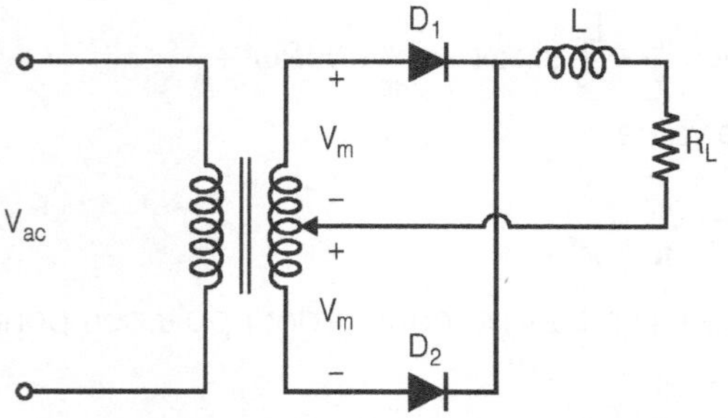


*Figure 8.11 FWR with inductor filter*

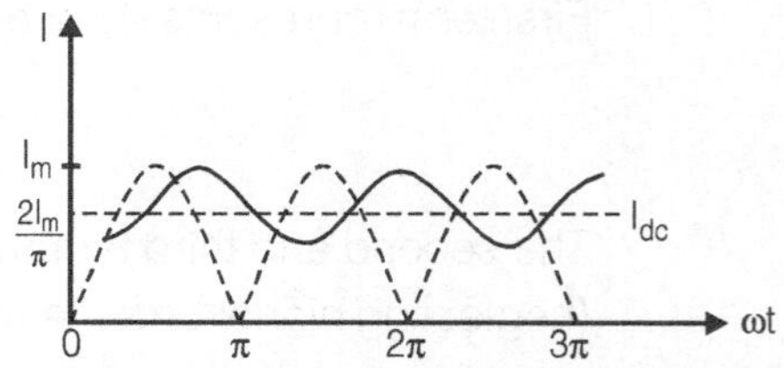


*Figure 8.12 Current wave-form of FWR with inductor filter*

The Fourier series equation for FWR is given by

$$I = \frac{2I_m}{\pi} - \frac{4I_m}{3\pi}\cos 2\omega t + \cdots$$

First term is dc component

$$I_{dc} = \frac{2I_m}{\pi} = \frac{2V_m}{\pi R_L} \qquad \text{Where } I_m = \frac{V_m}{R_L}$$

Second term which is ac component (second harmonic)

$$I_m \text{ of second harmonic} = \frac{4I_m}{3\pi} = \frac{4V_m}{3\pi\sqrt{R_L^2 + 4\omega^2 L^2}}$$

$$\text{Therefore} \quad I = \frac{2V_m}{\pi R_L} - \frac{4V_m}{3\pi\sqrt{R_L^2 + 4\omega^2 L^2}}\cos(2\omega t - \phi)$$

$$\text{where} \quad \phi = \tan^{-1}\frac{2\omega L}{R_L}$$

$$\text{ripple factor} \quad r = \frac{I_{rms}}{I_{dc}} = \frac{\dfrac{4V_m}{3\pi\sqrt{R_L^2 + 4\omega^2 L^2}\sqrt{2}}}{\dfrac{2V_m}{\pi R_L}}$$

$$r = \frac{2R_L}{3\sqrt{2}}\frac{1}{\sqrt{R_L^2 + 4\omega^2 L^2}} = \frac{2}{3\sqrt{2}}\frac{1}{\sqrt{1 + \dfrac{4\omega^2 L^2}{R_L^2}}}$$

If $R_L = \infty$

$$r = \frac{2}{3\sqrt{2}} = 0.472$$

If $\dfrac{4\omega^2 L^2}{R_L^2} >> 1$

$$r = \frac{2}{3\sqrt{2}}\frac{1}{\sqrt{\dfrac{4\omega^2 L^2}{R_L^2}}}$$

$$\boxed{r = \frac{R_L}{3\sqrt{2}\,\omega L}}$$

For a given load $R_L$, the ripple varies inversely as the magnitude of inductance. So, inductive filters are used for lighter load ($R_L$ is small) applications.

# 8.9 L Section Filter

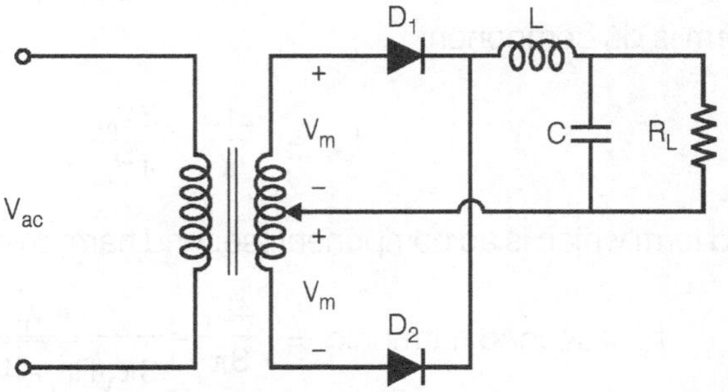


*Figure 8.13 L section filter*

Figure (8.13) shows an LC filter with a full-wave rectifier. It was found that ripple is reduced by decreasing the load resistance in an inductor and increase in a capacitor filter.

A combination of these two filters forms a LC filter. Its ripple is independent of load resistance $R_L$.

## Ripple Factor

Inorder the choke can suppress the ripple components, the reactance of the choke must be very large compared to the parallel impedance of the capacitor and the resistance.

i.e. $X_L >> X_C$

Therefore the total impedance is practically the choke alone.

$$X_L = 2\omega L$$

Ripple voltage has a magnitude $\frac{4V_m}{3\pi}$

Therefore ripple current passing through choke is

$$I_m = \frac{4V_m}{3\pi}\frac{1}{X_L}$$

rms value is $I_{rms} = \frac{I_m}{\sqrt{2}}$

$$I_{rms} = \frac{4}{\sqrt{2}}\cdot\frac{V_m}{3\pi}\cdot\frac{1}{X_L} = \frac{\sqrt{2}}{3}\cdot\frac{2V_m}{\pi}\cdot\frac{1}{X_L} = \frac{\sqrt{2}}{3}\cdot\frac{V_{dc}}{X_L}$$

Across the load $R_L$ the ac ripple voltage is same as across the capacitor

$$V_{rms} = I_{rms}\cdot X_C = \frac{\sqrt{2}}{3}\cdot\frac{V_{dc}}{X_L}\cdot X_C$$

Where $X_C = \frac{1}{2\omega C}$

ripple factor $r = \frac{V_{rms}}{V_{dc}} = \frac{\frac{\sqrt{2}\cdot V_{dc}\cdot X_C}{3\cdot X_L}}{V_{dc}} = \frac{\sqrt{2}\cdot X_C}{3\cdot X_L} = \frac{\sqrt{2}}{3}\cdot\frac{1}{2\omega C}\cdot\frac{1}{2\omega L} = \frac{\sqrt{2}}{12\omega^2 LC}$

If we assume $f = 50$ Hz then $\boxed{r = \frac{1.19}{LC}}$

L section filter is used for variable loads.

## 8.10 π Section Filter

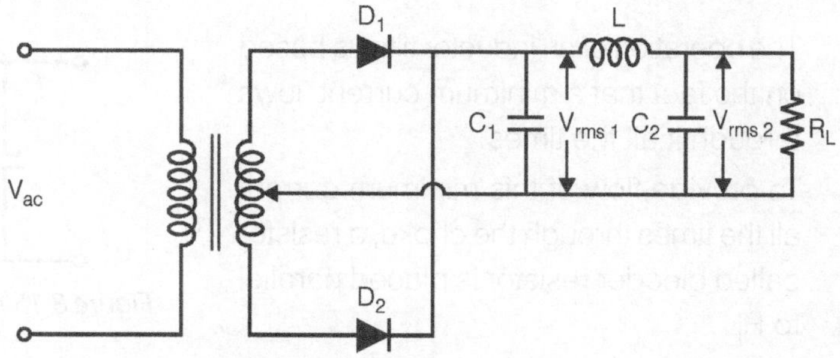


*Figure 8.14 π section filter*

- Figure (8.14) shows π-section filter which consists of two capacitor $C_1$ and $C_2$ and an inductor 'L'. This circuit is used where ever a low output current and a high dc output voltage is required.
- The capacitor $C_1$ offers a low reactance to ac ripple components of rectifier output, but it offers infinite resistance to the dc and low ripple component.
- a.c low frequency ripple move towards 'L'. The output of $C_1$ filter is a triangular in shape.
- When the capacitor $C_1$ output is fed to input of L section filter, inductor offers high reactance to ac ripple components, which could not by passed by $C_1$.
- The capacitor $C_2$ by passes the ac components which can not blocked by L and $C_1$.

From capacitor filter in FWR,

$$V_r = \frac{I_{dc}}{2fC}$$

Fourier series of triangular wave is

$$V = V_{dc} - \frac{V_r}{\pi}\left(\sin 2\omega t - \frac{\sin 4\omega t}{2} + \cdots\right)$$

The second order ripple component is

$$V_{rms\,1} = \frac{V_r}{\pi\sqrt{2}} = \frac{I_{dc}}{2\sqrt{2}\,\pi fC} = \sqrt{2}\,I_{dc} \cdot X_{C_1}$$

$$\text{Where} \quad X_{C_1} = \frac{1}{2\omega C_1} = \frac{1}{4\pi fC_1}$$

The output on LC filter is

$$V_{rms\,2} = V_{rms1} \cdot \frac{X_{C_2}}{X_{C_2} + X_L} = \frac{V_{rms1} \cdot X_{C_2}}{X_L} \qquad \text{Where } X_L >> X_{C_2}$$

$$= \frac{\sqrt{2}\,I_{dc} \cdot X_{C_1} \cdot X_{C_2}}{X_L} = \frac{\sqrt{2}\,V_{dc}\,X_{C_1} \cdot X_{C_2}}{X_L\,R_L}$$

$$r = \frac{V_{rms2}}{V_{dc}} = \frac{\sqrt{2} \cdot X_{C_1} \cdot X_{C_2}}{X_L \cdot R_L}$$

In case of π section filter the ripple factor depends on $R_L$.

# 8.11 Bleeder Resistor

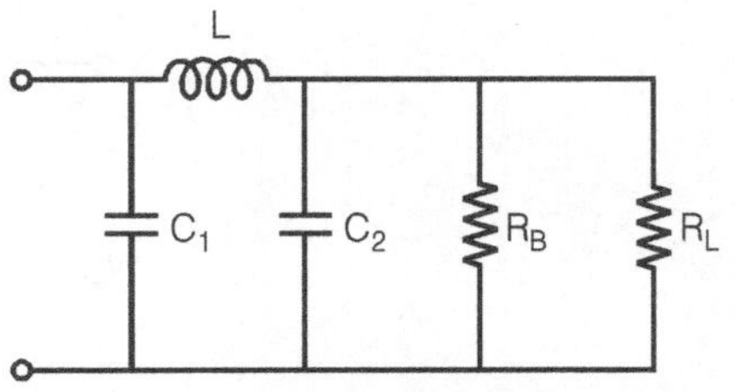

*Figure 8.15 π section filter with bleeder resistor $R_B$*

- The operation of an inductor filter is based on the fact that a minimum current flows through it all the times.
- To provide flow of this minimum current all the times through the choke, a resistor called bleeder resistor is placed parallel to $R_L$.
- It is used to maintain a certain minimum current through the choke, even the load resistor gets open circuited, and improves filtering action.
- The bleeder resistor can improve the voltage regulation. It provides safety to the operator by providing a discharge path to the capacitor, to avoid electrical shocks.

## VOLTAGE REGULATORS

# 8.12 Introduction

- The primary function of a voltage regulator is to maintain a constant dc output voltage. How ever, it also rejects ac ripple voltage that is not removed by the filter.
- A regulator may be constructed from a Zener diode or discrete transistor or integrated circuits. All voltage regulators must have a stable voltage reference source.
- Practically all electronic devices need dc supply. Although a battery serves the purpose. But in practice a rectifier and a suitable filter can act to give required dc voltage from ac supply lines.
- However, this does not ensure a dc voltage of constant magnitude.
- With change in supply voltage or load dc supply voltage tends to change.
- A regulated power supply is that the voltage at the input terminals of a device remain practically constant, even when ac supply voltage or load $R_L$ varies.

# 8.13 Block Diagram of Power Supply

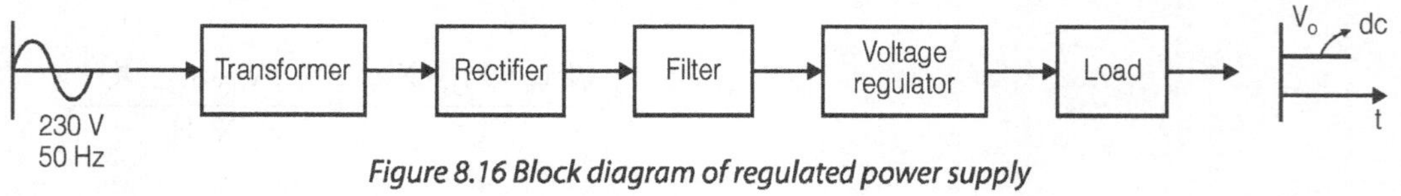

*Figure 8.16 Block diagram of regulated power supply*

- The input to the transformer is a ac voltage of 230 V 50 Hz supply.
- The transformer should be step down transformer and the rectifier converts ac into pulsating dc.

- As the output of rectifier is not pure dc. The ripple components are removed by filter.
- The output of the filter is pure dc which can not ensure a dc voltage of constant magnitude.
- Voltage regulator maintains a constant dc voltage due to the variations in the supply voltage and the load impedance.

## 8.14 Load Regulation

If the voltage regulator is not perfect, voltage across the load changes with variations in load current. As the load current increases load voltage diminishes.

"The change in the load voltage with change of load current is given by a parameter called as load regulation"

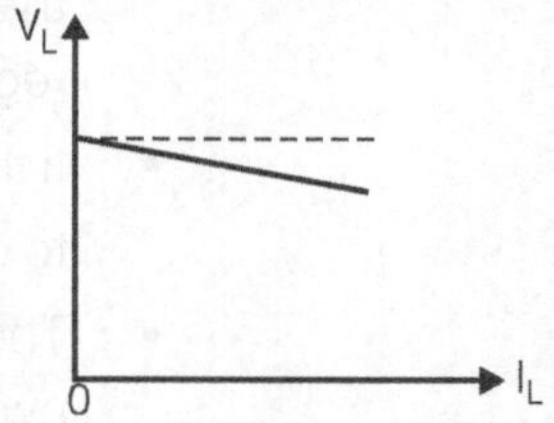


$$\% \text{ load regulation} = \frac{V_{dcNL} - V_{dcFL}}{V_{dcFL}} \times 100\%$$

## 8.15 Line Regulation

Variations of supply voltage at the transformer may also cause variation of load voltage. As the input voltage decreases (line voltage) will cause decrease of load voltage.

"The change of load voltage due to 10% change of line voltage is named as line regulation".

$$\therefore \quad \% \text{ line regulation} = \left[\frac{\text{Change in } V_o \text{ for 10\% change of } V_i}{V_o}\right] \times 100\%$$

## 8.16 Series Voltage Regulator

The block diagram of series voltage regulator is shown in Figure (8.17).

- The input to the control element (BJT) is un regulated dc voltage. The operation of control element is to control the output voltage.
- Sampling circuit samples the output voltage and if any deviation is there, it goes as a feedback voltage to the comparator which is compared with reference voltage.

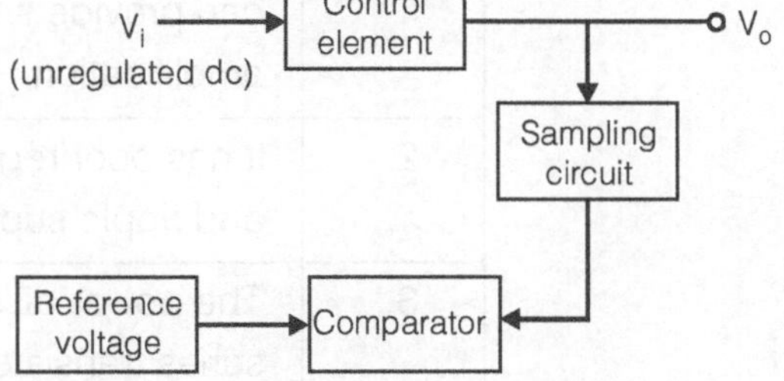


*Figure 8.17 Block diagram of series voltage regulator*

- The comparator provides control signal to the series element, which will decrease or increase according to the output voltage.
- This process continues till constant voltage is maintained at output.

## 8.17 Shunt Voltage Regulator

The block diagram of shunt voltage regulator is shown in Figure (8.18).

- The input unregulated voltage provides current to the load.
- Some of this current is shunted away by the control element to maintain the regulated output voltage across the load.
- If the output voltage tends to change due to change in load, the sampling circuit provides a feedback signal to a comparator circuit.
- The comparator circuit provides a control signal to vary the magnitude of current shunted away from the load.

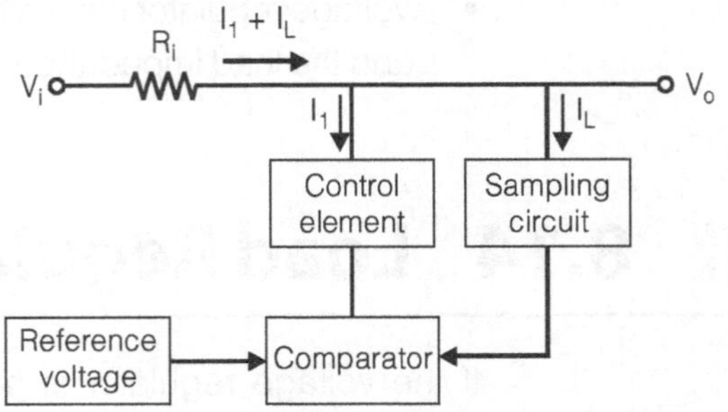


*Figure 8.18 Block diagram of shunt voltage regulator*

For example, when the output voltage tends to decrease, the sampling circuit provides a feedback signal to the comparator circuit which then provides a control signal to draw less shunt current, providing more load current, keeping voltage constant.

## 8.18 Limitations of Series and Shunt Voltage Regulators

**LIMITATIONS**

| S.No. | Series voltage regulator | Shunt voltage regulator |
|---|---|---|
| 1. | It can not provide good regulation at high currents because one transistor can provide small amplification. | There will be power loss in series resistor $R_S$. |
| 2. | It has poor regulation and and ripple suppression. | More current flows through the transistor than to load. |
| 3. | The power dissipation of series transistor is large ($V_{CC}$ $I_C$). | Over-voltage protection problems are there in this circuit. |
| 4. | Series voltage regulator is limited to low output voltages. | Compared to shunt voltage regulator, series regulator is preferred. |

## SUMMARY

- A rectifier converts bidirectional current to unidirectional current. There are three types of rectifiers.
  - half wave rectifier
  - full wave with centre tapped
  - bridge rectifier
- A rectifiers employs diodes for rectification. There are one, two and four diodes employed in half-wave, full-wave and bridge rectifiers respectively.
- The characteristics of rectifiers are:
  - Average current $I_{dc}$
  - DC output voltage $V_{dc}$
  - RMS load current $I_{rms}$
  - Ripple factor
  - Voltage regulation
  - Rectification efficiency
  - Transformer utilization factor
  - Peak inverse voltage
  - Form factor
  - Peak factor
- Half-wave rectifier is least used since its performance is poor. Full-wave and bridge rectifier are widely used.
- The ripple in the output for HWR is 121% and for FWR is 48%. The ripple can be reduced by a filter.
- There are various types of filters:
  - Capacitor filter
  - Inductor filter
  - L section filter
  - $\pi$ section filter
- Capacitor filter is used for heavier load ($R_L$ is more) applications.
- Inductor filter is used for lighter load ($R_L$ is less) applications.
- L section filter is used for variable load applications.
- $\pi$ section filter is dependent on $R_L$. So it can not be used for variable load applications.
- Regulated power supply basically converts ac supply to required dc power (constant magnitude).
- Regulated power supply consists of a step-down transformer, a rectifier, a filter and a voltage regulator.
- There are two types of voltage regulators:
  - Series voltage regulator
  - Shunt voltage regulator

## GATE AND IES QUESTIONS

**Q.1** In a half-wave rectifier. If an ac supply is 60 Hz, then what is the ac ripple at output?
(a) 30 Hz (b) 60 Hz
(c) 120 Hz (d) 15 Hz

**Exp.** (b)
The ripple frequency for half-wave rectifier is f.

**Q.2** A half-wave rectifier having a resistance load of 1 kΩ rectifiers an ac voltage of 325 V peak value and the diode has a forward resistance of 100 Ω. What is the RMS value of the current?
(a) 295.4 mA (b) 94 mA
(c) 147.7 mA (d) 208 mA

**Exp.** (c)

$$I_{rms} = \frac{I_m}{2} \text{ For HWR}$$

Where $$I_m = \frac{V_m}{R_f + R_L} = \frac{325}{100+1000} = 295.4 \quad ; \quad I_{rms} = \frac{295.4}{2} = 147.7 \text{ mA}$$

**Q.3** In a centre tap full-wave rectifier, 100 V is the peak voltage between the centre tap and one end of the secondary. What is the maximum voltage across the reverse biased diode?
(a) 200 V (b) 141 V
(c) 100 V (d) 86 V

**Exp.** (a)
The PIV rating of full-wave rectifier with centre tap is $2V_m = 2 \times 100 = 200$ V

**Q.4** The circuit shown below ie the Thevenins equivalent circuit of a centre tapped full-wave rectifier with diode forward resistance $R_f = 100\ \Omega$, transformer secondary coil resistance $R_s = 30\ \Omega$, peak input voltage $V_m = 10$ V. What are the values of V and $R_o$ respectively.
(a) 10 V, 100 Ω
(b) 6.36 V, 130 Ω
(c) 6.36 V, 115 Ω
(d) 4.54 V, 130 Ω

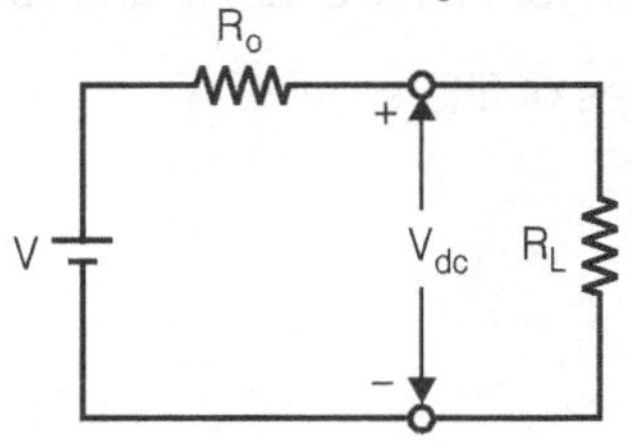


**Exp.** (c)

$$V = \frac{2V_m}{\pi} = 6.36 \text{ V} \quad ; \quad R_o = 100 + \frac{30}{2} = 115\ \Omega$$

**Q.5** Consider the following rectifier circuits for low load:
1. Half-wave rectifier without filter.
2. Full-wave rectifier without filter.
3. Full-wave rectifier with series inductor filter.
4. Full-wave rectifier with capacitor filter.

The sequence of these rectifier circuits in decreasing order of their ripple factor is
(a) 1, 2, 3, 4 (b) 3, 4, 1, 2
(c) 1, 2, 4, 3 (d) 3, 2, 1, 4

**Exp. (c)**

Half-wave rectifier without filter >
Full-wave rectifier without filter >
Full-wave rectifier with capacitor filter >
Full-wave rectifier with series inductor filter, for low load conditions.

**Q.6** A dc power supply has no load voltage of 30 V, and a full load voltage of 25 V at a full-load current of 1 A. Its output resistance and load regulation, respectively are

(a) 5 Ω and 20%
(b) 25 Ω and 20%
(c) 5 Ω and 16.7%
(d) 25 Ω AND 16.7%

**Exp. (b)**

$$R_L = \frac{V_{FL}}{I_L} = \frac{25\,V}{1A} = 25\,\Omega \;;\quad \%\,V_R = \frac{V_{dc\,NL} - V_{dc\,FL}}{V_{dc\,FL}} = \frac{30-25}{25} = 20\%$$

**Q.7** Which one of the following statements is correct?
An ideal regulated power supply should have

(a) 100% regulation
(b) 50% regulation
(c) 0% regulation
(d) 75% regulation

**Exp. (c)**

An ideal regulated power supply should have 0% regulation.

**Q.8** Which one of the following statements is correct? In case of load regulation.

(a) When the temperature changes, the output voltage remains constant.
(b) When the input voltage changes, the load current remains constant.
(c) When the load changes, the load current remains constant.
(d) When the load changes, the output voltage remains constant.

**Exp. (d)**

In case of load regulation, when the load changes, the output voltage remains constant.

**9.** Calculate the output voltage and Zener diode current in the regulator circuit of figure

Given $R = 1.8\ k\Omega$, $R_L = 2\ k\Omega$,
$V_z = 8.3\ V$, $\beta = 100$

(a) 7.6 V, 3.6842 mA
(b) 8.5 V, 3.6842 mA
(c) 7.6 V, 2.5 mA
(d) 7.6 V, 4.8 mA

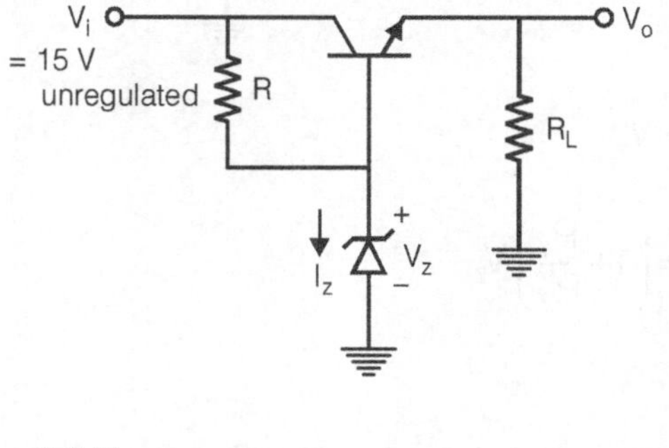


**Exp. (a)**

Output voltage $$V_o = V_z - V_{BE} = 8.3 - 0.7 = 7.6\ V$$

Current through $$R = \frac{V_i - V_z}{R} = \frac{15-8.3}{1.8}\ mA = 3.72\ mA$$

$$I_L = \frac{V_o}{R_L} = \frac{7.6}{2} = 3.8\ mA\ ;\ I_B = \frac{I_C}{\beta} = \frac{I_L}{\beta} = \frac{3.8}{100}\ mA = 0.038\ mA$$

$$I_z = I_R - I_B = 3.7222 - 0.038\ mA = 3.6842\ mA$$

10. What is the regulated output voltage of a series voltage regulator circuit shown in figure

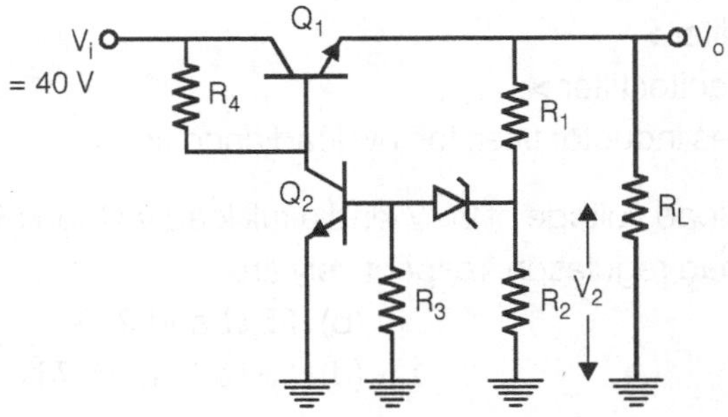


Given, $R_1 = 3.3\ k\Omega$, $R_2 = 2.2\ k\Omega$, $R_3 = 5\ k\Omega$, $R_4 = 10\ k\Omega$, $V_z = 10$ V

(a) 26.75 V (b) 24 V
(c) 25 V (d) 28.6 V

**Exp.** (a)

$$V_2 = V_{BE_2} + V_Z = 0.7 + 10 = 10.7\ \text{V} \quad ; \quad V_2 = V_o\left(\frac{R_2}{R_1+R_2}\right)$$

$$V_o = V_2\left(\frac{R_1+R_2}{R_2}\right) = 10.7\left(\frac{3.3+2.2}{2.2}\right) = 26.75\ \text{V}$$

11. Find $V_o$ for the circuit below

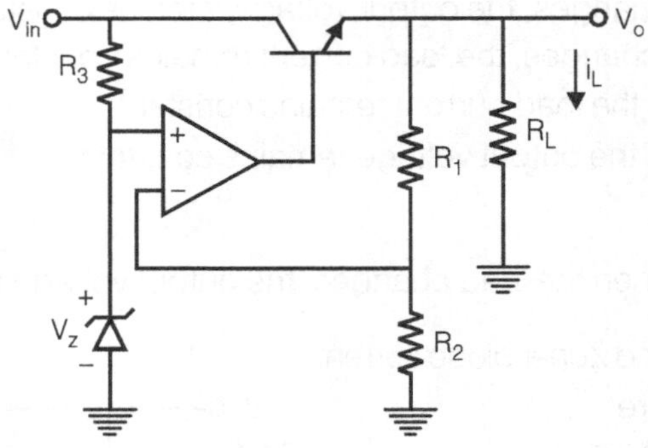


(a) $V_o = V_z$ (b) $V_o = -\frac{R_Z}{R_1}V_Z$

(c) $V_o = \left(1+\frac{R_1}{R_2}\right)V_Z$ (d) none

**Exp.** (c)

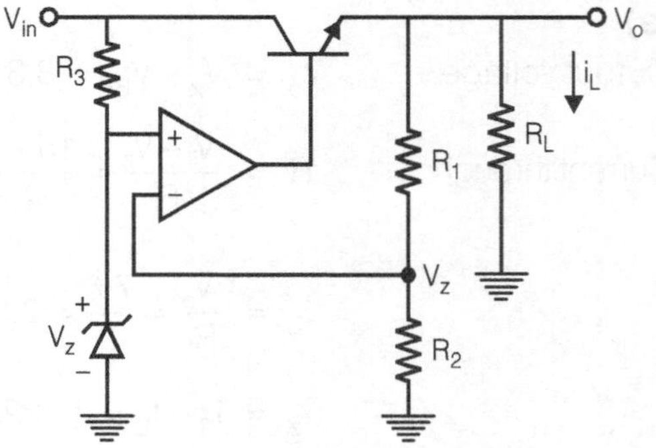


$$V_Z = \frac{V_o \times R_2}{R_1 + R_2}$$

$$\Rightarrow \quad V_o = V_Z\left(\frac{R_1+R_2}{R_2}\right) = V_Z\left(1+\frac{R_1}{R_2}\right)$$

12. Find output voltage of the shunt regulator shown

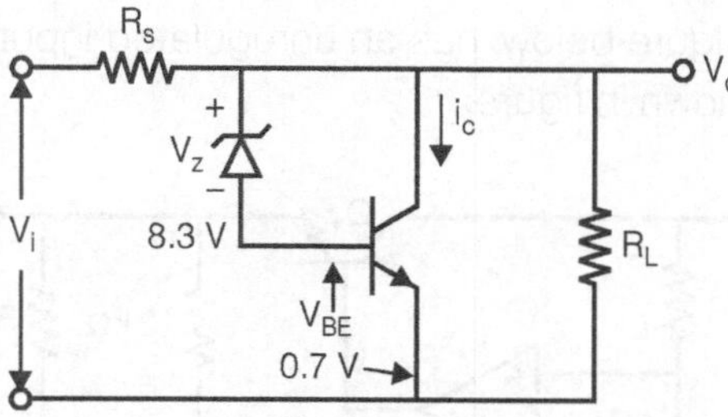


(a) 9 V
(b) 8.3 V
(c) 6 V
(d) 7.6 V

**Exp. (a)**

$$V_o = V_z + V_{BE} = 8.3\text{ V} + 0.7\text{ V} = 9\text{ V}$$

13. Find output voltage of a shunt regulator circuit shown:

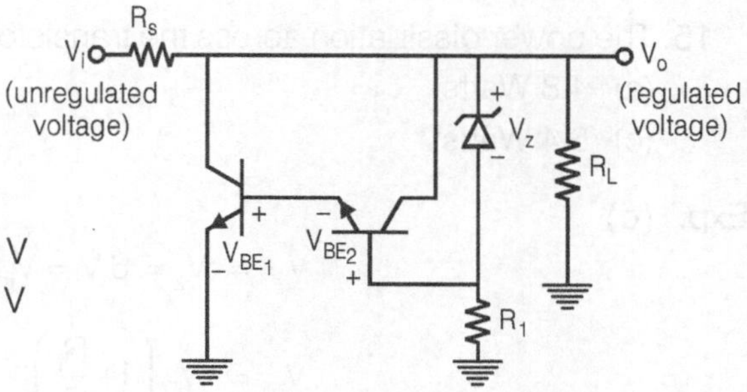


Given $V_{BE_1} = V_{BE_2} = 0.7\text{ V}$

$V_z = 6\text{ V}$

(a) 6 V
(b) 6.8 V
(c) 7.4 V
(d) 8.6 V

**Exp. (c)**

$$V_o = V_{BE_1} + V_{BE_2} + V_z = 6 + 0.7 + 0.7 = 7.4\text{ V}$$

14. In the following circuit the op-amp is ideal. If $\beta_f = 60$ then the total current supplied by 15 V source is

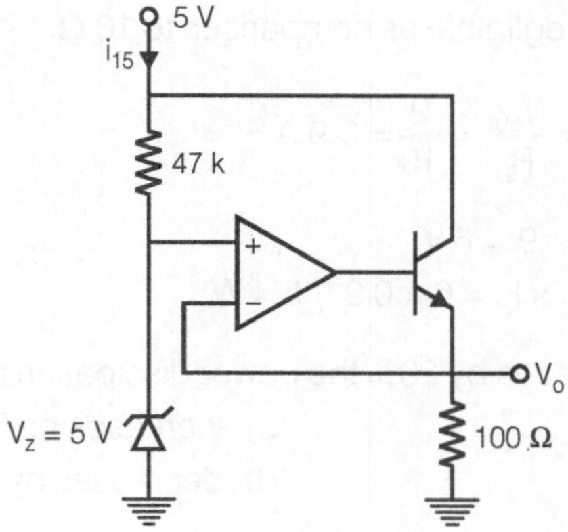


(a) 123.1 mA
(b) 98.3 mA
(c) 49.4 mA
(d) 168 mA

**Exp. (c)**

$$V_+ = 5\text{V}$$
$$V_- = V_E$$

The input current to op-amp is zero.

$$i_{15} = i_z + i_c = i_z + \alpha I_E$$
$$= \frac{15.5}{47\text{ k}\Omega} + \frac{60}{61}\left(\frac{5}{100}\right) = 49.4\text{ mA}$$

Common Data Q.15 and Q.16

A regulated power supply shown in figure below, has an unregulated input of 15 V and generates a regulated output $V_{out}$. The component valves shown in figure:

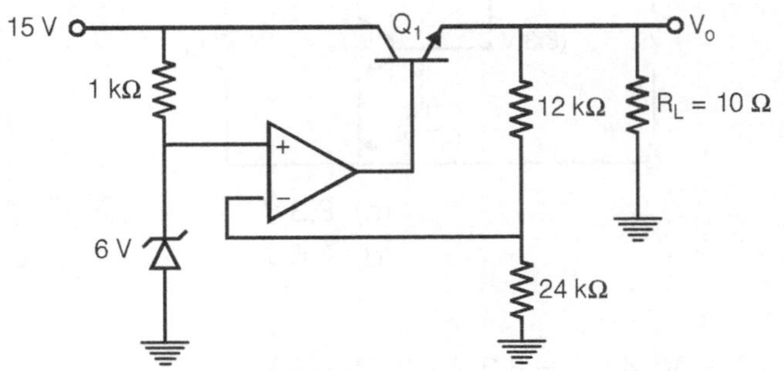


**15.** The power dissipation across the transistor $Q_1$ shown in figure
(a) 4.8 Watts
(b) 5.0 Watts
(c) 5.4 Watts
(d) 6.0 Watts

**Exp. (c)**

$$V_+ = V_z = 6\text{ V} = V_{in}$$

$$V_o = V_{in}\left(1+\frac{R_f}{R_1}\right)$$

$$V_o = 6\left(1+\frac{12\text{ k}\Omega}{24\text{ k}\Omega}\right) = 9\text{ V}$$

The current in 12 kΩ branch is negligible as comparison to 10 Ω.

Thus current $$I_C = I_E = \frac{V_{out}}{R_L} = \frac{9}{10} = 0.9\text{ A}$$

$$V_{CE} = 15 - 9 = 6\text{ V}$$

$$P_D = V_{CE} \times I_C = 6 \times 0.9 = 5.4\text{ W}$$

**16.** If the unregulated voltage increases by 20% the power dissipation across the transistor $Q_1$
(a) increases by 20%
(b) increases by 50%
(c) remains unchanged
(d) decreases by 20%

**Exp. (b)**

If unregulated voltage increase by 20% then the unregulated voltage is 18 V, but the $V_z = V_{in}$.

$$V_{CE} = 18 - 9 = 9\text{ V}$$

$$I_C = 0.9\text{ A}$$

$$P_D = V_{CE} \times I_C = 9 \times 0.9 = 8.1\text{ W}$$

% Power increase

$$\frac{8.1-5.4}{5.4} \times 100 = 50\%$$

**17.** The transistor shunt regulator shown in the figure has a regulated output voltage of 10 V, when the input varies from 20 V to 30 V. The relevant parameters for the Zener diode and the transistor are; $V_z = 9.5$ V, $V_{BE} = 0.3$ V, $\beta = 99$. Neglect the current through $R_B$. Then maximum power dissipated in the Zener diode ($P_z$) and the transistor ($P_T$) are

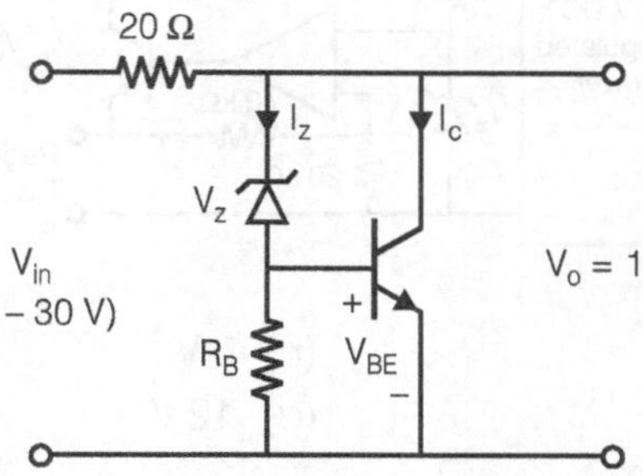


(a) $P_z = 75$ mW, $P_T = 7.9$ W
(b) $P_z = 85$ mW, $P_T = 8.9$ W
(c) $P_z = 95$ mW, $P_T = 9.9$ W
(d) $P_z = 115$ mW, $P_T = 11.9$ W

**Exp. (c)**

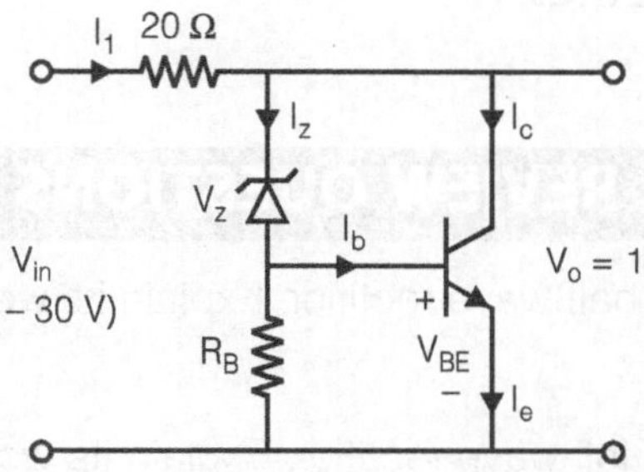


$$I_{1\,max} = \frac{V_{inmax} - V_o}{20}$$

$$I_1 = \frac{30-10}{20} = 1\,A \quad (I_z = 0)$$

$$I_E = I_C + I_Z$$

$$I_B = I_Z \text{ (as no current flows in } R_B)$$

$$\beta = \frac{I_C}{I_B} = \frac{I_C}{I_Z}$$

$$\Rightarrow \quad I_C = \beta I_Z$$

$$I_E = \beta I_Z + I_Z = (99 + 1)\,I_Z$$

$$I_E = 100\,I_Z$$

$$I_1 = I_E = 100\,I_Z$$

$$I_Z = \frac{I_1}{100} = \frac{1}{100} = 0.01\,A$$

$$P_Z = V_Z I_Z = 9.5 \times 0.01 = 95 \text{ mW}$$

$$I_C = 99\,I_Z = 99 \times 0.01 = 0.99\,A$$

$$P_T = V_C\,I_C = 10 \times 0.99 = 9.9\,W$$

18. Consider the circuit given below:

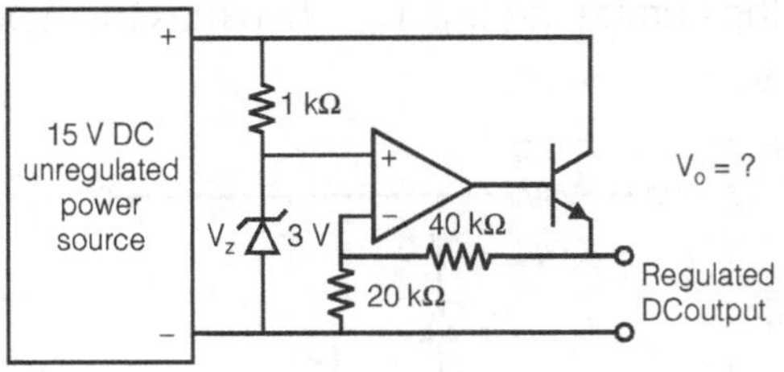


(a) 3 V
(b) 6 V
(c) 9 V
(d) 12 V

**Exp. (c)**

3 V at inverting terminal because of virtual ground

$$V_o = \frac{3\text{ V} \times 60\text{ k}\Omega}{20\text{ k}\Omega} = 9\text{ V}$$

## REVIEW QUESTIONS

1. Draw the circuit diagram of a half-wave rectifier. Explain its working. Sketch the input and output waveforms.
2. Draw the circuit diagram of a full-wave rectifier. Explain its working. Sketch the input and output waveforms.
3. Draw the circuit diagram of a bridge rectifier. Explain its working.
4. What is ripple? Derive the ripple factors of half-wave and full-wave rectifiers?
5. Explain the operation of full-wave rectifier with capacitor filter and derive the expression for ripple factor.
6. Explain the operation of full-wave rectifier with inductor filter and derive the expression for ripple factor.
7. Explain the operation of L-section filter and derive the expression for ripple factor.
8. Give the comparison of L-section and π-section filters.
9. What is a regulated power supply? Draw its block diagram and explain the function of each block.
10. Give various types of voltage regulators and discuss their performance.

○○○○

# CHAPTER 9

# OPERATIONAL AMPLIFIERS

## About This Chapter

In this part of the chapter, we study about

- Differential amplifiers
- Operational amplifiers
- Operational amplifier practical applications
- Active filters
- 555 timer

# 9 OPERATIONAL AMPLIFIERS

## 9.1 Introduction

- The differential amplifier is the basic building block of analog circuits.
- It is the input stage of operational amplifier.

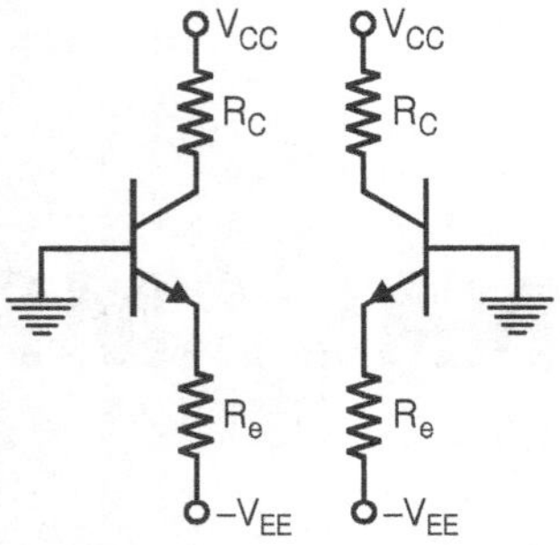


*Figure 9.1 Requirement for a differential amplifier*

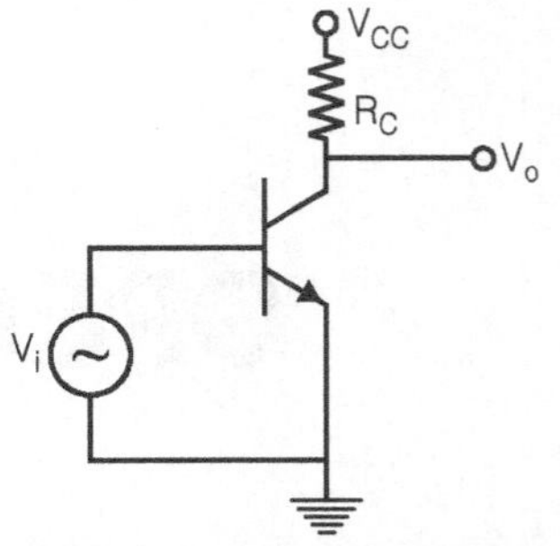


*Figure 9.2 Single ended amplifier*

The advantages of differential amplifier compared to single ended amplifiers are:

1. It is having good noise rejection capability.
2. There is no need of coupling or bypass capacitors.
3. The input impedance of a differential amplifier is high.

## 9.2 Differential Amplifier

The differential amplifier forms the input stage of an operational amplifier. It is used to achieve voltage gain, current gain or power gain. There are many different types of amplifiers, and one of the most important types is the differential amplifier.

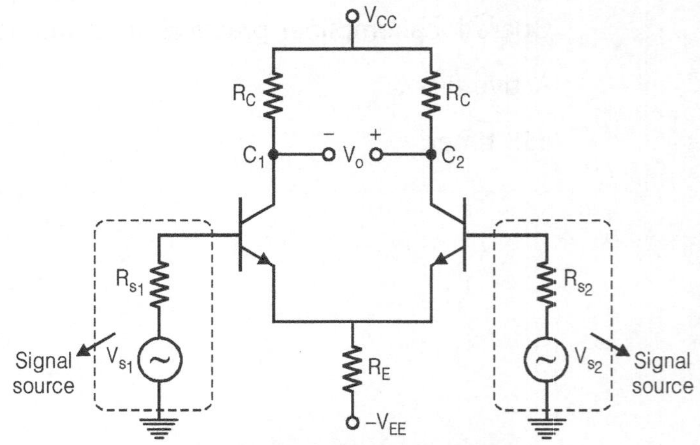


*Figure 9.3 Differential amplifier*

A differential amplifier mainly comprises of two transistors with their emitters connected as shown in the Figure (9.3). The transistors are direct-coupled. There are two inputs denoted by $V_{S_1}$ and $V_{S_2}$, output denoted as $V_o$.

If ideal conditions are assumed, then the two transistors are identical and their collector resistors are of equal magnitude.

If the two inputs $V_{S_1}$ and $V_{S_2}$ are equal, it is evident that the output voltage $V_o$ is zero. However the two inputs are unequal, an output voltage results and its polarity depends upon whether $V_{S_1} > V_{S_2}$ or $V_{S_2} > V_{S_1}$.

Of the two inputs, $V_{S_1}$ is termed as non-inverting input, because the output $V_o$ is in phase with the input voltage $V_{S_1}$. $V_{S_2}$ is termed as inverting input, since the output voltage is 180° out of phase with the input voltage $V_{S_2}$.

The differential amplifier of Figure (9.3) amplifies the voltage difference $(V_1 - V_2)$ and the output voltage $V_o$ is given as

$$\boxed{V_o = A(V_{S_1} - V_{S_2})}$$

If $V_{S_1} = 0$ i.e. if a voltage $V_{S_2}$ is applied at the inverting input terminal, keeping the non inverting input terminal grounded we have

$$V_o = -AV_{S_2}$$

That is, there is phase inversion between $V_o$ and $V_{S_2}$.

If $V_{S_2} = 0$ i.e. if a voltage $V_{S_1}$ is applied at the non-inverting input terminal, keeping inverting input terminal grounded, we have

$$V_o = AV_{S_1}$$

There is no phase inversion between $V_o$ and $V_{S_1}$.

## 9.3 Circuit Configurations

At the input side of the differential amplifier, we have

(i) Single input

(ii) Dual input

At the output side of the differential amplifier, we have

(i) Balanced output (output is taken across $C_1$ and $C_2$)

(ii) Unbalanced output (output is taken across $C_1$ to ground or $C_2$ to ground)

According to this classification, the circuit configurations are:

1. Single input balanced output.
2. Single input unbalanced output.
3. Dual input balanced output.
4. Dual input unbalanced output.

## 9.4 Circuit Diagrams

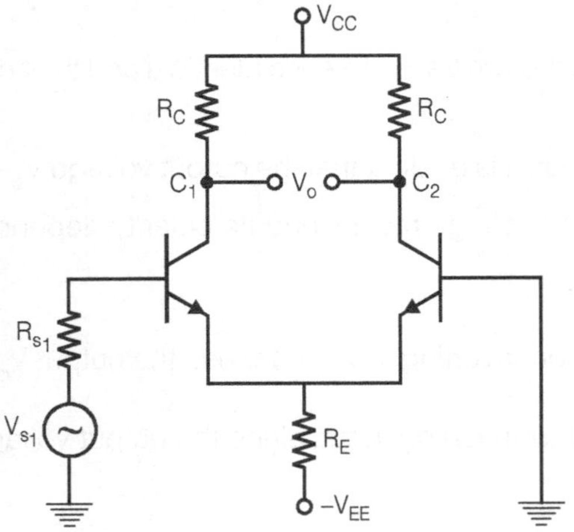


Figure 9.4 Single input balanced output

$A_v$ = Voltage gain = $\frac{R_c}{r_e}$

$R_{in}$ = Input resistance = $2\beta_{ac}\, r_e$

$R_o = R_c$

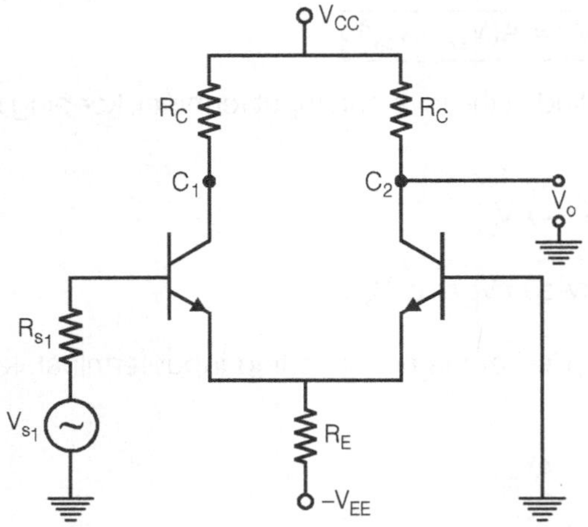


Figure 9.5 Single input unbalanced output

$A_v = \frac{R_c}{2r_e}$

$R_{in} = 2\beta_{ac}\, r_e$

$R_o = R_c$

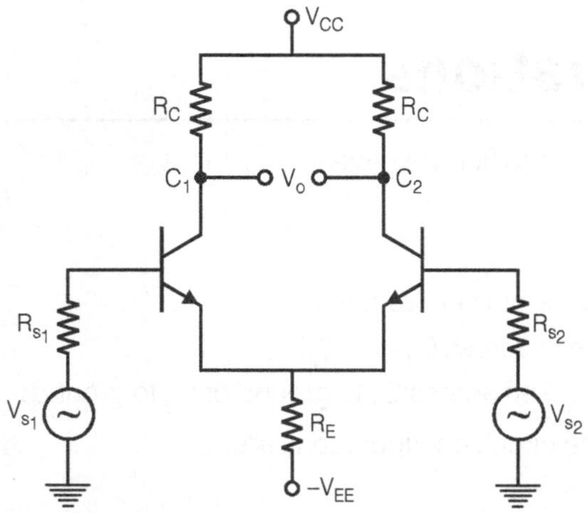


Figure 9.6 Dual input balanced output

$A_v = \frac{R_c}{r_e}$

$R_{in} = 2\beta_{ac}\, r_e$

$R_o = R_c$

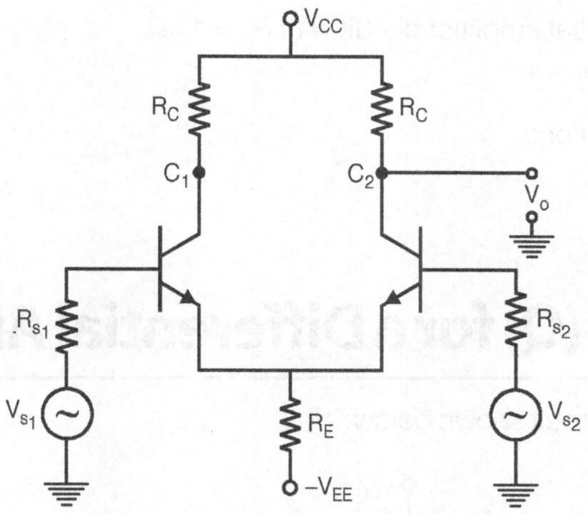


*Figure 9.7 Dual input unbalanced output*

$$A_v = \frac{R_c}{2r_e}$$

$$R_{in} = 2\beta_{ac} r_e$$

$$R_o = R_c$$

## PROBLEMS

1. Calculate the voltage gain for a single input balanced output or dual input balanced output differential amplifier circuit with $R_c$ = 1 kΩ and $I_E$ = 26 mA.

**Solution:**

For single input balanced output or dual input balanced output, the expression for voltage gain is same

i.e. $$A_V = \frac{R_c}{r_e}$$

Where $R_c$ is collector resistance
$r_e$ is emitter internal ac resistance

$$r_e = \frac{V_T}{I_E} = \frac{26\text{ mV}}{26\text{ mA}} = 1\,\Omega \qquad (V_T = \text{volt equivalent temperature})$$

$$A_v = \frac{1000\,\Omega}{1\Omega} = 1000$$

2. Calculate the voltage gain for a single input unbalanced output or dual input unbalanced output differential amplifier circuit with $R_c$ = 1 kΩ and $I_E$ = 26 mA.

**Solution:**

For single input unbalanced output or dual input unbalanced output, the expression for voltage gain is same.

i.e. $$A_v = \frac{1}{2}\frac{R_c}{r_e} = \frac{1000}{2\times 1} = 500$$

3. Calculate the input resistance of a differential amplifier circuit with $\beta_{ac}$ = 50 and $I_E$ = 26 mA.

**Solution:**

The formula is valid for all the four configurations

$$R_i = 2\beta_{ac} r_e$$

Where $$r_e = \frac{V_T}{I_E} = \frac{26\text{ mV}}{26\text{ mA}} = 1\,\Omega$$

$$R_i = 2 \times 50 \times 1\,\Omega = 100\,\Omega$$

4. Calculate the output resistance of a differential amplifier circuit with $R_c = 1\ k\Omega$.

Solution:

The formula is valid for all the four configurations

$$R_o = R_c = 1\ k\Omega$$

## 9.5 Operating Point (Q) for a Differential Amplifier

Consider a differential amplifier circuit shown below:

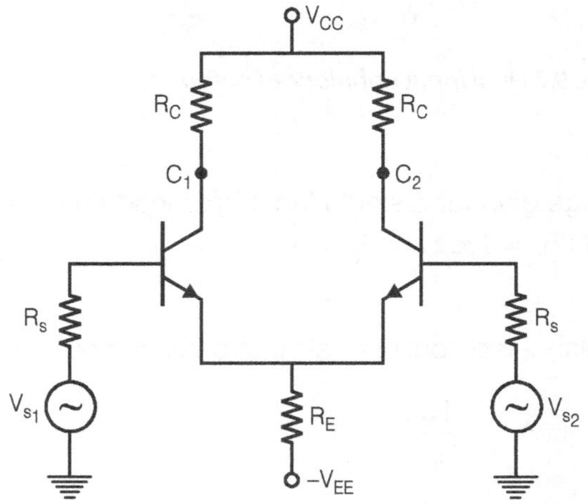


*Figure 9.8 Differential amplifier*

### ANALYSIS

To find the operating point (Q) ac should be grounded. Then the Figure (9.8) modify as

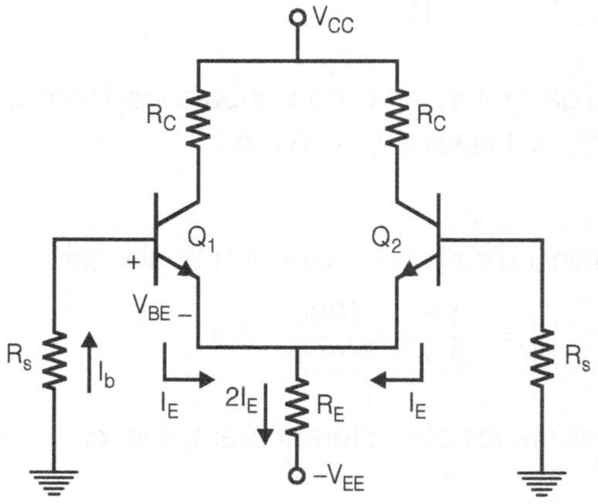


For the analysis of differential amplifier only one half section is enough.

For the first half section, input loop equation is

$$V_{EE} = I_b R_s + V_{BE} + 2I_E R_E$$

$$V_{EE} - I_b R_s - V_{BE} - 2I_E R_E = 0$$

We know
$$I_b = \frac{I_E}{1+\beta}$$

$$V_{EE} - \frac{I_E}{1+\beta}R_s - V_{BE} - 2I_E R_E = 0$$

$$I_E = \frac{V_{EE} - V_{BE}}{2R_E + \frac{R_s}{1+\beta}}$$

If $\beta$ is large

$$(I_C)_Q \simeq I_E = \frac{V_{EE} - V_{BE}}{2R_E}$$

$$(V_{CE})_Q = V_C - V_E$$
$$V_C = V_{CC} - (I_C)_Q R_C$$
$$V_E = -V_{BE}$$
$$(V_{CE})_Q = V_{CC} - (I_C)_Q R_C + V_{BE}$$

## 9.6 Practical Design Problems in Differential Amplifiers

### CASE-I

#### 1. Voltage gain

$$A_v = \frac{R_C}{r_e} \quad \text{For a differential amplifier}$$

In the expression of voltage gain $r_e$ is the dynamic resistance which makes the gain unstabilize.

#### 2. Input resistance

$$R_i = 2\beta_{ac} r_e$$

In previous problems we got $R_i = 100\ \Omega$ which is very less.
Generally the input resistance should be high.

**The technique used to improve gain stabilization and input resistance is "swamping resistor technique".**

### CASE-II

Operating point Q is not stable because $(I_C)_Q$ depends on emitter current $I_E$

$$\text{i.e. } (I_C)_Q = I_E = \frac{V_{EE} - V_{BE}}{2R_E}$$

$I_E$ current should be constant to make operating point Q stable.

The techniques used to make the operating point Q stable are:

**(i)** Constant current source

**(ii)** Current mirror

## 9.7 Swamping Resistor Technique

The modified differential amplifier circuit is shown in Figure (9.9) using swamping resistor.

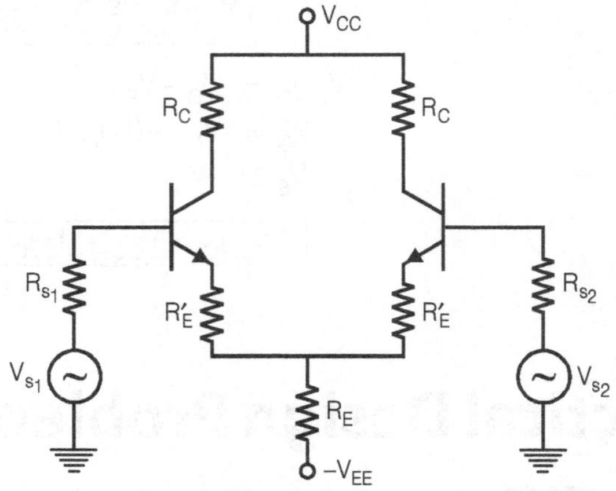


*Figure 9.9 Differential Amplifier with swamping resistor*

"$r_e$" model of a transistor BJT will be

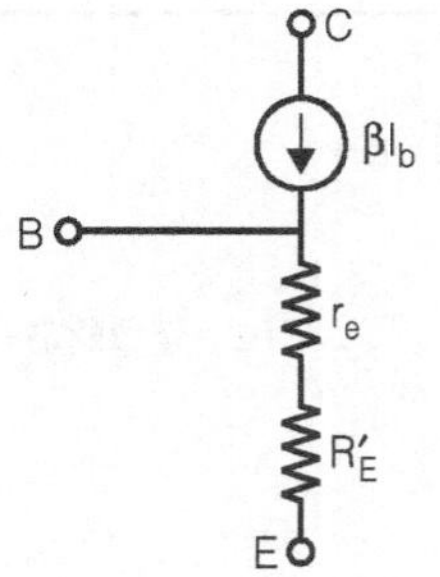


According to the "$r_e$" model external resistance $R'_E$ become in series with $r_e$. Therefore

The voltage gain $$A_v = \frac{R_c}{r_e + R'_E}$$

If $R'_E >> r_e$ then,

$$A_v = \frac{R_c}{R'_E}$$

∴ Gain is independent of $r_e$ (dynamic resistance of emitter). Therefore gain is stabilized.

Input resistance

$$R_i = 2\beta_{ac}(r_e + R'_E)$$

If $R'_E = 100\ \text{k}\Omega$

$r_e = 1\ \Omega$

$\beta_{ac} = 50$

Then

$$R_i = 2 \times 50\,(1\ \Omega + 100\ \text{k}\Omega)$$

$$\cong 100{,}00\ \text{k}\Omega$$

Therefore the input resistance is increased.

### Swamping Resistor Technique

{$r_e$ is replaced by $(r_e+R'_E)$}

| SNo | Configuration | $A_v$ | $R_i$ | $R_o$ |
|---|---|---|---|---|
| 1. | Single input balanced output | $\frac{R_C}{r_e+R'_E}$ | $2\beta_{ac}(r_e+R'_E)$ | $R_c$ |
| 2. | Single input unbalanced output | $\frac{R_C}{2(r_e+R'_E)}$ | $2\beta_{ac}(r_e+R'_E)$ | $R_c$ |
| 3. | Dual input balanced output | $\frac{R_e}{r_e+R'_E}$ | $2\beta_{ac}(r_e+R'_E)$ | $R_c$ |
| 4. | Dual input unbalanced output | $\frac{R_C}{2(r_e+R'_E)}$ | $2\beta_{ac}(r_e+R'_E)$ | $R_c$ |

# 9.8 Constant Current Source Circuits

An ideal current source provides a constant current regard-less of the load connected to it. There are many applications in electronics for a circuit that provides a constant current with a very high impedance. Constant-current sources can be built using FET devices, bipolar devices, and a combination of these components. Some of the forms will be considered here.

## JFET CONSTANT CURRENT SOURCE

Figure (9.10) shows a simple JFET current source.

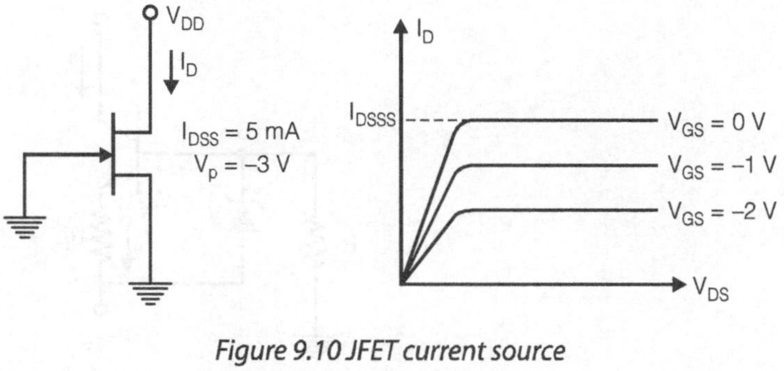


*Figure 9.10 JFET current source*

When $V_{GS}$ = 0 V, in JFET, the drain current $I_D$ is fixed at $I_{DSS}$. From the Figure (9.10), the device operates like a constant source of magnitude 5 mA at $V_{GS}$ = 0 V.

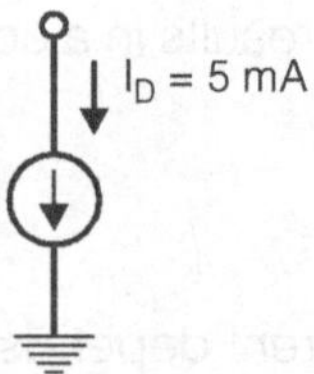

## BJT CONSTANT CURRENT SOURCE

Figure (9.11) shows a circuit using an npn transistor and a few resistors for operation as a constant-current source.

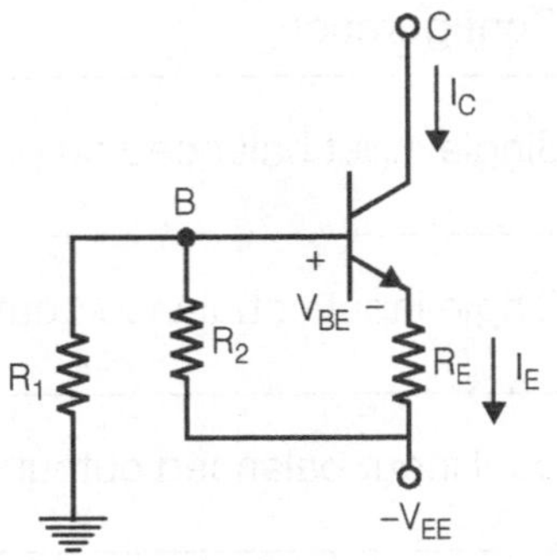


*Figure 9.11 Discrete constant current source*

Constant current $I_C$ is given by the following equation

$$I_C \approx I_E = \frac{V_E - (-V_{EE})}{R_E} = \frac{V_E + V_{EE}}{R_E}$$

Here $V_E = V_B - V_{BE}$

$$\text{Then } I_C = \frac{(V_B - V_{BE}) + V_{EE}}{R_E} = \frac{V_B - 0.7 + V_{EE}}{R_E}$$

$$\text{Where } V_B = \frac{R_1}{R_1 + R_2} \times (-V_{EE})$$

## BJT/ZENER CONSTANT CURRENT SOURCE

If resistor $R_2$ is replaced with a Zener diode in the circuit in Figure (9.11), an improved constant current source is obtained in Figure (9.12).

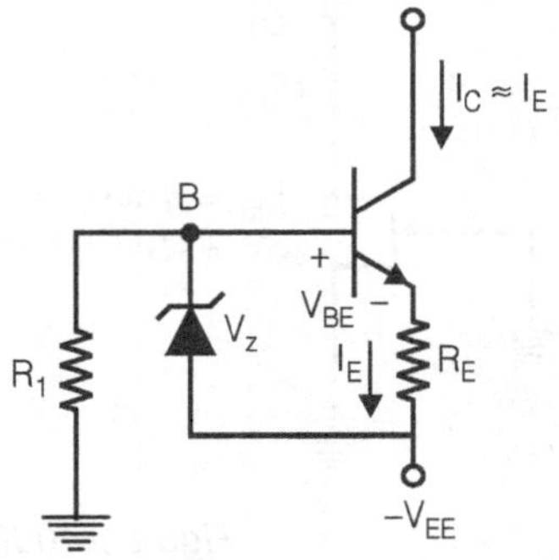


*Figure 9.12 BJT and Zener diode constant current source*

The Zener diode results in a constant current I given by the equation

$$I = I_E = \frac{V_B - V_{BE}}{R_E} = \frac{V_Z - V_{BE}}{R_E} = \frac{V_Z - 0.7}{R_E}$$

The constant current depends on Zener diode voltage, which remains quite constant and the emitter resistor $R_E$. The voltage supply $V_{EE}$ has no effect on the magnitude of current I delivered by the circuit.

## 9.9 Current Mirror Circuits

The two-transistor current source, also called a current mirror, is the basic building block in the design of integrated circuit current sources.

**MODEL-1**

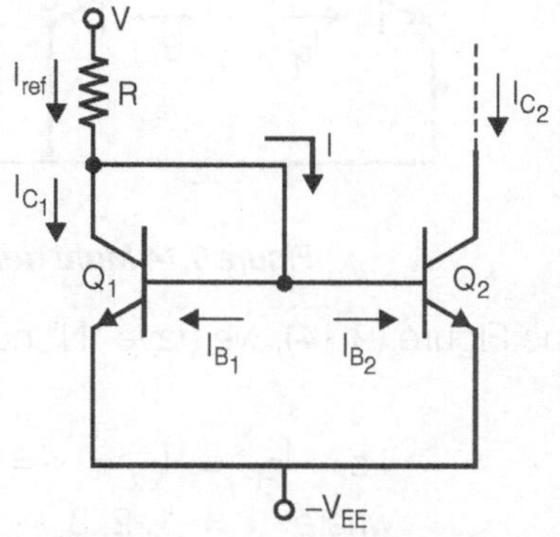


*Figure 9.13 Two transistor current mirror*

- The Figure (9.13) shows the basic current-source circuit, which consists of two matched or identical transistors $Q_1$ and $Q_2$, operating at the same temperature with their base terminals and emitter terminals connected together.
- The base-emitter voltage, $V_{BE}$ is therefore the same in the two transistors. Transistor $Q_1$ is connected as a diode, when supply voltage is applied. The base-emitter junction of transistor $Q_1$ is forward biased and a reference current $I_{ref}$ is established. Transistor $Q_1$ activate the transistor $Q_2$ and generates the output current. Which is a constant current source.

From Figure (9.13)

$$I_{ref} = I_{C_1} + I = I_{C_1} + I_{B_1} + I_{B_2}$$

According to current mirror concept

$$I_{C_1} = I_{C_2} \quad \text{and} \quad I_{B_1} = I_{B_2}$$

Therefore

$$I_{ref} = I_{C_2} + 2I_{B_2} = I_{C_2} + 2\frac{I_{C_2}}{\beta} \qquad \left[\because I_{C_2} = \beta I_{B_2}\right]$$

$$= I_{C_2}\left[1 + \frac{2}{\beta}\right]$$

$$\therefore I_{C_2} = \frac{I_{ref}}{1 + \frac{2}{\beta}}$$

As $\beta >> 2$

$$I_{C_2} \approx I_{ref}$$

i.e. the constant current provided at the collector of transistor $Q_2$ mirrors that of transistor $Q_1$.

Since

$$I_{ref} = \frac{V_{CC} - V_{BE}}{R}$$

## MODEL-2

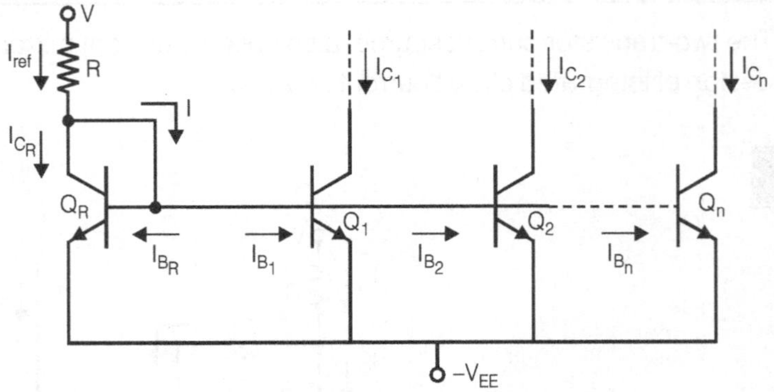


*Figure 9.14 Multi transistor current mirror*

According to the Figure (9.14), we have "N" number of transistors which are current mirror to transistor $Q_R$.

i.e. $I_{C_1} = I_{C_2} = \cdots = I_{C_n} = I_{C_R} = I_{C_i}$

where $i = 1, 2, 3, \cdots n$

$$I_{B_1} = I_{B_2} = \cdots = I_{B_n} = I_{B_R}$$

$$I_{ref} = I_{C_R} + I = I_{C_R} + I_{B_R} + I_{B_1} + I_{B_2} + \cdots + I_{B_n}$$

$$= I_{C_R} + I_{B_R}(1+N) = I_{C_R} + \frac{I_{C_R}}{\beta}(1+N) = I_{C_R}\left[1 + \frac{(1+N)}{\beta}\right]$$

$$I_{C_i} = I_{C_R} = \frac{I_{ref}}{1 + \dfrac{(1+N)}{\beta}}$$

Suppose If $N = 1$

$$I_{C_1} = I_{C_R} = \frac{I_{ref}}{1 + \dfrac{2}{\beta}}$$

## MODEL-3

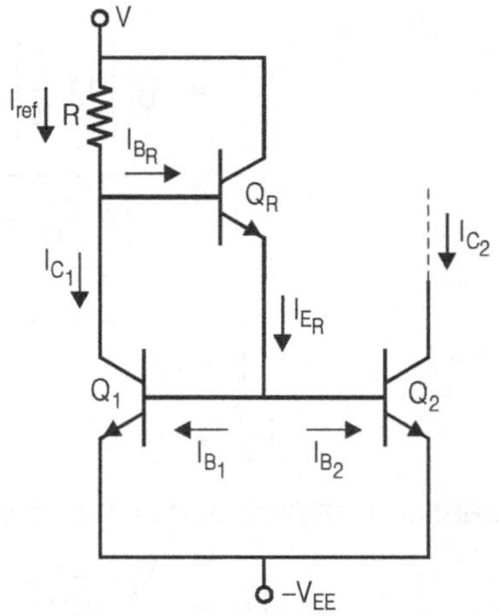


*Figure 9.15 Modified Current Mirror using two transistors*

In the above Figure (9.15), $Q_1$ and $Q_2$ are current mirror but not $Q_R$. The parameters of $Q_R$ are not identical to $Q_1$ and $Q_2$.

$$I_{ref} = I_{C_1} + I_{B_R} = I_{C_1} + \frac{I_{E_R}}{1+\beta_R}$$

$\beta_R$ is the current gain of transistor $Q_R$

$$I_{ref} = I_{C_1} + \frac{I_{B_1} + I_{B_2}}{(1+\beta_R)} = I_{C_2} + \frac{2I_{B_2}}{(1+\beta_R)} \quad (\because\ I_{C_1} = I_{C_2} \text{ and } I_{B_1} = I_{B_2})$$

$$= I_{C_2} + \frac{2I_{C_2}}{\beta(1+\beta_R)} = I_{C_2}\left[1 + \frac{2}{\beta(1+\beta_R)}\right]$$

$$\boxed{I_{C_2} = \frac{I_{ref}}{1 + \dfrac{2}{\beta(1+\beta_R)}}}$$

In this case $I_{C_2}$ is more approximate to $I_{ref}$.

## MODEL-4

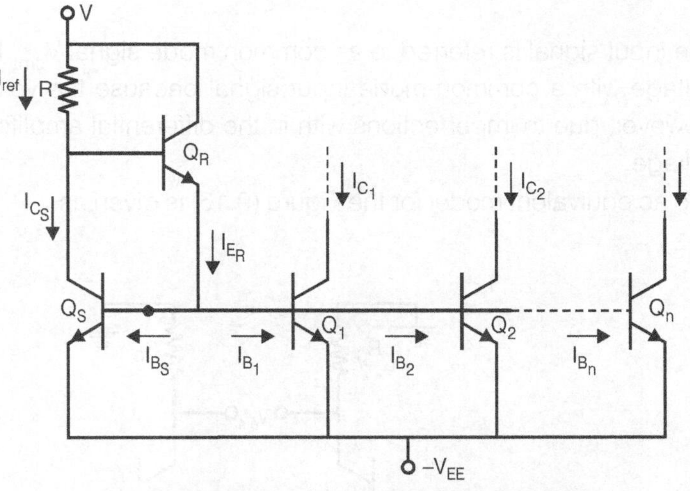


*Figure 9.16 Multi transistor current mirror*

According to the Figure (9.16)

$$I_{ref} = I_{C_S} + I_{B_R} = I_{C_S} + \frac{I_{E_R}}{1+\beta_R} = I_{C_S} + \frac{I_{B_S} + I_{B_1} + I_{B_2} + \cdots + I_{B_n}}{(1+\beta_R)}$$

$$= I_{C_S} + \frac{I_{B_S}[1+N]}{(1+\beta_R)} = I_{C_S} + \frac{I_{C_S}}{\beta}\frac{(1+N)}{(1+\beta_R)} = I_{C_S}\left[1 + \frac{(1+N)}{\beta(1+\beta_R)}\right]$$

$$\boxed{I_{C_S} = I_{C_i} = \frac{I_{ref}}{1 + \dfrac{(1+N)}{\beta(1+\beta_R)}}}$$

## 9.10 Common-Mode Voltage Gain

If in phase signals $V_{in}$ are applied to each base of the differential amplifier, as shown in Figure (9.16).

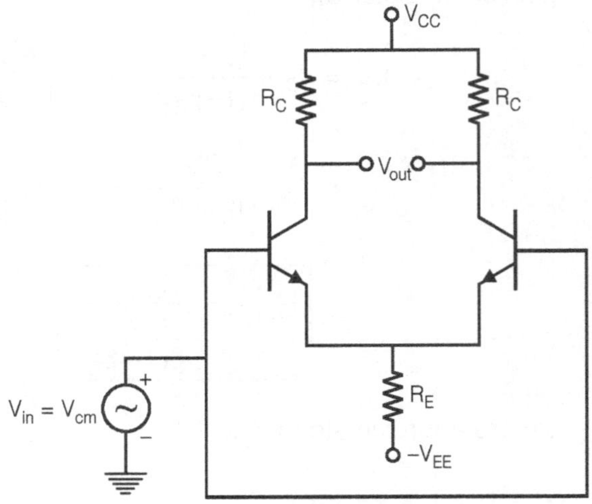


*Figure 9.16 Differential amplifier in common mode configuration*

The input signal is referred to as common mode signal $V_{cm}$. Ideally, there will be no ac output voltage with a common-mode input signal because the voltage between the bases is zero. However, due to imperfections with in the differential amplifier there will be a small ac output voltage.

The ac equivalent model for the Figure (9.16) is given as

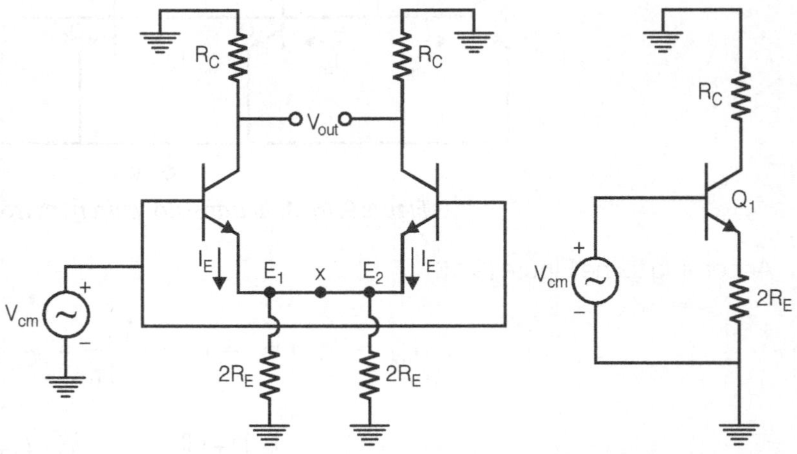


*Figure 9.17 AC equivalent model* *Independent Half section*

Using r-parameters, the common mode model for one half of the circuit is shown in Figure (9.18).

## Common Mode Voltage Gain $A_{cm}$

Emitter current $$I_E = \frac{V_{cm}}{2R_E + r_e}$$

Collector current $$I_C \approx I_E = \frac{V_{cm}}{2R_E + r_e}$$

Common mode output voltage is

$$V_{o\,cm} = I_C R_C = \frac{V_{cm} R_C}{2R_E + r_e}$$

The common mode voltage gain $A_{cm}$ is therefore

$$A_{cm} = \frac{V_{o\,cm}}{V_{cm}}$$

$$= \frac{R_C}{2R_E + r_e}$$

If $2R_E >> r_e$

$$A_{cm} = \frac{R_c}{2R_E}$$

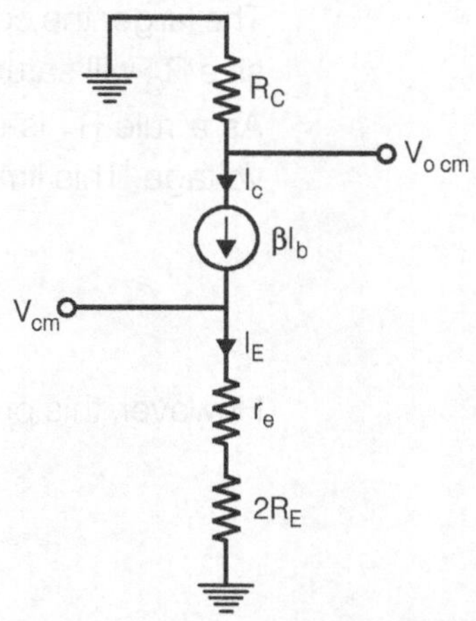


*Figure 9.18 $r_e$ model for one half of circuit*

## Common Mode Input Resistance $R_{in\,cm}$

Emitter current $$I_E = \frac{V_{cm}}{2R_E + r_e}$$

Collector current $$I_C = \frac{\beta}{\beta+1} I_E$$

$$= \frac{\beta}{\beta+1} \cdot \frac{V_{cm}}{2R_E + r_e}$$

Base current $$I_B = \frac{I_C}{\beta} = \frac{1}{1+\beta} \cdot \frac{V_{cm}}{2R_E + r_e}$$

Common mode input resistance

$$R_{in\,cm} = \frac{V_{cm}}{I_b} = \frac{V_{cm}}{\dfrac{V_{cm}}{(\beta+1)(2R_E + r_e)}}$$

$$R_{in\,cm} = (\beta + 1)(2R_E + r_e)$$

## 9.11 Differential Amplifier With Active Load

The larger the collector resistance $R_C$, the greater the differential voltage gain $A_d$, but a too large size $R_C$ will saturate the $Q_2$ transistor.

As a rule $R_C$ is selected to provide a quiescent voltage that is about half of collector supply voltage. This limits the size of $R_C$, which is limiting the voltage gain $A_d$.

$$A_d = \frac{R_C}{r_e}$$

However, this problem can be solved using an active load.

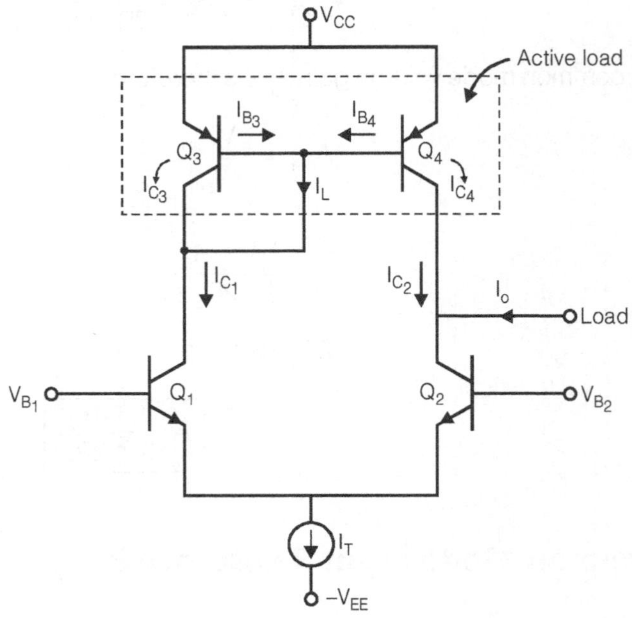


*Figure 9.19 Differential amplifier with active load*

If all the transistors are assumed identical, then a pure applied common mode voltage means that $V_{B_1} = V_{B_2}$ and current $I_T$ splits evenly between transistors $Q_1$ and $Q_2$. Neglecting base currents, $I_{C_3} = I_{C_4}$.

According to current source circuit, $I_{C_1} = I_{C_2} = I_{C_3} = I_{C_4} = \frac{I_T}{2}$ with no load connected at the output.

The quiescent load current may be determined as

$$I_L = I_{B_3} + I_{B_4} = \frac{1}{\beta}(I_{C_3} + I_{C_4})$$

$$= \frac{I_{C_1} + I_{C_2}}{\beta} = \frac{I_T}{\beta}$$

Since $\beta >> 1$, $I_L$ will be very small and thus there will be no loading problem.

# OPERATIONAL AMPLIFIER

## 9.12 Block Diagram of Operational Amplifier

- An operational amplifier, abbreviated as op-amp, is a very versatile electronic device which finds numerous practical applications.
- In analog computers, it is used to perform several mathematical operations like summing, integration, differentiation etc. and it gets its name from such use.
- Op-amps are used in many diverse fields like control systems, communications, instrumentation etc.

**"An operation amplifier is basically a very high gain, direct-coupled amplifier with high input impedance and low output impedance".**

The block diagram of a typical op-amp is as shown.

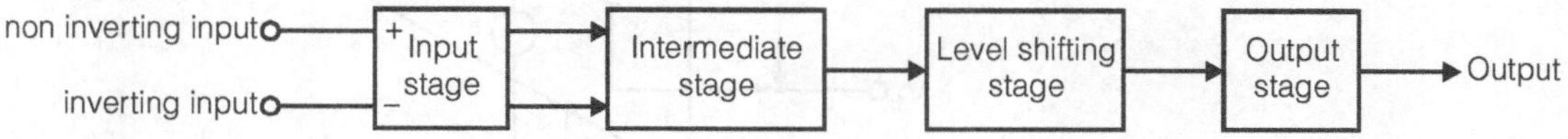


*Figure 9.20 Block diagram of op-amp*

- The input stage employs a dual-input, balanced output differential amplifier which provides a major part of the voltage gain and high input resistance.
- This drives the intermediate stage, which is another dual input, unbalanced output differential amplifier.
- Since, the input stage amplifier and the intermediate stage amplifier are direct-coupled. The dc voltage at the output of the intermediate stage tends to rise above the ground, which is not desirable.
- To make the dc voltage down to zero, a level shifter (or level translator) is employed. This usually is an emitter follower which also acts as a buffer, with very high input resistance and low output resistance.
- The output stage consists of a complementary symmetry class-B amplifier, which helps to increase the output voltage swing.

## 9.13 Symbol Representation of Op-Amp

The circuit symbol of an op-amp is as shown in Figure (9.21).

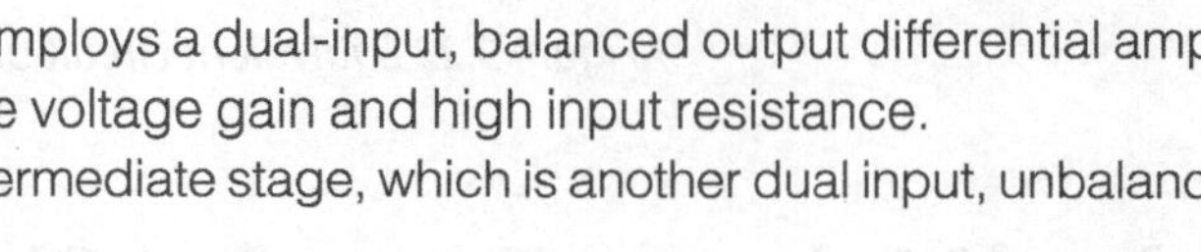


*Figure 9.21 Symbol of op-amp*

Let input voltages $V_1$ and $V_2$ be applied at the non-inverting and inverting terminals respectively of an op-amp. Let $V_o$ denote the output.
We have,
$V_o = A(V_1 - V_2)$ where A is the differential gain of the op-amp. Generally A is very high. (A is open loop dc gain).

# 9.14 Equivalent Circuit of an Op-Amp

Op-amp amplifies the difference between the two input signals applied at non inverting and inverting input terminals.

Let $V_1$ and $V_2$ denote the input signals.

$\therefore$ Difference input voltage $V_d = V_1 - V_2$

Let $A$ = Open loop gain of op-amp

We have, output voltage $V_o = A V_d$

Let $R_i$ = Differential input resistance

$R_o$ = Output resistance

The equivalent circuit of the op-amp is shown in Figure (9.22).

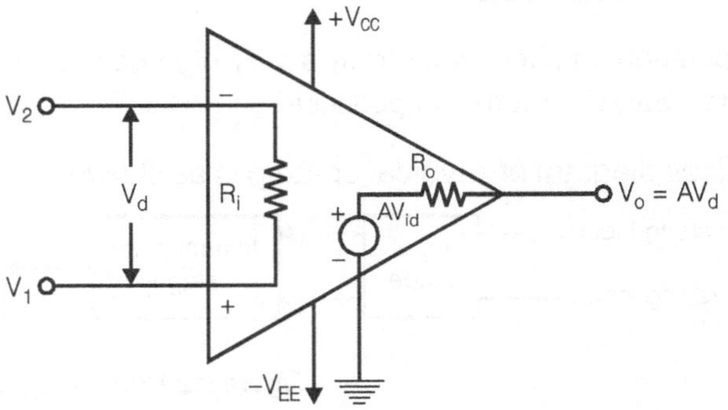


*Figure 9.22 Equivalent circuit of op-amp*

# 9.15 Ideal Voltage Transfer Characteristics

The voltage transfer curve is the graphical plot of the output voltage $V_o$ versus differential input voltage $V_d$.

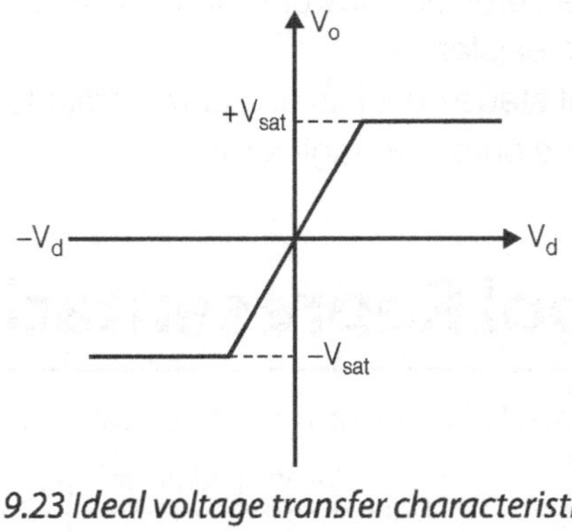


*Figure 9.23 Ideal voltage transfer characteristic graph*

The output voltage and the differential input voltage are related by the equation

$$V_o = A(V_1 - V_2) = AV_d$$

From the transfer characteristics

$V_o \propto V_d$ i.e. voltage gain A is assumed to be constant.

From the characteristics that the output voltage rises linearly as $V_d$ increases until it attains a maximum value slightly less than $V_{CC}$. For any further increase of $V_d$, there is no further increase of $V_o$. Similarly, for increasing negative values of $V_d$, $V_o$ rises linearly until it attains a negative maximum value, which is slightly less than $-V_{EE}$.

## 9.16 Pin Diagram of 741 IC

The pin diagram and schematic diagram of 741 IC operational amplifier is shown in Figure (9.24 and 9.25).

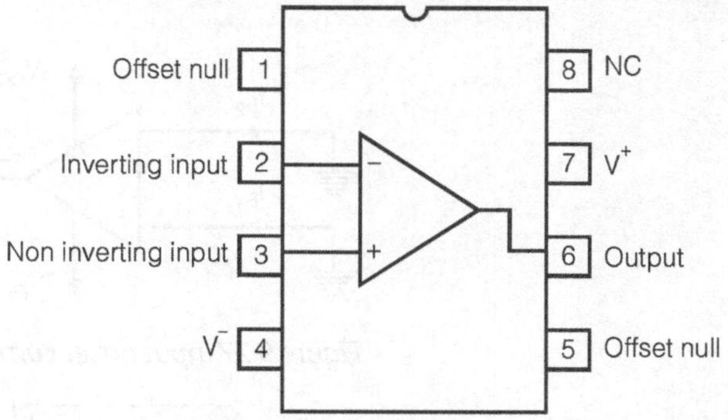


*Figure 9.24 Pin diagram of 741 IC*

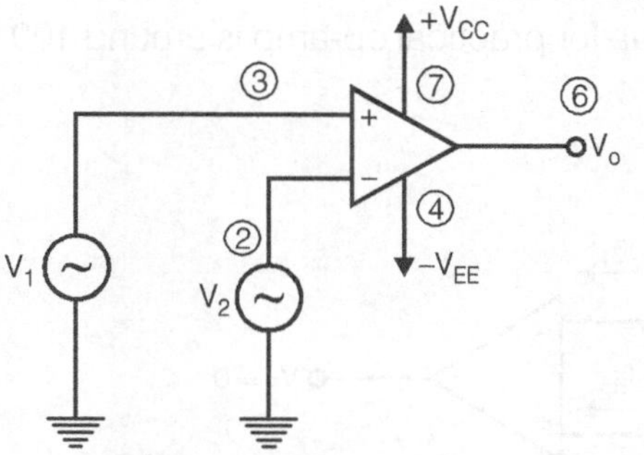


*Figure 9.25 Schematic diagram of 741 IC*

Where ② → Inverting input

③ → Non inverting input

④ → Negative supply

⑦ → Positive supply

⑥ → Output

## 9.17 Op-amp Characteristics

### INPUT OFFSET VOLTAGE ($V_{io}$)

If no external input signal is applied to the op-amp at the inverting and non inverting input terminals, the output must be zero.

But due to variation of transistors $Q_1$ and $Q_2$ in differential amplifier. The output of the op-amp will not be zero. This is known as offset.

**"It is the voltage that must be applied between the two input terminals of an op-amp to nullify output."**

$$V_{io} = (V_1 - V_2)$$

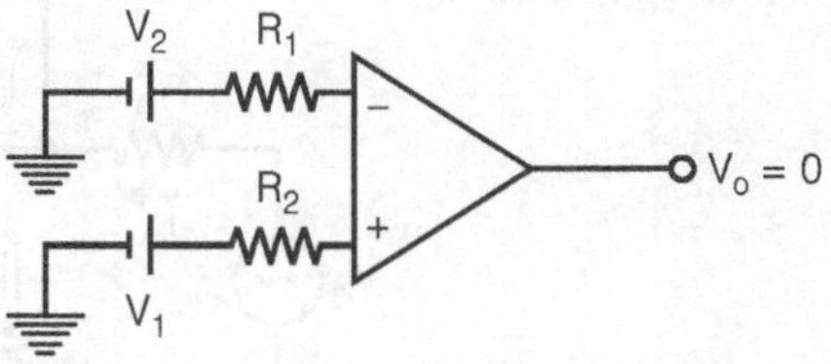


*Figure 9.26 Input offset voltage ($V_{io}$)*

The value for ideal op-amp is $V_{io}$ = 0 V practical value is around 100 μV (typical).

## INPUT OFFSET CURRENT ($I_{io}$)

Though for an ideal op-amp, the input impedance is infinite, it is not so practically, so the output current $I_o$ draws from the source, however small if may be.

**"The algebric difference between the currents into the inverting and non inverting terminals is referred to as input offset current $I_{io}$."**

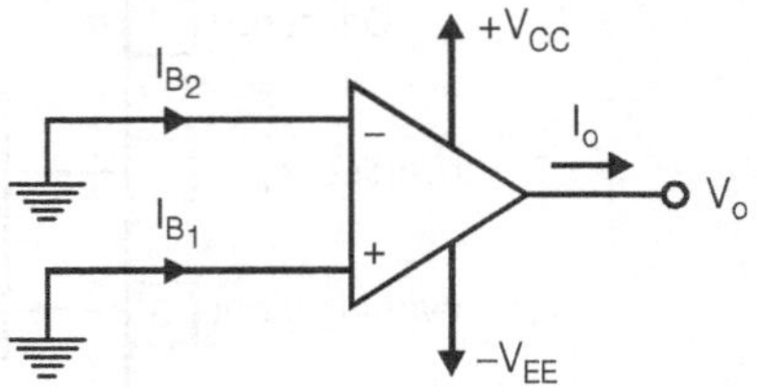


*Figure 9.27 Input offset current ($I_{io}$)*

$$I_{io} = |I_{B_1} - I_{B_2}|$$

For ideal op-amp, $I_{io}$ = 0, the typical value for practical op-amp is around 100 nA.

## INPUT BIAS CURRENT

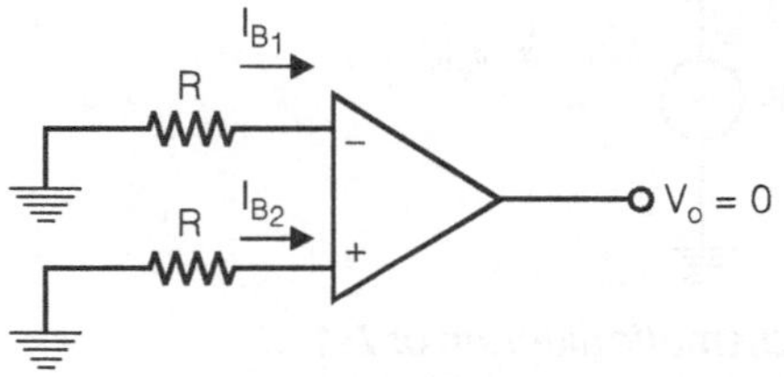


*Figure 9.28 Input bias current*

**"The average value of the two currents flowing into the op-amp input terminals is called as input bias current and is denoted as $I_B$."**

$$I_B = \frac{I_{B_1} + I_{B_2}}{2}$$

Even though very small, the input bias current $I_B$ can cause a significant output offset voltage in circuits using relatively large feedback resistors. This output offset voltage may not be as large as that caused by the input offset voltage.

To eliminate or reduce the output offset voltage due to input bias current $I_B$ some scheme is implemented in Figure (9.29).

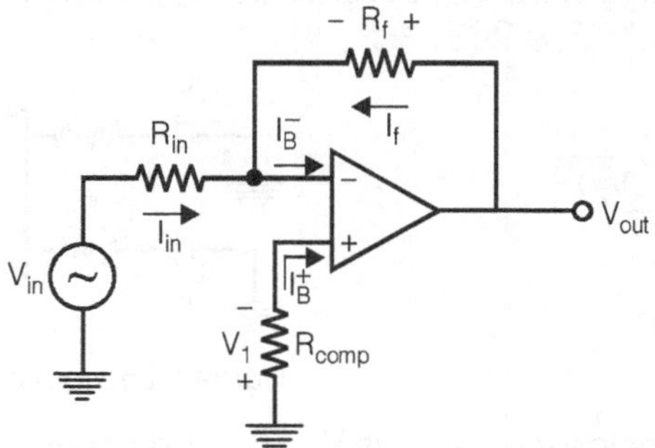


*Figure 9.29 Bias current compensation circuit*

A compensating resistor has been added between ground and the non inverting input current $I_B^+$ flowing through this resistor develops a voltage $V_1$.

Summing the voltages around the input feedback output loop, we have

$$-V_1 + 0 + V_f - V_{out} = 0$$

(or) $$V_{out} = V_f - V_1$$

If the $R_{comp}$ is selected of proper size, $V_1$ will cancel $V_f$ and output will be zero. The value of $R_{comp}$ is derived below:

$$V_1 = I_B^+ R_{comp}$$

(or) $$I_B^+ = \frac{V_1}{R_{comp}}$$

$$I_{in} = \frac{V_{in}}{R_{in}} \text{ and } I_f = \frac{V_f}{R_f} \qquad (\therefore \; V_{in} = V_1)$$

For compensation $$V_f = V_1$$

So $$I_f = \frac{V_f}{R_f} = \frac{V_1}{R_f}$$

The current $$I_B^- = I_f + I_{in} = \frac{V_1}{R_f} + \frac{V_1}{R_{in}} = V_1\left[\frac{1}{R_f} + \frac{1}{R_{in}}\right] = V_1\left[\frac{R_{in}+R_f}{R_f R_{in}}\right]$$

Assuming $$I_B^- = I_B^+$$

$$V_1\left[\frac{R_{in}+R_f}{R_{in} R_f}\right] = \frac{V_1}{R_{comp}}$$

or $$R_{comp} = \frac{R_{in} R_f}{R_{in} + R_f} = R_{in} \| R_f$$

So for compensating for bias current

$$\boxed{R_{comp} = R_{in} \| R_f}$$

# COMMON MODE REJECTION RATIO (CMRR)

Here the important point is that noise signals are not the signals that are desired to be amplified in the differential amplifier. Their distinguishing feature is that the noise signal appears equally at both inputs of the circuit. It means that any undesired (noise) signals that appear in polarity, or common to both input terminals, will be largely rejected, or cancelled out at the differential amplifier output.

**"A measure of rejection of noise signals common to both inputs is referred to as the common-mode rejection of the amplifier".**

CMRR is defined as the ratio of differential voltage gain to common-mode voltage gain and is given as

$$\boxed{\text{CMRR} = \frac{A_d}{A_{cm}}}$$

Consider the block diagram of op-amp.

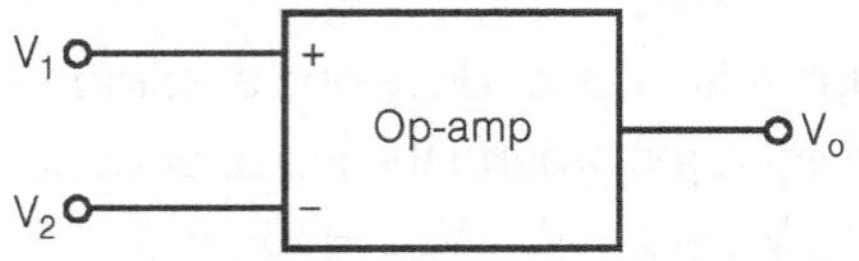


In ideal op-amp the output voltage depends only on the difference of the two input voltages hence

If $V_1 = V_2$ then $V_o$ reduces to zero.

In practical op-amp the output voltage depends on the difference and average of the two input voltages hence

If $V_1 = V_2$, then $V_o$ never reduces to zero.

By applying superposition principle

$$V_o = A_1 V_1 + A_2 V_2 \quad ...(i)$$

i.e. total response is sum of individual response.

$$A_1 = \text{gain for input } V_1 \quad \text{when } V_2 = 0$$

$$A_2 = \text{gain for input } V_2 \quad \text{when } V_1 = 0$$

Difference input signal voltage $V_d$

$$\boxed{V_d = V_1 - V_2} \quad ...(ii)$$

Common mode signal voltage

$$\boxed{V_C = \frac{V_1 + V_2}{2}} \quad ...(iii)$$

Solving equation (ii) and (iii) for $V_1$ and $V_2$, we get

$$V_d = V_1 - V_2 \quad ...(A)$$

$$2V_C = V_1 + V_2 \quad ...(B)$$

Adding A and B

$$V_d + 2V_C = 2V_1$$

$$V_1 = V_C + \frac{V_d}{2}$$

Subtracting A and B

$$V_d - 2V_C = -2V_2$$

$$V_2 = V_C - \frac{V_d}{2}$$

Substitute the values of $V_1$ and $V_2$ in equation (i)

$$V_o = A_1\left[V_C + \frac{V_d}{2}\right] + A_2\left[V_C - \frac{V_d}{2}\right] = V_C[A_1 + A_2] + \frac{V_d}{2}[A_1 - A_2]$$

i.e.

$$\boxed{V_o = A_c V_c + A_d V_d}$$

Where $A_C = A_1 + A_2$ (Common mode gain)

and $A_d = \frac{A_1 - A_2}{2}$ (Differential mode gain)

$$V_o = A_d V_d\left[1 + \frac{A_C}{A_d}\frac{V_C}{V_d}\right]$$

$$\boxed{V_o = A_d V_d\left[1 + \frac{1}{CMRR}\frac{V_C}{V_d}\right]}$$

The above equation gives the relationship between output voltage $V_o$ and CMRR.
For ideal op-amp common mode gain $A_C$ is zero.
i.e. for ideal op-amp CMRR is ∞.
In practical op-amp CMRR is $10^6$ or 120 dB.
CMRR can be expressed in decibel

$$\text{CMRR in dB} = 20\log_{10}\left|\frac{A_d}{A_c}\right|$$

$$(\text{CMRR})_{dB} = 20\log|A_d| - 20\log|A_c|$$

**CMRR is called the figure of merit for the op-amp.**

## SLEW RATE (SR)

**"Slew rate is defined as the maximum rate at which the amplifier output can change in volts/μs".**

$$SR = \left.\frac{\Delta V_o}{\Delta t}\right|_{\mu s} \text{ Volts/}\mu\text{sec}$$

Slew rate indicates how quick is the op-amp to respond to the sudden changes in the input parameters with respect to time interval of 1 μsec.

$$SR = \frac{\Delta V_o}{\Delta V_i} \times \frac{\Delta V_i}{\Delta t} \text{ V/}\mu\text{sec}$$

$$= |A_{CL}| \times \frac{\Delta V_i}{\Delta t} \text{ V/}\mu\text{sec}$$

$$\boxed{SR = |A_{CL}|\frac{\Delta V_i}{\Delta t}} \text{ V/}\mu\text{sec}$$

For the better performance of the op-amp slew rate must be larger.
For ideal op-amp slew rate is ∞. For practical op-amp slew rate is 80 V/μsec.
The maximum operating input signal frequency for the op-amp is limited by its slew rate and given by

$$\omega_{max} = \frac{SR}{K} \text{ rad/sec}$$

$$f_{max} = \frac{SR}{2\pi K} \text{ Hz}$$

Where k is a constant given by

$$k = |A_{CL}| \times \Delta V_i$$

**If the operating input signal frequency ω is greater than $\omega_{max}$ then the output of the op-amp will be distorted. So for distortionless output**

$$\omega \le \omega_{max}$$

## GAIN BANDWIDTH PRODUCT

**"This is the bandwidth (BW) of the op-amp when the voltage gain is unity".**

As the frequency increases, the gain decreases because as f increases.

$X_C = \dfrac{1}{2\pi fC}$ decreases, so $R_L$ also decreases.

Hence $V_o = I_L \times R_L$ also decreases. So the gain $A_V$ decreases.

The gain bandwidth product indicates the capability of the op-amp to amplify the input signals of larger frequency.

The other terms used to describe gain bandwidth are closed loop bandwidth and unity gain bandwidth.

Ideal value of gain bandwidth product is ∞.

Typical value is 1 MHz.

**Characteristics of op-amp in ideal and practical cases**

| S.No. | Characteristic | Ideal | Practical |
|---|---|---|---|
| 1. | CMRR | ∞ | $10^6$ or 120dB |
| 2. | Slew rate | ∞ | 80 V/μsec |
| 3. | Input resistance | ∞ | $10^6\Omega$ |
| 4. | Output resistance | 0 | 100Ω |
| 5. | Voltage gain $A_V$ | ∞ | $10^6$ |
| 6. | Bandwidth | ∞ | $10^6$Hz |
| 7. | Offset voltage | 0 | Negligible |
| 8. | Offset current | 0 | Negligible |

# 9.18 Importance of Negative Feedback in Op-Amp

- If gain in op-amp will be more then stability will be less, so to reduce the gain, we can apply negative feedback.
- Open loop gain of op-amp is very large and therefore the stability is less. To increase the stability a small amount of negative feedback is applied so that gain is reduced and frequency stability is increased.
- Closed loop gain $A_{CL}$ is always less than open loop gain ($A_{OL}$).
- Op-amp is an analog IC that it can amplify a DC signal along with a wide band of ac signals such as sine, square and triangular waves.
- Op-amp is basically linear analog IC linear means we can apply superposition principle.

## 9.19 Virtual Ground

This is an important concept for an op-amp, as it plays a vital role in deriving the expression for output voltage $V_o$.

Consider the op-amp circuit shown in Figure (9.30)

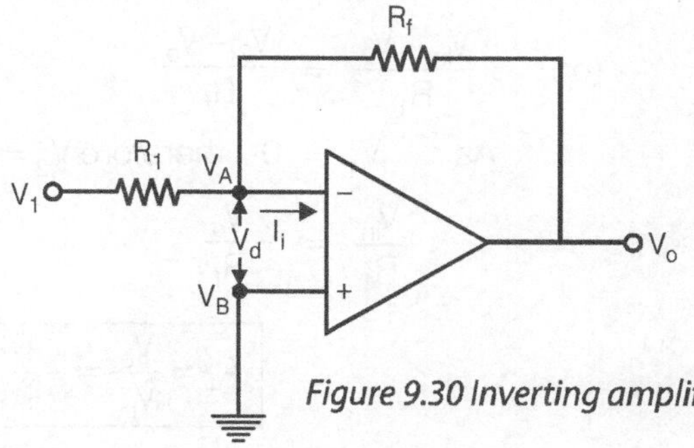


*Figure 9.30 Inverting amplifier*

$V_B$ is a potential defined at non-inverting input.

$V_A$ is a potential defined at feedback point of inverting input.

$V_d$ is difference between $V_B$ and $V_A$.

i.e. $$V_d = V_B - V_A$$

According to ideal characteristics of op-amp.

Input resistance $R_i$ is infinite, therefore the current entering into op-amp $I_i$ will be zero. That means differential voltage $V_d$ will be zero.

i.e. $$0 = V_B - V_A$$

$$\therefore \quad V_B = V_A$$

The node voltage $V_B$ at terminal positive is same as negative feedback node point $V_A$.

This concept is called as virtual ground concept.

# PRACTICAL APPLICATIONS OF OP-AMP

## 9.20 Inverting Amplifier

The Figure (9.31) shows the inverting amplifier. Its non inverting input terminal is grounded where as the external input signal $V_{in}$ is applied to the inverting input terminal through resistance $R_1$. A feedback resistor $R_f$ is connected from the output to the inverting input terminal of the op-amp.

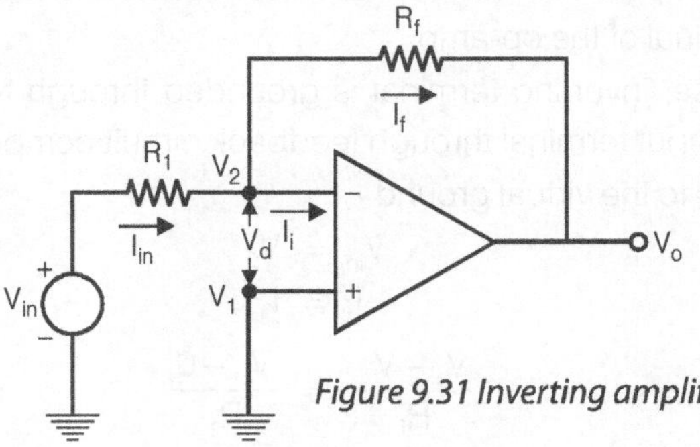


*Figure 9.31 Inverting amplifier*

This arrangement provides voltage shunt negative feedback. According to the virtual ground concept.

$$\therefore \quad V_1 = V_2$$

Applying Kirchhoffs current law at input node $V_2$ in Figure (9.31).

$$I_{in} = I_f + I_i$$

$$\text{and} \quad I_{in} = I_f \quad \text{since } I_i = 0$$

$$\text{(or)} \quad \frac{V_{in} - V_2}{R_1} = \frac{V_2 - V_o}{R_f}$$

$$\text{As} \quad V_1 = 0 \quad \text{therefore } V_2 = 0$$

$$\frac{V_{in}}{R_1} = -\frac{V_o}{R_f}$$

$$A_f = \frac{V_o}{V_{in}} = -\frac{R_f}{R_1}$$

The negative sign in above equation indicates that the input and output voltages are out of phase by 180°. Because of this phase inversion, this configuration is commonly called as inverting amplifier.

$R_1$ and $R_f$ are external resistances. Thus it is seen that, for this type of amplifier, the closed loop gain depends only upon the feedback resistor $R_f$ and $R_1$, and by properly choosing $R_f$ and $R_1$, any desired gain can be obtained.

## 9.21 Non-Inverting Amplifier

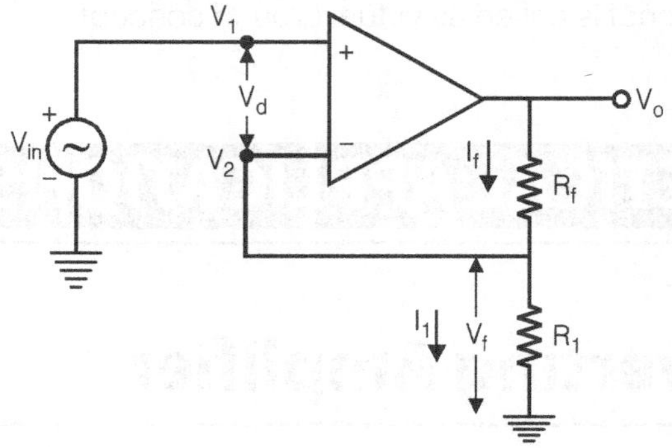


*Figure 9.32 Non-inverting amplifier*

The circuit is commonly known as non-inverting amplifier with feedback or closed loop non inverting amplifier because it uses a feedback and input signal is applied to the non inverting input terminal of the op-amp.

In this case, inverting terminal is grounded through resistor $R_1$ and output is applied to the inverting input terminal through feedback circuit composed of two resistors $R_1$ and $R_f$.

According to the virtual ground

$$V_{in} = V_2$$

$$I_f = I_1$$

$$\frac{V_o - V_{in}}{R_f} = \frac{V_{in} - 0}{R_1}$$

$$A_V = \frac{V_o}{V_{in}} = \left[1 + \frac{R_f}{R_1}\right]$$

$R_f$ and $R_1$ are both positive. Hence gain is positive. Also there is no phase inversion.

The expression for closed loop gain reveals that the voltage gain is always greater than unity, whatever the values of $R_1$ and $R_f$. Also, the closed loop is independent of the open loop gain, but depends only upon the external resistors $R_1$ and $R_f$.

## 9.22 Phase Shifter

In an inverting amplifier if $R_1 = R_f = R$ then

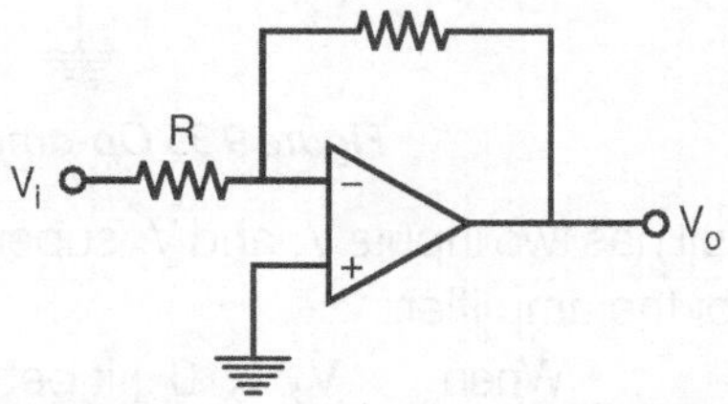


*Figure 9.33 Inverting amplifier with $R_1 = R_f = R$*

$$\text{Closed loop gain } A_V = -\frac{R_f}{R_1} = -\frac{R}{R} = -1$$

Negative sign indicates that the input and output voltages are out of phase by 180°, but magnitude is same. Therefore it can work like a phase shifter.

## 9.23 Voltage Follower

In non inverting amplifier, when the resistor $R_f$ is set equal to zero or $R_1$ is made ∞ by keeping it open-circuited.

The output voltage $V_{out} = V_{in}$

i.e. the output voltage of the op-amp exactly tracks the input voltage both in sign and magnitude. This is the reason that this circuit is called a voltage follower.

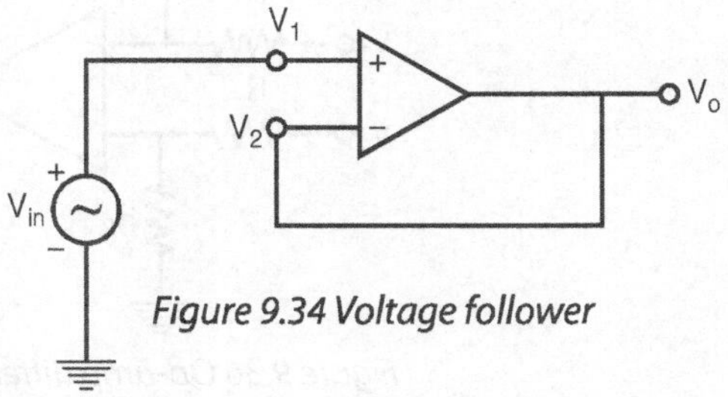


*Figure 9.34 Voltage follower*

The direct connection of the output voltage to the inverting terminal of the op-amp represents the case of 100% negative feedback of the output to the input.

From the Figure (9.34), $V_2$ is equal to $V_0$ and $V_{in}$ is equal to $V_1$.

$$\therefore V_o = V_{in}$$

## 9.24 Differential Amplifier

Some times it is necessary to amplify the voltage difference between two input lines neither of which is grounded. In this case, the amplifier is called a differential amplifier.
The circuit of an op-amp differential amplifier is shown in Figure (9.35).

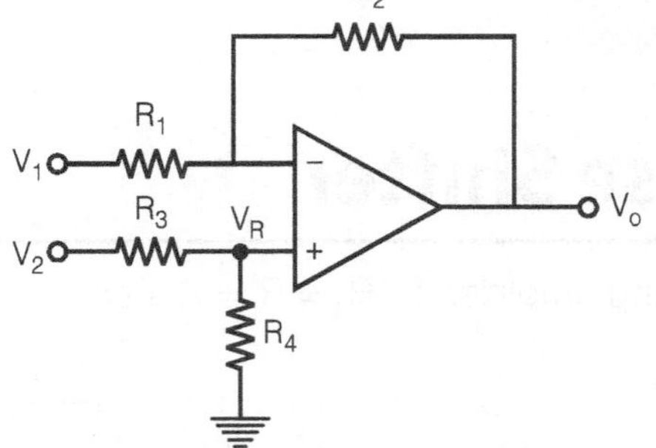


*Figure 9.35 Op-amp differential amplifier*

Since the circuit has two inputs $V_1$ and $V_2$ superposition theorem will be used for determination of voltage gain of the amplifier.

When $V_2 = 0$ ; it behaves like a inverting mode

$$\therefore \quad V_{01} = -\frac{R_2}{R_1} V_1$$

When $V_1 = 0$ ; it behaves like a non-inverting mode.

$$V_{02} = \left(1+\frac{R_2}{R_1}\right) V_R$$

Where $$V_R = \frac{V_2 \times R_4}{R_3 + R_4}$$

$$\therefore \quad V_O = V_{01} + V_{02}$$

$$V_o = -\frac{R_2}{R_1} V_1 + \left(1+\frac{R_2}{R_1}\right)\frac{V_2 \times R_4}{R_3 + R_4}$$

If suppose the differential amplifier with $R_3 = R_1$ and $R_4 = R_2$ then

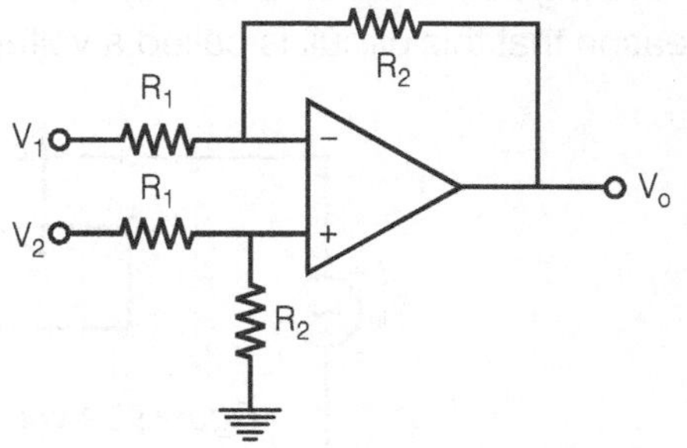


*Figure 9.36 Op-amp differential amplifier*

$$V_o = -\frac{R_2}{R_1} V_1 + \left(1+\frac{R_2}{R_1}\right) V_2 \times \frac{R_4}{R_3 + R_4} = -\frac{R_2}{R_1} V_1 + \left(1+\frac{R_2}{R_1}\right) V_2 \times \frac{R_2}{R_1 + R_2} = \frac{R_2}{R_1}(V_2 - V_1)$$

$$\therefore \quad V_o = \frac{R_2}{R_1}(V_2 - V_1)$$

# 9.25 Subtracter

If $R_1 = R_2 = R_3 = R_4 = R$ in op-amp differential amplifier then

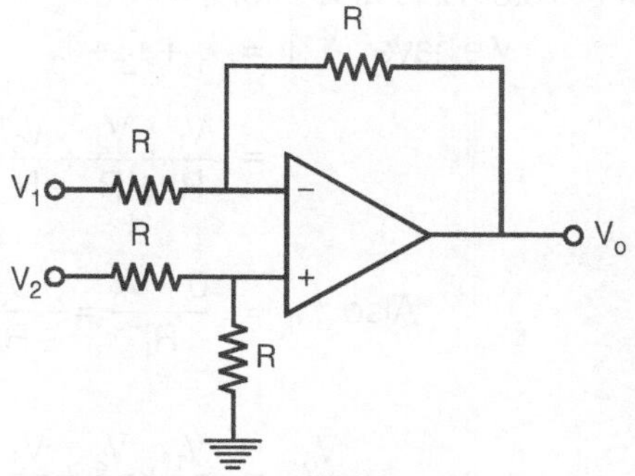


*Figure 9.37 Subtractor*

We have $$V_o = \frac{R_2}{R_1}(V_2 - V_1)$$

Here $$R_1 = R_2 = R$$

$$V_o = \frac{R}{R}(V_2 - V_1)$$

$$\boxed{V_o = V_2 - V_1}$$

Which is the practical application of op-amp as a subtracter.

# 9.26 Inverting Adder

Inverting adder can also be called as summing amplifier. The output voltage is the sum of all input voltages with negative sign.

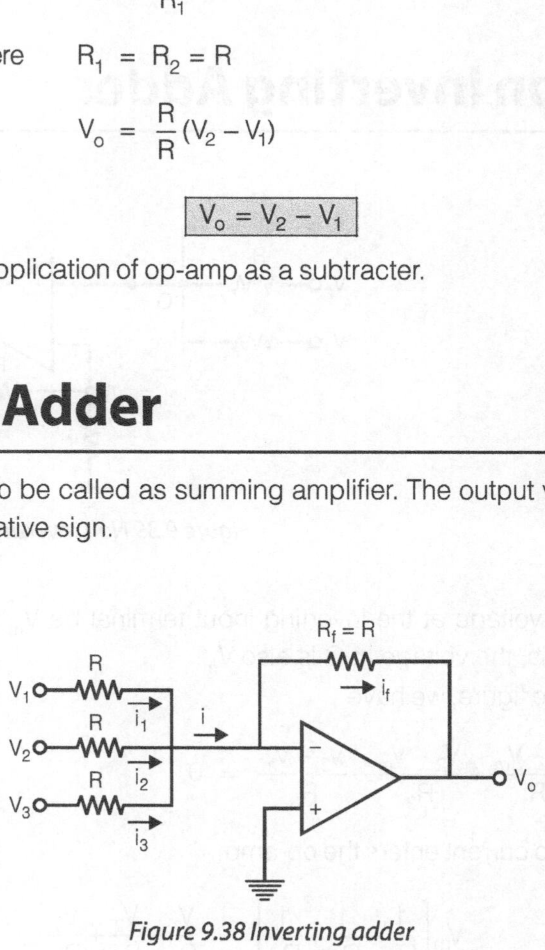


*Figure 9.38 Inverting adder*

Several input voltages $V_1$, $V_2$ and $V_3$ are applied to the inverting input of the op-amp through resistors $R_1$, $R_2$ and $R_3$ as shown, keeping the non inverting input grounded. If i denotes the input current, we have

$$i = i_1 + i_2 + i_3$$

Because of virtual ground, practically no current flows into the op-amp, so that current through the feedback resistor $R_f$ is also i i.e. $i_f = i$

$$\text{we have} \qquad i = i_1 + i_2 + i_3$$

$$= \frac{V_1}{R} + \frac{V_2}{R} + \frac{V_3}{R}$$

$$\text{Also} \quad i_f = \frac{0 - V_o}{R_f} = -\frac{V_o}{R_f}$$

$$\therefore \qquad -\frac{V_o}{R} = \frac{V_1}{R} + \frac{V_2}{R} + \frac{V_3}{R}$$

$$\boxed{V_o = -(V_1 + V_2 + V_3)}$$

i.e. the output voltage is the sum of the input voltages, with negative sign.

## 9.27 Non Inverting Adder

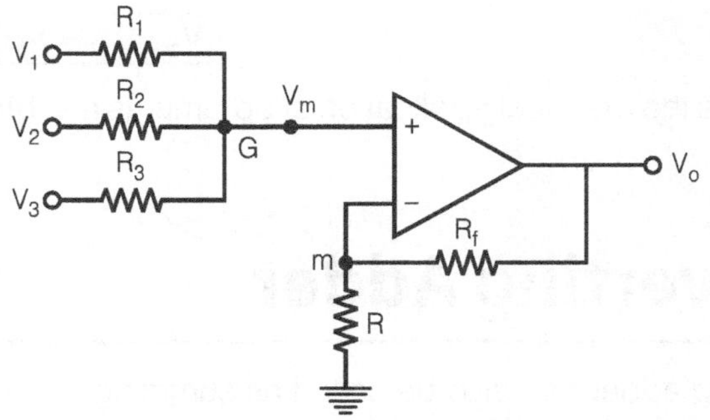


*Figure 9.39 Non inverting adder*

Let the voltage at the inverting input terminal be $V_m$. Because of virtual ground at the input terminals, the voltage at G is also $V_m$.

From the figure, we have

$$\frac{V_1 - V_m}{R_1} + \frac{V_2 - V_m}{R_2} + \frac{V_3 - V_m}{R_3} = 0,$$

Since no current enters the op-amp.

$$\text{(or)} \qquad V_m\left[\frac{1}{R_1} + \frac{1}{R_2} + \frac{1}{R_3}\right] = \frac{V_1}{R_1} + \frac{V_2}{R_2} + \frac{V_3}{R_3}$$

(or) $$V_m = \frac{\frac{V_1}{R_1}+\frac{V_2}{R_2}+\frac{V_3}{R_3}}{\frac{1}{R_1}+\frac{1}{R_2}+\frac{1}{R_3}}$$

As it is a non inverting amplifier, closed loop voltage gain.

$$\frac{V_o}{V_m} = \left(1+\frac{R_f}{R}\right)$$

$$\therefore \quad V_o = V_m\left[1+\frac{R_f}{R}\right]$$

Where $$V_m = \frac{\frac{V_1}{R_1}+\frac{V_2}{R_2}+\frac{V_3}{R_3}}{\frac{1}{R_1}+\frac{1}{R_2}+\frac{1}{R_3}}$$

If $$R = R_1 = R_2 = R_3 = \frac{R_f}{2}$$

We have $$\boxed{V_o = V_1 + V_2 + V_3}$$

Hence, the output is the sum of the input voltages, without change of sign.

## 9.28 Current to Voltage Converter

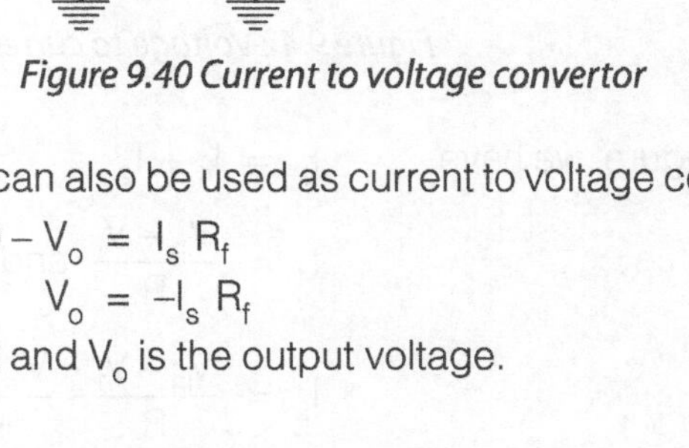


*Figure 9.40 Current to voltage convertor*

Op-amp inverting amplifier can also be used as current to voltage converter

$$0 - V_o = I_s R_f$$
$$V_o = -I_s R_f$$

Where $I_s$ is the input current and $V_o$ is the output voltage.

It is seen that $V_o \propto I_s$.

Hence an input current is converted into an output voltage.

## 9.29 Voltage to Current Converter

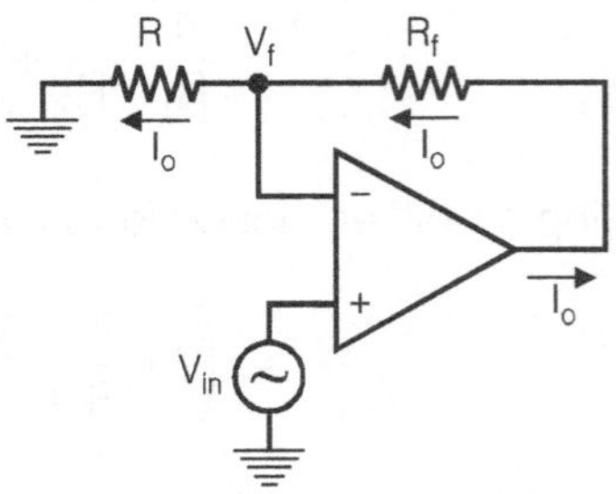


*Figure 9.41 Voltage to current converter*

The input voltage $V_{in}$ is applied to the non inverting input terminal of the op-amp. Because of virtual ground

$$V_{in} = V_f$$
$$V_f = I_o R$$
$$V_{in} = I_o R$$

$$I_o = \frac{V_{in}}{R}$$

Hence the input voltage $V_{in}$ gets converted into an output current $I_o$.

## 9.30 Voltage to Current Converter (with load)

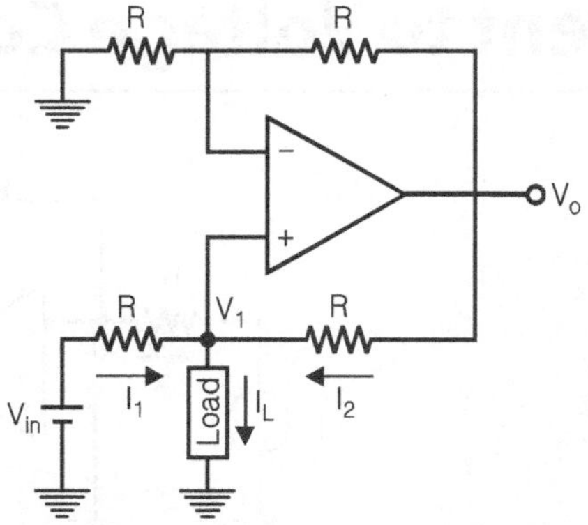


*Figure 9.42 Voltage to current converter with load*

From the figure, we have $\quad I_L = I_1 + I_2$

$$I_1 = \frac{V_{in} - V_1}{R} \text{ and } I_2 = \frac{V_o - V_1}{R}$$

$$I_L = \frac{V_{in} - V_1}{R} + \frac{V_o - V_1}{R} = \frac{V_{in} + V_o - 2V_1}{R}$$

$$V_1 = \frac{V_{in} + V_o - I_L R}{2}$$

We know that the closed loop gain of the non inverting op-amp is given as

$\left(1+\frac{R_f}{R}\right)$ Where $R_f$ = resistance of feedback path

$$\text{But} \quad R_f = R$$

$$\text{Overall gain} = 1+\frac{R}{R} = 2$$

$$\text{Output voltage} \quad V_o = 2V_1$$

$$\text{Where} \quad V_1 = \frac{V_{in}+V_o-I_L R}{2}$$

$$V_o = V_{in}+V_o-I_L R$$

$$I_L R = V_{in}$$

$$\boxed{I_L = \frac{V_{in}}{R}}$$

The input voltage gets converted into output load current $I_L$,

# 9.31 Op-Amp Integrator

An integrator is a circuit that performs a mathematical operation called integration. Integration is a process of continuous addition.

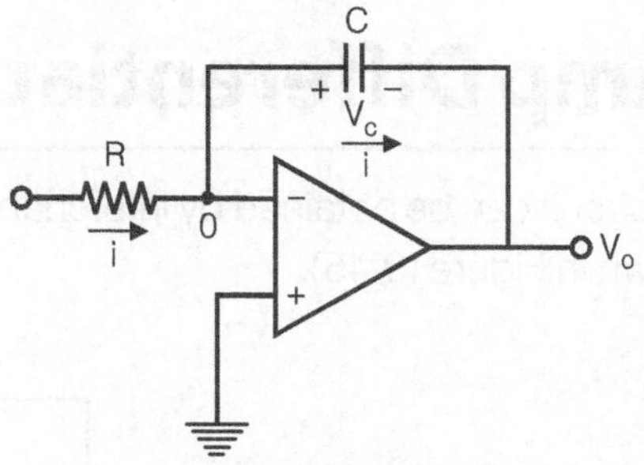


*Figure 9.43 Integrator*

The most popular application of an integrator is to produce a ramp of output voltage, which is a linearly increasing or decreasing voltage. The feedback is through capacitor C instead of resistor $R_f$.

From the figure, according to the virtual ground concept

$$0 - V_o = V_c$$

$$V_o = -V_c$$

$$= -\frac{1}{C}\int i\,dt = -\frac{1}{C}\int \frac{V_{in}}{R}dt$$

$$V_o = -\frac{1}{RC}\int V_{in}\,dt$$

$$\boxed{V_o = -\frac{1}{\tau}\int V_{in}\,dt}$$

$$\text{Where} \quad \tau = RC \qquad \text{(time constant)}$$

- If the input voltage is a step voltage, then the output voltage will be a ramp or linearly changing voltage.
- If the input voltage is a square-wave, the output voltage will be a triangular wave.
- If the input voltage is a sinusoidal one, the output voltage will be cosine wave.

Integrators are widely used in ramp or sweep generators, filters, analog computers etc.

## SUMMING INTEGRATOR

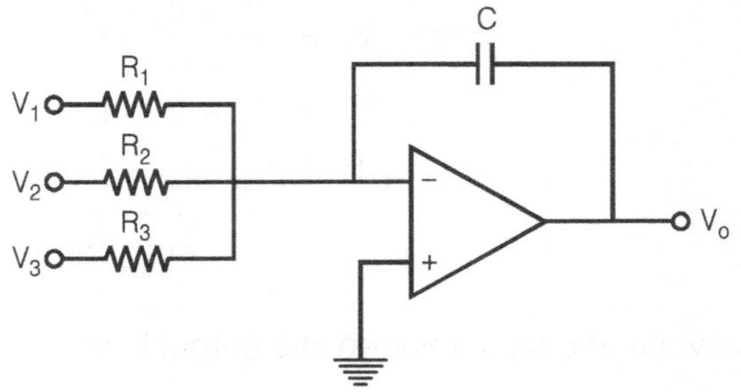


*Figure 9.44 Summing integrator circuit*

Generally, summing integrator circuits are used in analog computers.

The output $$V_o = -\left[\frac{1}{R_1C}\int V_1\,dt + \frac{1}{R_2C}\int V_2\,dt + \frac{1}{R_3C}\int V_3\,dt\right]$$

# 9.32 Op-Amp Differentiator

Differentiator circuit can be obtained by interchanging the resistor and capacitor of the integrator circuit as shown in Figure (9.45).

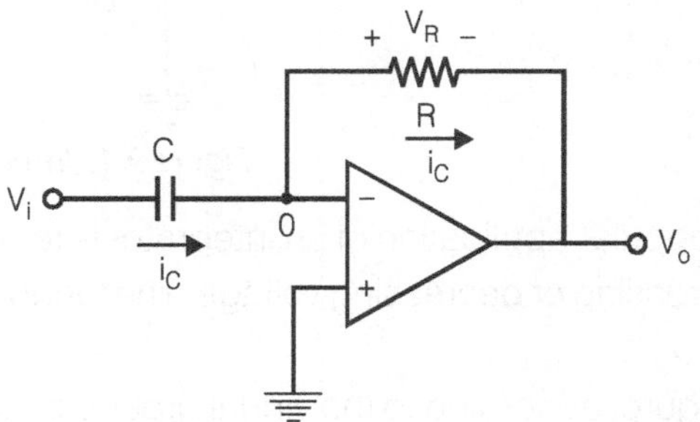


*Figure 9.45 Differentiator*

We have from figure $$i_c = C\frac{dv_i}{dt}$$

$$0 - V_o = V_R$$

$$V_o = -V_R$$

$$V_o = -i_C R = -RC\frac{dv_i}{dt}$$

$$V_o = -RC\frac{dv_i}{dt}$$

$$V_o = -\tau \frac{dv_i}{dt}$$

Where $\tau$ = RC (time constant)

Its function is to provide an output voltage proportional to the rate of change of the input voltage. It is an inverse mathematical operation to that of an integrator.

- The common application of the op-amp differentiator is to produce very narrow spikes.
- A cosine wave input will generate a sine wave output.
- A triangular input will produce a square wave output.

But the above circuit shown in Figure (9.45) will not be able to do so because of some practical problems like unstabilising and susceptibility to high-frequency noise.

## PRACTICAL OP-AMP DIFFERENTIATOR

Both of the problems of unstability and susceptibility to high frequency noise can be overcome by modifying the circuit given below in Figure (9.46).

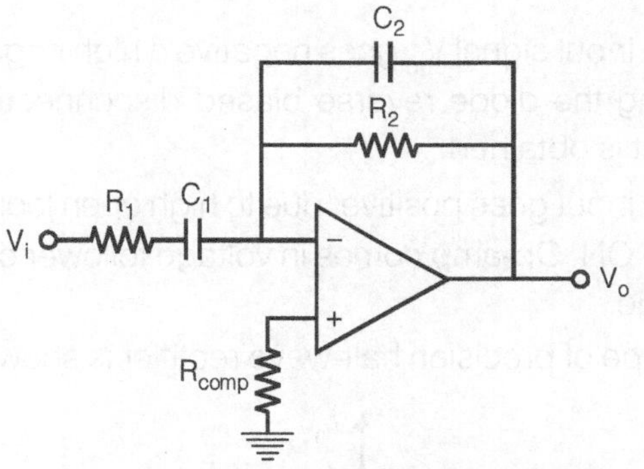


*Figure 9.46 Practical differentiator*

# 9.33 Precision Rectifiers

In the context of rectification using semiconductor diodes, we studied that a diode readily conductor when it is forward biased and the bias voltage is more than the cut in voltage, which is 0.7 V for silicon diodes and 0.3 V for germanium diodes. Thus a diode can not conduct if the biasing voltage is less than the cut in voltage.

If voltages less than the threshold voltage, as for example, a few milli volts or micro volts are to be rectified, precision rectifiers are needed. An op-amp ideally fulfils this need.

## CIRCUIT-1

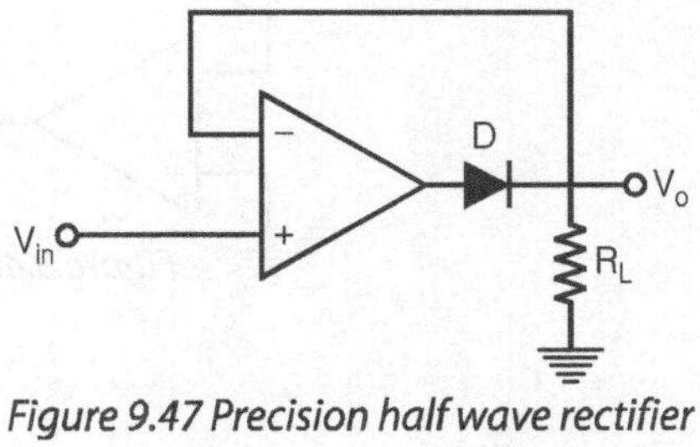


*Figure 9.47 Precision half wave rectifier*

In the circuit shown diode is used in non inverting configuration. Diode is driven by the op-amp output.

- When diode is OFF, op-amp acts in open loop configuration.

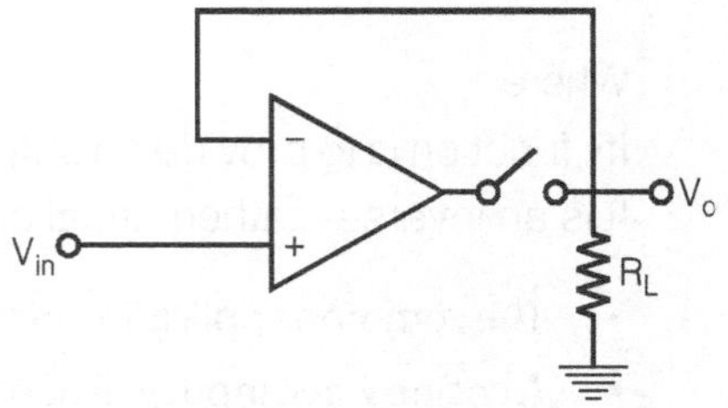


- When diode is ON, op-amp acts in voltage follower configuration.

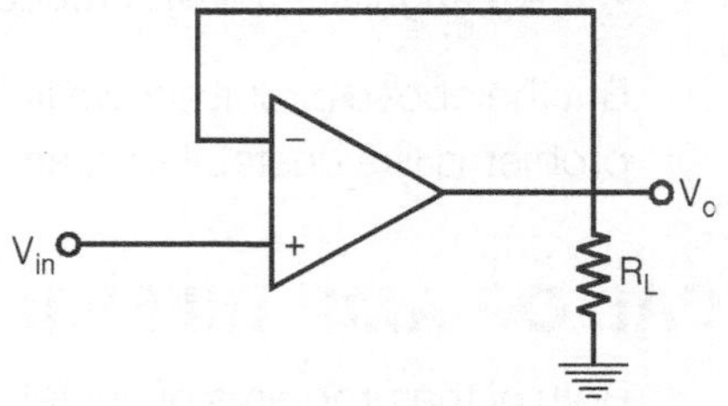


## Operation

- When input signal $V_{in}$ goes negative a high negative signal appears at the output of op-amp making the diode reverse biased disconnecting $V_o$. Thus during negative half cycle no output is obtained.
- When input goes positive, due to high open loop gain op-amps output goes high driving the diode ON. Op-amp comes in voltage follower configuration the output voltage equals input voltage.

Wave shape of precision half-wave rectifier is shown.

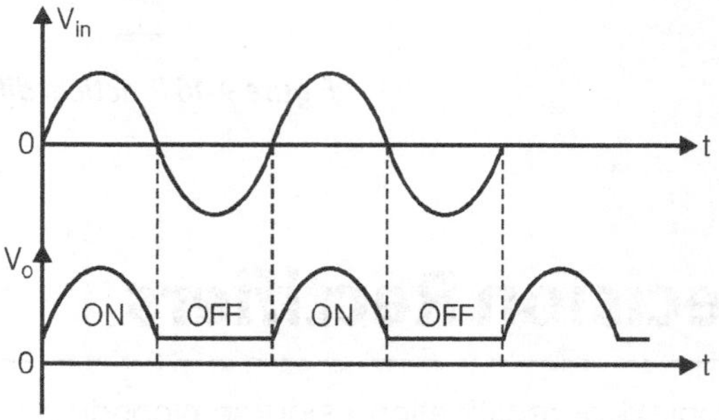


## CIRCUIT-2

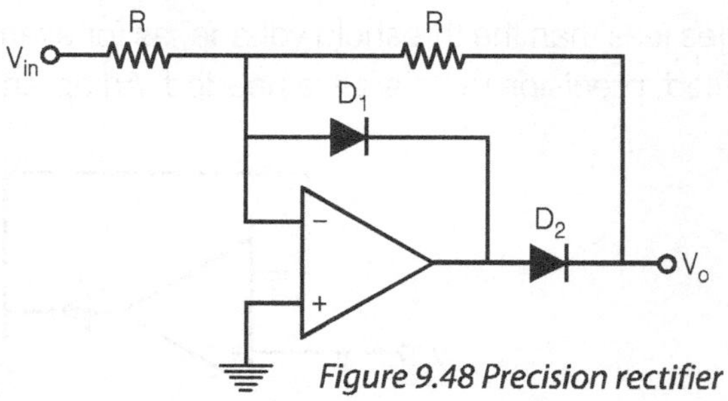


*Figure 9.48 Precision rectifier*

In positive half cycle, $D_1$ diode is forward bias and $D_2$ diode is reverse bias. The equivalent circuit is shown below:

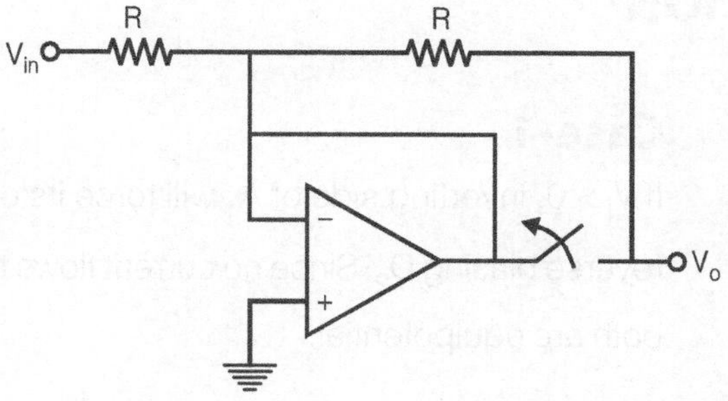


$\therefore \quad V_o = 0$

In negative half cycle $D_1$ diode is reverse bias and $D_2$ diode is forward bias. The equivalent circuit is shown below:

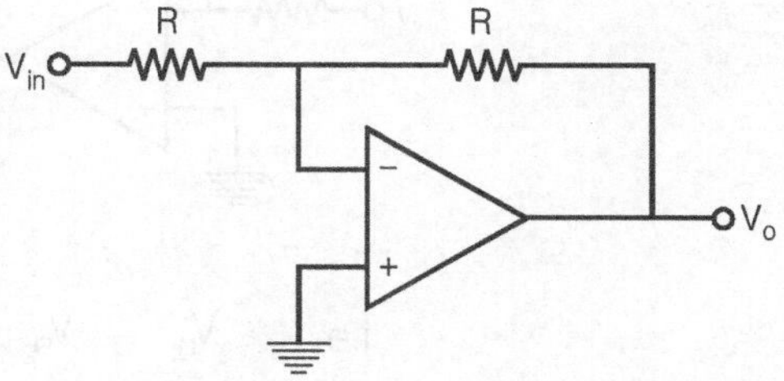


$\therefore \quad V_o = -V_i$ (Phase shifter)

Wave shape for this circuit is shown below:

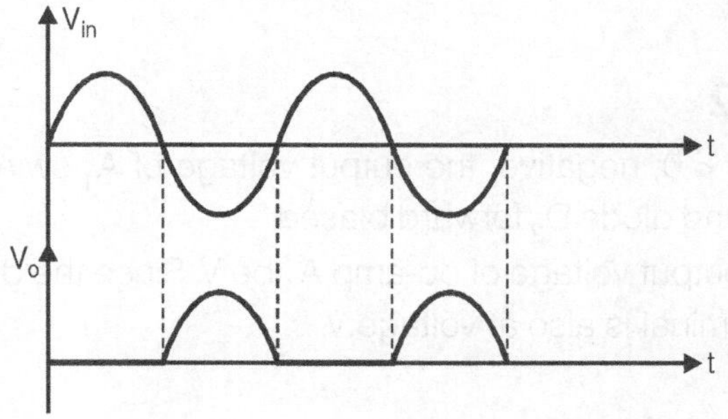


## CIRCUIT-3

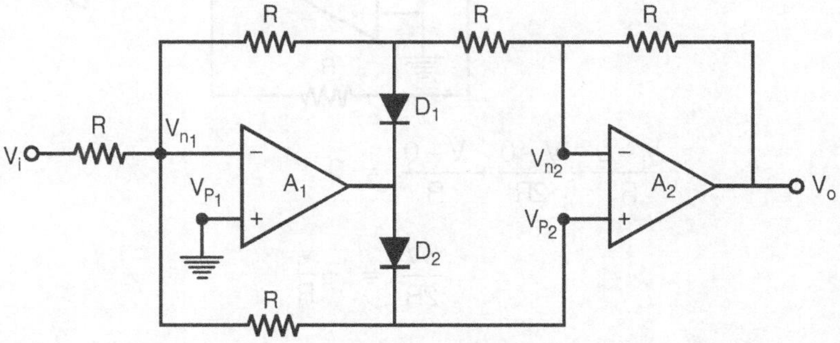


*Figure 9.49 Precision full-wave rectifier*

# OPERATION

## Case-1

If $V_i > 0$, inverting side of $A_1$ will force its output to swing negative, thus forward biasing $D_1$ and reverse biasing $D_2$. Since no current flows through resistance R connected between $V_{n_1}$ and $V_{p_1}$, both are equipotential.

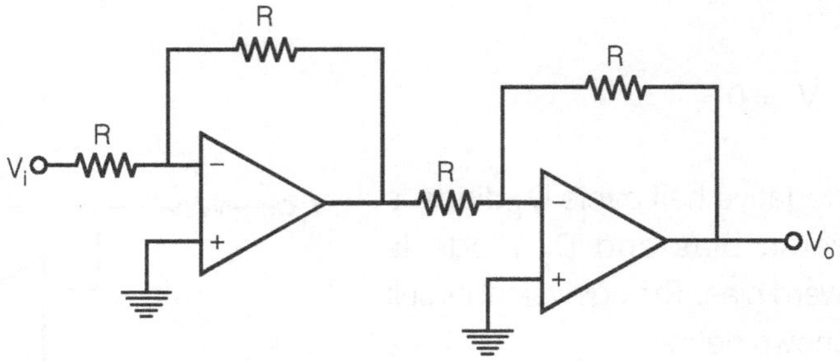


i.e. $V_{n_1} = V_{p_2} = 0\,V$

The output voltage can be given as

$$V_o = \left(-\frac{R}{R}\right)\left(-\frac{R}{R}\right)V_i = V_i$$

$$\therefore \quad V_o = V_i$$

## Case-2

When $V_i < 0$, negative, the output voltage of $A_1$ swings to positive, making diode $D_1$ reverse biased and diode $D_2$ forward biased.

Let the output voltage of op-amp $A_1$ be V. Since the differential input to $A_2$ is zero, the inverting input terminal is also at voltage V.

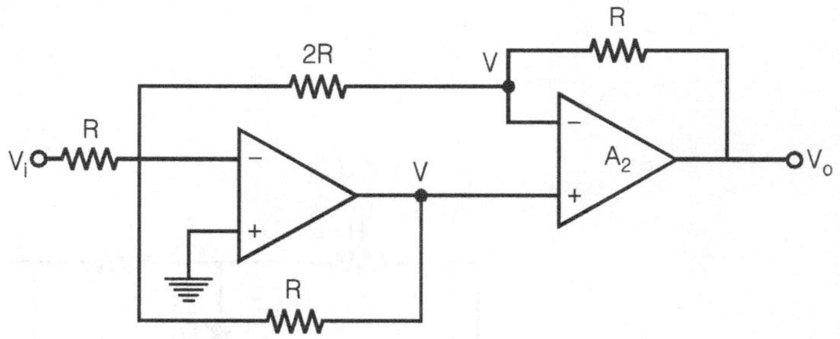


$$\frac{V_i - 0}{R} + \frac{V - 0}{2R} + \frac{V - 0}{R} = 0$$

$$\frac{3V}{2R} = -\frac{V_i}{R}$$

$$V = -\frac{2}{3}V_i$$

Consider a part of circuit above

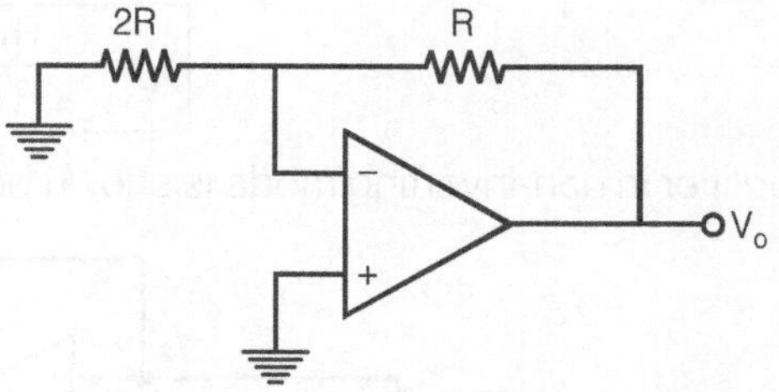


To find $V_o$ in terms of V, we concentrate on the equivalent circuit of $A_2$ as shown

$$V_o = \left[1+\frac{R}{2R}\right]V = \left[\frac{2R+R}{2R}\right]V = \frac{3}{2}V$$

Substituting value of V in above equation

$$V_o = \frac{3}{2}\left[-\frac{2}{3}V_i\right] = -V_i$$

$$\therefore \quad V_o = -V_i$$

The waveform for the precision full-wave rectifier is shown below:

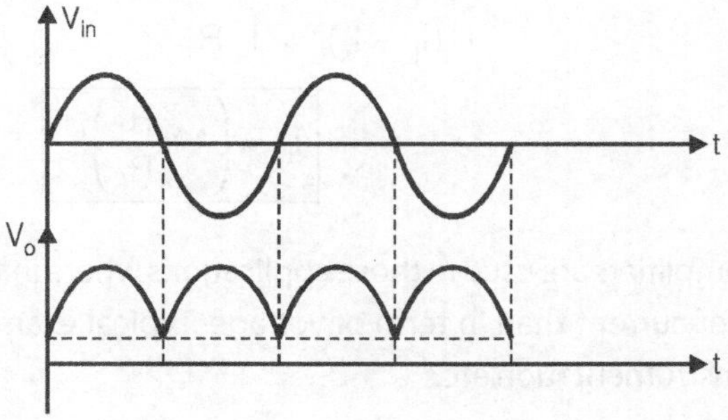


## 9.34 Current Amplifiers

Though op-amp is basically a voltage amplifier, it can be configured to act as a current amplifier. Current amplifier in inverting mode is shown below:

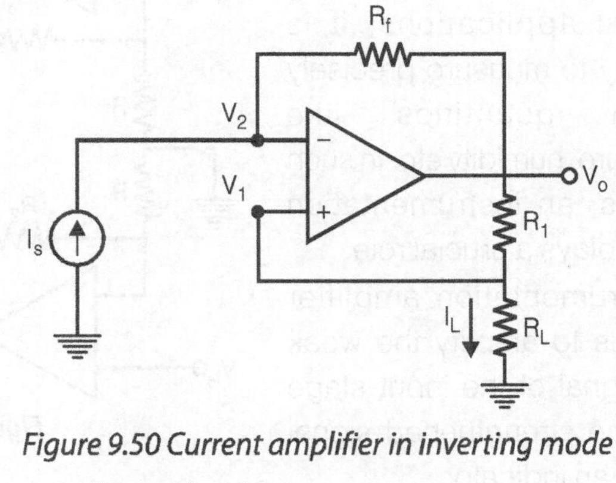


*Figure 9.50 Current amplifier in inverting mode*

$$I_s R_f = V_2 - V_o$$

and $$R_1 I_L = V_o - V_1$$

Since op-amp keeps $V_2 = V_1$, the above equation becomes

$$I_L = -\frac{R_f}{R_1} I_s$$

Current amplifier in non-inverting mode is shown below

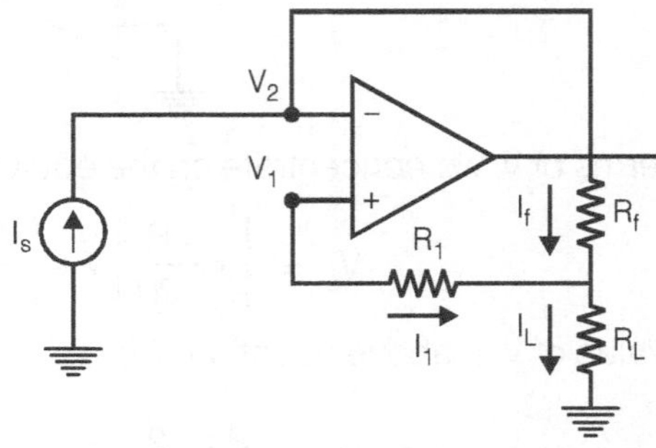


*Figure 9.51 Current amplifier in inverting mode*

Here $V_2 = V_1$ by op-amp action, this implies from the figure that

$$I_f R_f = I_1 R_1$$
$$R_f(I_L - I_1) = I_1 R_1$$
$$R_f(I_L - I_s) = I_s R_1 \qquad (\because I_1 = I_s)$$

or

$$I_L = \left(1 + \frac{R_1}{R_f}\right) I_s$$

**Current amplifiers are used in those applications where information is more conveniently represented in terms of current than in term of voltage. Typical examples are photo detector, two-wire remote sensing instrumentation etc.**

## 9.35 Instrumentation Amplifier

The main use of instrumentation amplifier is in industry. In many industrial applications, it is necessary to measure precisely physical quantities like temperature, humidity etc. In such situations, an instrumentation amplifier plays a crucial role.

The instrumentation amplifier function is to amplify the weak output signal of the input stage so that the strengthened signal can drive an indicator.

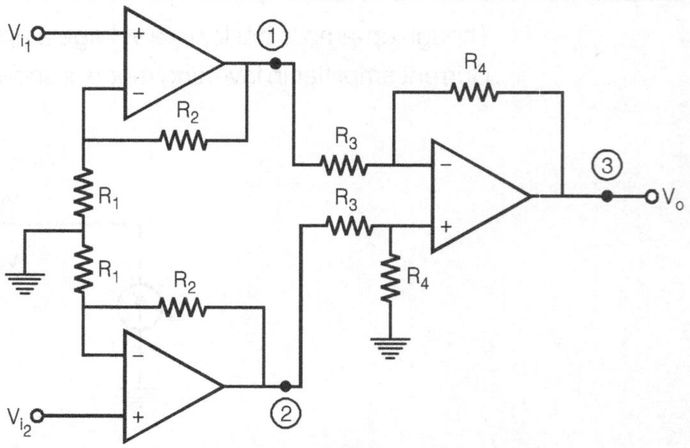


*Figure 9.52 Instrumentation amplifier*

The instrumentation amplifier

circuit is shown below:

The output of node ① is $\left(1+\frac{R_2}{R_1}\right)V_{i_1}$

The output of node ② is $\left(1+\frac{R_2}{R_1}\right)V_{i_2}$

The output of node ③ is

$$V_o = \frac{R_4}{R_3}\left(1+\frac{R_2}{R_1}\right)(V_{i_2} - V_{i_1})$$

Where $V_{i_2} - V_{i_1} = V_{id}$

**Instrumentation amplifier is used to improve the input impedance and voltage gain.**

# 9.36 Voltage Limiters

Voltage limiters are used to limit a part of the signal input $V_i$.

## CIRCUIT-1

In the positive half cycle, $Z_1$ is FB which will have $V_{D_1}$ drop and $Z_2$ is RB which will have $V_z$ drop. As it is in inverting mode $V_o = -(V_{D_1} + V_Z)$.

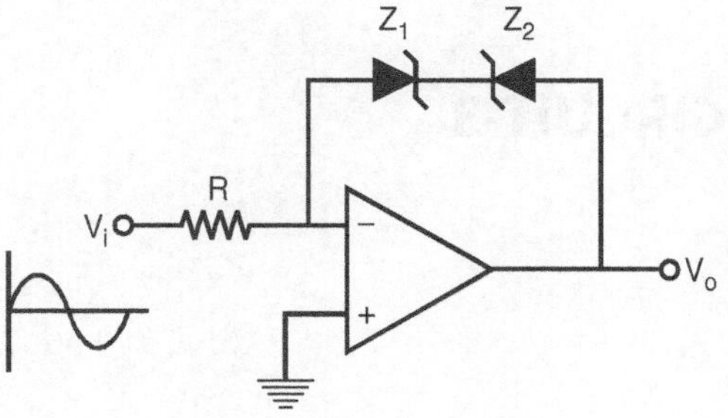


In the negative half cycle, $Z_1$ is RB which haves a drop of $V_Z$ and $Z_2$ is FB which haves a drop of $V_{D_2}$.

$V_o = (V_Z + V_{D_2})$

The output waveform is shown below:

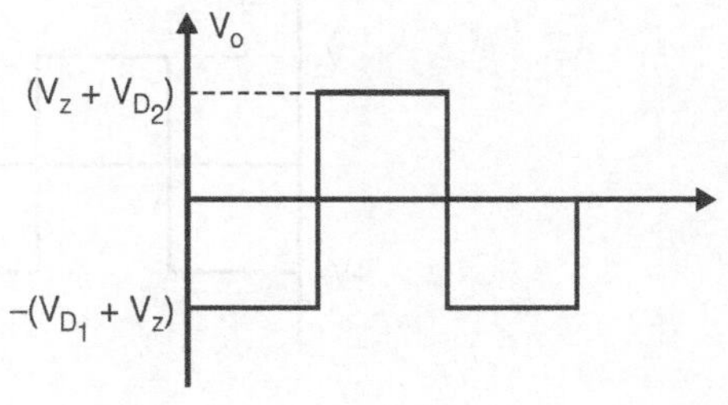

## CIRCUIT-2

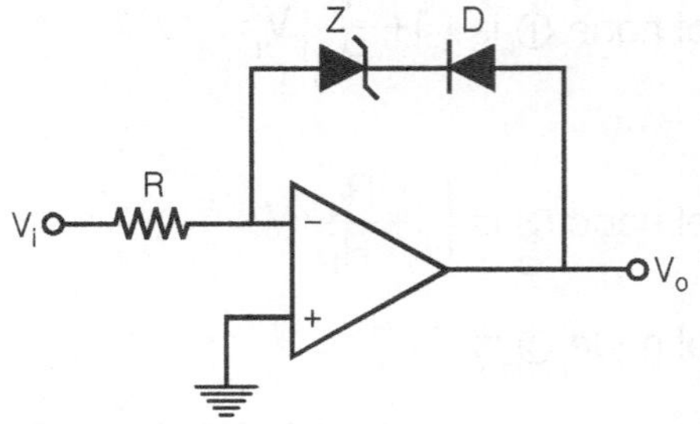


In positive half cycle, Zener diode is FB which haves a drop of $V_D$ and diode is RB which is open circuited. Therefore the circuit acts in open loop system, where the gain is very high. So output saturates to $-V_{sat}$.

$$V_o = -V_{sat}$$

In negative half cycle, Zener diode is RB which haves a drop of $V_z$ and diode is FB which haves a drop of $V_D$. Therefore

$$V_o = (V_z + V_D)$$

The output waveform is shown below:

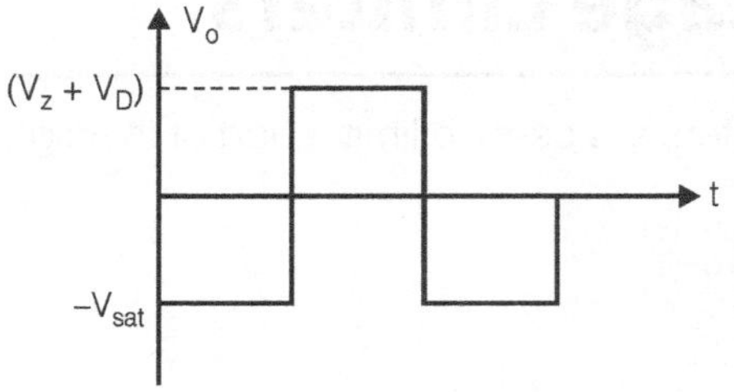


## CIRCUIT-3

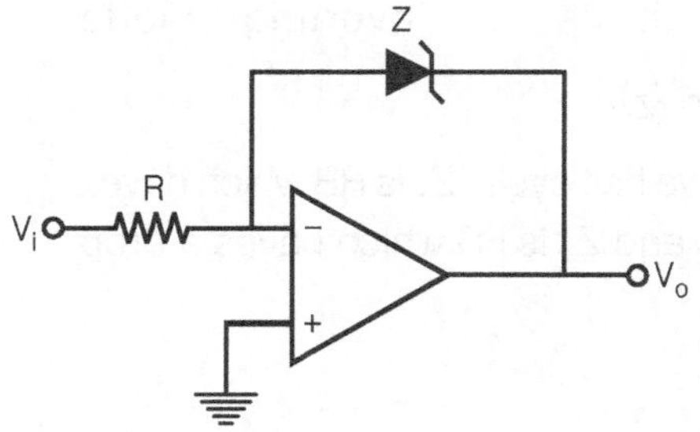


The output $V_o$ for the above circuit is

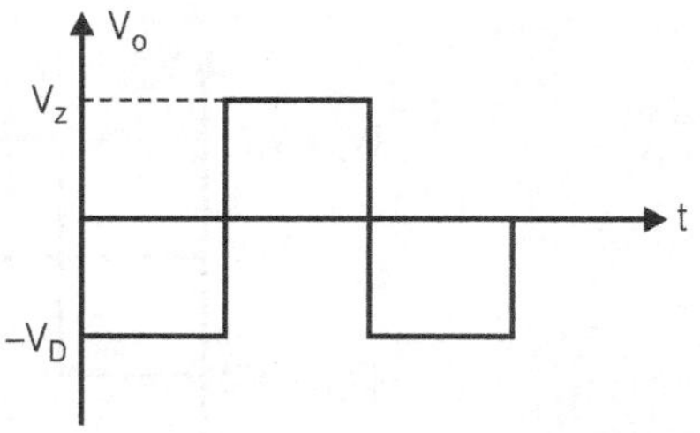

## 9.37 Log Amplifiers

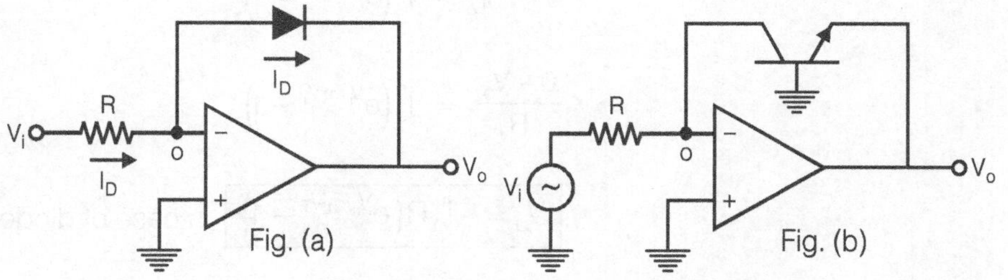


*Figure 9.53 Log amplifiers using diode (or) BJT*

Assume Fig. 9.53 (a)

We know diode current equation

$$I_D = I_o\left(e^{V_D/\eta V_T} - 1\right)$$

As $V_D$ is positive, we can neglect one.

$$I_D = I_o\, e^{V_D/\eta V_T}$$

$$V_D = \eta V_T \, ln\left[\frac{I_D}{I_o}\right]$$

$$0 - V_o = V_D$$

$$V_o = -V_D$$

$$V_o = \eta V_T \, ln\left(\frac{I_o}{I_D}\right)$$

Where $I_D = \dfrac{V_i}{R}$

$$V_o = \eta V_T \, ln\left(\frac{I_o R}{V_i}\right) \quad \text{in case of diode}$$

$$V_o = \eta V_T \, ln\left(\frac{I_{CO} R}{V_i}\right) \quad \text{in case of BJT}$$

The relation between input and output gives logarithmic operation.

## 9.38 Anti Log Amplifiers

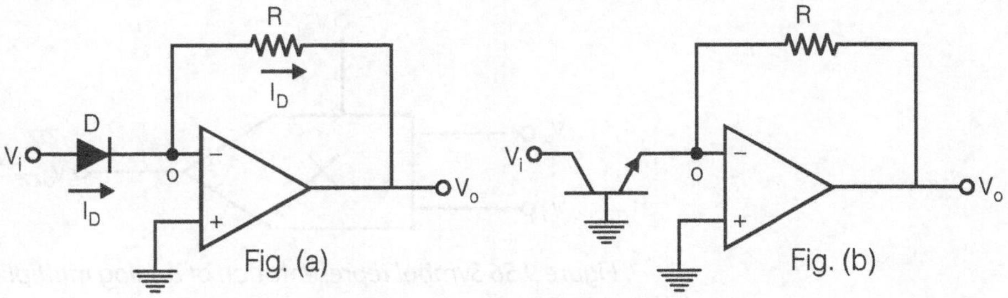


*Figure 9.54 Antilog amplifiers diode or BJT*

Consider Figure 9.54 (a)
The diode current equation is

$$I_D = I_o\left(e^{V_i/\eta V_T} - 1\right)$$

$$\frac{0 - V_o}{R} = I_o\left(e^{V_i/\eta V_T} - 1\right)$$

$$\boxed{V_o = -I_o R\left(e^{V_i/\eta V_T} - 1\right)} \text{ in case of diode}$$

$$\boxed{V_o = -I_{C_o} R\left(e^{V_i/\eta V_T} - 1\right)} \text{ in case of BJT}$$

## 9.39 Analog Multiplier

Analog multiplier finds a large number of practical applications like multiplication of two signals, division of signals, squaring of signals, frequency doubling, frequency shifting etc.
The block diagram of a log-antilog multiplier based on the above principle.

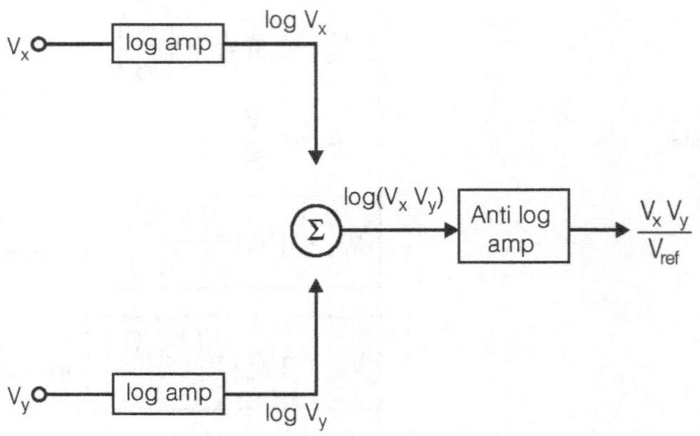


*Figure 9.55 Analog multiplier*

The symbol representation of analog multiplier is given by

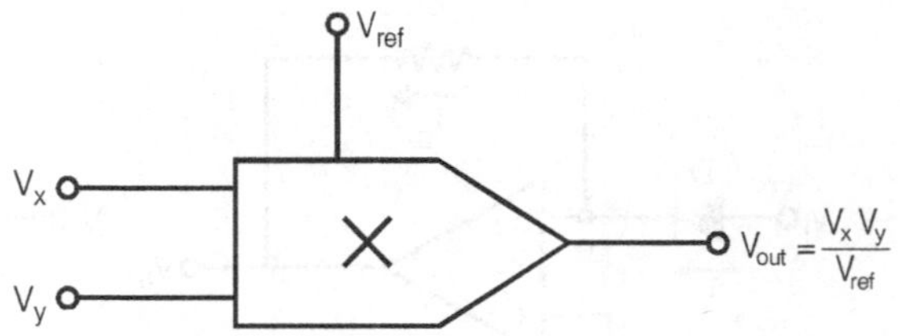


*Figure 9.56 Symbol representation of analog multiplier*

# 9.40 Analog Divider

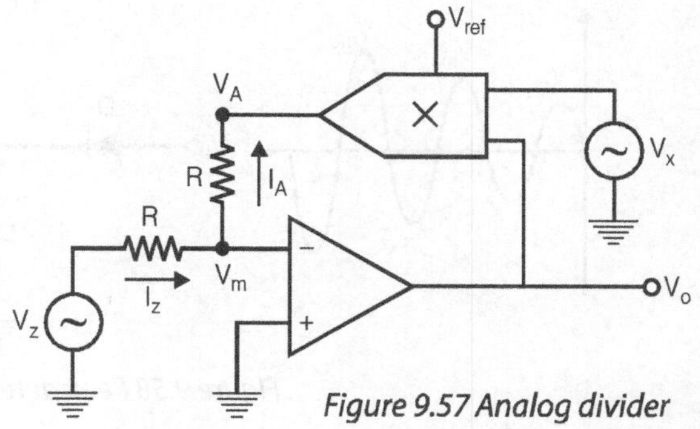


*Figure 9.57 Analog divider*

In the above circuit $V_z$ is dividend and $V_x$ is divisor.

$$\text{output voltage } V_o \ \alpha \ \frac{V_z}{V_x}$$

From the Figure (9.57), analog multiplier is placed in the feedback path of the op-amp. Signal $V_z$ which is the divided is applied to the inverting terminal of op-amp, keeping non inverting terminal grounded.

Because virtual ground concept at the input terminals of op-amp, no current enters the op-amp then

$$I_z = I_A$$

$$\text{Where } I_z = \frac{V_z - V_m}{R} = \frac{V_z}{R} \quad (\because V_m = 0)$$

The input signals to analog multiplier are $V_x$ and $V_0$.

$$\text{Output of multiplier is} \quad V_A = \frac{V_x V_o}{V_{ref}}$$

$$V_A = -I_A R$$

$$I_A = -\frac{V_A}{R} = -\frac{V_x V_o}{V_{ref} R}$$

$$\text{We have} \quad I_z = I_A$$

$$I_z = -\frac{V_x V_o}{V_{ref} R}$$

$$V_z = I_z R$$

$$V_z = -\frac{V_x V_o R}{V_{ref} R} = -\frac{V_x V_o}{V_{ref}}$$

$$\boxed{V_o = -V_{ref} \frac{V_z}{V_x}}$$

$$\text{Since } V_{ref} \text{ is fixed} \quad V_o \ \alpha \ \frac{V_z}{V_x}$$

# 9.41 Peak Detector

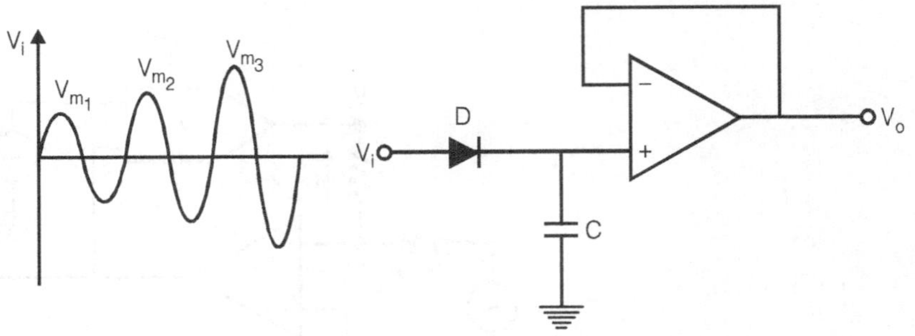


*Figure 9.58 Peak detector*

If the input is having different peak values $V_{m_1}$, $V_{m_2}$ and $V_{m_3}$. To detect the highest peak in the waveform peak detector is used.

For the first half cycle, capacitor charges to $V_{m_1}$ and in negative half cycle capacitor will not have a path to discharge, therefore $V_o = V_{m_1}$.

Then if $V_{m_3}$ is the highest peak which will be charged by the capacitor and remains same in negative half cycle and finally $V_o = V_{m_3}$ (highest peak detected).

# 9.42 Sample and Hold Circuit

A sample and hold circuit is a device which finds very wide application in analog to digital and pulse modulation systems. As the name implies, the circuit samples as input signal, and holds on to it i.e. retains the last sampled value until it again samples the input signal.

The basic circuit of a sample and hold device using op-amp and enhancement mode MOSFET is shown in the Figure (9.59).

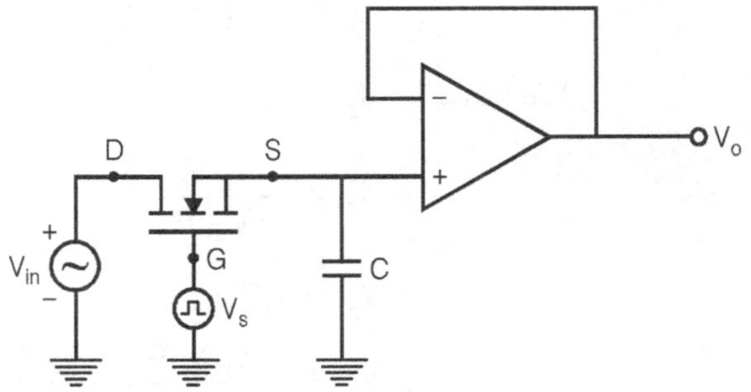


*Figure 9.59 Sample and hold circuit*

## OPERATION

The sample and hold control voltage $V_s$ is applied to the gate of the MOSFET and this voltage controls the operation of MOSFET as a switch.

The input signal $V_{in}$ i.e. the voltage to be sampled is applied to the drain terminal D. Capacitor C acts as a storage element.

When the control voltage $V_s$ is positive, it is seen that the MOSFET readily conducts, acting as a closed switch. The input voltage $V_{in}$ charges the capacitor to voltage $V_C$. Since the op-amp is a voltage follower, this voltage $V_c$ appears at the output terminal i.e. $V_o = V_c$.

When $V_s = 0$, MOSFET becomes OFF, thus acting as an open switch. Capacitor C must now discharge through the op-amp, since this is the only available path. But it can not discharge (because the input resistance of the op-amp is very high). Hence the capacitor retains the charge and output voltage remains at $V_c$.

The waveforms for sample and hold circuit are shown below:

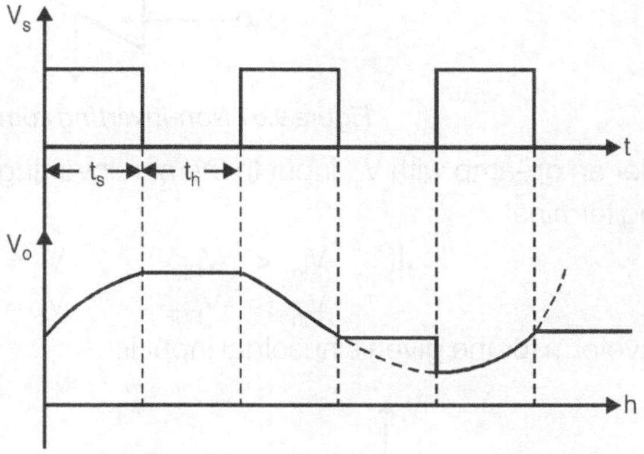


Where $t_s \rightarrow$ sample period and $t_h \rightarrow$ hold period.

## 9.43 Op-amp Comparator

Op-amp comparator is a circuit with two inputs and one output. The two inputs can be compared with each other i.e. one of them can be considered as a reference voltage, $V_{REF}$

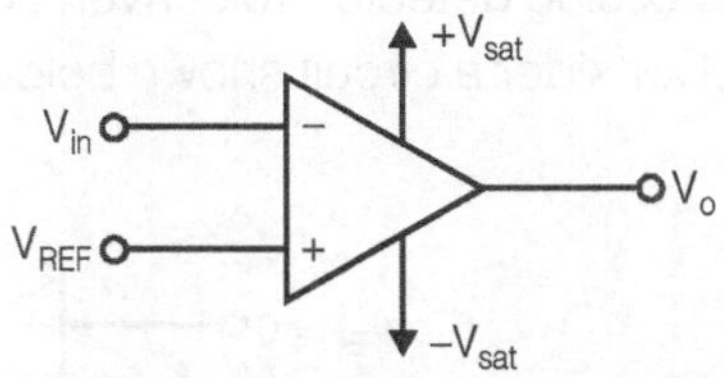


*Figure 9.60 Inverting comparator circuit*

Consider an op-amp with $V_{in}$ input to the inverting terminal, and $V_{REF}$ is given to non inverting terminal.

$$\text{If} \quad V_{in} < V_{REF} \qquad V_o = +V_{sat}$$
$$V_{in} > V_{REF} \qquad V_o = -V_{sat}$$

The waveform for the given sinusoidal input is

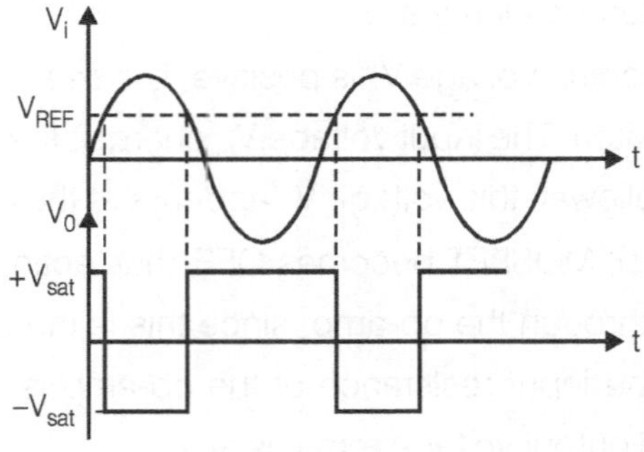


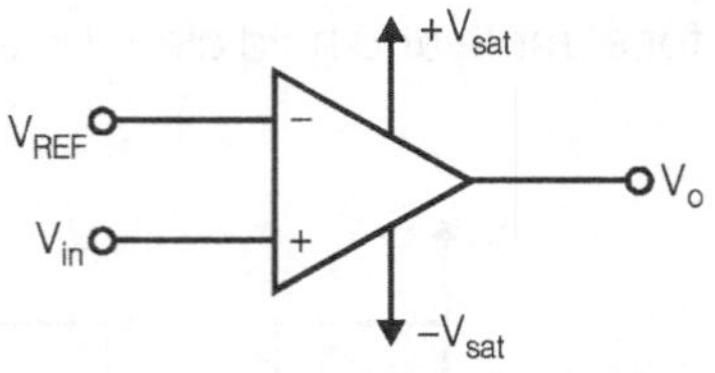


*Figure 9.61 Non-inverting comparator circuit*

Consider an op-amp with $V_{in}$ input to the non inverting terminal and $V_{REF}$ (negative) is given to inverting terminal.

If $V_{in} < -V_{REF}$ $\quad V_o = -V_{sat}$

$V_{in} > -V_{REF}$ $\quad V_o = +V_{sat}$

The waveform for the given sinusoidal input is

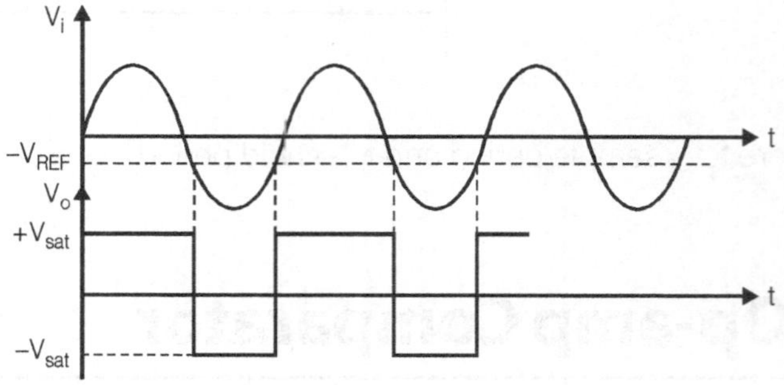


## APPLICATION OF COMPARATOR AS ZCD

ZCD means zero crossing detector. To convert comparator as zero crossing detector we have to assume $V_{REF} = 0$. Consider a circuit shown below:

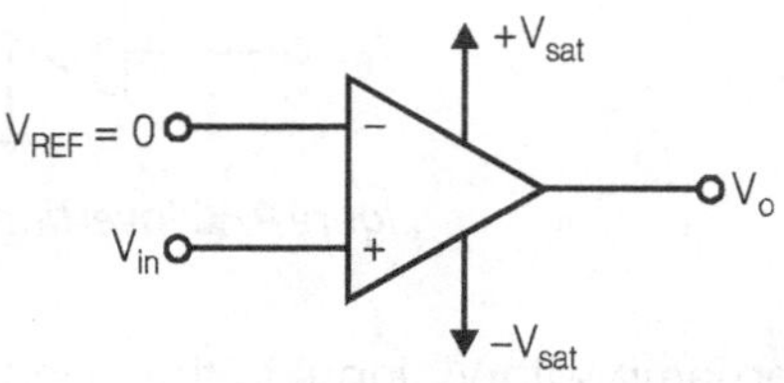


*Figure 9.62 Zero crossing detector*

If $V_{in} < V_{REF}$ $\quad V_o = -V_{sat}$

$V_{in} > V_{REF}$ $\quad V_o = +V_{sat}$

The waveform for the given sinusoidal input is

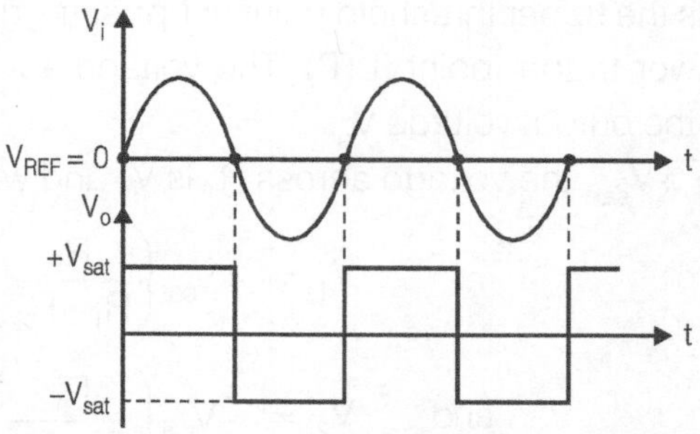


- For a zero crossing detector always the output will be square wave.
- The zero crossing detector can not give square wave output for any other arbitrary waveform which is not sinusoidal. To avoid this draw back we have a special circuit called as "Schmitt trigger".
- Always comparator works under open loop systems.

### Practical Comparator Circuit

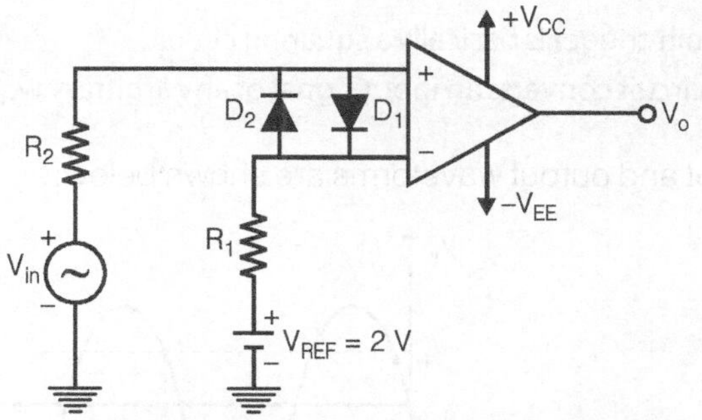


*Figure 9.63 Non inverting practical comparator circuit*

Diodes $D_1$ and $D_2$ are provided in the circuit to protect the op-amp against damage due to excessive input voltage. Because of these diodes, the differential input voltage $V_d$ is clamped to either +0.7 V or –0.7 V, hence the diodes are called as clamp diodes.

## 9.44 Schmitt Trigger

The basic principle of Schmitt trigger is that whenever the input voltage crosses certain voltage levels termed as upper trigger point (UTP) and lower trigger point (LTP), the state of the output changes. In the process, the output voltage takes on the shape of a square wave.

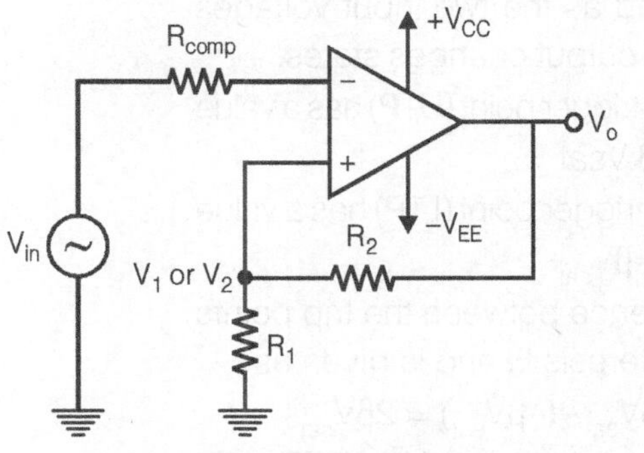


*Figure 9.64 Schmitt trigger*

From the Figure (9.64), the op-amp provided with positive feedback can function as Schmitt trigger $V_1$ is the upper threshold point or upper trigger point (UTP) and $V_2$ is the lower threshold point or lower trigger point (LTP). The voltage across $R_1$ depends upon the magnitude and polarity of the output voltage $V_o$.

When $V_o = +V_{sat}$, the voltage across $R_1$ is $V_1$ and when $V_o = -V_{sat}$, the voltage across $R_1$ is $V_2$.

$$V_1 = +V_{sat}\left(\frac{R_1}{R_1+R_2}\right)$$

and $$V_2 = -V_{sat}\left(\frac{R_1}{R_1+R_2}\right)$$

When the input voltage $V_{in}$ becomes slightly greater i.e. more positive than the UTP ($V_1$). The output voltage $V_0$ changes from $+V_{sat}$ to $-V_{sat}$, and it remains at that level i.e. $-V_{sat}$ until $V_{in}$ becomes equal to $V_2$ (LTP).

When $V_{in}$ becomes slightly more negative than $V_2$, $V_o$ switches instantaneously from $-V_{sat}$ to $+V_{sat}$, until $V_{in}$ again reaches the value $V_1$.

Because of positive feedback, the output voltage $V_o$ is enabled to switch rapidly from $+V_{sat}$ to $-V_{sat}$.

- **Schmitt trigger is basically a squaring circuit.**
- **The circuit converts an input signal of any arbitrary waveform into square wave.**

The input and output waveforms are shown below:

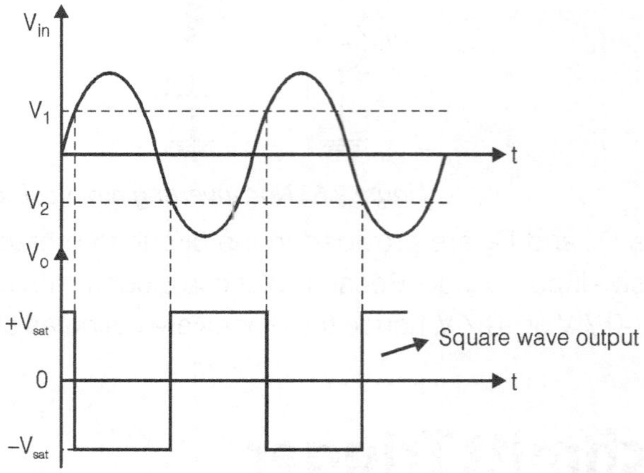


## Hysteresis

From the Figure (9.65), the trip points are defined as the two input voltages where the output changes states.

The upper trigger point (UTP) has a value

$$UTP = \beta\ Vsat$$

and lower trigger point (LTP) has a value

$$LTP = -\beta V_{sat}$$

The difference between the trip points is the hysteresis H and is given as

$$H = +\beta V_{sat} - (-\beta V_{sat}) = 2\beta V_{sat}$$

i.e. Hysteresis voltage $V_H = UTP - LTP$

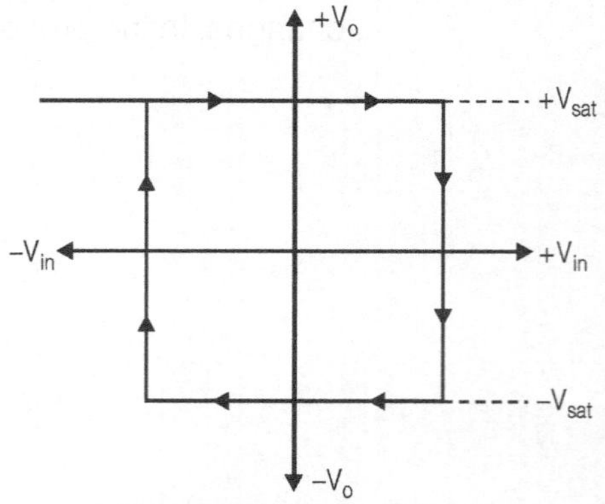


*Figure 9.65 Hysteresis loop*

## 9.45 Astable Multivibrator (free-running oscillator)

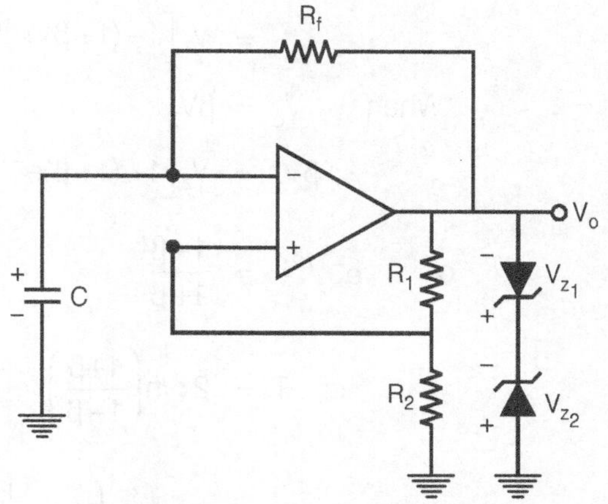


*Figure 9.66 Astable multivibrator*

In the op-amp square wave generator circuit given in Figure (9.66), the output voltage $V_{out}$ is shunted to ground by two Zener diodes $Z_1$ and $Z_2$ connected back-to-back and is limited to either $V_{Z_2}$ or $-V_{Z_1}$.

A fraction of the output is feedback to the non inverting input terminal. Combination of $R_f$ and C acting as a low-pass R-C circuit is used to integrate the output voltage $V_o$ and the capacitor voltage $V_c$ is applied to the inventing input terminal in place of external signal.
The differential input voltage is

$$V_{in} = V_c - \beta V_o \qquad \text{Where } \beta = \frac{R_2}{R_1 + R_2}$$

Where $V_{in}$ is positive, $V_o = -V_{Z_1}$ and when $V_{in}$ is negative, $V_o = +V_{Z_2}$

The waveforms of the capacitor voltage $V_c$ and output voltage $V_o$ (or $V_z$) are shown:

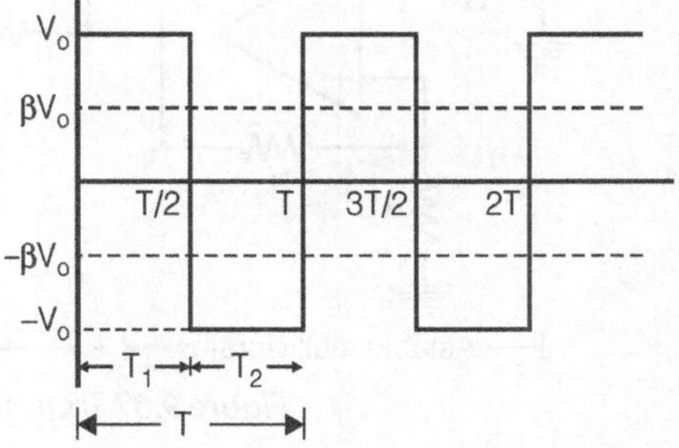


$$\text{When} \quad t = \frac{T}{2}, \; V_c = +\beta V_z \text{ or } + \beta V_o$$

∴ The voltage across the capacitor, $V_c$ when it is charging from $-\beta V_z$ to $+V_z$ is given by

$$V_c = V_z\left[1-(1+\beta)e^{-t/R_fC}\right]$$

$$= V_z\left[1-(1+\beta)e^{-t/\tau}\right]$$

When $V_c = \beta V_z$

$$\beta V_z = V_z\left[1-(1+\beta)e^{-T/2\tau}\right]$$

or $$e^{-T/2\tau} = \frac{1-\beta}{1+\beta}$$

or $$T = 2\tau\, ln\left(\frac{1+\beta}{1-\beta}\right)$$

$$= 2R_fC\, ln\left(\frac{1+\frac{R_2}{R_1+R_2}}{1-\frac{R_2}{R_1+R_2}}\right)$$

$$T = 2R_fC\, ln\left(1+\frac{2R_2}{R_1}\right)$$

The frequency f = 1/T, of the square-wave is independent of output voltage $V_o$. This circuit is also known as free running or astable multivibrator.

## 9.46 Triangular-Wave Generator

The op-amp triangular wave generator is another example of a relaxation oscillator.

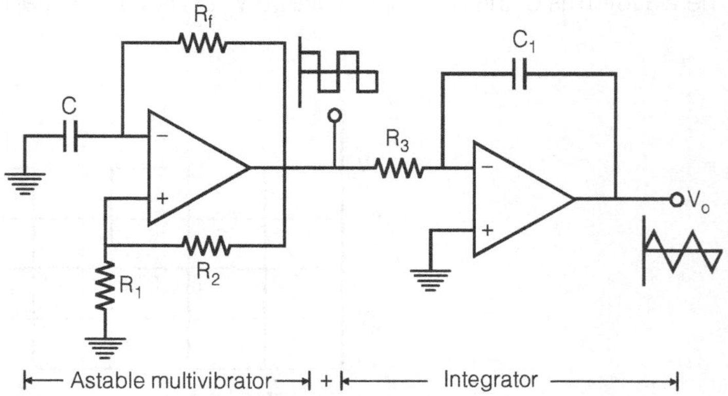


*Figure 9.67 Triangular wave generator*

We know that the integrator output waveform will be triangular if the input is a square-wave. It means that a triangular wave generator can be formed by simply cascading an integrator and a square-wave generator as shown in Figure (9.67).

## 9.47 Monostable Multivibrator (pulse generator)

A monostable multivibrator has one stable state and one quasi stable state. The circuit remains in its stable state till an external triggering pulse causes a transition to the quasi stable state.

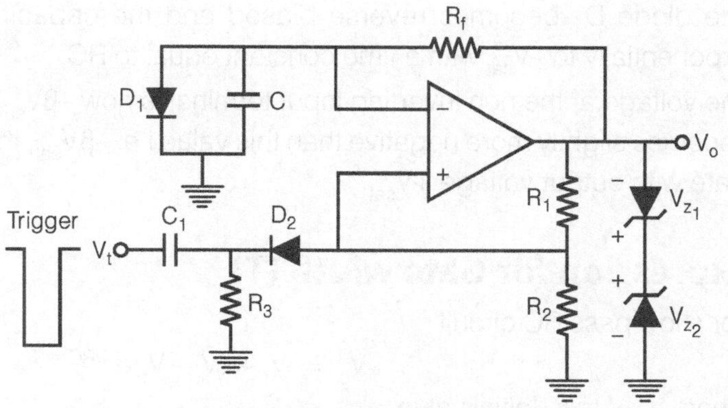


*Figure 9.68 Monostable Multivibrator*

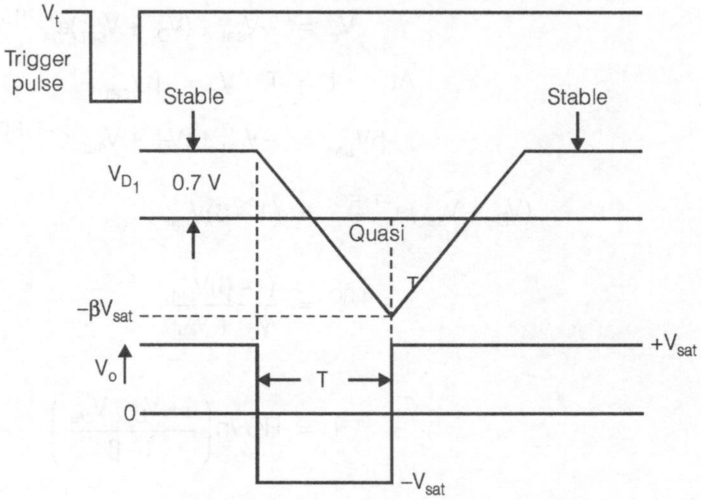


The circuit uses two diodes $D_1$ and $D_2$, two Zener diodes, a capacitor C and several resistors with an op-amp.

Diode $D_1$ acts as clamping diode, since it is connected across the capacitor C, it clamps the capacitor voltage to 0.7 V, when the output voltage is at positive saturation level.

The triggering of the multivibrator is by a negative pulse $V_t$ applied to the non inverting terminal of the op-amp through diode $D_2$.

Let us assume that it is in the stable state, so that the output voltage is at the positive saturation level $V_{sat}$.

Diode $D_1$ is forward biased and hence conduct. The capacitor gets charged and a voltage develops across it. But this voltage is clamped to +0.7 V, since the volt drop across the diode is 0.7 V.

A negative triggering pulse of amplitude $V_t$ is applied to the non inverting terminal such that the net voltage at this terminal is $(\beta V_{sat} - V_t)$ which is less than 0.7 V.

The output voltage immediately switches to $-V_{sat}$.

The diode $D_1$ becomes reverse biased and the capacitor discharges through resistor R exponentially to $-V_{sat}$ with a time constant equal to RC.

The voltage at the non inverting input terminal is now $-\beta V_{sat}$. As soon as the capacitor voltage becomes slightly more negative than this value i.e. $-\beta V_{sat}$, the multi switches back to the stable state with output voltage $+V_{sat}$.

## Expression for Gate width (T)

For a low pass RC circuit

$$V_o = V_f + (V_i - V_f)e^{-t/RC}$$

Where $V_i \rightarrow$ Initial value

$V_f \rightarrow$ Final value of capacitor voltage

We have $$V_i = V_{D_1} \text{ and } V_f = -V_{sat}$$

Capacitor voltage $$V_c = V_o$$

$$V_c = -V_{sat} + (V_{D_1} + V_{sat})e^{-t/RC}$$

At $$t = T, \quad V_c = -\beta V_{sat}$$

$$-\beta V_{sat} = -V_{sat} + (V_{D_1} + V_{sat})e^{-T/RC}$$

$$(V_{D_1} + V_{sat})e^{-T/RC} = (1 - \beta)V_{sat}$$

$$e^{-T/RC} = \frac{(1-\beta)V_{sat}}{V_{D_1} + V_{sat}}$$

$$T = RC\, ln\left(\frac{1 + V_{D_1}/V_{sat}}{1-\beta}\right)$$

We have $$\beta = \frac{R_2}{R_1 + R_2}$$

If $$R_1 = R_2 \text{ and } V_{sat} >> V_{D_1}$$

$$T = RC\, ln\left(\frac{1}{1-1/2}\right)$$

$$= RC\, ln\, 2$$

$$\therefore T = 0.693\, RC$$

# ACTIVE FILTERS

## 9.48 Introduction

- A filter is an electric network which has the ability to transmit a certain range of frequencies, and suppress or attenuate all other frequencies.
- A filter circuit, therefore, possesses atleast one pass band - a band of frequencies in which the output is approximately equal to the input and an attenuation band in which output is zero.
- The frequencies that separate the different pass and attenuation bands are called the cut-off frequencies.
- An ideal filter should pass the band of frequencies which is designed to accept, without attenuation, while at the same time, it should attenuate, without fail, all frequencies outside this band.
- Thus, in the pass band, attenuation must be zero, whereas in the stop band attenuation must be infinite.
- Electrical filters are used in practically all circuits which require separation of signals according to their frequencies.

## 9.49 Classification of Active Filters

Practical filters are classified into five groups, depending upon the frequency range, they accept for transmission without attenuation.

1. Low pass filter
2. High pass filter
3. Band pass filter
4. Band stop filter
5. All pass filter.

### 1. Low Pass Filter

A low pass filter transmits, without attenuation, all frequencies below a certain frequency termed as cut-off frequency. This band of frequencies is called pass band. The band of frequencies beyond the cut-off frequency, which are attenuated is called stop band.

The frequency response characteristic of low pass filter is as shown.

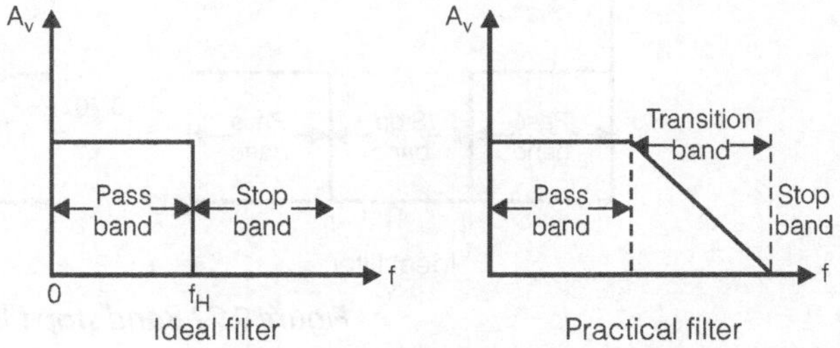


*Figure 9.69 Low pass filter response*

## 2. High Pass Filter

A high pass filter transmits, without attenuation, all frequencies above the frequency $f_L$, called lower cut-off frequency, and attenuates all frequencies below the frequency $f_L$.

The frequency response characteristic of a high pass filter is as shown:

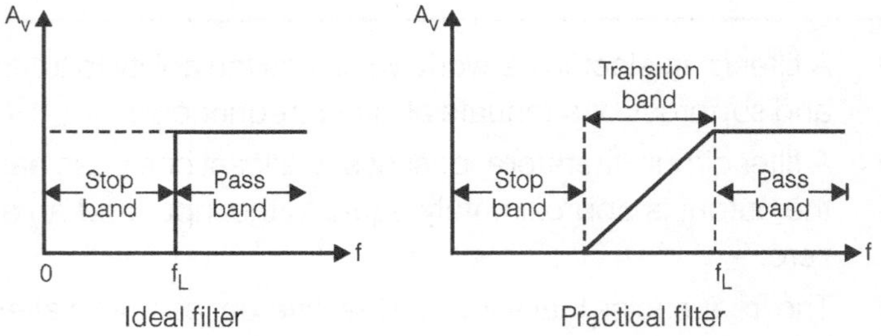


*Figure 9.70 High pass filter response*

## 3. Band Pass Filter

Band pass filter transmits, with out attenuation, all frequencies lying in between two cut-off frequencies $f_L$, the lower cut-off frequency, and $f_H$, the higher cut-off frequency and attenuates all frequencies below $f_L$ and beyond $f_H$.

The frequency response shown below:

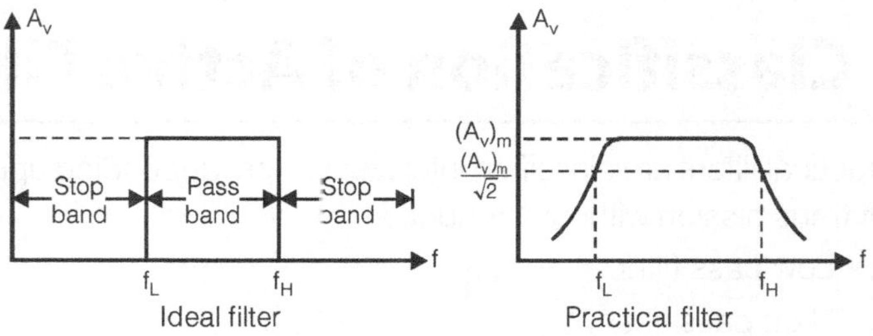


*Figure 9.71 Band pass filter response*

## 4. Band Stop Filter

This filter is also called as band reject filter and band elimination filter. This filter transmits all frequencies below $f_L$ and all frequencies above $f_H$, but attenuates all frequencies lying in between $f_L$ and $f_H$.

The frequency response is shown below:

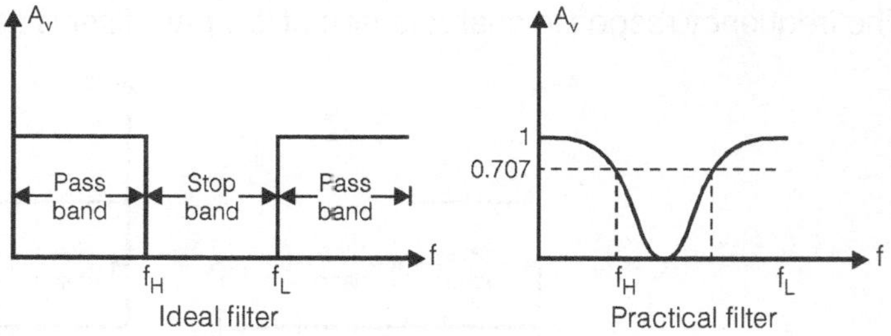


*Figure 9.72 Band stop filter*

This filter is also termed as Notch filter.

### 5. All Pass Filter

All pass filter transmits all frequencies without attenuation, but it brings about the necessary phase shift between the output and input signals. Thus there is no attenuation band. The gain of the filter remains the same.

The graph of voltage versus time of all pass filter is as shown:

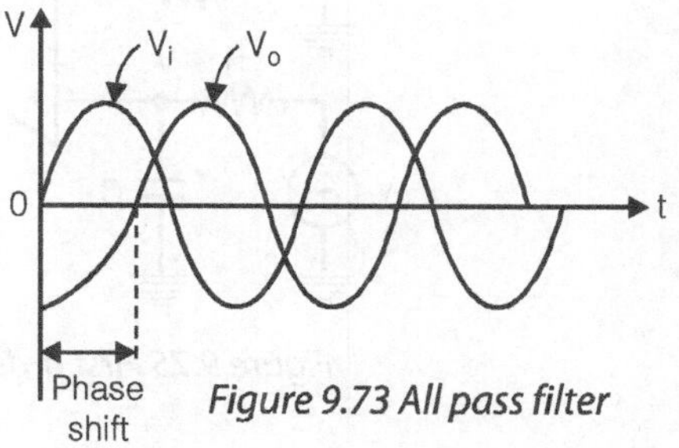


*Figure 9.73 All pass filter*

An all pass filter is also termed as delay equaliser or phase corrector.

## 9.50 Butterworth Filter

- This filter is also called a maximally flat or flat-flat filter. Ideal and practical frequency responses for three types of butterworth low pass and high pass filters are depicted in Figure. 9.74 (a) & (b) by solid line and dashed lines respectively.

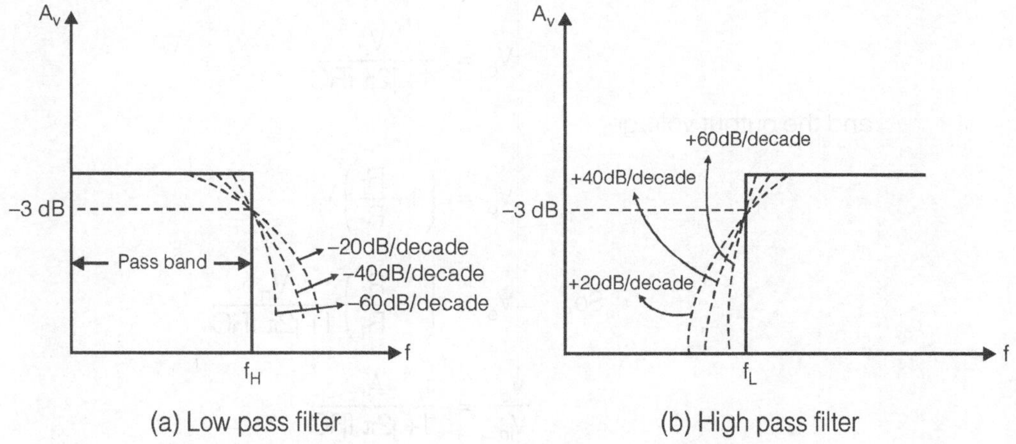


*Figure 9.74 Low pass and High pass filters*

- In many low-pass and high-pass filter applications it is necessary that the closed loop gain is an close to unity as possible with in the pass band.
- Butterworth low pass and high pass filters are simple in design, and are the most popular active filters. In this section, we discuss only Butterworth filters.

## 9.51 First Order Low Pass Butterworth Filter

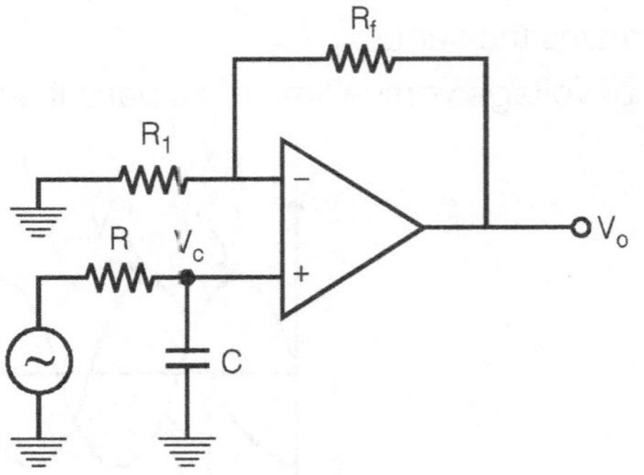


*Figure 9.75 First order low pass filter*

A first order low pass Butterworth filter using an R-C network for filtering is shown in Figure (9.75).

The op-amp is in the non inverting mode of operation, and hence its voltage gain $A_V = \left(1 + \frac{R_f}{R_1}\right)$.

Voltage at the non inverting (+) input terminal is equal to voltage across the capacitor C is given by

$$V_c = \frac{-jX_C}{R - jX_C} V_{in}$$

Substitute $-jX_C = \frac{1}{j2\pi fC}$ in above equation

$$V_c = \frac{V_{in}}{1 + j2\pi fRC}$$

and the output voltage,

$$V_o = \left(1 + \frac{R_f}{R_1}\right) V_C$$

So, $$V_o = \left(1 + \frac{R_f}{R_1}\right) \frac{V_{in}}{1 + j2\pi fRC}$$

$$\frac{V_o}{V_{in}} = \frac{A_f}{1 + j2\pi fRC}$$

Where $A_f = \left(1 + \frac{R_f}{R_1}\right)$, pass band gain of the filter

$$\frac{V_o}{V_{in}} = \frac{A_f}{1 + j\left(\frac{f}{f_H}\right)}$$

Where $f_H = \frac{1}{2\pi RC}$ (higher cut-off frequency).

The gain magnitude and phase angle can be determined by applying modulus to the above equation.

$$\left|\frac{V_o}{V_{in}}\right| = \frac{A_f}{\sqrt{1+\left(\frac{f}{f_H}\right)^2}}$$

and phase angle $$\phi = -\tan^{-1}\left(\frac{f}{f_H}\right)$$

## Frequency Response Characteristic

We have, $$\frac{V_o}{V_i} = \frac{A_f}{1+j\left(\frac{f}{f_H}\right)}$$

- Af $f = 0$, $\left|\frac{V_o}{V_i}\right| = A_f$.
- For very low frequency i.e. $f << f_H$; $\frac{V_o}{V_i} = A_f$
- For frequency equal to high cut-off frequency i.e. $f = f_H$;

$$\frac{V_o}{V_i} = \frac{A_f}{\sqrt{2}} = 0.707\, A_f$$

- For frequencies greater than high cut-off frequency i.e. $f >> f_H$

$$\frac{V_o}{V_i} < A_f.$$

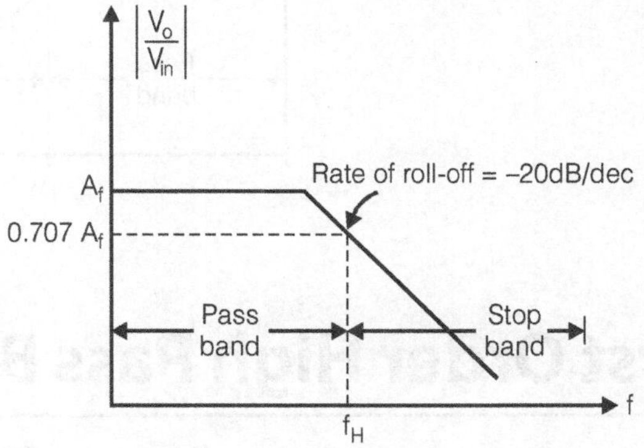


## Frequency Scaling

The procedure of converting a given cut-off frequency to a new cut-off frequency is called the frequency scaling.

Usually a standard capacitor is used for the R-C circuit, and the required value of the resistor is computed as

$$R_{required} = R_{selected}\left(\frac{f_H}{f'_H}\right)$$

## SECOND ORDER LOW PASS FILTER

The rate of roll-off of the gain in the stop band of a second order filter is –40dB/decade since the filter is of the second order, its frequency response is nearer to the ideal response.

A first order low pass filter can be converted into a second order filter by the addition of another R-C network, and modified network is shown in Figure (9.76).

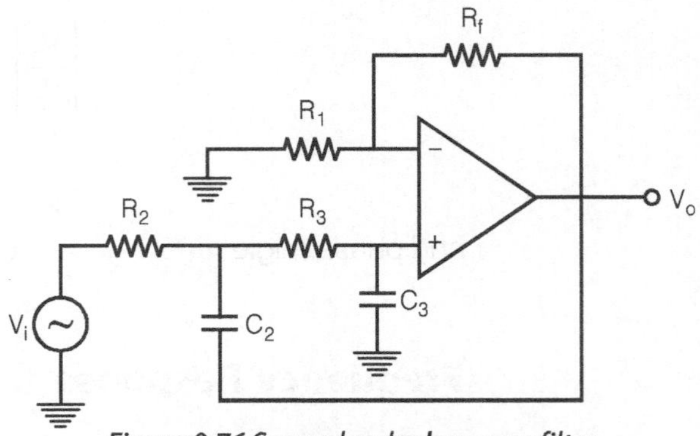


*Figure 9.76 Second order low pass filter*

The higher cut-off frequency $f_H$ is given by

$$f_H = \frac{1}{2\pi\sqrt{R_2 R_3 C_2 C_3}}$$

The voltage gain of the second order filter is given in magnitude as

$$\left|\frac{V_o}{V_i}\right| = \frac{A_f}{\sqrt{1+\left(\frac{f}{f_H}\right)^4}}$$

The frequency response characteristic is as shown

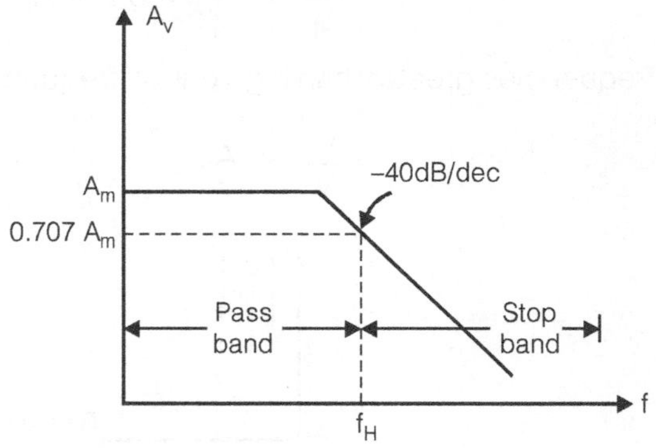


# 9.52 First Order High Pass Butterworth Filter

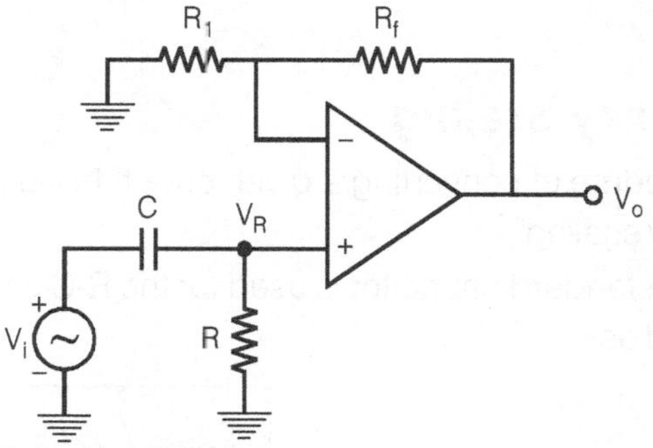


*Figure 9.77 First order high pass filter*

The first-order high-pass filter is formed from a first-order low pass filter by interchanging components R and C as shown in Figure (9.77). Such circuit passes the high frequencies but blocks the low frequencies.

The voltage across resistor R forms the input at the non inverting input terminal of the op-amp.

Voltage across R is $V_R = \left(\frac{R}{R - jX_C}\right)V_i$

Where $X_c = \frac{1}{2\pi fC}$

$$\therefore \quad V_R = \frac{R}{\left(R - j\frac{1}{2\pi fC}\right)}V_i$$

$$V_R = \frac{RV_i}{-jX_C\left(-\frac{R}{jX_C} + 1\right)}$$

$$\text{Where} \quad -\frac{1}{jX_C} = \frac{j}{(1/2\pi\, fC)} = j2\pi\, fC$$

$$-\frac{1}{j} = j \text{ and } X_C = \frac{1}{2\pi fC}$$

$$\therefore \quad V_R = \left(\frac{j2\pi\, fRC}{1 + j2\pi\, fRC}\right)V_i$$

$$\text{We have} \quad \frac{1}{2\pi\, RC} = f_L \text{ (lower cut-off frequency)}$$

$$V_R = V_i\left(\frac{j(f/f_L)}{1 + j(f/f_L)}\right)$$

$$\text{We know,} \quad V_o = A_f V_R$$

$$\therefore \quad V_o = A_f V_i\left(\frac{j(f/f_L)}{1 + j(f/f_L)}\right)$$

$$\frac{V_o}{V_i} = A_f\left(\frac{j(f/f_L)}{1 + j(f/f_L)}\right)$$

Magnitude of the voltage transfer function

$$A_v = \left|\frac{V_o}{V_i}\right| = \frac{A_f(f/f_L)}{\sqrt{1 + (f/f_L)^2}}$$

Since high pass filters are built from low pass filters simply by exchanging the place of resistors and capacitors in the frequency determining section of the filter. The design and frequency scaling procedures are the same as for low pass filters.

## Frequency Response Characteristic

- At f = 0, $\left|\frac{V_o}{V_i}\right| = 0.$
- At low frequencies i.e. $f < f_L$

$$\left|\frac{V_o}{V_i}\right| < A_f.$$

- At $f = f_L$, $\left|\frac{V_o}{V_i}\right| = \frac{A_f}{\sqrt{2}} = 0.707\ A_f.$
- At high frequencies i.e. $f > f_L$.

$$\left|\frac{V_o}{V_i}\right| = A_f.$$

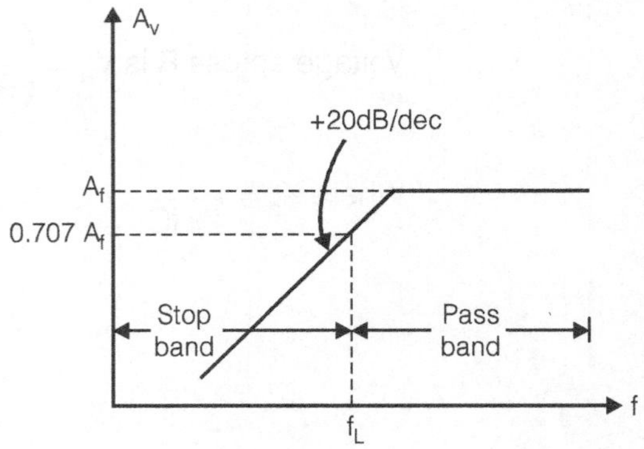


## Second Order High Pass Filter

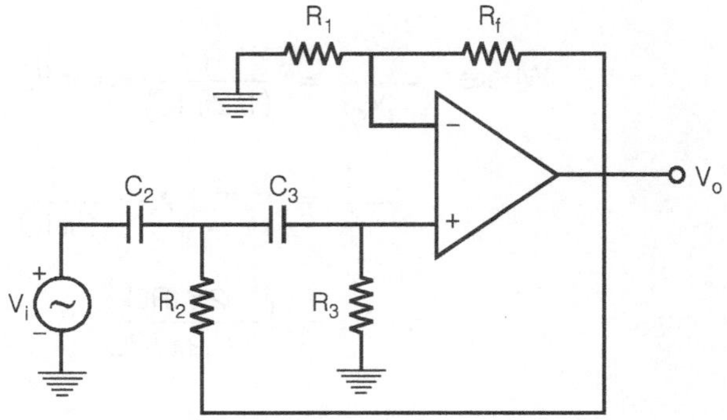


*Figure 9.78 Second order high pass filter*

The rate of roll-off of the gain in the stop band of a second order filter is +40dB/dec since the filter is of the second order, its frequency response is nearer to the ideal response.

A first order high pass filter can be converted into a second order filter by the addition of another R-C network and modified network is shown in Figure (9.78).

The voltage gain magnitude is given by

$$\left|\frac{V_o}{V_{in}}\right| = \frac{A_f}{\sqrt{1+(f_L/f)^4}}$$

The frequency response is given by

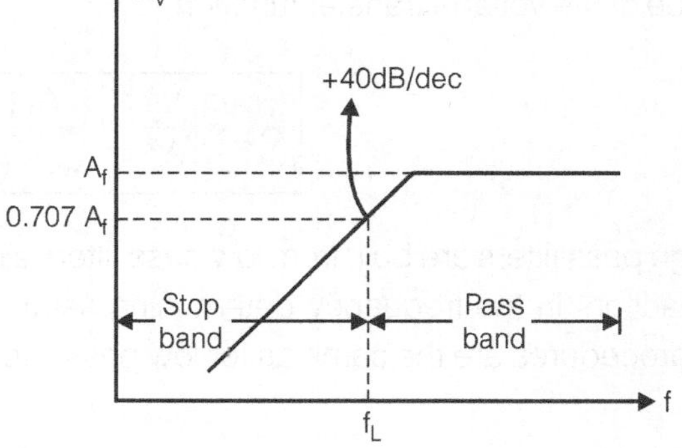

# 9.53 Band Pass Filters

A band pass filter is a circuit which is designed to pass signals only in a certain band of frequencies while attenuating all signals outside this band.

The parameters of importance in a band-pass filter are the high and low cut-off frequencies ($f_H$ and $f_L$), the bandwidth, centre frequency $f_C$, and selectivity or Q.

There are basically two types of band-pass filters

(i) Wide band-pass filter.

(ii) Narrow band-pass filter.

A band pass filter is defined as a wide band pass if its figure of merit or quality factor Q is less than 10. While the band pass filters with Q > 10 are called the narrow band pass filters.

Thus Q is a measure of selectivity, meaning the higher the value of Q the more selective is the filter, or the narrower is the bandwidth.

The relation between bandwidth, quality factor, and centre frequency $f_C$ is given by

$$Q = \frac{f_C}{BW} = \frac{f_C}{f_H - f_L}$$

For wide band pass filter the centre frequency can be defined as

$$f_C = \sqrt{f_H f_L}$$

Where $f_H$ and $f_L$ are respectively the high and low cut-off frequencies in Hz.

In a narrow band pass filter, the output voltage peaks at the centre frequency $f_C$.

## (i) Wide Band Pass Filter

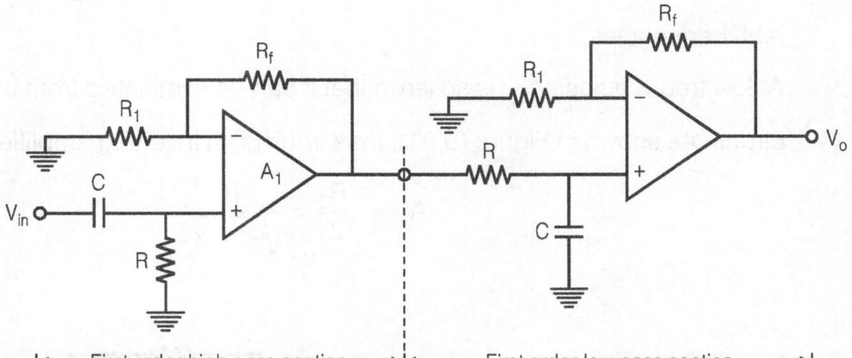


*Figure 9.79 Wideband pass filter*

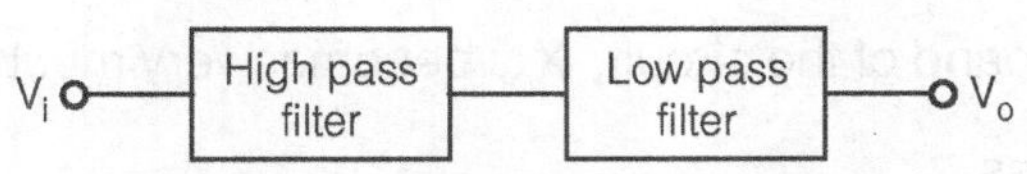


*(a) Cascaded high pass and low pass filters*

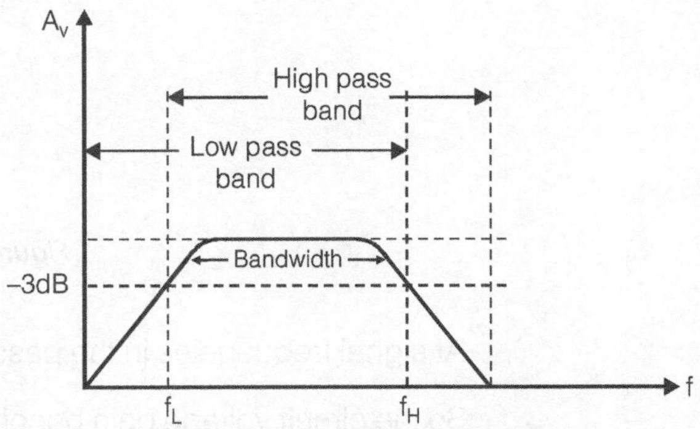


*(b) Wideband pass frequency response*

A band-pass filter can be constructed simply by connecting low-pass and high-pass filters in cascade. For example, suppose a low pass filter with $f_C$ = 100 kHz is cascaded with a high-pass circuit which has $f_C$ = 10 kHz. The low pass circuit will pass all frequencies upto 100 kHz, while the high pass circuit will block all frequencies below 10 kHz. Consequently the combination gives a filter with a pass band from 10 kHz to 100 kHz.

## Single-Stage Wide band Pass Filter

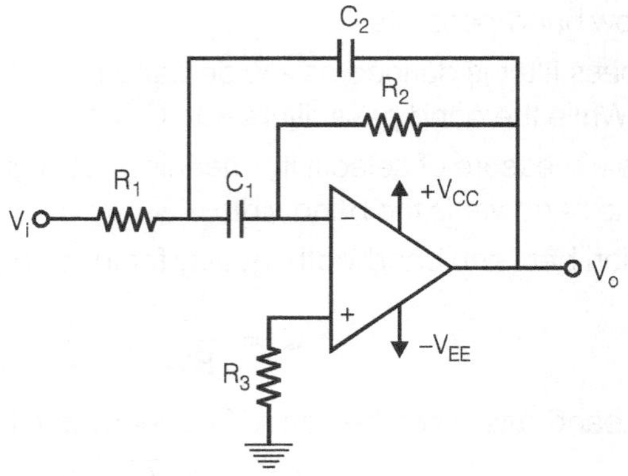


*Figure 9.80 Single stage band pass filter*

A single stage band pass filter circuit is shown in Figure (9.80). If the capacitors were not present, this circuit would look like an inverting amplifier. The capacitors are selected to have $X_{C_2}$ large enough to be neglected at low frequencies and $X_{C_1}$ small enough to be neglected at high frequencies.

At low frequencies, $X_{C_2}$ is so large that it can be eliminated from the low frequency equivalent circuit. As shown in Figure (9.81), the circuit is an inverting amplifier with a voltage gain of

$$A_v = \frac{R_2}{Z_1} = \frac{R_2}{\sqrt{R_1^2 + X_{C1}^2}} \qquad \text{...(i)}$$

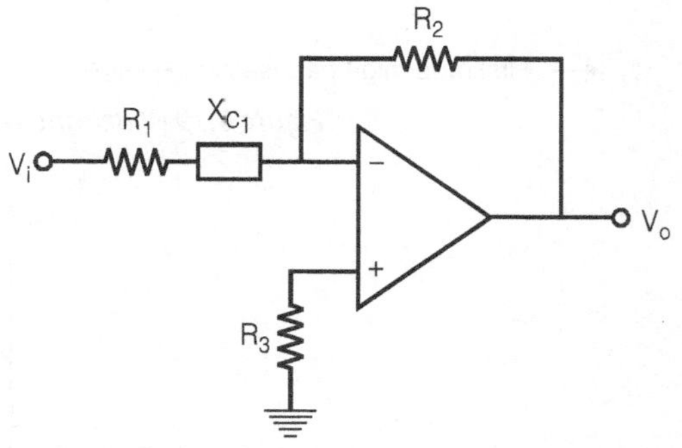


*Figure 9.81 Low frequency equivalent circuit*

At signal frequencies in the pass band of the circuit, $X_{C_1}$ becomes very much smaller than $R_1$.

So the circuit voltage gain becomes

$$A_v \approx \frac{R_2}{R_1}$$

The equation (i) shows that for the voltage gain to be down by 3dB

$$X_{C_1} = R_1 \text{ at } f_L$$

Where $f_L$ is the lower cut-off frequency.

At high frequencies, $X_{C_1}$ becomes so small compared to $R_1$ that $C_1$ can be eliminated from the high frequency equivalent circuit. But $X_{C_2}$ is longer large enough to be neglected. So the high frequency equivalent circuit is as shown in Figure (9.82).

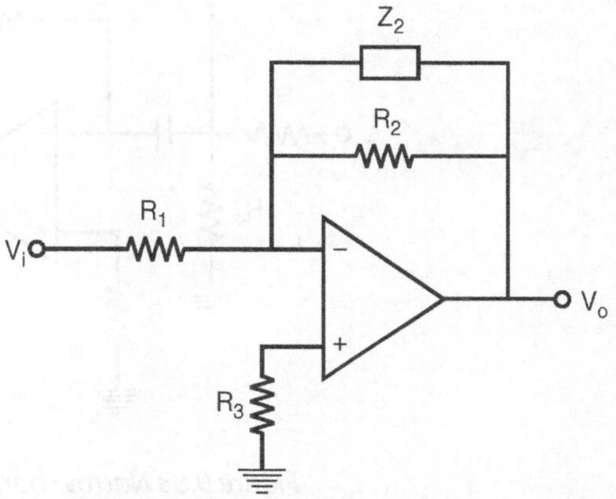


*Figure 9.82 High-frequency equivalent circuit*

The circuit functions as an inverting amplifier, and its voltage gain is

$$A_v = \frac{X_{C_2} \| R_2}{R_1} = \frac{R_2}{R_1\sqrt{1+\left(\frac{R_2}{X_{C_2}}\right)^2}} \qquad \text{...(ii)}$$

At frequencies in the passband, $X_{C_2}$ is very much larger than $R_2$, and so the circuit voltage gain is

$$A_v \approx \frac{R_2}{R_1}$$

From equation (ii) the voltage gain is down by 3dB from its mid frequency value when

$$X_{C_2} = R_2 \text{ at } f_2$$

Where $f_2$ is the high cut-off frequency.

It is seen that at the circuit behaves as a high-pass filter for low signal frequencies, as an inverting amplifier for frequencies in the pass band, and as a low-pass filter for high frequencies.

The quailing factor Q is the relationship of the centre frequency $f_o$ to the bandwidth.

$$Q = \frac{f_o}{BW}$$

The centre frequency of the filter can be determined from

$$f_o = \sqrt{f_{-}\, f_2}$$

## (ii) Narrow-band Pass Filter

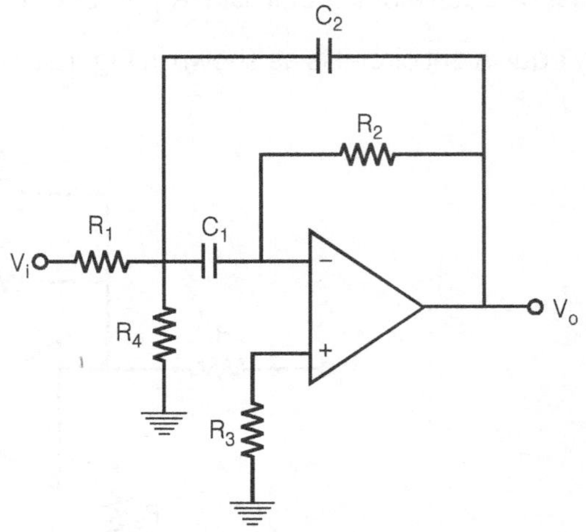


*Figure 9.83 Narrow band pass filter*

From the Figure (9.83), it is reconsidered, that if $f_L$ and $f_H$ are brought closer together to narrow the bandwidth, $C_2$ will affect the low cut-off frequency, and $C_1$ will interfere with the high cut-off frequency. To counter this, another resistor ($R_4$) is included at the circuit input. This modification produces larger Q factors and narrower bandwidths. The additional resistor ($R_4$) makes the circuit analysis quite complex.

How ever the following equations can be shown to relate the component values to the circuit performance.

$$C_1 = C_2 = C$$

$$R_1 = 0.5\, R_2$$

$$R_2 = 2Q X_C \text{ at } f_o$$

$$Q^2 = \frac{R_1 + R_4}{2R_4}$$

$$A_f = \frac{R_2}{2R_1}$$

Where $A_f$ is the gain at centre frequency.

## 9.54 Band Stop Filter

The band-pass filter passes one set of frequencies while rejecting all others. The band stop filter does just the opposite. It rejects a band of frequencies, while passing all others. This is also called a band reject or band-elimination filter.

Like band-pass filters, band stop filters may also be classified as

(i) Wide band stop filter.

(ii) Narrow band stop filter.

The narrow band reject filter is also called a notch filter. Because of its higher Q, which exceeds 10, the bandwidth of the narrow band reject filter is much smaller than that of a wide band reject filter.

### Wide band Stop Filter

A wide band stop filter is the inverse of a band-pass filter. Its function is to block a band of signal frequencies. Other names are band elimination and band reject filter.

The Figure 9.84 (a & b) shows the block diagram and circuit diagram of wide band stop filter.

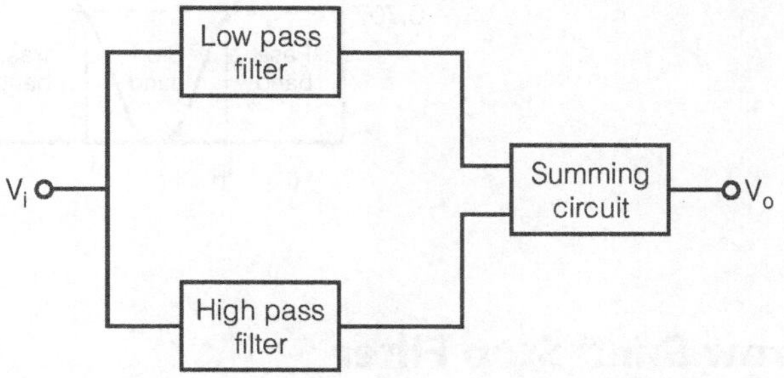


*(a) Block diagram*

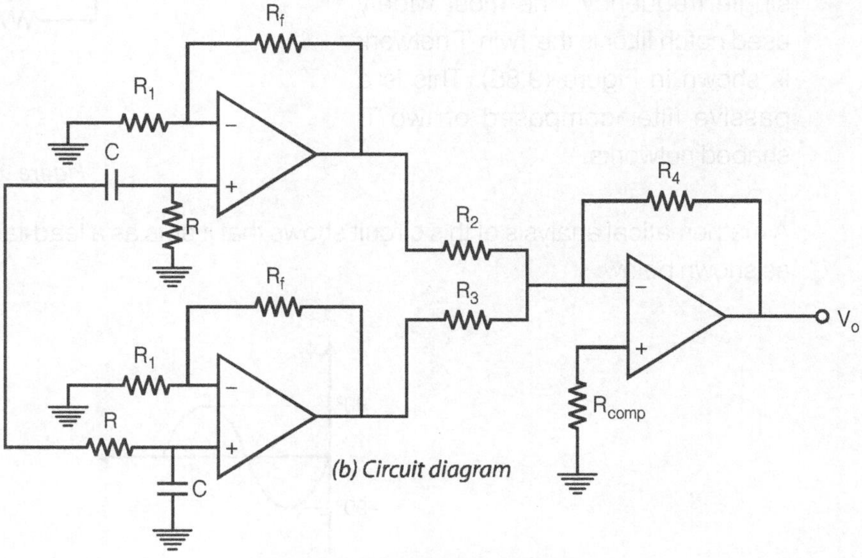


*(b) Circuit diagram*

*Figure 9.84 Wideband stop filter*

A wide band stop filter is a combinaticn of a low pass filter, a high pass filter and summing amplifier.

The filter operates provided,

**(i)** the low pass and high pass filter sections have equal pass band gain and

**(ii)** The lower cut-off frequency $f_L$ of the high pass filter is greater than the higher cut-off frequency $f_H$ of the low pass filter.

In practice,

$$R_2 = R_3 = R_4 = R$$

Compensating resistor $$R_{comp} = R_2 \| R_3 \| R_4 \| = \frac{R}{3}$$

The frequency response characteristic is as shown:

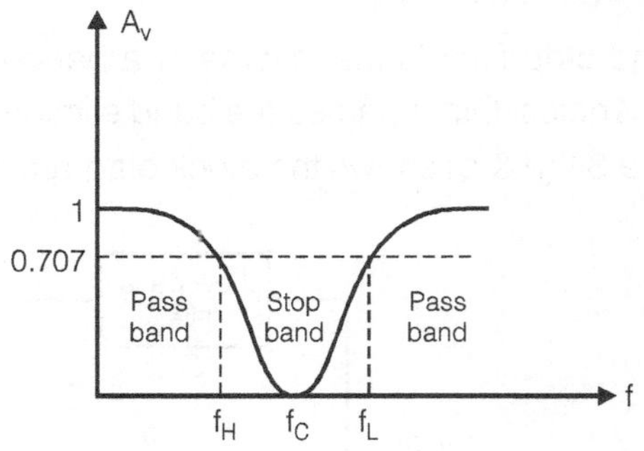


## Narrow Band Stop Filter

This filter is referred as notch filter, it is commonly used for attenuation of a single frequency. The most widely used notch filter is the Twin-T network is shown in Figure (9.85). This is a passive filter composed of two T-shaped networks.

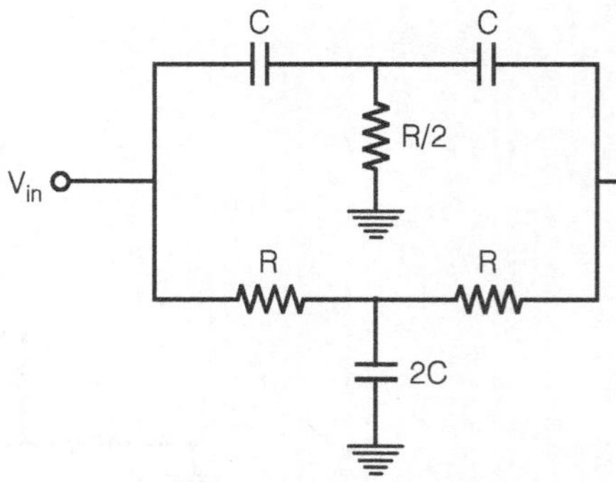


*Figure 9.85 Twin-T network*

A mathematical analysis of this circuit shows that it acts as a lead-lag circuit with a phase angle as shown below:

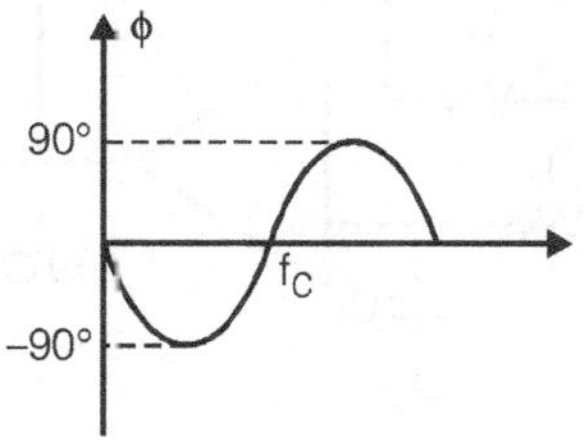


There is a frequency $f_c$ at which the phase shift is equal to 0°.

The voltage gain versus frequency graph is shown below:

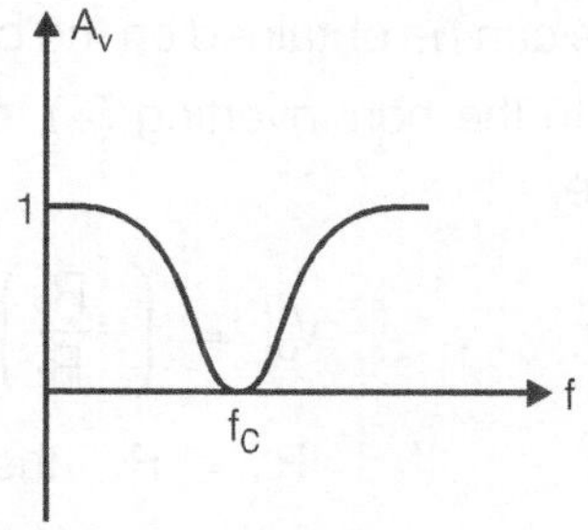


The voltage gain is equal to 1 at low and high frequencies. In between, there is a frequency $f_c$ at which voltage gain drops to zero. thus such a filter notches out, or blocks frequencies near $f_C$. The frequency at which maximum attenuation occurs is called notch out frequency given by

$$f_N = f_c = \frac{1}{2\pi RC}$$

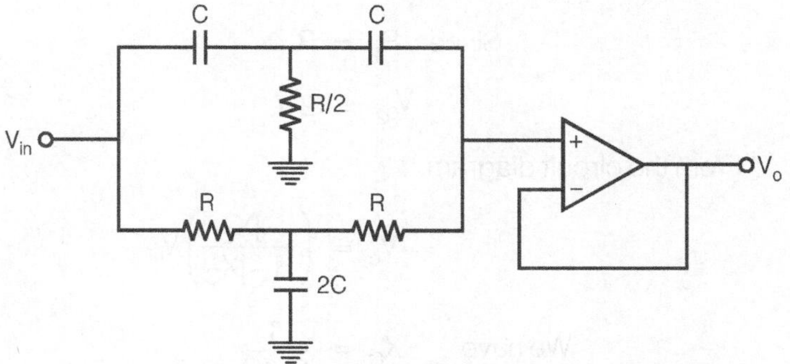


*Figure 9.86 Notch filter*

One drawback of above notch filter is that it has relatively low figure of merit Q. However, Q of the network can be increased significantly if it is used with the voltage follower.

Notch filters are most commonly used in communications and biomedical instrumentation for elimination of undesired frequencies.

# 9.55 All Pass Filter

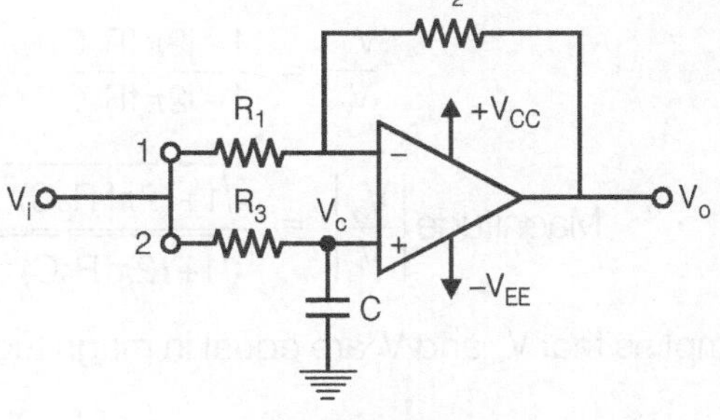


*Figure 9.87 All pass filter*

An all-pass filter is a circuit that will pass all frequencies without attenuation but with phase shifts that vary with the signal frequency.

A phase lag can also be termed as phase delay, so an all-pass phase-lag circuit is also known as a delay filter.

## To Find the Output Voltage $V_o$

The output voltage can be obtained on the basis of superposition theorem.

**(i)** Let the input to the non inverting (+) terminal to be zero. Then the op-amp operates in inverting mode.

$$\therefore \quad V_{O_1} = \left(-\frac{R_2}{R_1}\right)V_i$$

$$\text{If} \quad R_2 = R_1 \quad \text{then}$$

$$\therefore \quad V_{O_1} = -V_i \quad \text{...(i)}$$

**(ii)** Let the input at the inverting terminal be zero.

The op-amp operates as a non inverting amplifier.

$V_{O_2} = \left(1+\frac{R_2}{R_1}\right)V_c$, where $V_c$ is the input at the non inverting terminal.

$$\text{Since} \quad R_2 = R_1$$

$$V_{O_2} = 2V_c \quad \text{...(ii)}$$

From the circuit diagram,

$$V_c = \left(\frac{-jX_C}{R-jX_C}\right)V_i$$

$$\text{We have} \quad X_C = \frac{1}{2\pi fC}$$

Substituting this value for $X_C$ and simplifying, we get

$$V_C = V_i\left(\frac{1}{1+j2\pi fR_3C}\right)$$

Substitute $V_C$ in equation (ii), we get

$$V_{O_2} = 2V_i\left(\frac{1}{1+j2\pi fR_3C}\right)$$

$$V_o = V_{O_1}+V_{O_2} = -V_i+2V_i\left(\frac{1}{1+j2\pi fR_3C}\right) = V_i\left(-1+\frac{2}{1+j2\pi fR_3C}\right)$$

$$\frac{V_o}{V_i} = \frac{1-j2\pi fR_3C}{1+j2\pi fR_3C}$$

$$\text{Magnitude}\left|\frac{V_o}{V_i}\right| = \frac{\sqrt{1+(2\pi fR_3C)^2}}{\sqrt{1+(2\pi fR_3C)^2}} = 1$$

This implies that $V_o$ and $V_i$ are equal in magnitude.

$$\text{Phase angle}\ \phi = -2\tan^{-1}\left(\frac{R_3}{X_C}\right) = -2\tan^{-1}\left(\frac{2\pi fRC}{1}\right)$$

This implies that the output voltage $V_o$ lags behind the input voltage $V_i$ by angle $\phi$.

## Phase Versus Frequency Graph

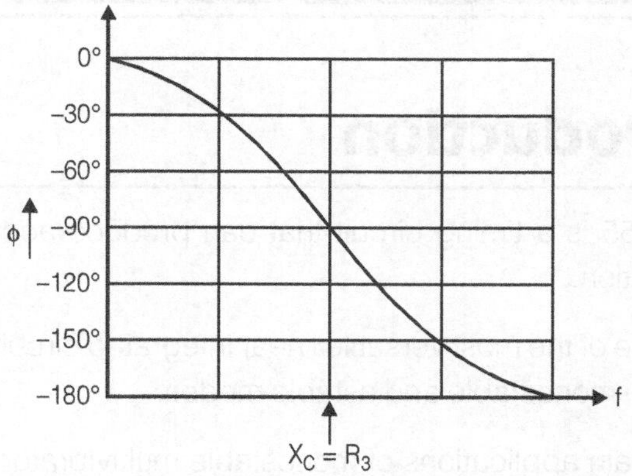


*Figure 9.88 Phase versus frequency for phase-lag circuit*

When $X_C = R_3, \phi = -90°$

At high signal frequencies, when $X_{C_1} << R_3$

$$\phi \approx -180°$$

At low signal frequencies, when $X_{C_1} >> R_3$

$$\phi \approx 0°$$

## Phase-Lead Circuit

The phase-lead circuit in Figure (9.89) is similar to the phase-lag circuit, except that $R_3$ and C are interchanged.

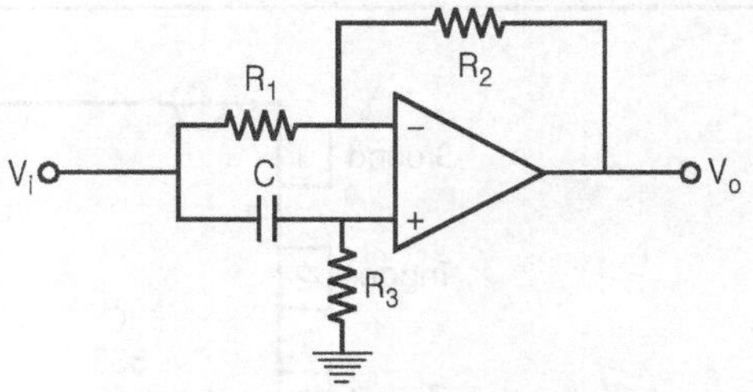


*Figure 9.89 All pass phase lead circuit*

This has the effect of causing the output to lead the input by a phase angle dependent on the relationship between $X_C$ and $R_3$.

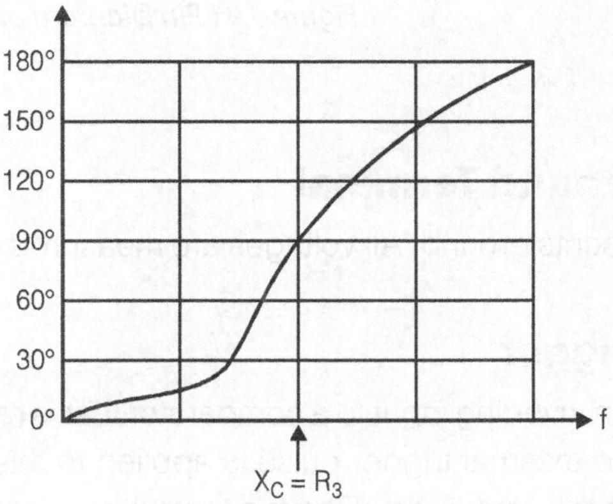


*Figure 9.90 Phase versus frequency plot for phase lead circuit*

# 555 TIMER

## 9.56 Introduction

- The 555 is a timing circuit that can produce accurate and highly stable time delays or oscillation.
- It is one of the most versatile linear integrated circuit devices in practical use. It can operate in both monostable and astable modes.
- The main applications of monostable multivibrator are pulse width modulator, linear ramp generator, frequency divider etc.
- The main applications of astable multivibrator are square wave generator, voltage control oscillator (VCO), FSK generator.
- This section deals with the basic features of the timer, block diagram, 555 timer in two modes i.e. mono stable and astable mode.
- In monostable mode, it can produce accurate time delays from micro seconds to hours. In the astable mode it can produce rectangular waves with a variable duty cycle.

## 9.57 555 Timer Pin Configuration

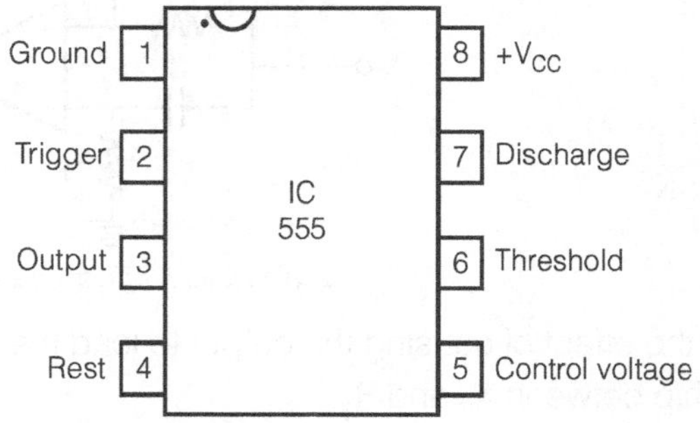


*Figure 9.91 Pin Diagram of 555 Timer*

### Pin-1: Ground Terminal

Pin-1 represents ground. All voltages are measured with respect to this terminal.

### Pin-2: Trigger

This pin is an inverting input to a comparator that is responsible for transition of flip-flop from set to reset. The external trigger pulse is applied to this terminal. The output voltage of the timer depends on the amplitude of the trigger pulse.

### Pin-3: Output

A load may be connected between this pin and the ground pin.

### Pin-4: Reset

To reset or disable the timer a negative pulse is applied to this pin due to which it is referred as reset terminal. When this pin is not to be used for reset purpose, it should be connected to $+V_{cc}$ to avoid any possibility of false triggering.

### Pin-5: Control Voltage

When an external voltage is applied to control voltage, both trigger and threshold voltages vary, so that output voltage pulse width changes.

### Pin-6: Threshold Voltage

This is the non inverting input terminal of comparator-1. When the voltage at this terminal exceeds the control voltage 2/3 $V_{CC}$, the output of compartor-1 goes high, as a result the timer output goes low.

### Pin-7: Discharge

This pin is connected to the collector terminal of the discharge transistor $Q_1$.
During operation, when the timer output is low, $Q_1$ is saturated and it short circuits the capacitor connected across it externally. When the timer output is high, $Q_1$ is cut-off.

### Pin-8: +$V_{CC}$ Supply

A supply voltage of +5 V to +18 V is applied to this terminal with respect to ground (pin-1).

## 9.58 Block Diagram of 555 Timer

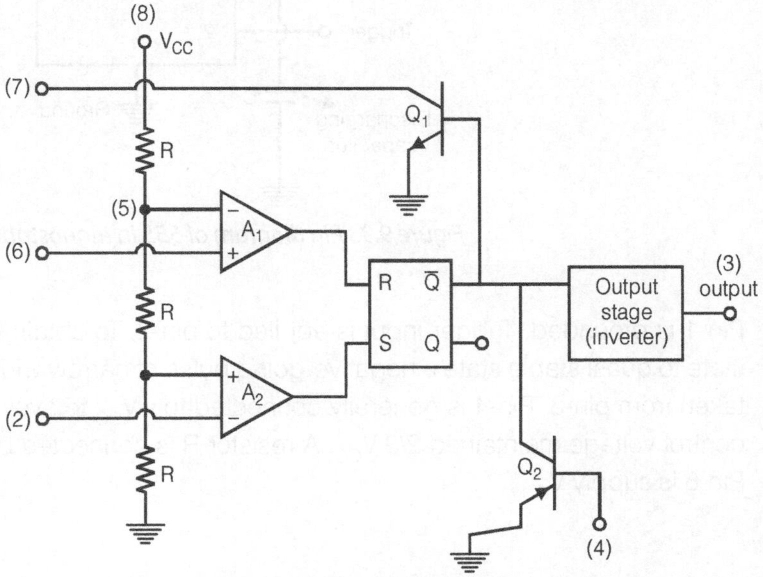


*Figure 9.92 Block diagram of 555 timer*

- As shown in the block diagram, a 555 timer consists of two comparators, an R-S flip-flop, two transistors (npn and pnp) and a resistive network.
- Resistive network consists of three equal resistors of 5 k. So it is named as 555 timer.
- Comparator-1 compares threshold voltage (pin-6) with a control voltage (pin-5) +2/3 $V_{CC}$. Comparator-2 compares the trigger voltage (pin-2) with a reference voltage +1/3 $V_{CC}$. The two comparators are connected to R-S flip flop.
- One of the two transistors is a discharge transistor of which collector is connected to pin-7. This npn transistor $Q_1$ saturates or cut-off according to the output state of the flip-flop.
- Base of another transistor $Q_2$ (pnp) is connected to a reset terminal (pin-4). When a pulse is applied to this terminal resets the whole timer irrespective of any input.
- We have to discuss the operation of 555 timer as monostable and astable multivibrator.

## 9.59 555 Timer as Monostable Multivibrator

A monostable multivibrator is also called as one-shot multivibrator. In such a multivibrator, one state of output is stable while the other is quasi-stable state.

The pin diagram of 555 timer in the monostable mode of operation is as shown in Figure (9.93).

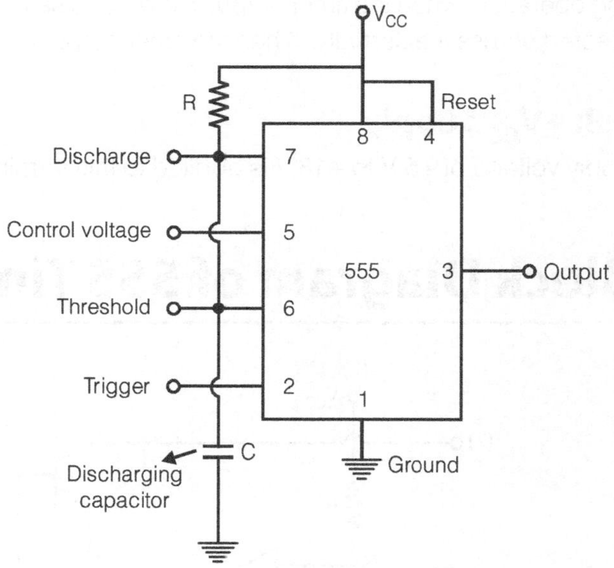


*Figure 9.93 Pin diagram of 555 in monostable mode*

Pin-1 is grounded. Trigger input is applied to pin-2. To obtain transition of output from stable state to quasi stable state a negative-going pulse of narrow width is applied to pin-2. Output is taken from pin-3. Pin-4 is generally connected to +$V_{CC}$ to avoid unnecessary reset. Pin-5 is a control voltage maintained 2/3 $V_{CC}$. A resistor R is connected between pin-8, pin-7 and pin-6. Pin-8 is supply $V_{CC}$.

## Mono-stable Operation

The block diagram of 555 timer in mono-stable mode of operation is as shown in Figure (9.94).

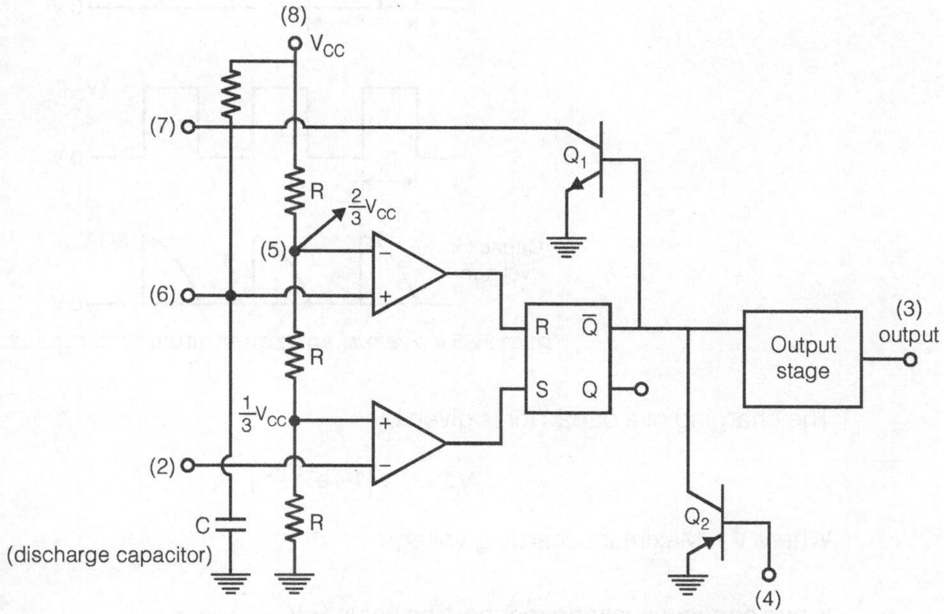


*Figure 9.94 Block diagram of 555 timer in mono-stable mode*

- Initially, when the output at pin-3 is low i.e. the circuit is in a stable state. The transistor $Q_1$ is on and capacitor C is shorted to grounded.
- When a negative pulse is applied to pin-2. the trigger input falls below $+1/3\ V_{CC}$. The output of comparator-2 goes high which sets the flip-flop and consequently the transistor $Q_1$ turns-off and the output at pin-3 goes high. This is the transition of the output from stable to quasi-stable state.
- As the transistor $Q_1$ is cut-off, the capacitor C begins charging toward $+V_{CC}$ through resistance R with a time constant equal to RC.
- When the increasing capacitor voltage becomes slightly greater than $+2/3\ V_{CC}$, the output of comparator-1 goes high, which resets the flip-flop. The transistor goes to saturation, there by discharging the capacitor C and the output of the timer goes low.
- Thus the output returns back to stable state from quasi-stable state. The output of multivibrator remains low until a trigger pulse is applied again.

| Condition | Q | $\bar{Q}$ | Output | Transistor Position |
|---|---|---|---|---|
| 1. Stand by | 0 | 1 | 0 | Discharging (on) |
| 2. Trigger $< \frac{1}{3}V_{CC}$ | 1 | 0 | 1 | Charging (off) |
| 3. Threshold $> \frac{2}{3}V_{CC}$ | 0 | 1 | 0 | Discharging (on) |

The wave forms of triggering signal, output pulse and capacitor voltage is shown below:

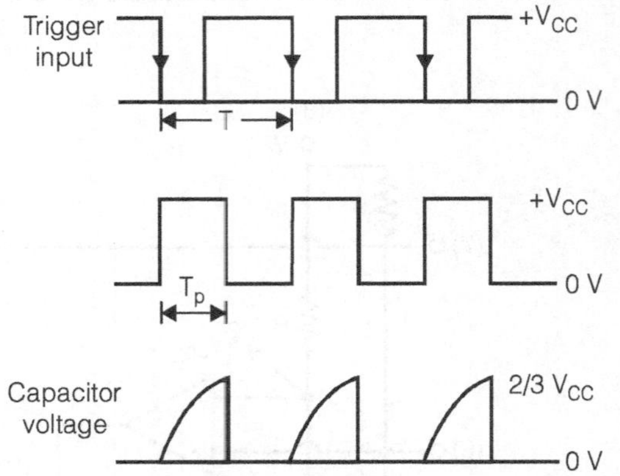


*Figure 9.95 Waveforms of trigger, output and capacitor*

The charging of a capacitor is given by

$$V_C = V\left(1 - e^{-t/RC}\right)$$

Where V = Maximum charging voltage.

When capacitor voltage reached the value $\frac{2}{3}V_{CC}$.

$$\text{We have} \quad V_c = \frac{2}{3}V_{CC}$$

$$\text{Also} \quad V = V_{CC}$$

$$\frac{2}{3}V_{CC} = V_{CC}\left[1 - e^{-t/RC}\right]$$

$$\frac{2}{3} - 1 = -e^{-t/RC}$$

$$\frac{1}{3} = -e^{t/RC}$$

$$t = -RC\, ln\left(\frac{1}{3}\right)$$

$$\boxed{t = 1.1RC}$$

Where RC is time constant.

The pulse width of output waveform is

$$\boxed{T_p = 1.1RC}$$

# 9.60 555 Timer as Astable Multivibrator

- Astable multivibrator can also be called as free running oscillator. An astable multivibrator has two output states and both these states are quasi-stable states.
- The multi keeps on switching between these two state by itself, and it does not need any external triggering.
- The pin diagram of 555 Timer in the astable mode of operation is shown in Figure (9.96).

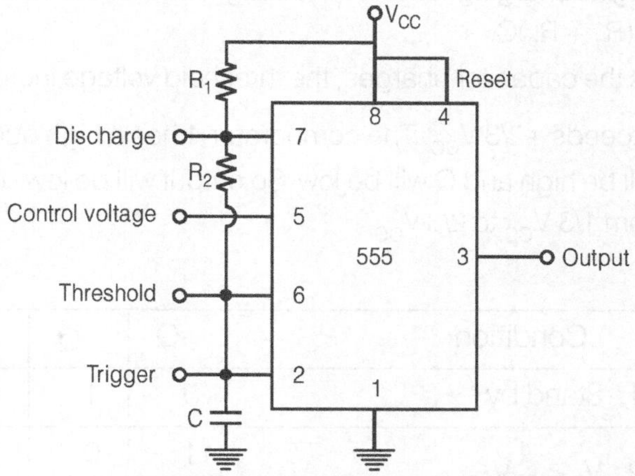


*Figure 9.96 Pin diagram of 555 in astable mode*

Pin-1 is grounded. Here capacitor is connected to pin-2 which becomes internal trigger input. Output is taken from pin-3. Pin-4 is usually connected to $+V_{CC}$ to avoid unnecessary reset. Pin-5 is a control voltage maintained 2/3 $V_{CC}$. A resistor $R_1$ is connected between pin-8 and pin-7. Another resistor $R_2$ is connected between pin-7 and pin-6. Pin-8 is supply $V_{CC}$.

## Astable Mode of Operation

The block diagram of 555 timer in astable mode of operation is as shown in Figure (9.97).

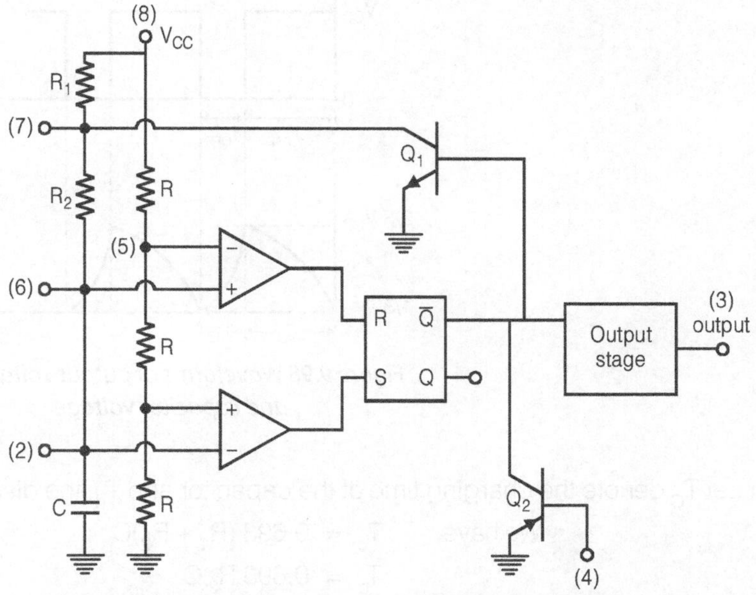


*Figure 9.97 Block diagram of astable mode of 555*

- When $\bar{Q}$ is high and the timer output is low, the transistor $Q_1$ saturates and pin-7 grounds, so that the capacitor C discharges through resistance $R_2$ with a time constant $R_2C$.
- With the discharging of capacitor, trigger voltage (capacitor voltage) at inverting input of comparator 2 decreases. When it drops below 1/3 $V_{CC}$. The output of comparator-2 goes high and this sets the flip-flop so that Q is high and $\bar{Q}$ is low. The timer output is high.
- When $\bar{Q}$ is low or output is high, the discharging transistor is cut-off and the capacitor C begins charging toward $V_{CC}$ through resistances $R_1$ and $R_2$. Because of this the time constant is $(R_1 + R_2)C$.
- As the capacitor charges, the threshold voltage increases. Eventually, the threshold voltage exceeds +2/3 $V_{CC}$. The comparator-1 has a high output and triggers the flip-flop so that $\bar{Q}$ will be high and Q will be low. So output will be low. This proves the auto-transition in output from 1/3 $V_{CC}$ to 2/3 $V_{CC}$.

| Condition | Q | $\bar{Q}$ | Output | Transistor Position |
|---|---|---|---|---|
| 1. Stand by | 0 | 1 | 0 | Discharging (on) |
| 2. $V_C < \frac{1}{3}V_{CC}$ | 1 | 0 | 1 | Charging (off) |
| 3. Threshold voltage > $\frac{2}{3}V_{CC}$ | 0 | 1 | 0 | Discharging (on) |
| | ⋮ | ⋮ | ⋮ | ⋮ |

The waveforms of output voltage and capacitor voltage is shown below:

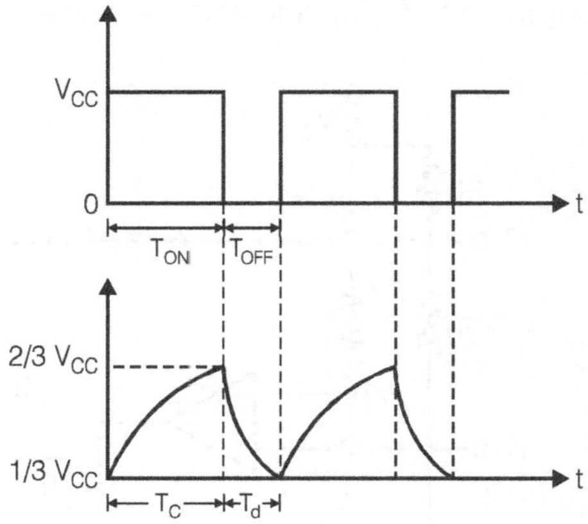


*Figure 9.98 Waveforms of output voltage and capacitor voltage*

Let $T_C$ denote the charging time of the capacitor and $T_d$, the discharging time.

We have,
$$T_c = 0.693\,(R_1 + R_2)C$$
$$T_d = 0.693\,R_2\,C$$

$$\text{Total time } T = T_c + T_d$$
$$= 0.693\,(R_1 + R_2)C + 0.693\,R_2C$$
$$= 0.693\,(R_1 + 2R_2)C$$

The frequency of oscillation is given by

$$f = \frac{1}{T} = \frac{1}{0.693\,(R_1 + 2R_2)C}$$

$$\boxed{f = \frac{1.443}{(R_1 + 2R_2)\,C}}$$

Duty cycle is given by $\quad D.C. = \dfrac{T_c}{T_c + T_d}$

$$= \frac{0.693\,(R_1 + R_2)C}{0.693(R_1 + 2R_2)C} = \frac{R_1 + R_2}{R_1 + 2R_2}$$

To convert pulse generator into square-wave generator. The modification of the astable multivibrator pin diagram is shown in Figure (9.99).

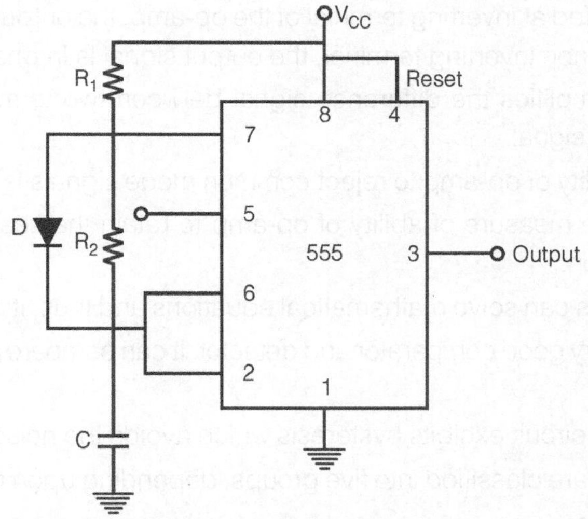


*Figure 9.99 Square wave generator*

According to this figure,

$$\text{Charging time } T_C = 0.693\,(R_1 + R_2)C$$
$$\text{Apply condition } R_2 = 0 \text{ (diode is parallel to } R_2)$$
$$T_C = 0.693\,R_1C$$
$$\text{Discharging time } T_d = 0.693\,R_2C$$

$$\text{Duty cycle} = \frac{T_c}{T_c + T_d} = \frac{0.693\,R_1C}{0.693\,(R_1 + R_2)C} = \frac{R_1}{R_1 + R_2}$$

$$\text{If } R_1 = R_2 = R$$

$$\text{Duty cycle} = \frac{R}{2R} = \frac{1}{2} = 50\%$$

That means, a symmetrical square wave can be obtained if a diode is connected across resistor $R_2$.

## SUMMARY

- Differential amplifier is the basic building block of analog circuits and it is the input stage of operational amplifier.
- There are four types of configurations in differential amplifiers:
  - Single input balanced output.
  - Single input unbalanaced output.
  - Dual input balanced output.
  - Dual input unbalanced output.
- The technique used to improve gain stabilization and input resistance is "swamping resistor technique".
- The techniques used to make the operating point Q stable are:
  - Constant current source.
  - Current mirror.
- Differential amplifier with active load is used to improve CMRR.
- Op-amp can be used over a very wide frequencies starting from dc signals.
- Virtual ground is nothing but zero potential across infinite impedance.
- When signal is fed at inverting terminal of the op-amp, the output signal is out of phase whereas when signal is fed at non inverting terminal, the output signal is in phase with the input.
- The op-amp amplifies the difference signal between two terminals of the input while it rejects the common mode signal.
- Measure of ability of op-amp to reject common mode signals is CMRR.
- Slew rate is the measure of ability of op-amp to retain the shape of sinusoidal signal at the output without distortion.
- Op-amp circuits can solve mathematical equations and thus, it can be referred as analog computer.
- Op-amp is a very good comparator and detector. It can compare and detect any signal to any reference per specified.
- Schmitt trigger circuit exhibits hysteresis which avoids the noise present in the input signal.
- Practical filters are classified into five groups, depending upon the frequency range:
  - Low pass filter
  - High pass filter
  - Band pass filter
  - Band stop filter
  - All pass filter.
- An ideal filter should pass the band of frequencies which is designed to accept, with out attenuation, while at the same time, it should attenuate, without fail, all frequencies outside this band.
- 555 timer is a timing circuit that can produce accurate and highly stable time delays or oscillation.
- 555 timer can operate in monostable and astable modes.
- The main applications of monostable multivibrator are pulse width modulator, linear ramp generator, frequency divider etc.
- The main applications of astable multivibrator are square wave generator, voltage control oscillator, FSK generator.

## GATE AND IES QUESTIONS

**Q.1** For the op-amp, determine output voltage. Assume ideal conditions.

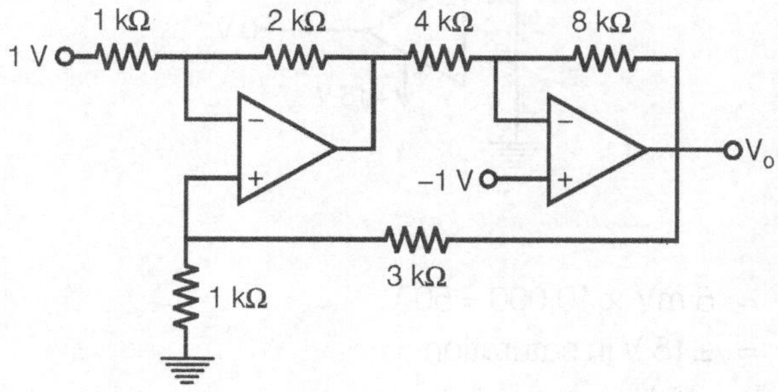


(a) $-\frac{8}{7}V$ (b) $-\frac{20}{7}V$

(c) −10 V (d) None of these

**Exp. (b)**

$$\frac{V_1 - 1}{1} + \frac{V_1 - V_2}{2} = 0 \quad \text{(First op-amp)}$$

$$3V_1 - V_2 = 2$$

Op-amp (Positive terminal of op-amp).

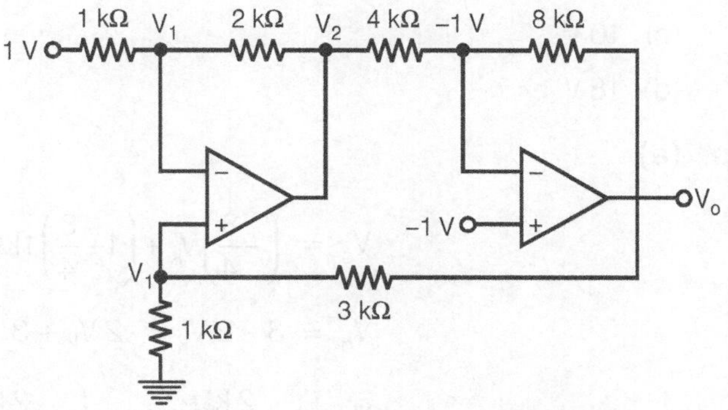


$$\frac{V_1 - V_o}{3} + \frac{V_1 - 0}{1} = 0$$

At negative terminal op-amp 2.

$$\frac{-1 - V_2}{4} + \left(\frac{-1 - V_o}{8}\right) = 0$$

$$-2 - 2V_2 - 1 - V_o = 0$$

$$V_o + 2V_2 = -3$$

$$3\frac{V_o}{4} - 2V_2 = 1$$

$$\frac{3}{4}V_o - 2(-3 - V_o) = 1$$

$$\frac{3}{4}V_o + V_o = -5$$

$$\frac{7}{4}V_o = 5 \; ; \; V_o = -\frac{20}{7}V$$

**Q.2** In the circuit of figure, $V_o$ is

(a) −1 V

(b) 2 V

(c) +1 V

(d) +15 V

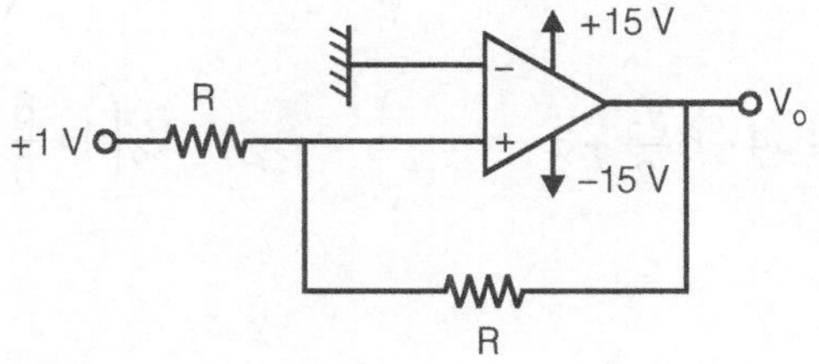


**Exp. (d)**

It is a positive feedback $V_o = +V_{sat} = +15$ V

**Q.3** The op-amp in the figure has an input offset voltage of 5 mV and an open loop voltage gain of 10000, then $V_o$ will be

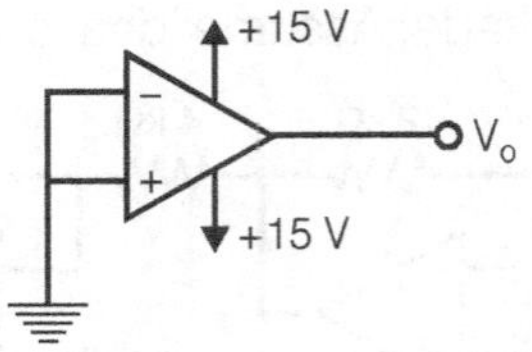


(a) 0
(b) 5 mV
(c) +15 V or −15 V
(d) +50 V or − 50 V

**Exp. (c)**

$$V_o = 5\text{ mV} \times 10{,}000 = 50\text{ V}$$

But $V_o = \pm 15$ V in saturation

So it can never be exceeds ±15 V

$$V_o = \pm V_{sat} = \pm 15\text{ V}$$

**Q.4** Find the output voltage of the following circuit assume ideal op-amp

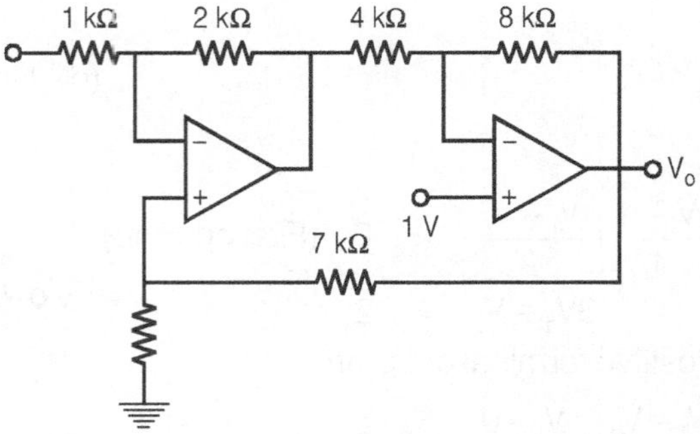


(a) 6 V
(b) 15 V
(c) 10 V
(d) 18 V

**Exp. (a)**

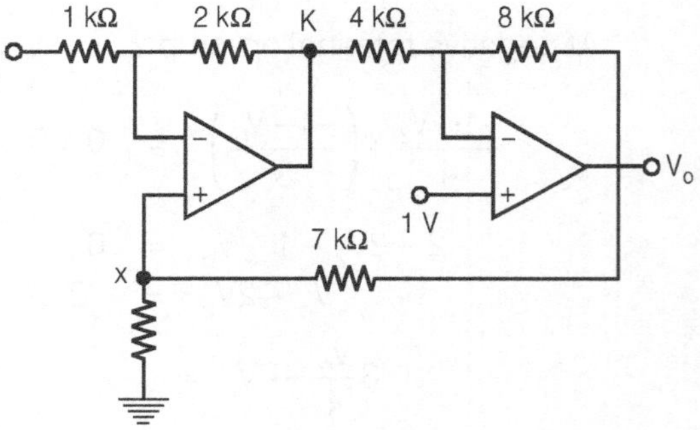


$$V_o = \left(-\frac{8}{4}\right)V_k + \left(1+\frac{8}{4}\right)1\text{k}\Omega$$

$$V_o = 3 - 2V_k = -2V_k + 3$$

$$V_k = -\frac{2\text{ k}\Omega}{1\text{k}\Omega}(1\text{ V}) + \left(1+\frac{2\text{ k}\Omega}{1\text{k}\Omega}\right)V_x$$

$$= -2 + 3V_x$$

$$\frac{0-V_x}{1\text{k}\Omega} = \frac{V_x - V_o}{7\text{ k}\Omega}$$

$$-7V_x = V_x - V_o$$

$$V_x = \frac{V_o}{8}$$

$$V_k = -2 + 3V_x = -2+3\frac{V_o}{8} \quad ; \quad V_o = -2\left(-2+3\frac{V_o}{8}\right)+3$$

$$V_o = 4-6\frac{V_o}{8}+3 \quad ; \quad V_o\left(1+\frac{6}{8}\right) = 4+3$$

$$\Rightarrow \quad V_o\left(\frac{14}{8}\right) = 7$$

$$V_o = 4\text{ V}$$

**Q.5** Consider the circuit shown below:

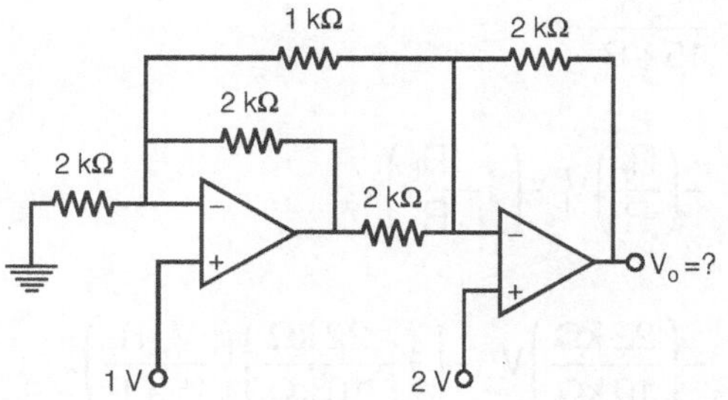


(a) 6 V (b) 15 V
(c) 10 V (d) 18 V

**Exp. (a)**

Due to virtual ground

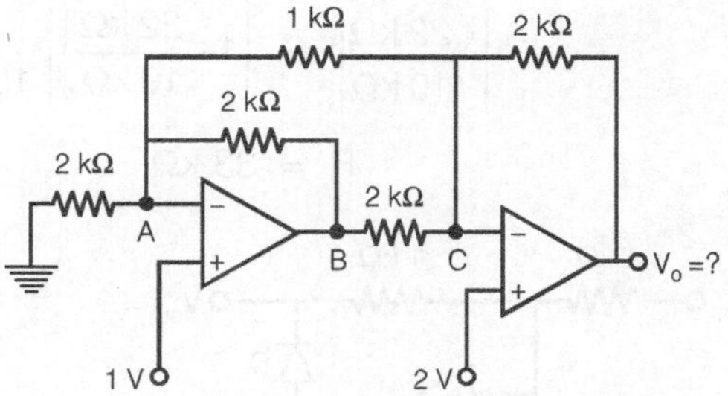


$$V_A = 1\text{ V}; V_c = 2\text{ V}$$

KCL at node A
$$\frac{0-1}{2\,k\Omega} = \frac{1-V_B}{2\,k\Omega}+\frac{1-2}{1k\Omega}$$

$$V_B = 0\text{ V} \quad \text{KCL at node C}$$

$$\frac{V_B - V_C}{2\,k\Omega}+\frac{V_A - V_C}{7\,k\Omega} = \frac{V_C - V_o}{2\,k\Omega}$$

$$\frac{0-2}{2\,k\Omega}+\frac{1-2}{1k\Omega} = \frac{2-V_o}{2\,k\Omega}$$

$$-\frac{1}{1k\Omega}-\frac{1}{1k\Omega} = \frac{1}{1k\Omega}-\frac{V_o}{2\,k\Omega}$$

$$\frac{V_o}{2\,k\Omega} = \frac{1}{1k\Omega}+\frac{1}{1k\Omega}+\frac{1}{1k\Omega}$$

$$V_o = \frac{3\times 2\,k\Omega}{1k\Omega}$$

$$= 6$$

**Q.6** In the given circuit, if the voltage $V_-$ and $V_+$ are to be amplified by the same amplification, the value of R should be

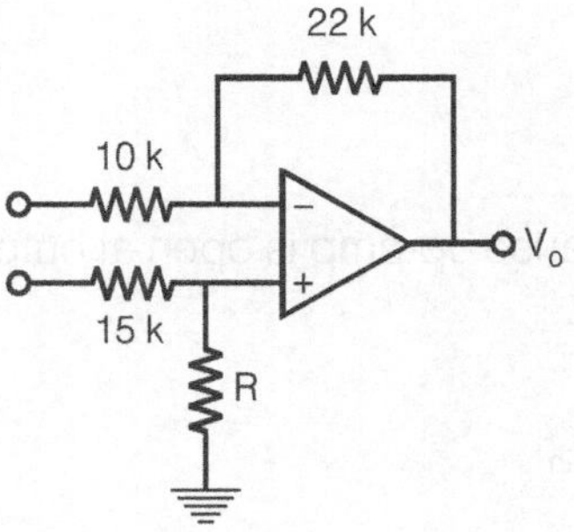


(a) 33 k (b) 22.3 k
(c) 43 k (d) none

Exp. (a)

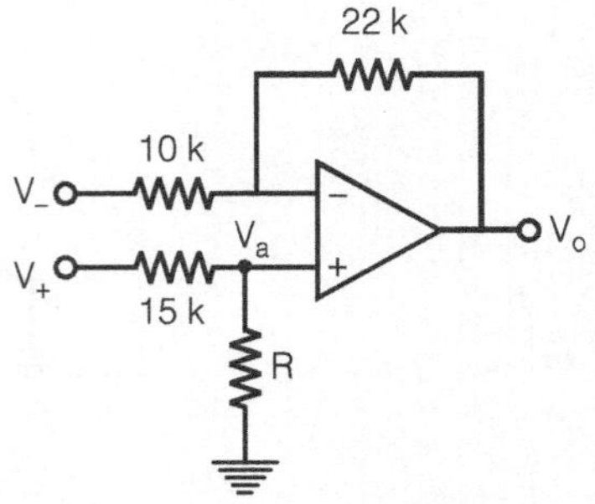


$$V_a = \frac{V_+ R}{15+R}$$

$$V_o = -\left(\frac{R_f}{R}\right)V_i + \left(1+\frac{R_f}{R_1}\right)V_a$$

$$= -\left(\frac{22\ k\Omega}{10\ k\Omega}\right)V_- + \left(1+\frac{22\ k\Omega}{10\ k\Omega}\right)\left(\frac{V_+ R}{15+R}\right)$$

$$\left(-\frac{R_f}{R}\right) = \left(1+\frac{R_f}{R_1}\right)\left(\frac{R}{15+R}\right)$$

$$\left|-\frac{22\ k\Omega}{10\ k\Omega}\right| = \left|1+\frac{22\ k\Omega}{10\ k\Omega}\right|\left|\frac{R}{15+R}\right|$$

$$R = 33\ k\Omega$$

Q.7

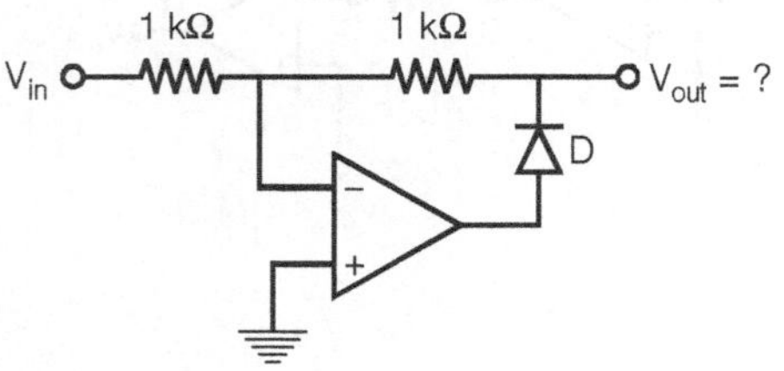


(a)

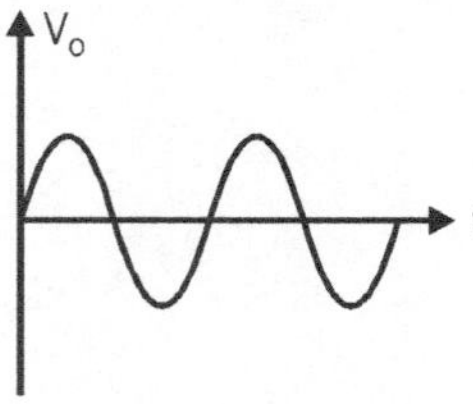


(b)

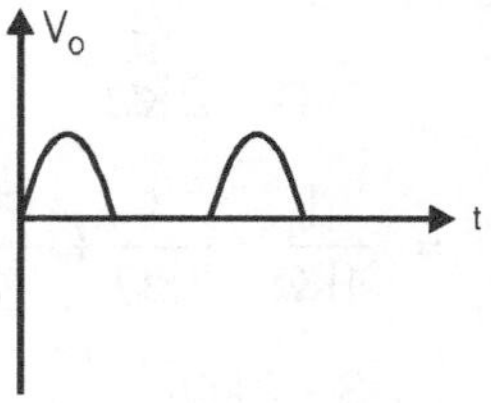


(c)

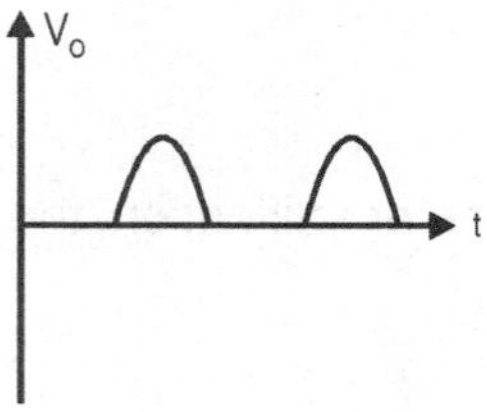


(d) none

Exp. (c)

In positive half cycle $D_1$ is RB

$V_o$ = 0 because op-amp is open at output.

In negative half cycle, $D_1$ is FB

$$V_o = -\frac{R_f}{R_1} V_{in}$$

$$V_o = -V_{in}$$

**Q.8** The output voltage $V_o$ of the circuit is

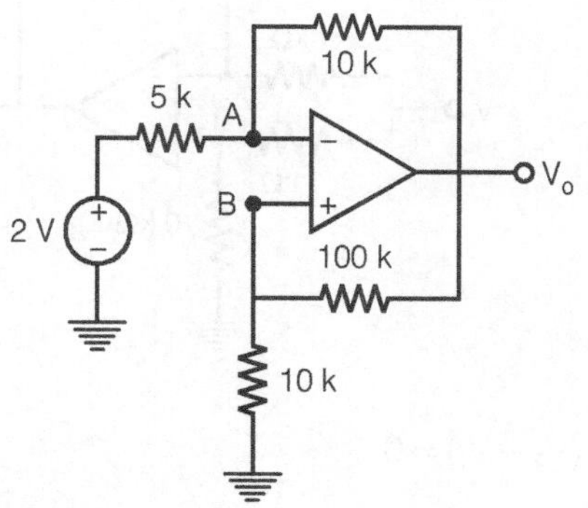


(a) –4 V
(b) 6 V
(c) 5 V
(d) –5.5 V

**Exp. (d)**

$$\text{Voltage at B} = \frac{V_o \times 10}{110} = \frac{V_o}{11}$$

$$\text{Voltage at A} = \text{Voltage at B} = \frac{V_o}{11}$$

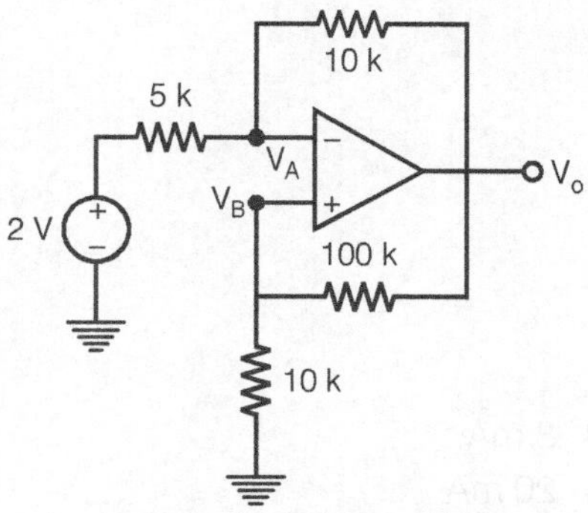


Using KCL at A

$$\frac{\frac{V_o}{11} - 2}{5} + \frac{\frac{V_o}{11} - V_o}{10} = 0$$

$$V_o = -5.5 \text{ V}$$

**Q.9** The voltage gain versus frequency of an op-amp is shown in figure

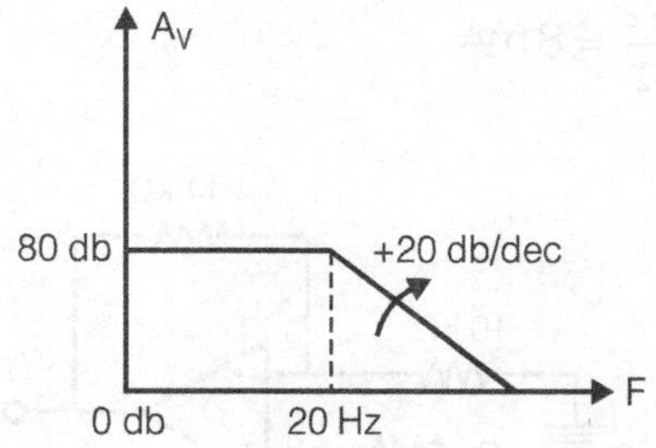


The gain bandwidth product of

(a) 200 Hz
(b) 200 MHz
(c) 200 kHz
(d) 2 MHz

**Exp. (c)**

$$\text{Gain BW product} = \text{Constant}$$
$$80 \text{ dB} = 20 \log A_V$$
$$A_V = 10^4$$
$$\text{GBW} = 10^4 \times 20 = 200 \text{ kHz}$$

**Q.10** The $V_o$ of the op-amp circuit shown is

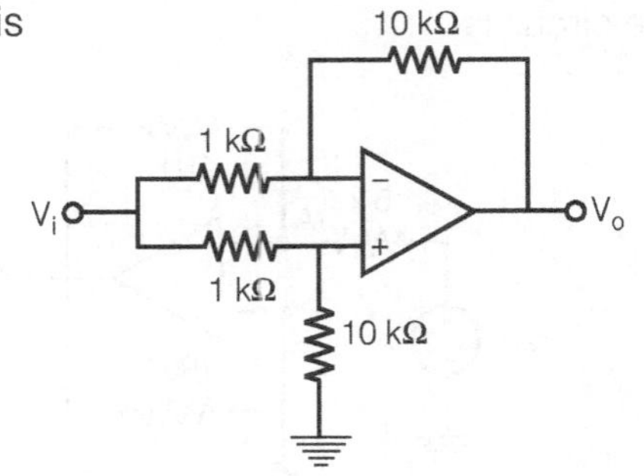


(a) $11\ V_i$

(b) $10\ V_i$

(c) $V_i$

(d) zero

**Exp. (d)**

$$V_o = \frac{R_2}{R_1}(V_2 - V_1) = 0$$

**Q.11** In the circuit shown in the figure the current flowing through resistance of 100 Ω would be

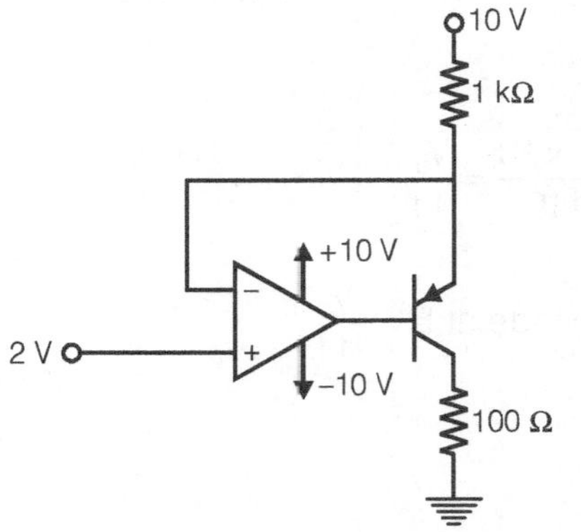


(a) 8 mA
(b) 10 mA
(c) 20 mA
(d) 100 mA

**Exp. (a)**

Current flowing through 100 Ω

$$= \frac{10-2}{1\text{k}\Omega} = 8\,\text{mA}$$

**Q.12** In the circuit shown in $V_o$ is

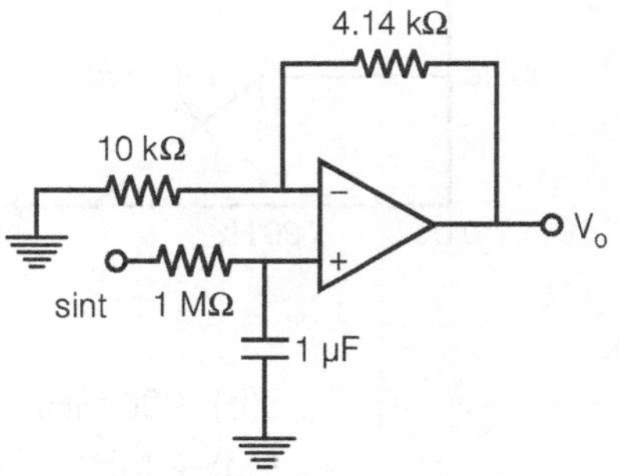


(a) $\sin t\left(\frac{t-\pi}{4}\right)$

(b) $\sin t\left(\frac{t+\pi}{4}\right)$

(c) sin t

(d) cos t

**Exp. (a)**

$$\frac{V_o}{V_i} = \left(\frac{1}{1+10^6 \times 10^{-6}s}\right)\left(1+\frac{4.14}{10}\right) = \left(\frac{1}{1+s}\right)1.414 = \frac{\sqrt{2}}{\sqrt{2}}\angle -45°$$

$$= 1\angle -45° = \sin\left(t-\frac{\pi}{4}\right)$$

**Q.13** Consider the circuit shown below:

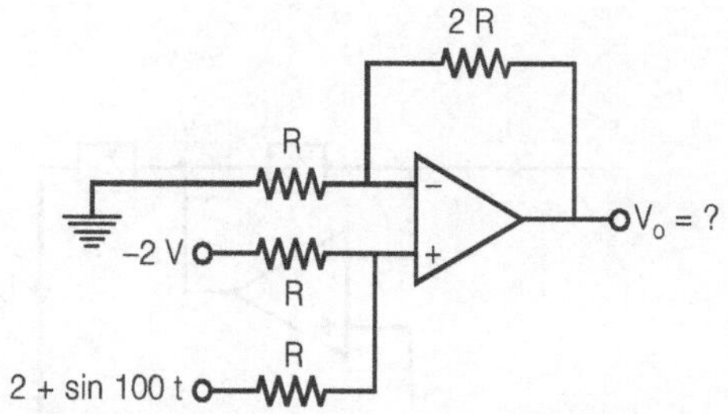


(a) $\frac{3}{2}\sin 100t$

(b) $3\sin 100t$

(c) $2\sin 100t$

(d) $3\sin 100t + \frac{1}{2}$

**Exp. (a)**

$$V_{01} = -2\left(\frac{R}{R+R}\right)\left(1+\frac{2R}{R}\right) = -2\left(\frac{1}{2}\right)(3) = -3\text{ V}$$

$$V_{02} = (2+\sin 100\,t)\left(1+\frac{2R}{R}\right)\left(\frac{R}{R+R}\right) = \frac{3}{2}(2+\sin 100t)$$

$$V_o = -3+3+\frac{3}{2}\sin 100t$$

$$V_o = \frac{3}{2}\sin 100t$$

**Q.14** A differential amplifier has inputs $V_1$ = 1050 μV and $V_2$ = 950 μV with CMRR = 1000, what is the error in differential output?

(a) 10%

(b) 1%

(c) 0.1%

(d) 0.01%

**Exp. (b)**

$$V_o = A_d V_d + A_c V_c$$

$$= A_d V_d\left(1+\frac{A_c V_c}{A_d V_d}\right) = A_d V_d\left(1+\frac{1}{CMRR}\frac{V_c}{V_d}\right)$$

Error in differential output.

$$= \frac{V_c}{CMRR \times V_d}\times 100$$

$$V_c = \frac{V_1+V_2}{2} = \frac{1050+950}{2}\,\mu V = 1000\,\mu V$$

$$V_d = V_1 - V_2$$

$$= (1050-950)\,\mu V = 100\,\mu V$$

$$\text{Error} = \frac{1000}{1000\times 100}\times 100 = 1\%$$

**Q.15** In the circuit shown above, what is the value of transfer function $I_o/I_i$?

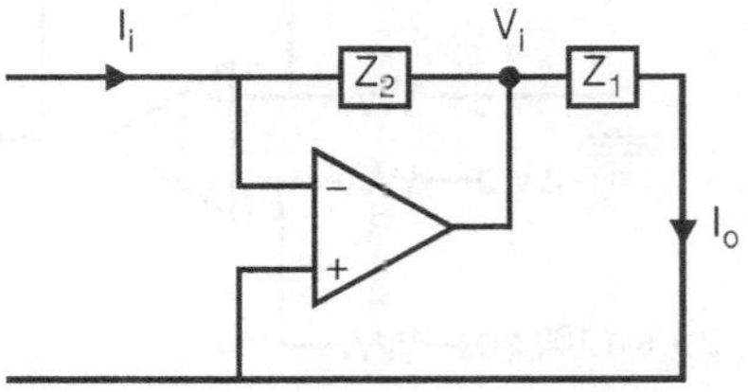


(a) $-\frac{Z_2}{Z_1}$

(b) $-\frac{Z_1}{Z_2}$

(c) $1+\frac{Z_2}{Z_1}$

(d) $1+\left(\frac{Z_1}{Z_2}\right)$

**Exp. (a)**

$$V_i = -I_i Z_2$$
$$V_i = Z_1 I_o$$
$$Z_1 I_o = -I_1 Z_2$$
$$\frac{I_o}{I_i} = -\frac{Z_2}{Z_1}$$

**Q.16** Assume the op-amp to be ideal. The current I through 1 kΩ resistor is

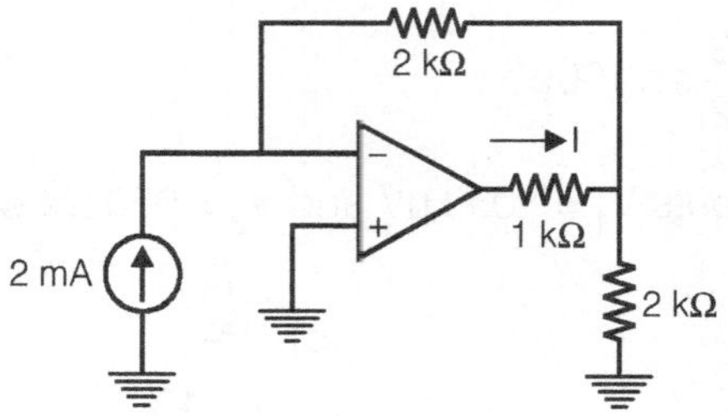


(a) –2 mA

(b) –4 mA

(c) –6 mA

(d) –8 mA

**Exp. (b)**

$$I_f = 2\text{ mA}$$
$$0 - V_o = 2\text{ mA} \times 2\text{ k}\Omega$$
$$V_o = -4\text{ V}$$
$$I_L = \frac{V_o}{2\text{ k}\Omega} = \frac{-4}{2\text{ k}\Omega} = -2\text{ mA}$$
$$I_f + I = I_L$$
$$I = -2\text{ mA} - 2\text{ mA} = -4\text{ mA}$$

**Q.17** With ideal operational amplifiers, the circuit is figure simulates the equation Vo(t) = .......

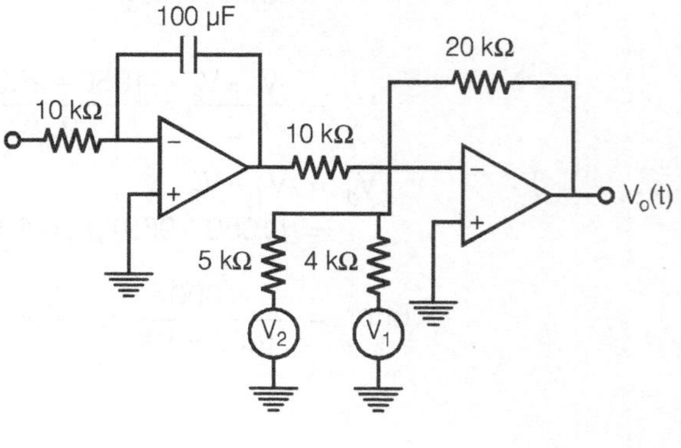


(a) $2V_1(t) + 3V_2 - 2V_3$

(b) $2V_1(t) - 4V_2 - 5V_3$

(c) $2V_1(t) - 3V_2 - 2V_3$

(d) $2V_1(t) + 4V_2 + 5V_3$

**Exp. (b)**

$$\text{Op-amp 1 output} = -\frac{V_1(t)}{RC} = -\frac{V_1(t)}{10\times10^3\times100\times10^{-6}} = -V_1(t)$$

$$V_o(t) = -V_1(t)\times\frac{-20\,k\Omega}{10\,k\Omega} - V_2\times\frac{20\,k\Omega}{5\,k\Omega} - V_3\times\frac{20\,k\Omega}{4\,k\Omega}$$

$$= 2V_1(t) - 4\,V_2 - 5\,V_3$$

**Q.18**

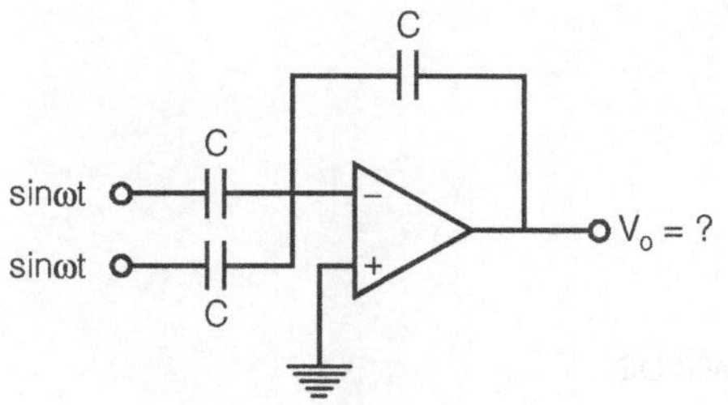


(a) zero
(b) $(V_1 - V_2)\sin\omega t$
(c) $-(V_1 + V_2)\sin\omega t$
(d) $(V_1 + V_2)\sin\omega t$

**Exp. (c)**

$$V_o = \left(\frac{\omega C}{\omega C}\right)V_1\sin\omega t - \left(\frac{\omega C}{\omega C}\right)V_2\sin\omega t$$

$$= -V_1\sin\omega t - V_2\sin\omega t$$

$$= -(V_1 + V_2)\sin\omega t$$

**Q.19** A change in the value of emitter resistance $R_e$ in a difference amplifier
(a) affects $A_d$
(b) affects
(c) affects both $A_d$ and $A_c$
(d) does not affect either $A_d$ or $A_c$

**Exp. (b)**

$$A_{CM} = -\frac{R_L}{2R_E}$$

**Q.20** In a differential amplifier, CMRR can be improved by using an increased
(a) emitter resistance
(b) collector resistance
(c) power supply voltage
(d) source resistance

**Exp. (a)**

$$\text{CMRR}\uparrow = \frac{A_d}{A_c} = \frac{A_d}{-R_L}\times 2R_E\uparrow$$

**Q.21** The op-amp of figure has very poor open loop voltage gain of 45, but is otherwise ideal. The gain of the amplifier equals

(a) 15
(b) 20
(c) 4
(d) 4.5

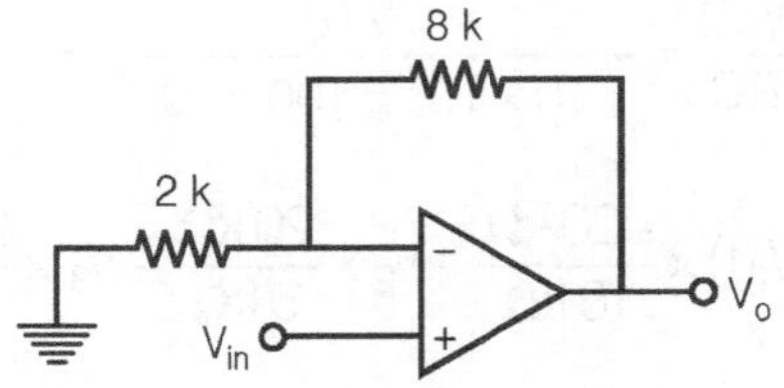


**Exp. (d)**

$$\beta = \frac{R_1}{R_1 + R_f} = \frac{2}{2+8} = \frac{1}{5}$$

$$A_f = \frac{A}{1 + A\beta} = \frac{45}{1 + 45 \times \frac{1}{5}} = \frac{45}{10} = 4.5$$

**Q.22** If the input to the circuit of figure is a sine wave. The output will be

(a) a half wave rectified sine wave
(b) a full wave rectified sine wave
(c) triangular wave
(d) square wave

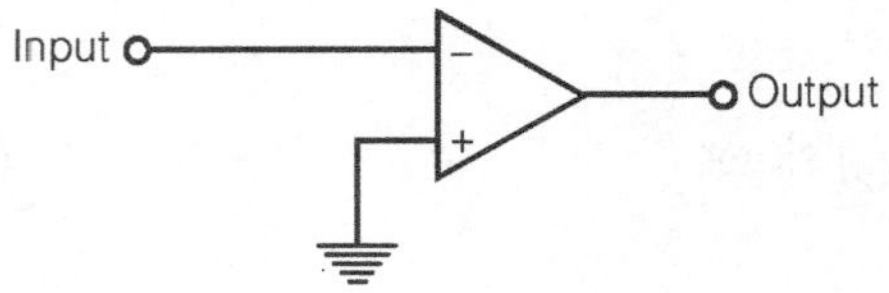


**Exp. (d)**

Output of op-amp will be square wave.

**Q.23** Frequency compensation is used in op-amp to increase its

(a) voltage gain
(b) input resistance
(c) bandwidth
(d) none

**Exp. (c)**

Frequency compensation is a method use in op-amp to increase the bandwidth.

**Q.24** In a practical op-amp $A_d$ = 100, $A_c$ = 0.1 is operated with input signal voltage of 1050 µr and 950 µv. Find $V_o$. Also calculate CMMR in dB

(a) 10.1 mV, 1000
(b) 9 V, 100
(c) 5 V, 1000
(d) none

**Exp. (a)**

$$V_o = A_c V_c + A_d V_d$$

$$V_c = \frac{V_1 + V_2}{2}$$

$$V_c = 1000 \times 10^{-6}\ V = 10^{-3}\ V$$
$$V_d = V_1 - V_2 = 100\ \mu V$$
$$V_o = 0.1 \times 10^{-3} + 100 \times 10^{-6} \times 100$$
$$= 10.1\ mV$$

$$(CMRR)_{dB} = 20 \log \frac{A_d}{A_c} = 20 \log \frac{100}{0.1} = 1000$$

**Q.25** Calculate CMRR for the op-amp circuit

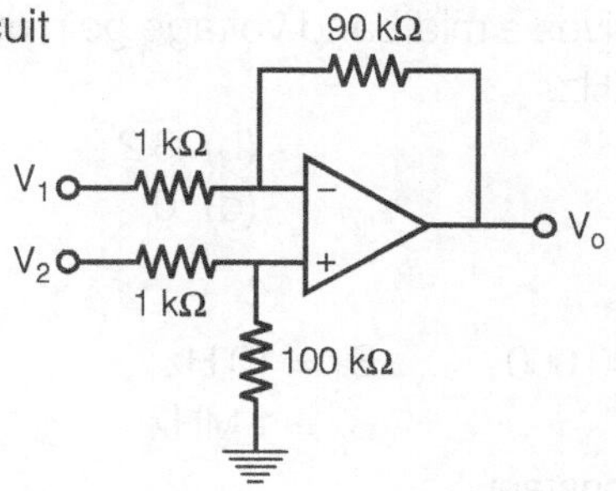


(a) 1000.5

(b) 900

(c) 825

(d) none

**Exp.** (a)

$$V_o = -\frac{90\,k\Omega}{1k\Omega}V_1 + \left(1+\frac{90\,k\Omega}{1k\Omega}\right)\frac{V_2 \times 100\,k\Omega}{1k\Omega + 100\,k\Omega}$$

$$V_o = -90\,V_1 + 90.09\,V_2$$

$$A_1 = -90$$

$$A_2 = 90.09$$

$$A_c = A_1 + A_2$$

$$A_d = \frac{A_1 - A_2}{2}$$

$$\text{CMRR} = \frac{A_d}{A_c} = 1000.5$$

**Q.26** The common mode voltage is completely attenuated at the output of the differential given below:

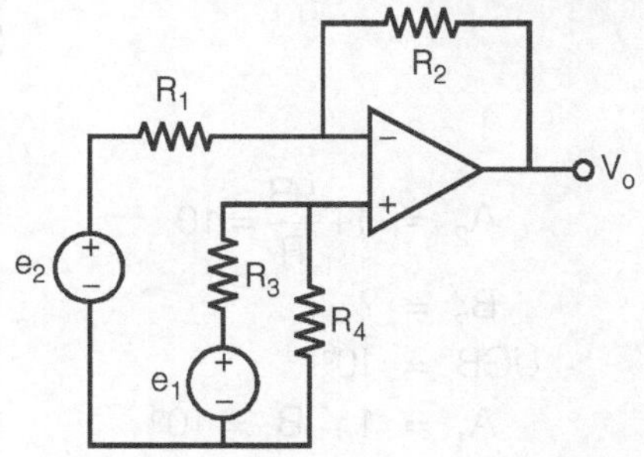


(a) $R_1 + R_2 = R_3 + R_4$

(b) $R_1 - R_2 = R_3 - R_4$

(c) $\frac{R_1}{R_2} = \frac{R_3}{R_4}$

(d) $\frac{R_2}{R_1} = \frac{R_3}{R_4}$

**Exp.** (c)

$$V_o = -\frac{R_2}{R_1}e_2 + \left(1+\frac{R_2}{R_1}\right)\frac{e_1 R_4}{R_3 + R_4}$$

$$A_2 = -\frac{R_2}{R_1}$$

$$A_1 = \frac{R_4}{R_1}\left(\frac{R_1 + R_2}{R_3 + R_4}\right)$$

$$A_c = A_1 + A_2 = 0$$

$$\Rightarrow \quad A_1 = -A_2$$

$$\frac{R_4}{R_1}\left(\frac{R_1+R_2}{R_3+R_4}\right) = \frac{R_2}{R_1}$$

$$R_4(R_1 + R_2) = R_2(R_3 + R_4)$$

$$R_1\,R_4 = R_2\,R_3$$

$$\frac{R_1}{R_2} = \frac{R_3}{R_4}$$

**Q.27** The data sheet of an op-amp gives a mid band voltage gain of 200000 with a cut-off frequency of 10 Hz. What is the voltage gain at 1 MHz.

(a) 2
(b) 3.2
(c) 1
(d) 0

**Exp. (a)**

$$A_1 = 200{,}000 \qquad B_1 = 10\text{ Hz}$$
$$A_2 = ? \qquad B_2 = 1\text{ MHz}$$
$$\text{GBW} = \text{Constant}$$
$$A_1 B_1 = A_2 B_2$$
$$A_2 = \frac{A_1 B_1}{B_2} = 2$$

**Q.28** The bandwidth of an amplifier circuit given below is. It is given that UGB product of the op-amp is 1 MHz.

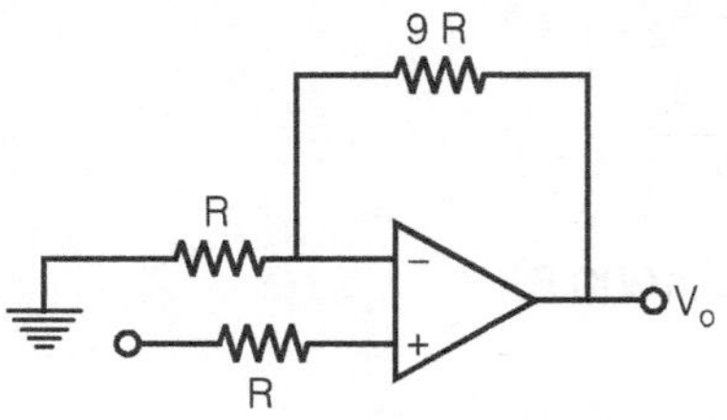


(a) $10^5$ Hz
(b) $10^7$ Hz
(c) $10^4$ Hz
(d) none

**Exp. (a)**

$$A_2 = 1+\frac{9R}{R} = 10$$
$$B_2 = ?$$
$$\text{UGB} = 10^6$$
$$A_1 = 1, \quad B_1 = 10^6$$
$$A_1 B_1 = A_2 B_2$$
$$B_2 = 10^5\text{ Hz}$$

**Q.29** For the op-amp circuit given below determine the maximum input signal frequency in rad/sec. that can be used to get an distortionless output by taking SR 0.5 V/μsec.

(a) $1.04 \times 10^6$ rad/sec
(b) $2.04 \times 10^6$ radsec
(c) $1.32 \times 10^4$ rad/sec
(d) $0.26 \times 10^5$ rad/sec

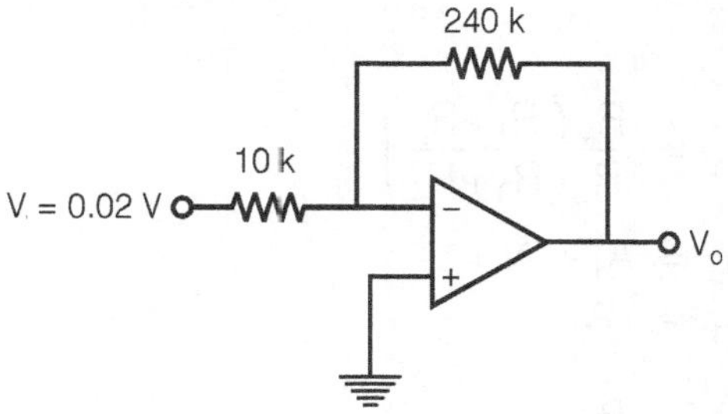


**Exp. (a)**

$$\omega_{max} = \frac{SR}{K}$$
$$K = |A_{CL}|\ \Delta V_i = \left|\frac{240}{10}\right| 0.02 = 24 \times 0.02 = 0.48$$
$$\omega_{max} = \frac{0.5\times10^{-6}}{24\times0.02} = 1.04\times10^6\text{ rad/sec}$$

**Q.30** Keep LED on $V_i$ must be

(a) > 12 V

(b) < 12 V

(c) > 6 V

(d) < 6 V

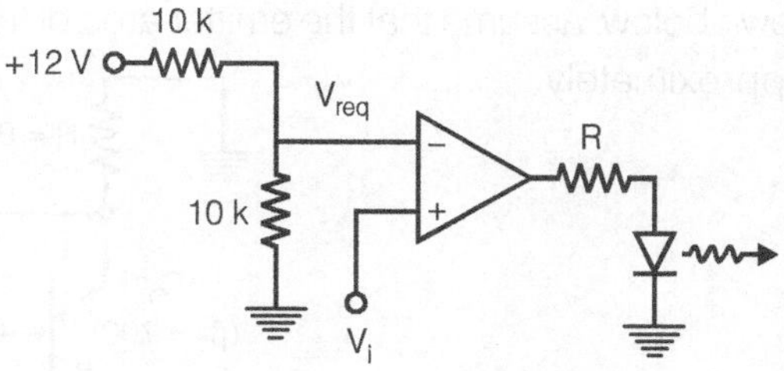


**Exp.** (c)

$$V_{ref} = \frac{12 \times 10\,k\Omega}{10\,k\Omega + 10\,k\Omega} = 6\,V$$

$$V_i > V_{req}$$

$$Vi > 6\,V$$

**Q.31** An op-amp has an offset voltage 1 mV and is ideal in all other aspects. If this op-amp is used in the circuit shown in figure. The output voltage will be

(a) 1 mV

(b) 1 V

(c) ±1 V

(d) 0 V

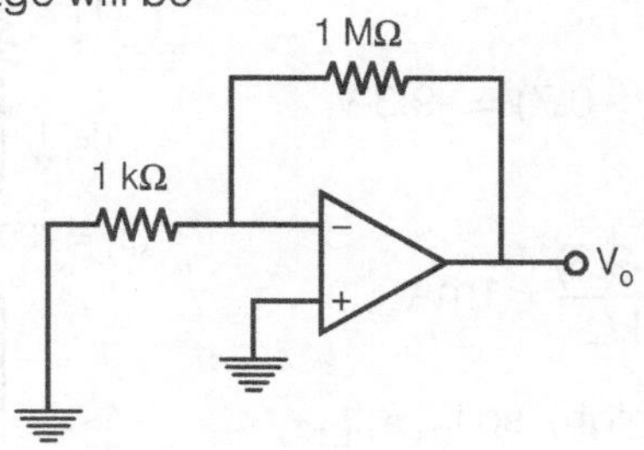


**Exp.** (b)

$$V_o = \text{offset voltage} \times \text{gain} = 10^{-3} \times \frac{10^6}{10^3} = 1\,V$$

**Q.32** Consider the figure $V_o = ?$

(a) $-V_{ref}\dfrac{V_Z}{V_X}$

(b) $-V_{ref}\dfrac{V_X}{V_Z}$

(c) $-V_{ref}\,V_Z$

(d) $-V_{ref}\,V_X$

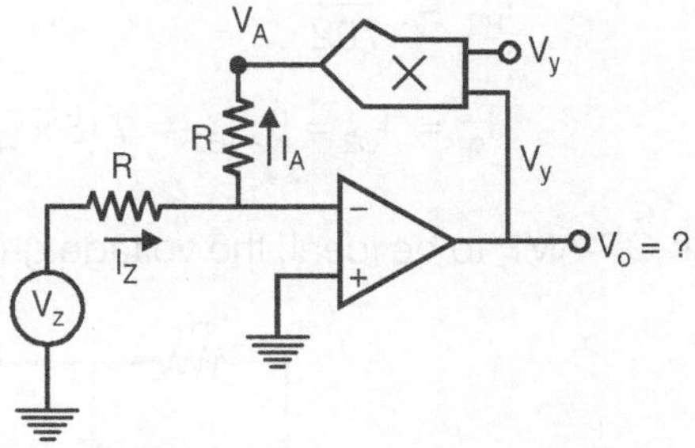


**Exp.** (a)

$$I_Z = \frac{V_Z - 0}{R} \;;\; V_A = \frac{V_X\,V_Y}{V_{ref}}$$

$$V_A = \frac{V_X\,V_o}{V_{ref}} \;;\; I_A = -\frac{V_A}{R} = \frac{-V_X\,V_o}{V_{ref}\,R}$$

$$I_A = I_Z \;;\; V_Z = I_Z R = \frac{V_X\,V_o}{V_{ref}}$$

$$V_o = -V_{ref}\frac{V_Z}{V_X}$$

**Q.33** In the silicon BJT circuit shown below, assume that the emitter area of transistor $Q_1$ is half that of transistor $Q_2$ The value of current $I_o$ is approximately

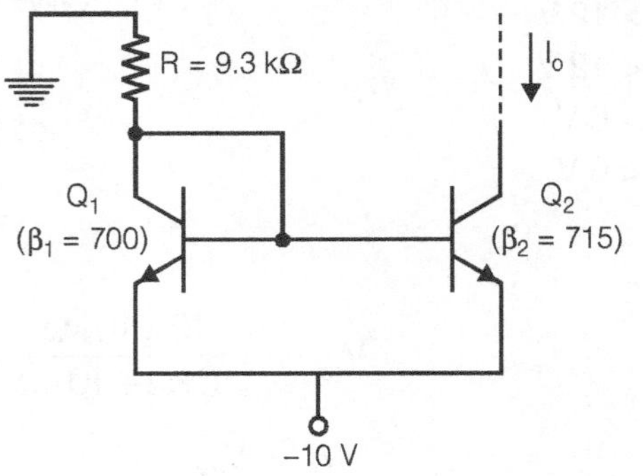


(a) 0.5 mA

(b) 2 mA

(c) 9.3 mA

(d) 15 mA

**Exp. (b)**

Since emitter area of transistor $Q_1$ is half of transistor $Q_2$, so current

$$I_{E_1} = \frac{1}{2}I_{E_2} \quad \text{and} \quad I_{B_1} = \frac{1}{2}I_{B_2}$$

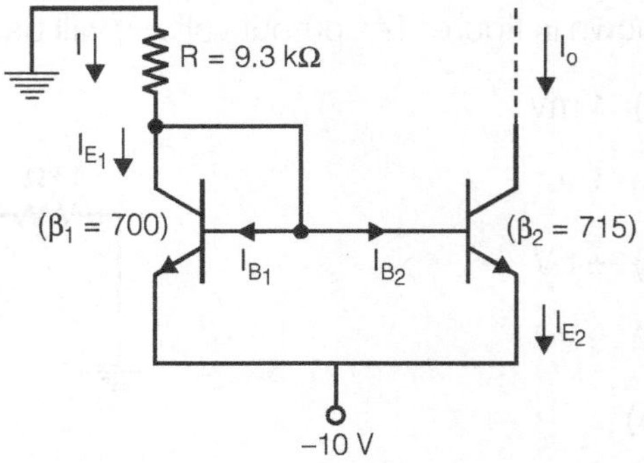


In the circuit

$$V_B = -10-(-0.7) = -9.3 \text{ V}$$

Collector current

$$I = \frac{0-(-9.3)}{9.3\text{k}\Omega} = 1\text{mA}$$

$$\beta_1 = 700 \text{ (high) so } I_{C1} \approx I_{E}.$$

Applying KCL at base

$$I - I_{E1} = I_{B1} + I_{B2}$$

$$1-(\beta_1+1)I_{B1} = I_{B1} + I_{B2}$$

$$1 = (700+1+1)\frac{I_{B2}}{2} + I_{B2}$$

$$I_{B2} \approx \frac{2}{702}$$

$$I_o = I_{C2} = \beta_2 I_{B2} = 715 \times \frac{2}{702} \cong 2 \text{ mA}$$

**Q.34** Assuming the OP-AMP to be ideal, the voltage gain of the amplifier shown below is

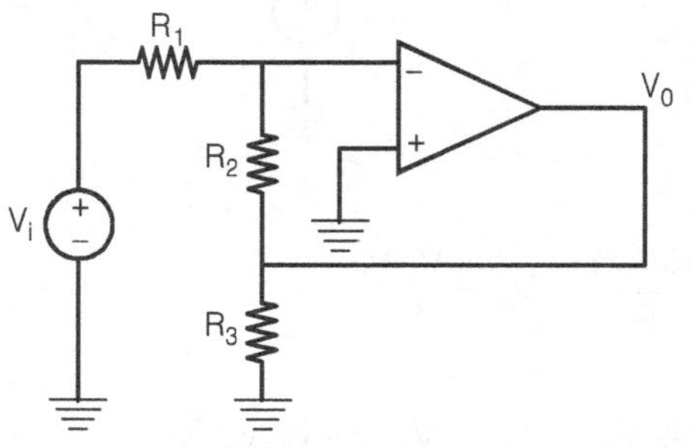


(a) $-\frac{R_2}{R_1}$

(b) $-\frac{R_3}{R_1}$

(c) $-\left(\frac{R_2 \| R_3}{R_1}\right)$

(d) $-\left(\frac{R_2+R_3}{R_1}\right)$

**Exp.** (c)

The proper design is

It is a inverting amplifier

$$\frac{V_o}{V_i} = -\left(\frac{R_2 \parallel R_3}{R_1}\right)$$

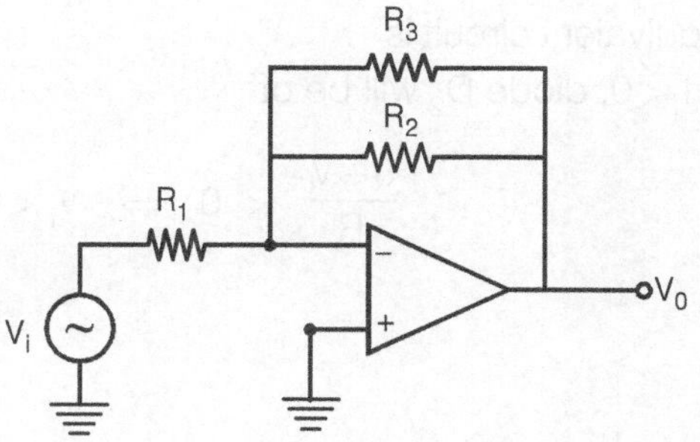


**Q.35** The transfer characteristic for the precision rectifier circuit shown below is (assume ideal OP-AMP and practical diodes)

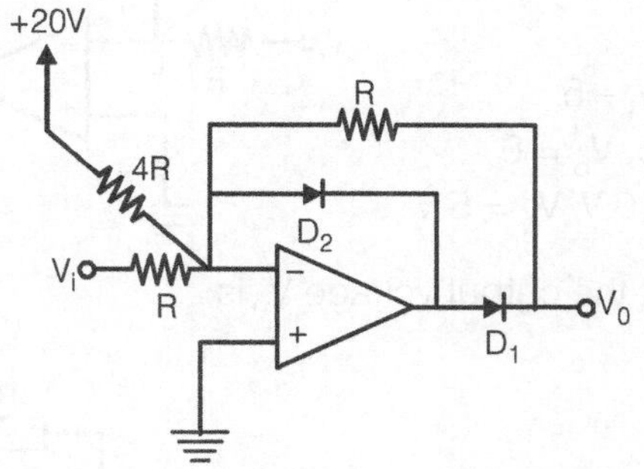


(a)

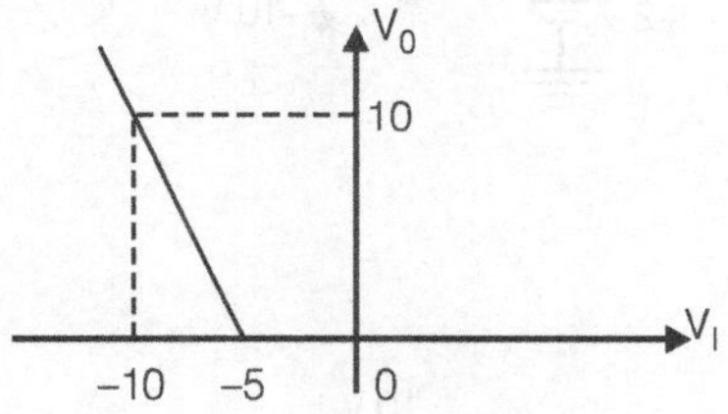


(b)

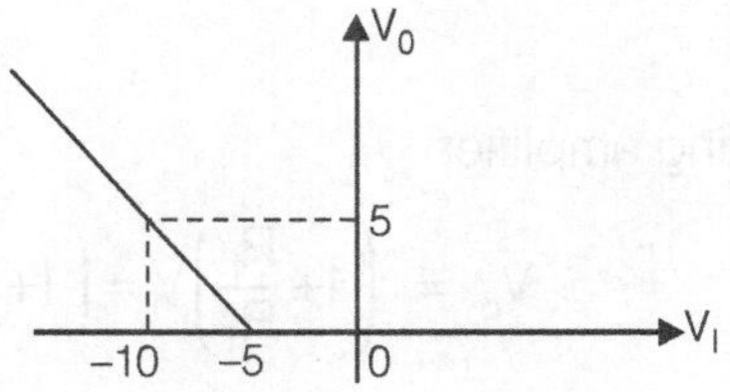


(c)

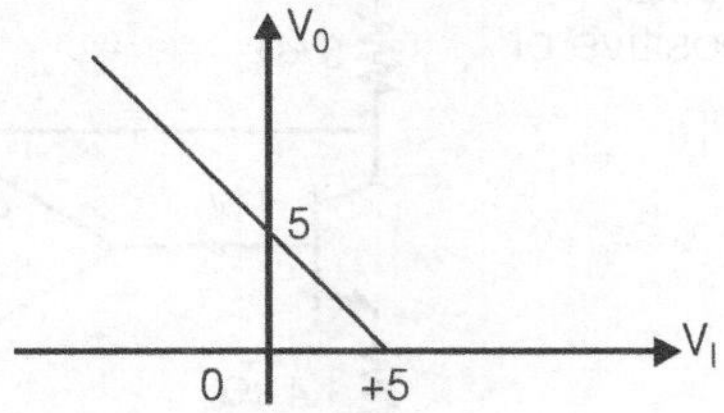


(d)

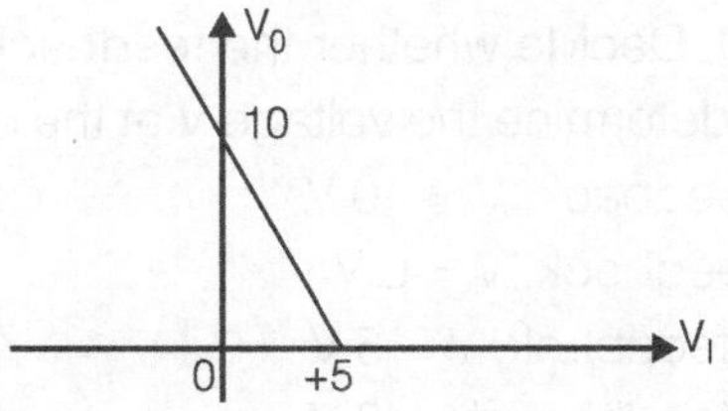


**Exp.** (b)

In the circuit

Current $\quad I = \frac{20}{4R} + \frac{V_i}{R} = \frac{5+V_i}{R}$

If $I > 0$, diode $D_2$ conducts

So for $\quad \frac{5+V_i}{R} > 0$

$V_i > -5$, $D_2$ conducts

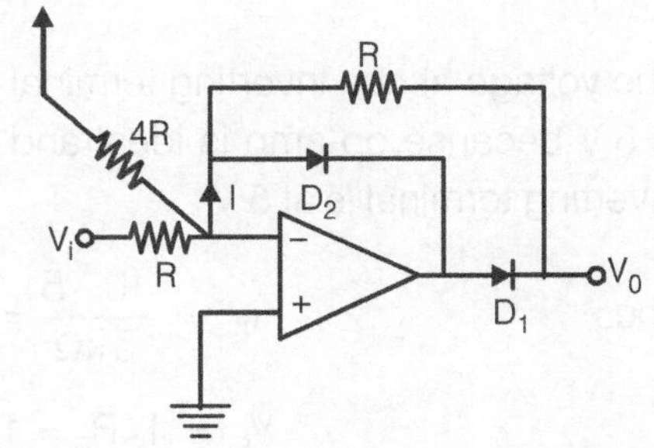

Equivalent circuit is

If I < 0, diode $D_2$ will be off

$$\frac{5+V_i}{R} < 0 \Rightarrow V_i < -5, D_2 \text{ is off}$$

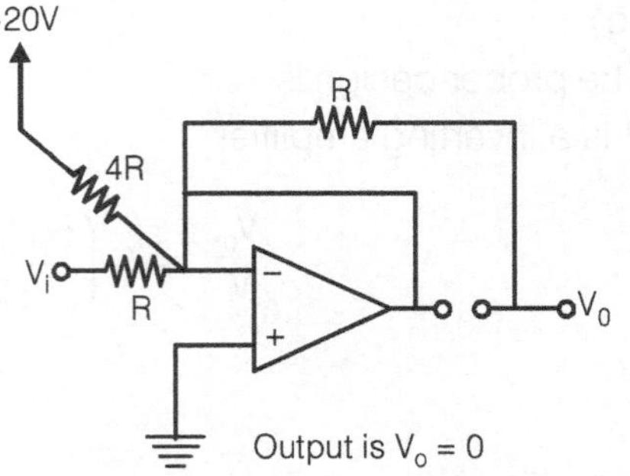


Output is $V_o = 0$

The circuit is

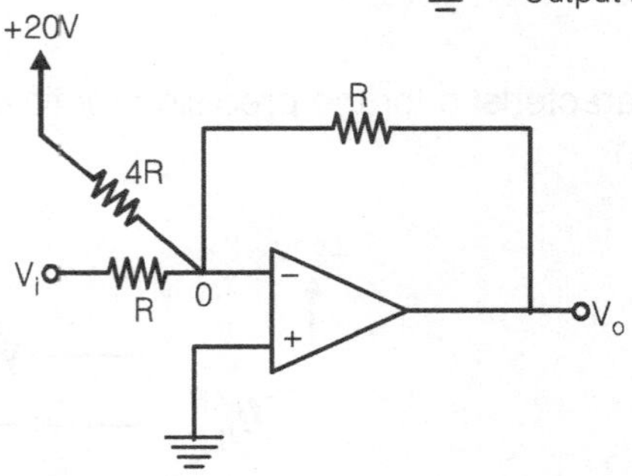


$$\frac{0-V_i}{R}+\frac{0-20}{4R}+\frac{0-V_o}{R} = 0$$

$$V_o = -V_i - 5$$

$$V_i = -5, V_o = 0$$

$$V_i = -10 \text{ V}, V_o = 5 \text{ V}$$

**Q.36** Given that the op-amp is ideal, the output voltage $V_0$ is

(a) 4 V

(b) 6 V

(c) 7.5 V

(d) 12.12 V

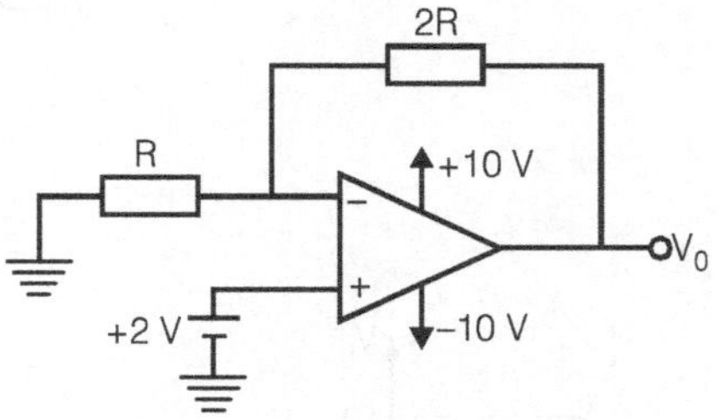


**Exp. (b)**

It is non inverting amplifier

$$V_o = \left(1+\frac{R_f}{R_1}\right)V_i = \left(1+\frac{2R}{R}\right)2 = 3\times 2 = 6\text{V}$$

**Q.37** In the circuit shown below, the op-amp is ideal, the transistor has $V_{BE} = 0.6$ V and $\beta = 150$. Decide whether the feedback in the circuit is positive or negative and determine the voltage V at the output of the op-amp.

(a) Positive feedback, V = 10 V

(b) Positive feedback, V = 0 V

(c) Negative feedback, V = 5 V

(d) Negative feedback, V = 2 V

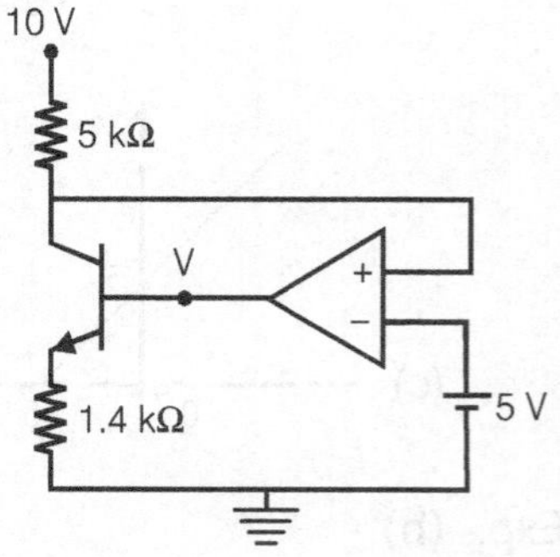


**Exp. (d)**

The circuit is

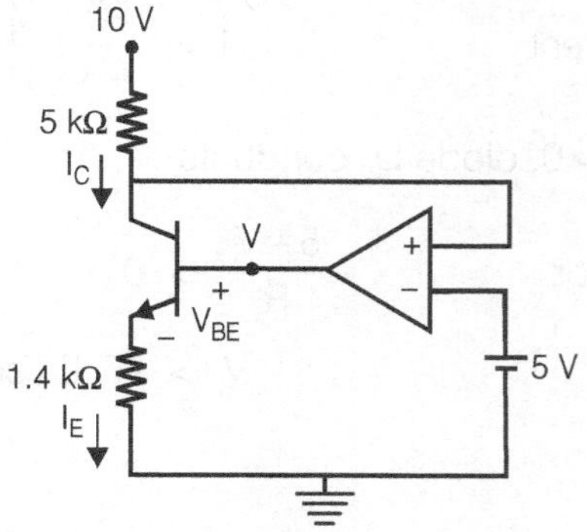


The voltage at non inverting terminal is 5 V because op-amp is ideal and inverting terminal is at 5 V

Thus $$I_L = \frac{10-5}{5\text{k}\Omega} = 1\text{mA}$$

$$V_E = I_E R_E = 1 \text{ mA} \times 1.4 \text{ k} = 1.4 \text{ V}$$

$$V = V_{BE} + V_E = 0.6 + 1.4 = 2 \text{ V}$$

Thus feedback is negative and output voltage is V = 2 V

**Q.38** The nature of feedback in the op-amp circuit shown is

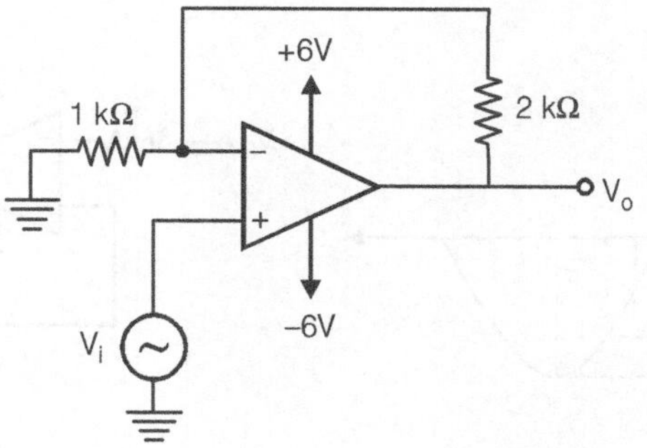


(a) Current – Current feedback
(b) Voltage – Voltage feedback
(c) Current – Voltage feedback
(d) Voltage – Current feedback

**Exp. (b)**

This is voltage series feedback

Then it can be called as voltage-voltage feedback.

**Q.39** The following circuit has R = 10 kΩ, C = 10 μF. The input voltage is a sinusoidal at 50 Hz with an rms value of 10 V. Under ideal conditions, the current $I_s$ from the source is

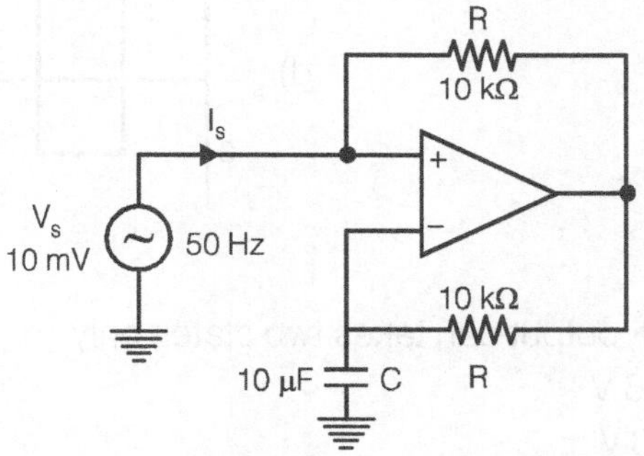


(a) 10πmA leading by 90°
(b) 20πmA leading by 90°
(c) 10πmA lagging by 90°
(d) 10πmA lagging by 90°

**Exp. (c)**

Let voltages at positive and negative terminals of op-amp are $V_+$ and $V_-$

$$V_+ = V_- = V_s \text{ (ideal op-amp)} \quad ; \quad \frac{V_- - 0}{\frac{1}{Cs}} + \frac{V_- - V_o(s)}{R} = 0$$

$$(RCs)\,V_- + V_- - V_o(s) = 0$$

$$(1 + RCs)\,V_s = V_o(s)$$

Current

$$I_s = \frac{V_s - V_o}{R} \quad ; \quad I_s = \frac{-RCs}{R} V_s \quad ; \quad I_s = -j\omega C\, V_s$$

$$I_s = 2\pi f \times 10 \times 10^{-6} \times 10 \angle -90°$$

$$|I_s| = 2 \times \pi \times 50 \times 10 \times 10^{-6} \times 10$$

$$|I_s| = 10\pi \text{ mA, lagging by } 90°$$

**Q.40** An ideal opamp circuit and its input waveform are shown in the figures. The output waveform of this circuit will be

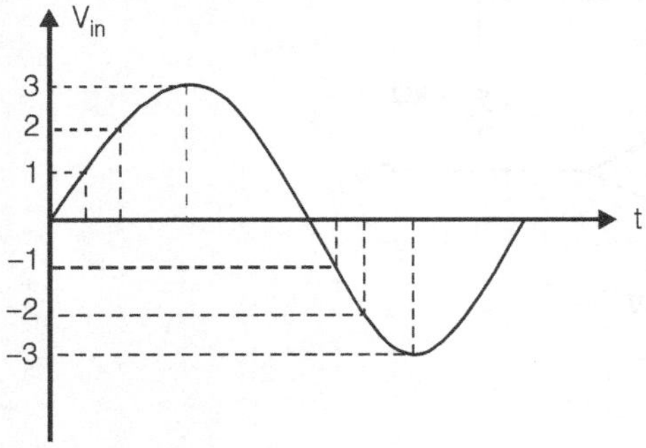


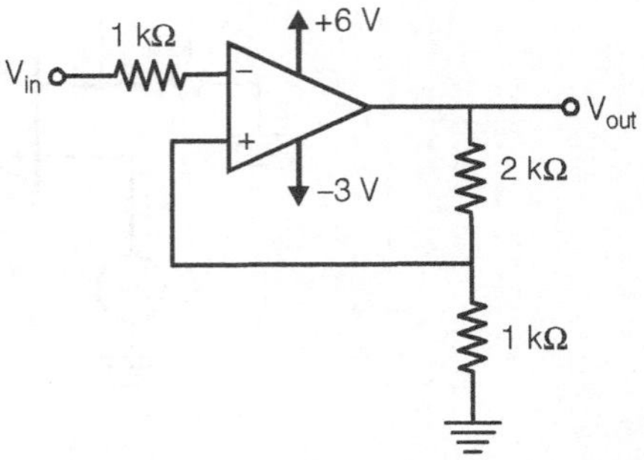


(a)

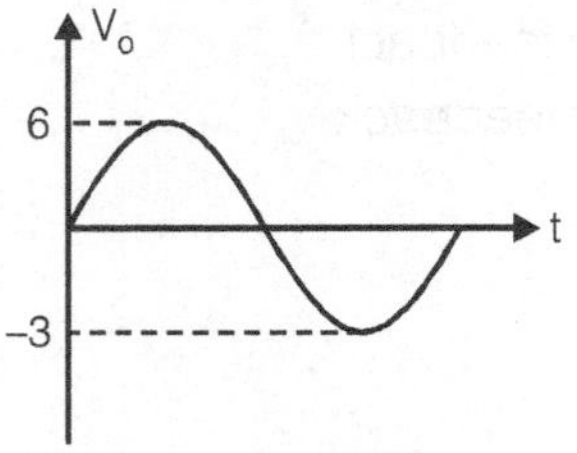


(b)

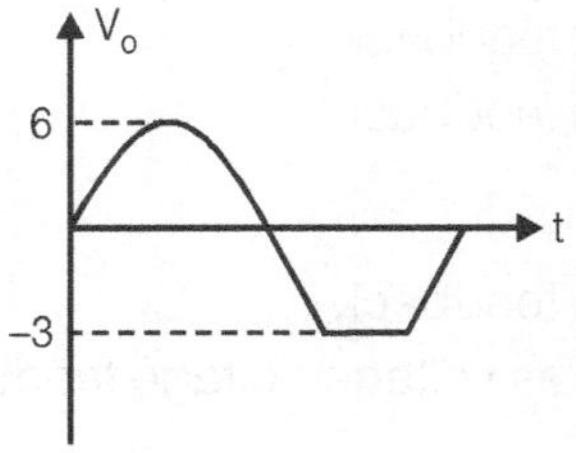


(c)

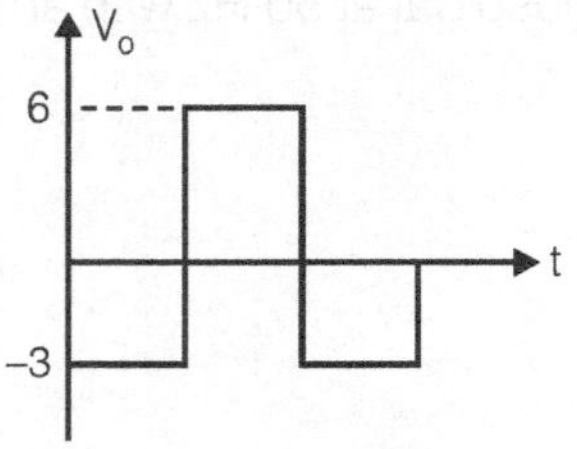


(d)

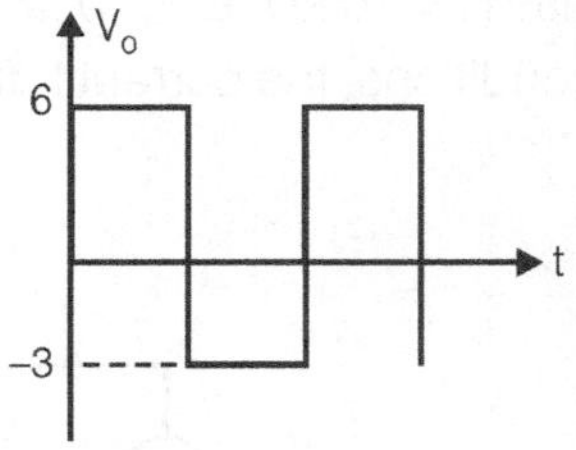


**Exp.** **(d)**

This is a Schmitt trigger circuit, output can takes two states only

$$V_{OH} = +6\text{ V}$$

$$V_{OL} = -3\text{ V}$$

Threshold voltages at non inverting terminals of op-amp is

$$\frac{V_{Th}-6}{2}+\frac{V_{Th}-0}{1} = 0$$

$$3V_{Th}-6 = 0$$

$$V_{Th} = 2\text{V (UTP)}$$

Similarly

$$\frac{V_{Th}-(-3)}{2}+\frac{V_{Th}}{1} = 0$$

$$3V_{Th}+3 = 0$$

$$V_{TL} = -1\text{ V (LTP)}$$

$V_{in} < 2$ V, $V_o = +6$ V

$V_{in} > 2$ V, $V_o = -3$ V

$V_{in} < -1$ V, $V_o = +6$ V

$V_{in} > -1$ V, $V_o = -3$ V

Output wave form

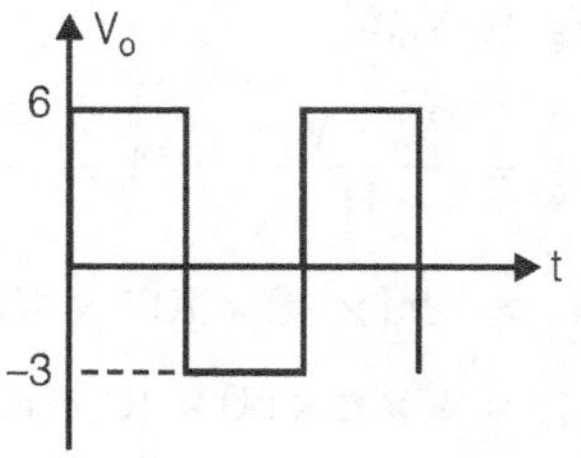

**Q.41** Consider the following circuit using an ideal OP-Amp. The I-V characteristics of the diode is described by the relation $I = I_0\left(e^{\frac{V}{V_T}} - 1\right)$ where $V_T$ = 25 mV, $I_0$ = 1 μA and V is the voltage across the diode (taken as positive for forward bias).

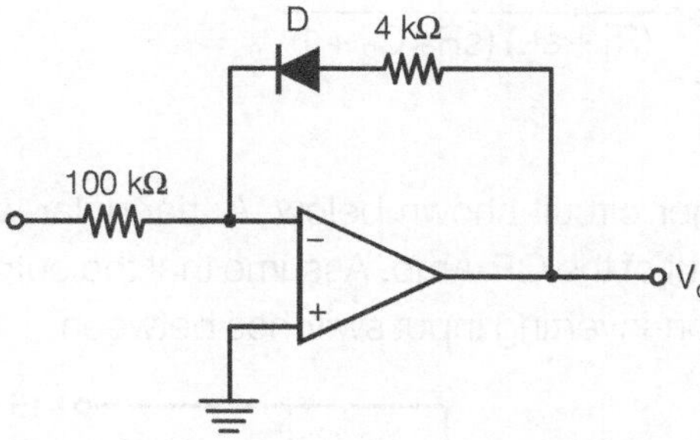


For an input voltage $V_i$ = –1V, the output voltage $V_0$ is

(a) 0 V
(b) 0.1 V
(c) 0.7 V
(d) 1.1 V

**Exp.** (b)

The circuit is using ideal op-amp. The non inverting terminal of op-amp is at ground. Thus inverting terminal is also at virtual ground.

Thus current will flow from –ve terminal (0 V) to –1 V source. The current I is

$$I = \frac{0-(-1)}{100\,k\Omega} = \frac{1}{100\,k\Omega}$$

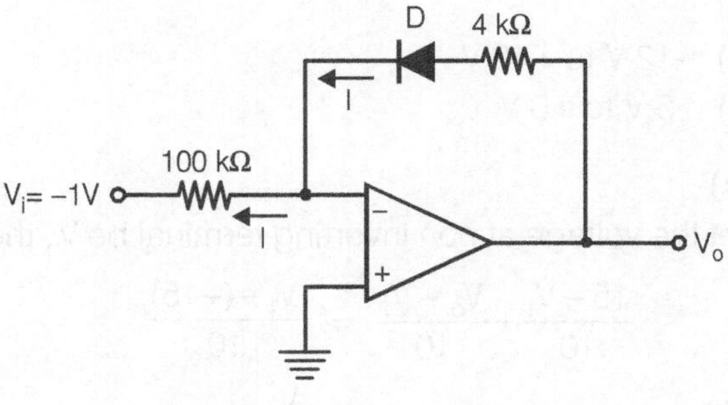


The current through diode is

$$I = I_o\,(e^{V/V_t} - 1)$$

Now $V_T$ = 25 mV and $I_o$ = 1 μA

Thus $I = 10^{-6}\left[e^{V/25\times10^{-3}} - 1\right] = \frac{1}{10^5}$

$$V = 0.06\text{ V}$$

$$V_o = I \times 4\,k\Omega + V = \frac{1}{100\,k\Omega} \times 4\,k\Omega + 0.06$$

$$V_o = 0.1\text{ V}$$

**Q.42** The OP-Amp circuit shown below represents a

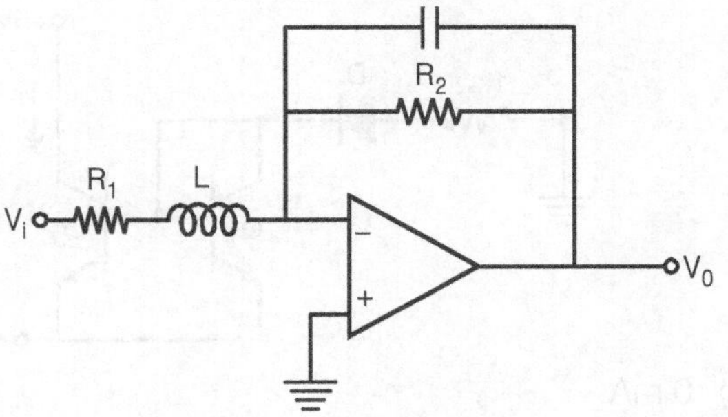


(a) high pass filter

(b) low pass filter

(c) band pass filter

(d) band reject filter

Exp. (b)

$$\frac{V_i - 0}{R_1 + sL} = \frac{V_o - 0}{\frac{R_2}{sR_2C_2} + 1}$$

$$\frac{V_o}{V_i} = \frac{R_2}{(R_1 + sL)(sR_2 C_2 + 1)}$$

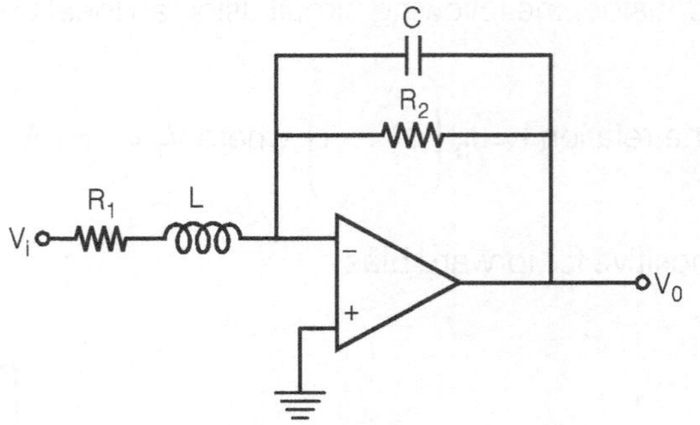


This is a low pass filter.

**Q.43** Consider the Schmitt trigger circuit shown below. A triangular wave which goes from −12 V to 12 V is applied to the inverting input of the OP-Amp. Assume that the output of the OP-Amp swings from +15 V to −15 V. The voltage at the non-inverting input switches between

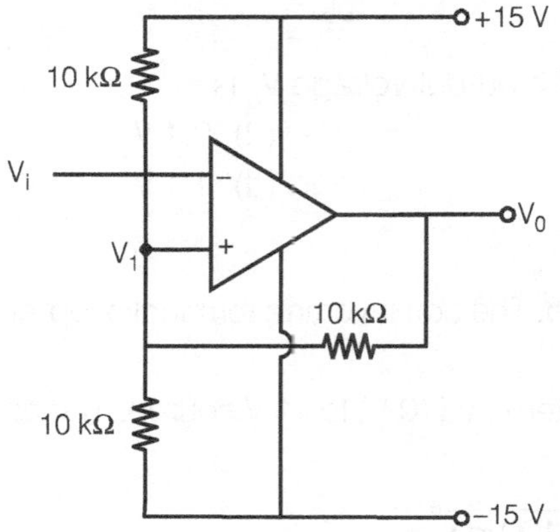


(a) −12 V to +12 V
(b) −7.5 to +7.5 V
(c) −5 V to +5 V
(d) 0 V and 5 V

Exp. (c)

Let the voltage at non inverting terminal be $V_1$ then

$$\frac{15 - V_1}{10} + \frac{V_o - V_1}{10} = \frac{V_1 - (-15)}{10}$$

$$V_1 = \frac{V_o}{3}$$

Hence $V_o$ swings from −15 V to +15 V.
Then $V_1$ swings between −5 V to + 5 V.

**Q.44** Two perfectly matched silicon transistor are connected as shown in the figure. Assuming the β of the transistors to be very high and the forward voltage drop in diodes to be 0.7 V, the value of current I is

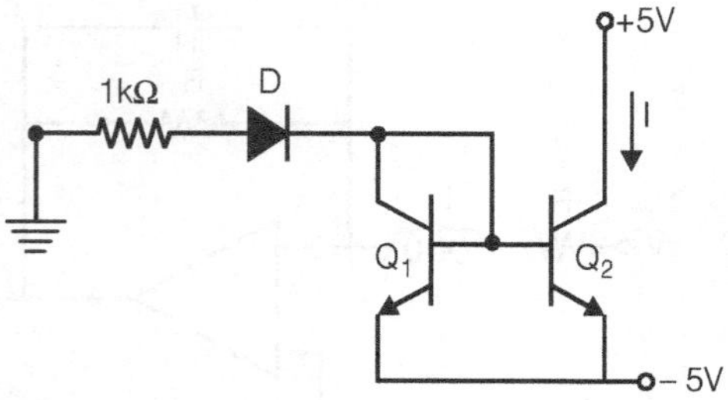


(a) 0 mA
(b) 3.6 mA
(c) 4.3 mA
(d) 5.7 mA

**Exp. (b)**

This is a current mirror circuit, since β is high $I_{C1} = I_{C2}$, $I_{B1} = I_{B2}$

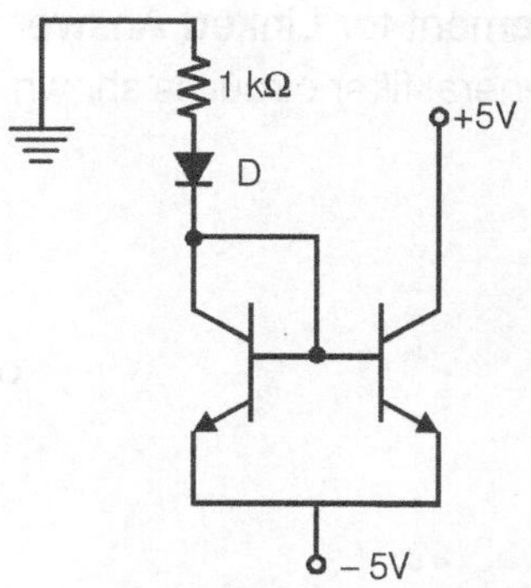


$$V_B = (-5 + 0.7)$$
$$= -4.3 \text{ V}$$

Diode $D_1$ is forward bias

$$\text{Current I} = \frac{0 - 0.7 - (-4.3)}{1\text{k}\Omega}$$
$$= 3.6 \text{ mA}$$

**Q.45** A waveform generator circuit using OPAMPs is shown in the figure. It produces a triangular wave at point 'P' with a peak to peak voltage of 5 V for $V_i = 0$ V.

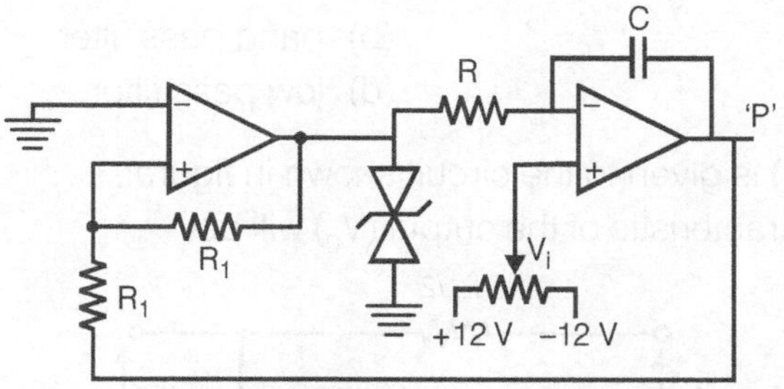


If the voltage $V_i$ is made +2.5 V, the voltage waveform at point 'P' will become

(a)
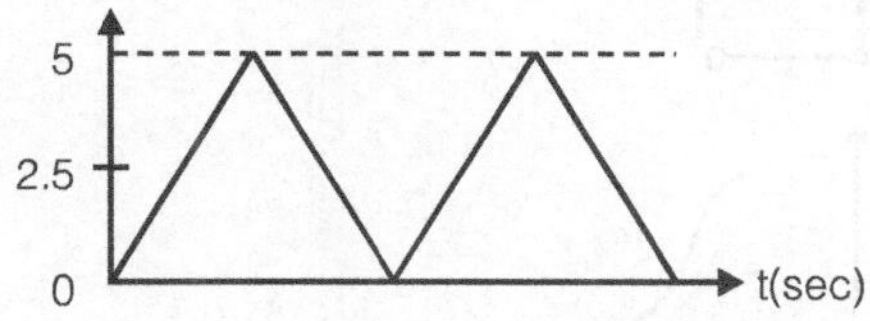


(b)
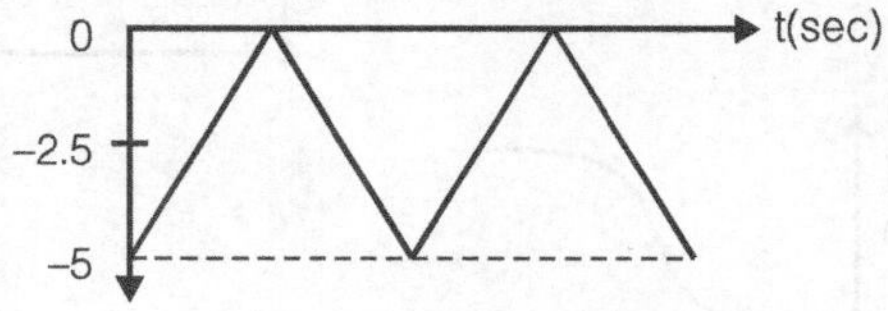


(c)
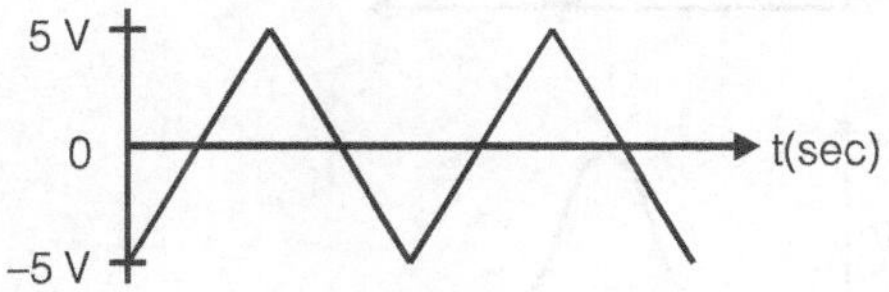


(d)
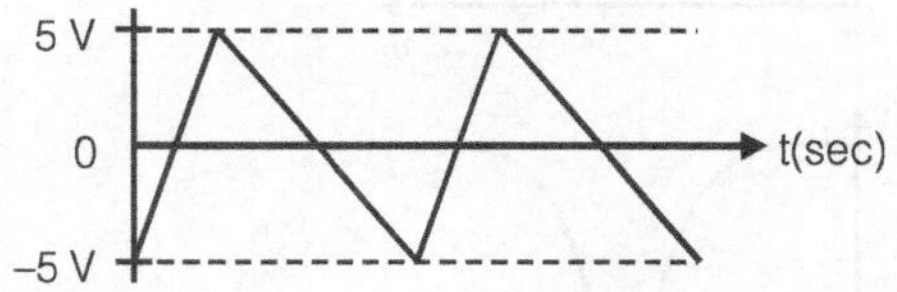


**Exp. (a)**

$V_i = 0$ V

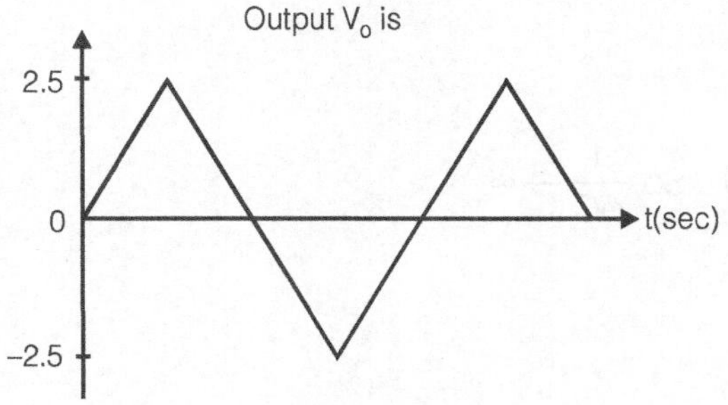


$V_i = 2.5$ V

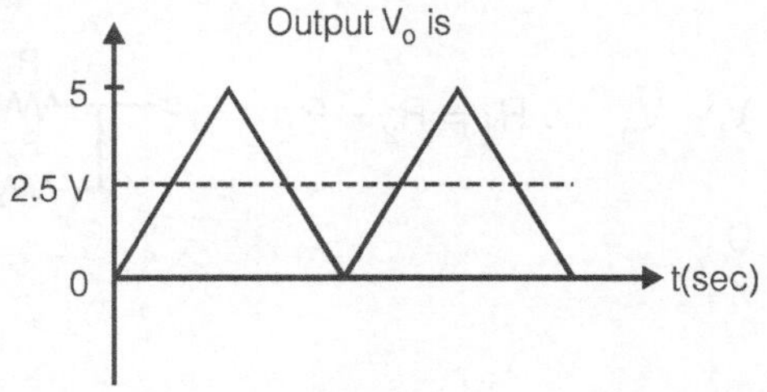

**Q.46 Statement for Linked Answer Q. (i) and Q. (ii)**

A general filter circuit is shown in the figure:

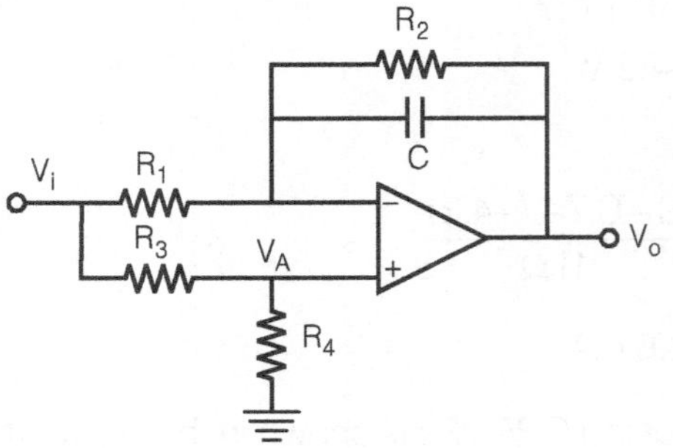


**(i)** If $R_1 = R_2 = R_A$ and $R_3 = R_4 = R_B$, the circuit acts as a

(a) all pass filter
(b) band pass filter
(c) high pass filter
(d) low pass filter

**(ii)** The output of the filter in **(i)** is given to the circuit shown in figure:

The gain vs frequency characteristic of the output ($V_0$) will be

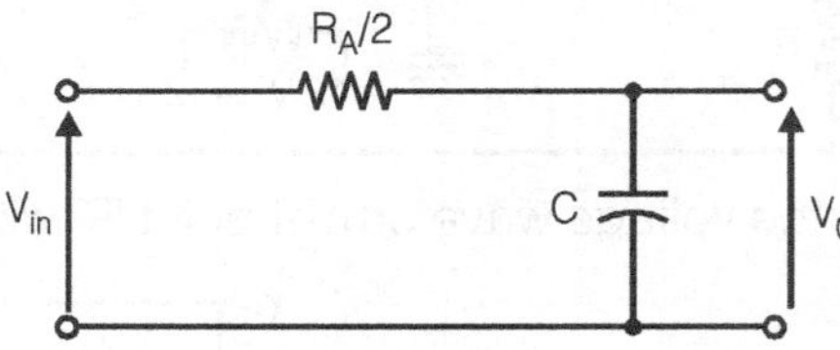


(a) 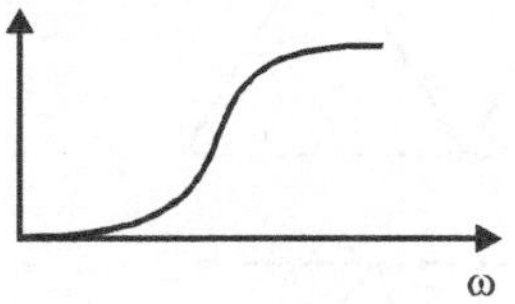


(b) 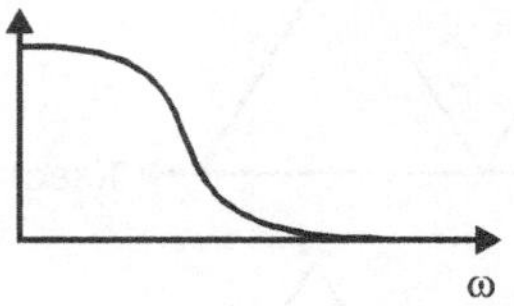


(c) 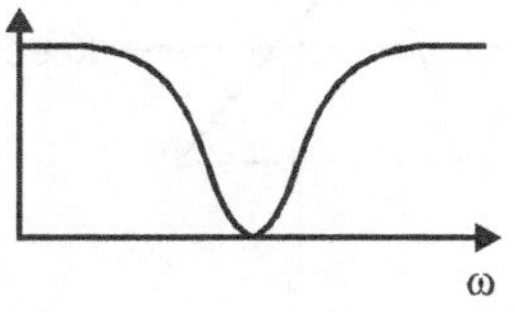


(d) 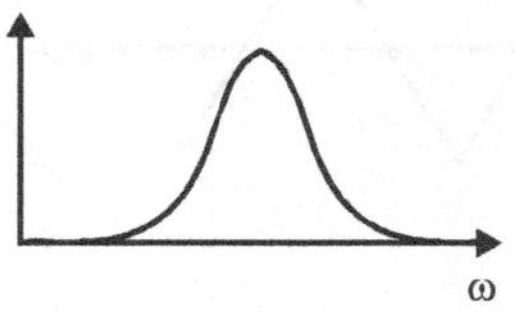


**Exp. (i). (c)**

For low frequencies

$$\omega \to 0 \quad \text{So } \frac{1}{\omega C} \to \infty$$

Equivalent circuit is

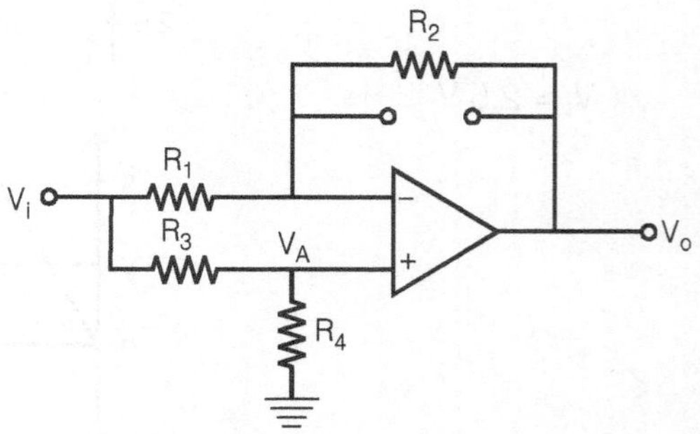


$$\frac{V_A - V_i}{R_1} + \frac{V_A - V_o}{R_2} = 0$$

$$2V_A = V_i + V_o \quad \therefore R_1 = R_2 = R_A$$

Similarly, $$\frac{V_A - V_i}{R_3} + \frac{V_A - 0}{R_4} = 0$$

$$2V_A = V_{in} \quad \therefore \quad R_3 = R_4 = R_B$$

So $\quad V_o = 0$

It will stop low frequency signals.

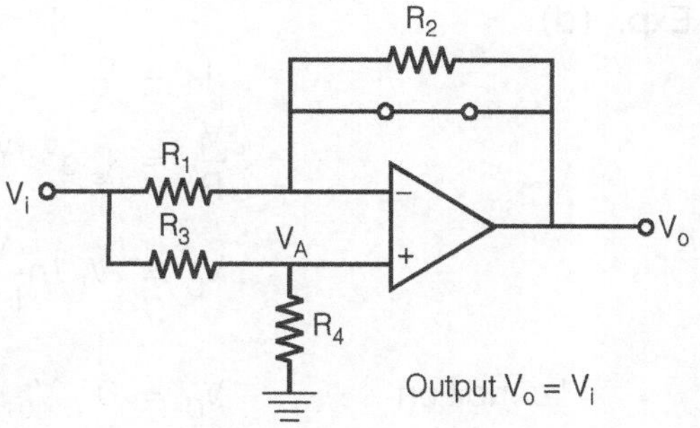


For high frequencies $\quad \omega \to \infty, \quad$ Then $\frac{1}{\omega C} \to 0$

Equivalent circuit is

So it will pass high frequency signal. This is a high pass filter.

**Exp. (ii). (d)**

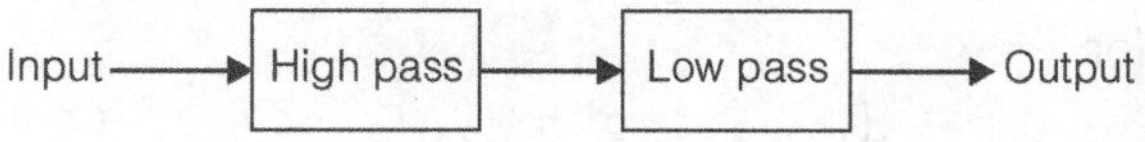


So it is band pass filter

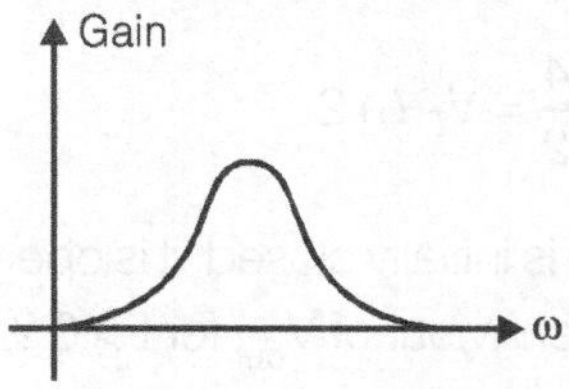


**Q.47** For the Op-Amp circuit shown in the figure, $V_0$ is

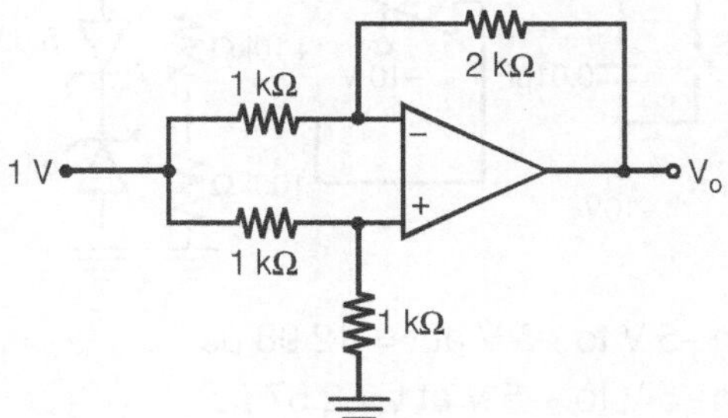


(a) –2 V
(b) –1 V
(c) –0.5 V
(d) 0.5 V

**Exp. (c)**

$$V_o = -\frac{2}{1}\times 1 + \left(1+\frac{2}{1}\right)\times\frac{1\times 1}{2} = -2+1.5 = -0.5\text{ V}$$

**Q.48** In the op-amp circuit shown, assume that the diode current follows the equation $I = I_s \exp(V/V_T)$. For $V_i = 2$ V, $V_0 = V_{01}$, and for $V_i = 4$ V, $V_0 = V_{02}$. The relationship between $V_{01}$ and $V_{02}$ is

(a) $V_{02} = \sqrt{2}\ V_{01}$

(b) $V_{02} = e^2\ V_{01}$

(c) $V_{02} = V_{01}\ ln\ 2$

(d) $V_{01} - V_{02} = V_T\ ln\ 2$

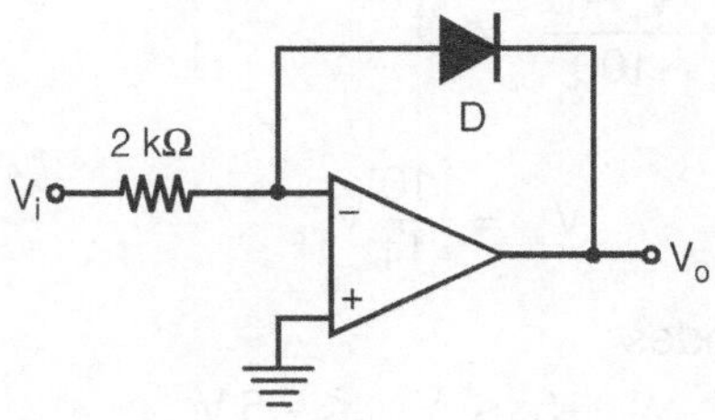

**Exp. (d)**

$$I_R = I_D$$

$$\frac{V_i}{R} = I_s e^{V_D/V_T}$$

$$V_D = V_T \ln\frac{V_i}{I_s R}$$

$I^{st}$ condition $\quad V_D = 0 - V_{01} = V_T \ln\frac{2}{I_s R}$

$II^{nd}$ condition $\quad V_D = 0 - V_{02} = V_T \ln\frac{4}{I_s R}$

Subtracting above equation

$$V_{01} - V_{02} = V_T \ln\frac{4}{I_s R} - V_T \ln\frac{2}{I_s R}$$

$$V_{01} - V_{02} = V_T \ln\frac{4}{2} = V_T \ln 2$$

**Q.49** The switch S in the circuit of the figure is initially closed, it is opened at time t = 0. You may neglect the Zener diode forward voltage drops. What is the behaviour of $V_{out}$ for $t > 0$?

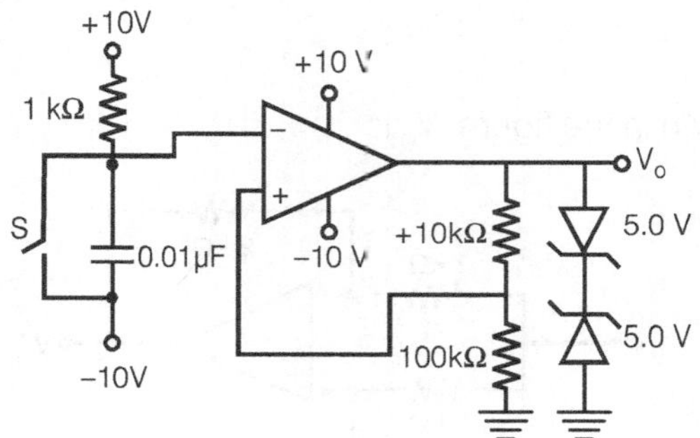


(a) It makes a transition from –5 V to +5 V at t = 12.98 μs
(b) It makes a transition from –5 V to + 5 V at t = 2.57 μs
(c) It makes a transition from +5 V to –5V at t=12.98 μs
(d) It makes a transition from + 5 V to – 5 V at t= 2.57 μs

**Exp. (b)**

In the circuit the capacitor starts charging from 0 V towards a steady state value of 20 V.

So at any time t, voltage across capacitor is given by

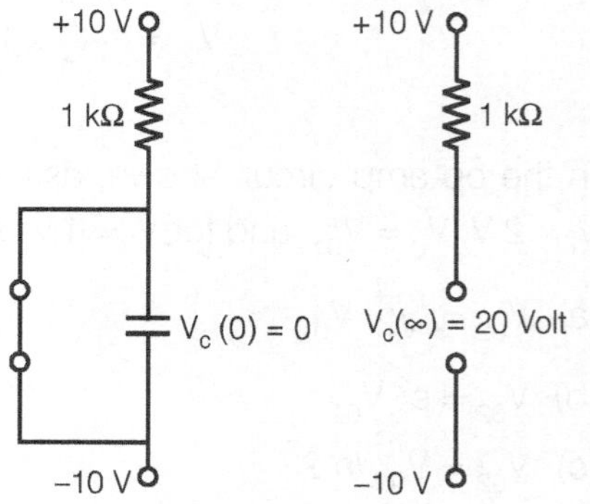


$$V_c(t) = V_c(\infty) + (V_c(0) - V_c(\infty))e^{-t/RC}$$

$$V_c(t) = 20(1 - e^{-t/RC})$$

Voltage at positive terminal of op-amp

$$\frac{V_+ - V_{out}}{10} + \frac{V_+ - 0}{100} = 0$$

$$V_+ = \frac{10}{11} V_{out}$$

Due to Zener diodes

$$-5 \le V_{out} \le +5\text{ V}$$

So, $$V_+ = \frac{10}{11}(5)\ \text{V}$$

Transistor from –5 V to 5 V occurs when capacitor charges upto $V_+$.

So, $$20(1-e^{-t/RC}) = \frac{10\times 5}{11}$$

$$1-e^{-t/RC} = \frac{5}{22}$$

$$\frac{17}{22} = e^{-t/RC}$$

$$t = RC\,ln\left(\frac{22}{17}\right) = 1\times 10^3 \times 0.01\times 10^{-6}\times 0.0257 = 2.57\ \mu\text{sec}$$

**Q.50** IC 555 in the figure is configured as an astable multivibrator. It is enabled to oscillate at t = 0 by applying a high input to pin 4. The pin description is: 1 and 8 – supply; 2-trigger; 4-reset; 6-threshold 7-discharge. The waveform appearing across the capacitor starting from t = 0, as observed on a storage CRO is

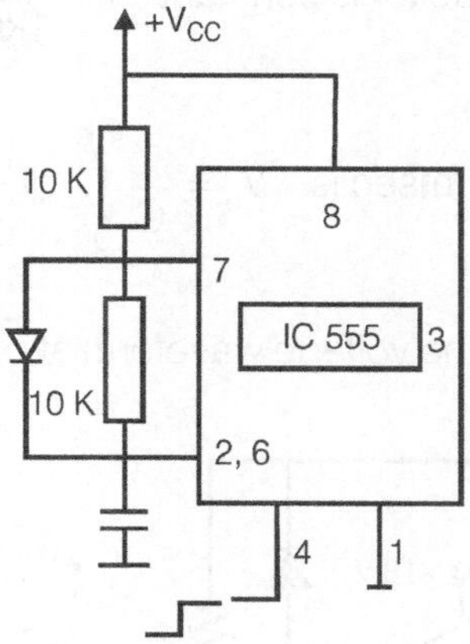


(a) 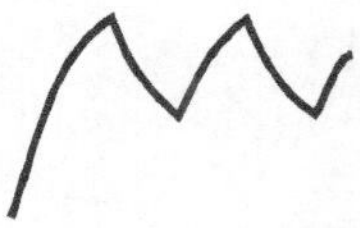 (b) 

(c) 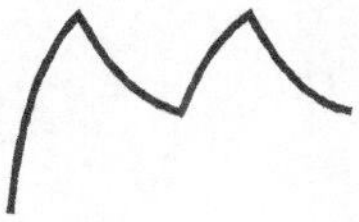 (d) 

**Exp.** **(b)**

In a 555 astable multivibrator circuit, charging of capacitor occurs through resistor ($R_A + R_B$) and discharging through resistor $R_B$ only. Time for charging and discharging is

$$T_C = 0.693\,(R_A + R_B)\,C$$
$$T_d = 0.693\,R_B\,C$$

But in the given circuit the diode will go in the forward bias during charging, so the capacitor will charge through resistor $R_A$ only and discharge through $R_B$ only

$$R_A = R_B$$
$$T_C = T_D$$

**Q.51** For the circuit shown in the following figure, the capacitor C is initially uncharged. At t = 0, the switch S is closed. The voltage $V_C$ across the capacitor at t = 1 millisecond is

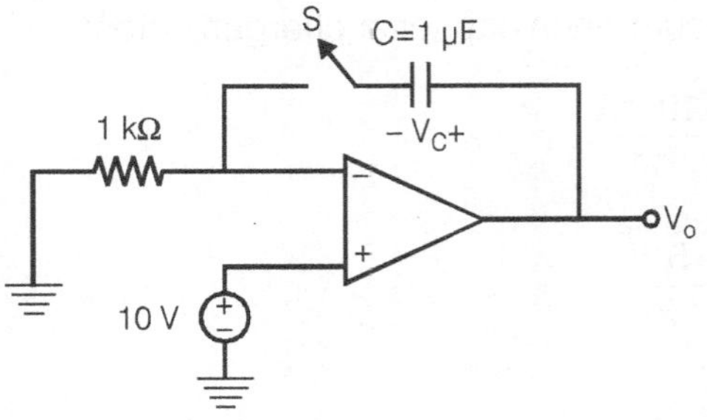


In the figure shown above, the OP AMP is supplied with ±15V.

(a) 0 Volt
(b) 6.3 Volts
(c) 9.45 Volts
(d) 10 Volts

**Exp. (d)**

The voltage at inverting terminal is

$$V_- = V_+ = 10 \text{ V}$$

Here not the current through the capacitor is constant $I = \dfrac{V_-}{1\text{k}\Omega} = \dfrac{10}{1\text{k}\Omega} = 10 \text{ mA}$

The voltage across capacitor at t = 1 msec is $V_c = \dfrac{1}{C}\int_0^{1\text{ms}} I\,dt = \dfrac{1}{1\mu}\int_0^{1\text{ms}} 10\,\text{m}dt = 10^4 \int_0^{1\text{ms}} dt = 10 \text{ V}$

**Q.52** For a given sinusoidal input voltage, the voltage waveform at point P of the clamper circuit shown in figure will be

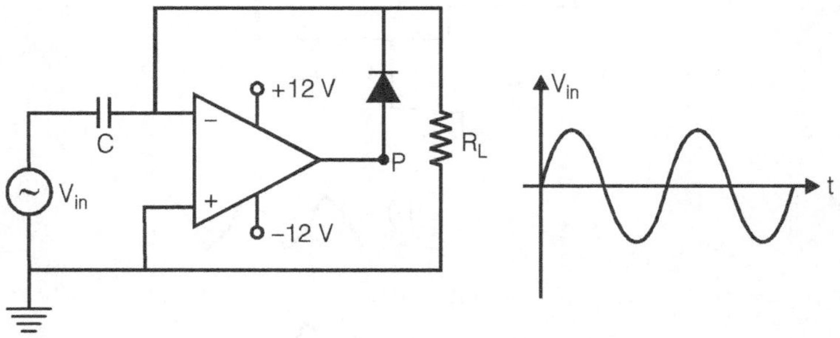


(a)
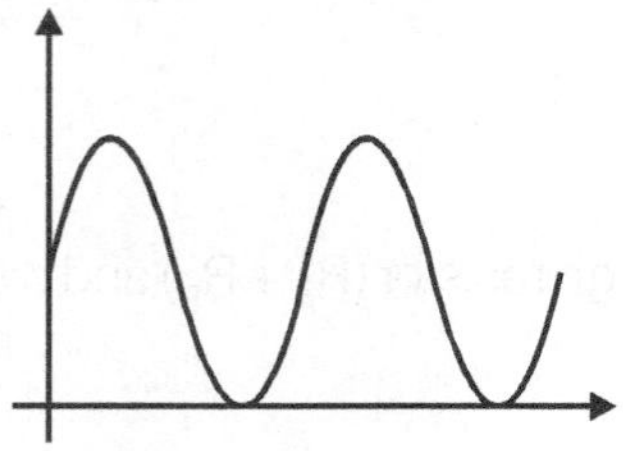

(b)
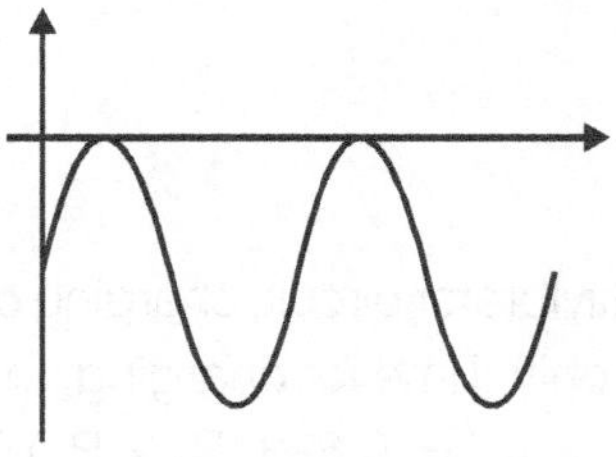

(c)
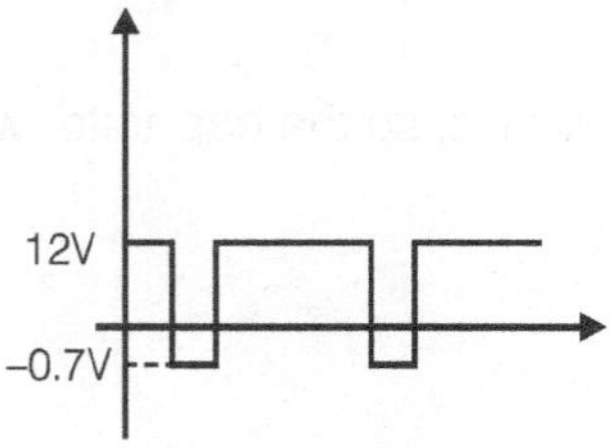


(d)
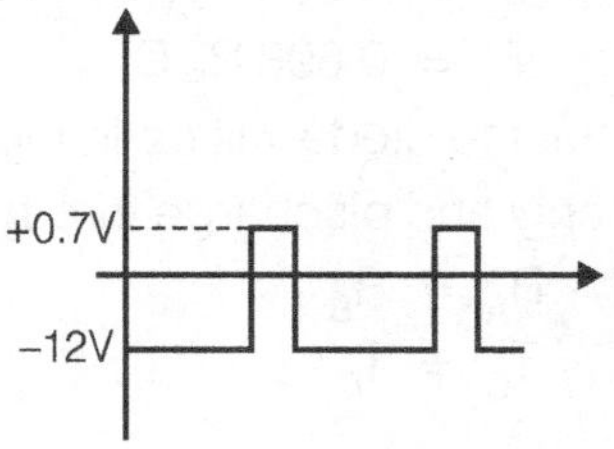

Exp. (d)

In the positive half cycle of input, diode $D_1$ will be RB and equivalent circuit is

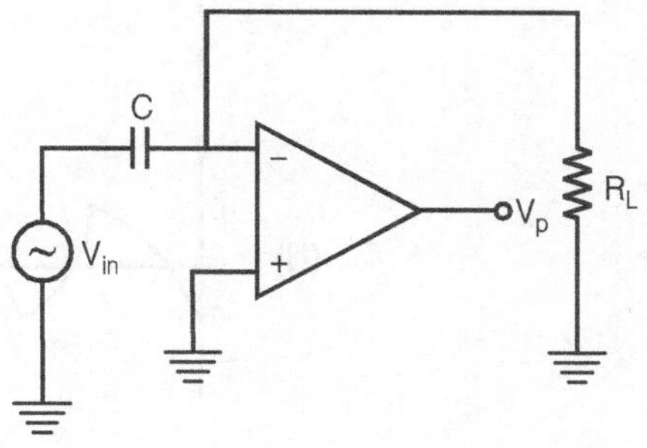


Since there is no feedback to the op-amp and op-amp has a high open loop gain so it goes in saturation. Input is applied at inverting terminal so

$$V_p = -V_{CC} = -12\text{ V}$$

In negative half cycle of input, diode $D_1$ is in forward bias and equivalent circuit is

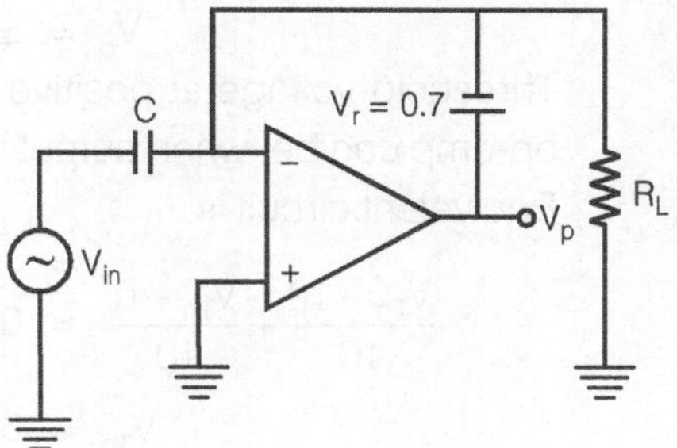


$$V_P = V_\gamma + V_-$$

Op-amp is at virtual ground so

$$V_+ = V_- = 0$$

$$V_p = V_\gamma = 0.7\text{ V}$$

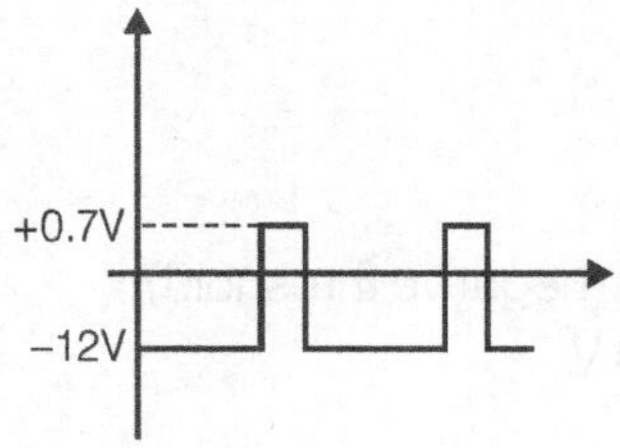


**Q.53** A relaxation oscillator is made using OPAMP as shown in figure. The supply voltages of the OPAMP are ± 12V. The voltage waveform at point P will be

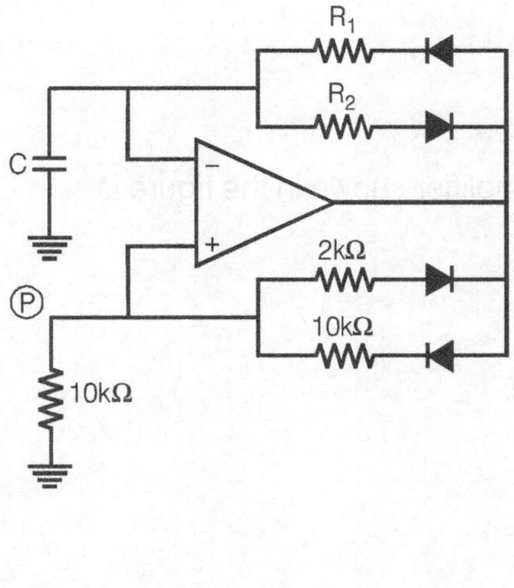

(a)

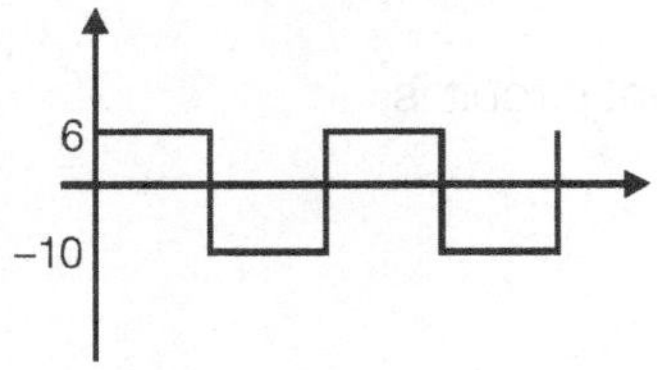


(b)

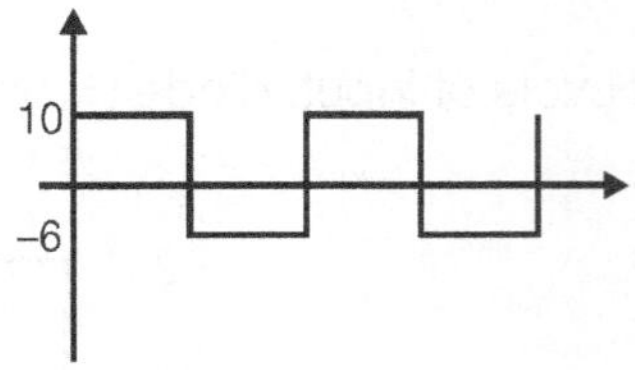


(c)

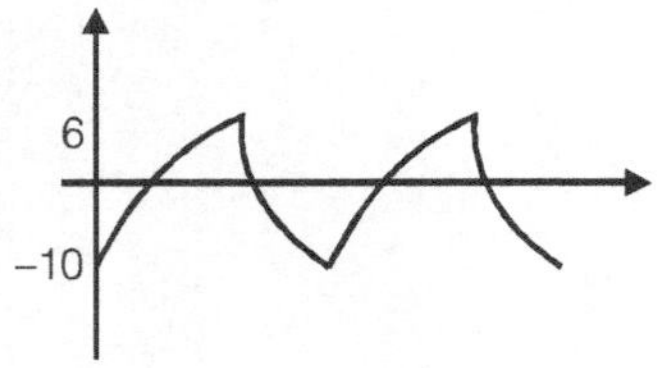


(d)

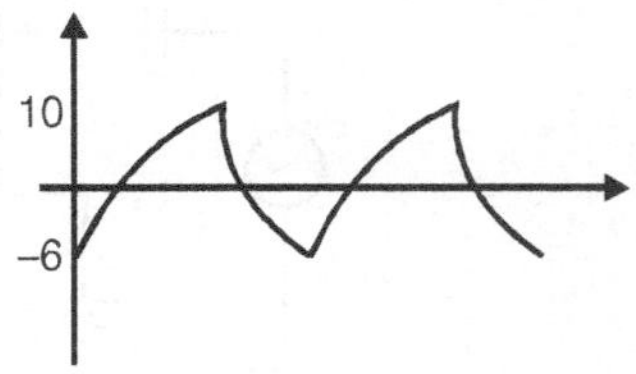


**Exp. (c)**

Here output of multivibrator is

$$V_o = \pm 12\text{ V}$$

Threshold voltage at positive terminal of op-amp can be, when output $V_o = +12$ V

Equivalent circuit is

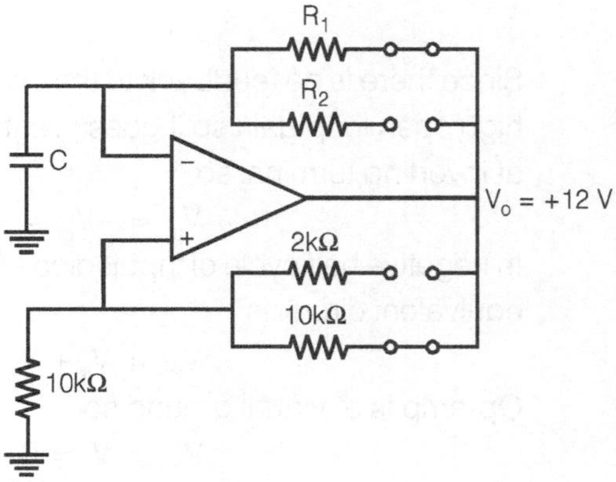


$$\frac{V_{Th} - 12}{10} + \frac{V_{Th} - 0}{10} = 0$$

$$V_{Th} = 6\text{ V} \quad \text{(positive threshold)}$$

So capacitor can charge upto 6 V.

When output $V_o = -12$ V, the equivalent circuit is

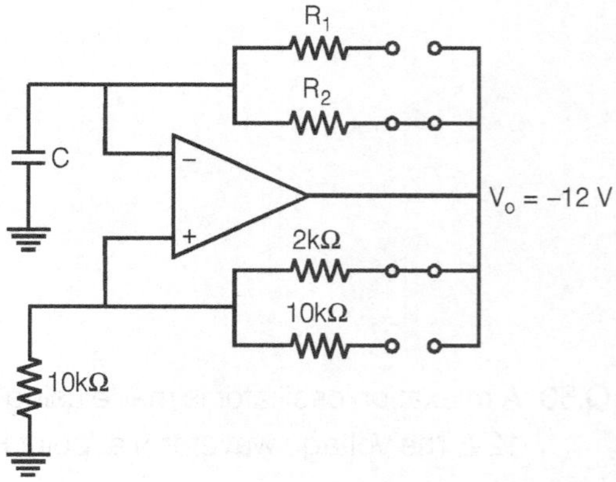


$$\frac{V_{Th} + 12}{2} + \frac{V_{Th} - 0}{10} = 0$$

$$5 V_{Th} + 60 + V_{Th} = 0$$

$$V_{Th} = -10\text{ V} \quad \text{(negative threshold)}$$

So capacitor will discharge upto −10 V.

At terminal p voltage waveform is

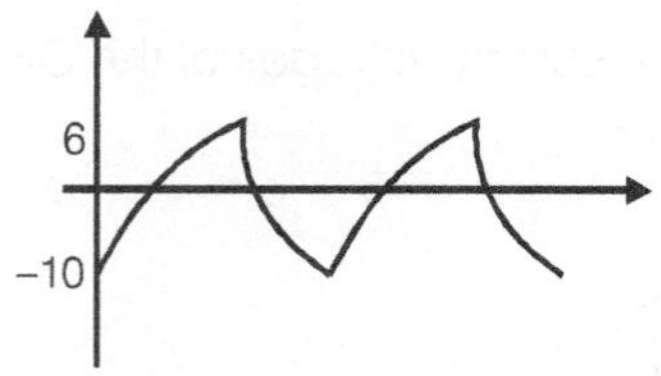


**Q.54** The input resistance $R_i$ of the amplifier shown in the figure is

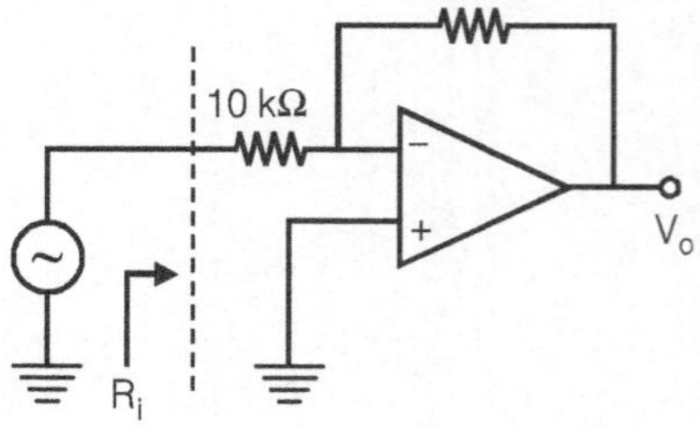


(a) $\frac{30}{4}$ kΩ

(b) 10 kΩ

(c) 40 kΩ

(d) infinite

**Exp. (b)**

Since the inverting terminal is at virtual ground the current flowing through the voltage source is

$$I_s = \frac{V_s}{10\,k\Omega}$$

$$\frac{V_s}{I_s} = 10\,k\Omega = R_{in}$$

**Q.55** The voltage $e_0$ indicated in the figure has been measured by an ideal voltmeter. Which of the following can be calculated?

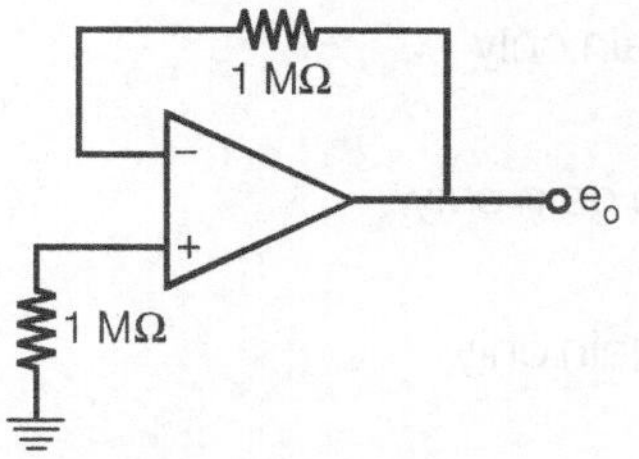


(a) Bias current of the inverting input only
(b) Bias current of the inverting and non-inverting inputs only
(c) Input offset current only
(d) Both the bias currents and the input offset current

**Exp. (c)**

The circuit as shown

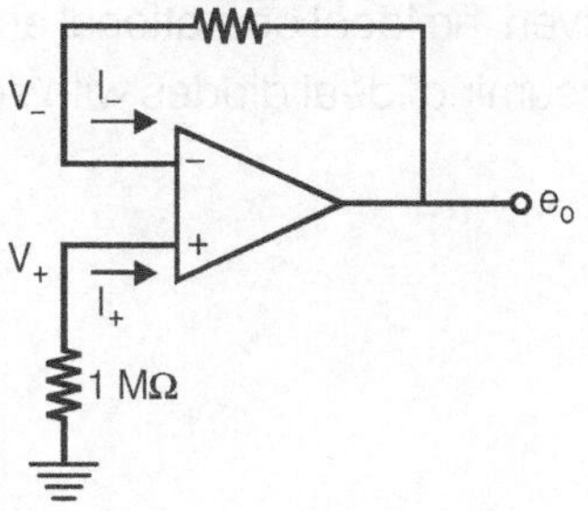


$$\frac{e_0 - V_-}{1M\Omega} = I_-$$

$$e_0 = I_-(1\,M\Omega) + V_-$$

$$\frac{0 - V_+}{1M\Omega} = I_+$$

$$V_+ = -I_+(1\,M\Omega)$$

Since ideal op-amp

$$V_- = V_+$$

$$e_0 = I_-(1\,M\Omega) - 1_+(1\,M\Omega) = (I_- - 1_+)\,1M\Omega = I_{os}\,(1\,M\Omega)$$

Thus if $e_0$ has been measured, we can calculate input offset current $I_{os}$ only.

**Q.56** The OP-amp circuit shown in the figure is a filter. The type of filter and its cut-off frequency are respectively

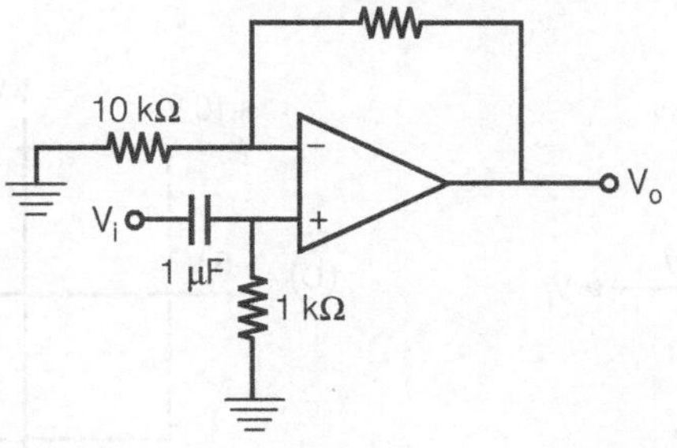


(a) high pass, 1000 rad/sec.
(b) low pass, 1000 rad/sec.
(c) high pass, 10000 rad/sec.
(d) low pass, 10000 rad/sec.

Exp. (a)

At low frequency, capacitor is open circuit and voltage across non inverting terminal is zero. At high frequency, capacitor act as short circuit and all input voltage appear at non inverting terminal. Thus, this is high pass circuit.

The frequency is $\omega = \frac{1}{RC} = \frac{1}{1\times10^3\times1\times10^{-6}} = 1000$ rad/sec

**Q.57** In an ideal differential amplifier shown in the figure, a large value of $R_E$.

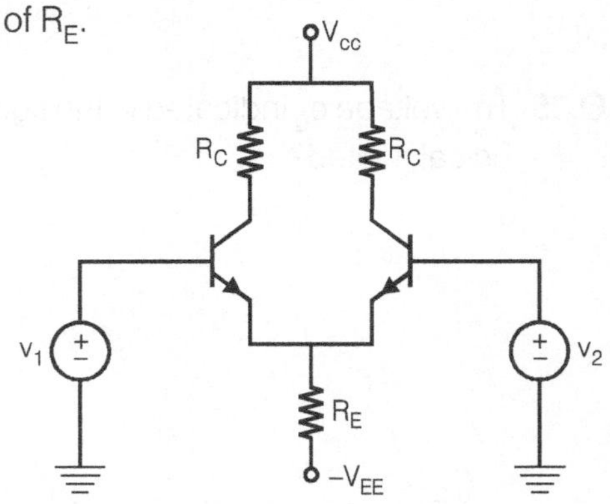


(a) increases both the differential and common-mode gains

(b) increases the common-mode gain only

(c) decreases the differential-mode gain only

(d) decreases the common-mode gain only

Exp. (d)

Common mode gain $A_{cm} = -\frac{R_C}{2R_E}$

and differential mode gain $A_{dm} = \frac{R_c}{r_e}$

Thus only common mode gain depends on $R_E$ and for large value of $R_E$ it decreases.

**Q.58** Given the ideal operational amplifier circuit shown in the figure indicate the correct transfer characteristics assuming ideal diodes with zero cut-in voltage.

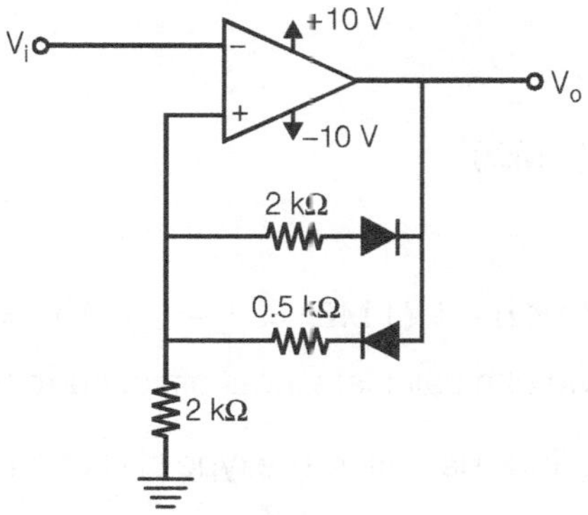


(a)

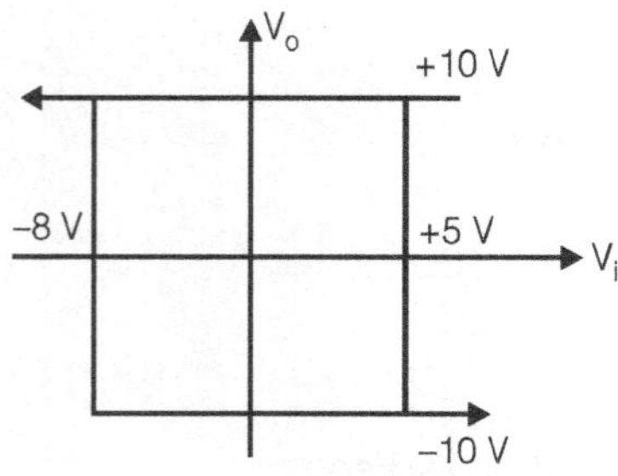


(b)

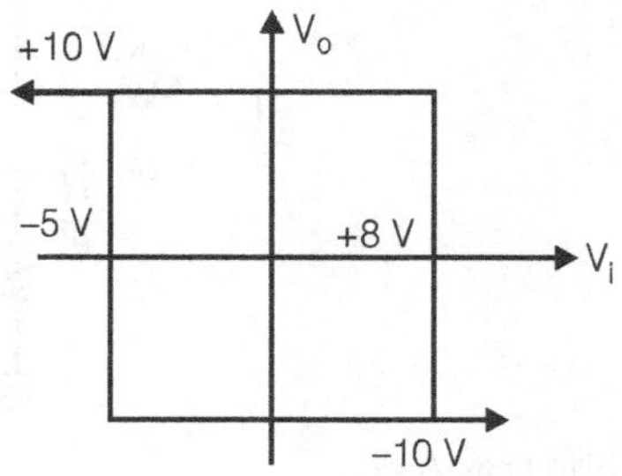

(c) 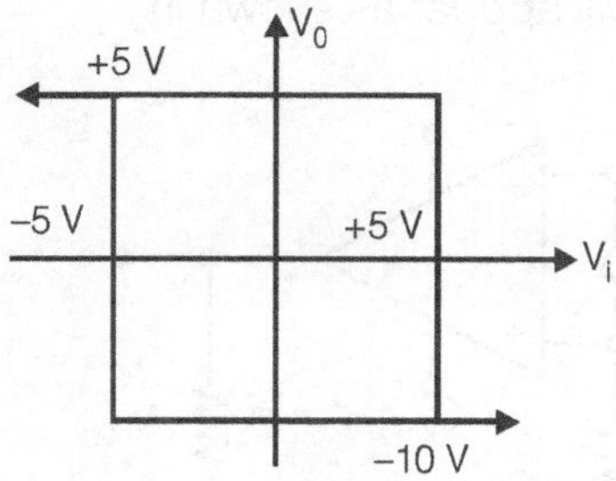


(d) 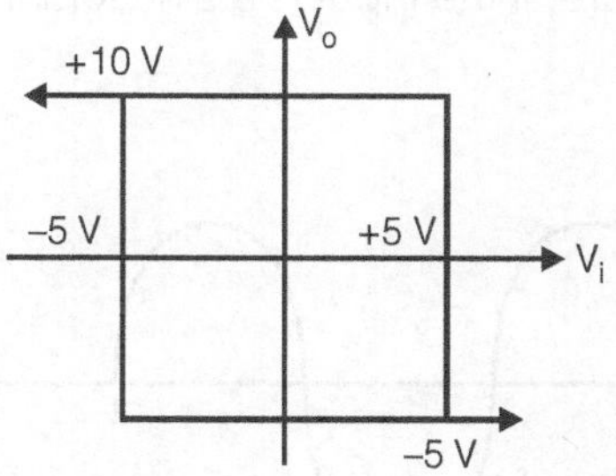


**Exp. (b)**

Only one diode will be in on condition when lower diode is on condition

$$V_L = \frac{2\,k\Omega}{2.5\,k\Omega}\cdot V_{sat} = \frac{2}{2.5}\times 10 = 8\text{ V}$$

When upper diode is on condition

$$V_U = \frac{2\,k\Omega}{4\,k\Omega}\times(-10) = -5\text{ V}$$

**Q.59** Consider the inverting amplifier, using an ideal operational amplifier shown in the figure. The designer wishes to realize the input resistance seen by the small-signal source to be as large as possible, while keeping the voltage gain between −10 and −25. The upper limit on $R_F$ is 1 MΩ. The value of $R_1$ should be

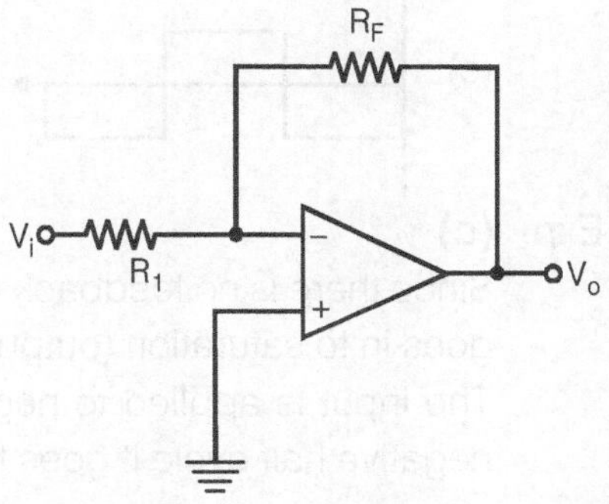


(a) Infinity
(b) 1 MΩ
(c) 100 kΩ
(d) 40 kΩ

**Exp. (c)**

Gain of inverting amplifier is

$$A_V = -\frac{R_f}{R_1}$$

$$A_V = -\frac{1\times 10^6}{R_1}$$

$$\therefore\quad R_f = 1\text{ M}\Omega \quad ; \quad R_1 = -\frac{1\times 10^6}{A_V}$$

$$A_V = -10 \text{ to } -25 \text{ so value of } R_1$$

$$R_1 = \frac{10^6}{10} = 100\text{ k}\Omega \qquad A_V = -10$$

$$R_1 = \frac{10^6}{25} = 40\text{ k}\Omega \qquad A_v = -25$$

$R_1$ should be as large as possible so

$$R_1 = 100\text{ k}\Omega$$

**Q.60** In the given figure, if the input is a sinusoidal signal, the output will appear as shown in

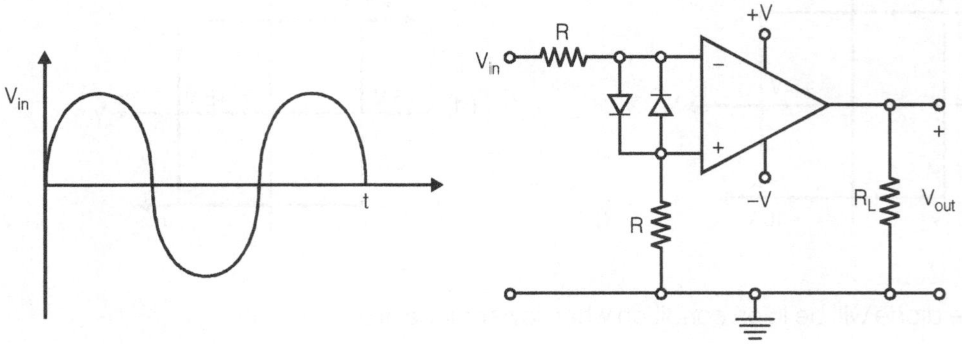


(a)

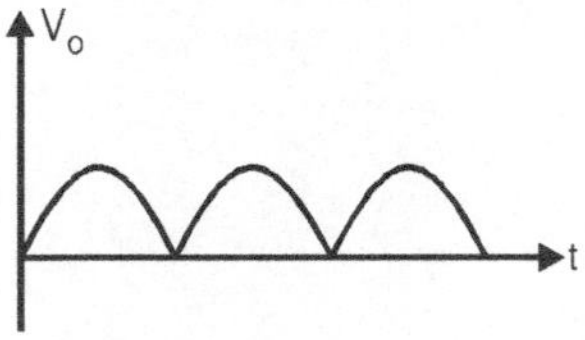


(b)

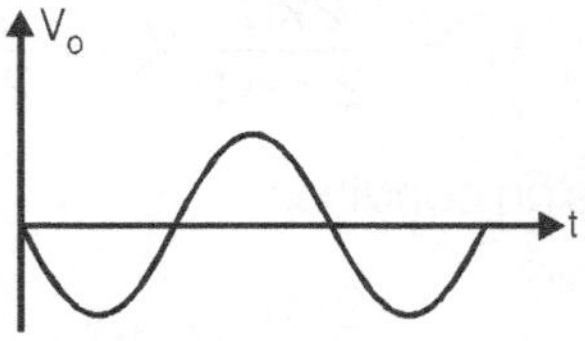


(c)

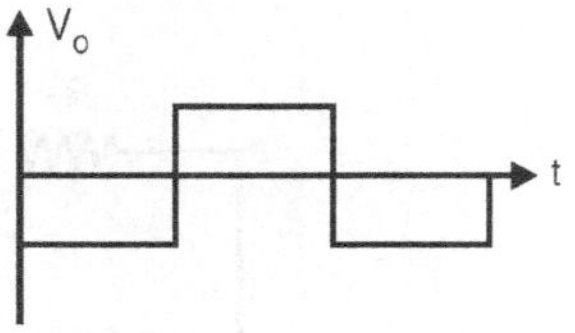


(d)

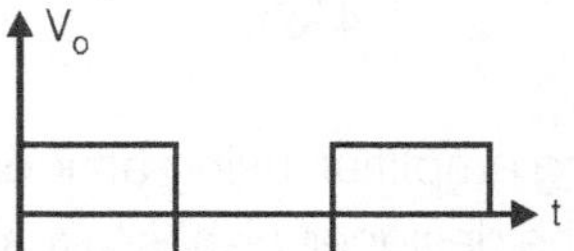


**Exp. (c)**

Since there is no feedback in the circuit and ideally op-amp has a very high value of open loop gain, so it goes in to saturation (output is either +V or –V) for small values of input.

The input is applied to negative terminal of op-amp, so in positive half cycle it saturates to –V and in negative half cycle it goes to +V.

**Q.61** An ideal op-amp is an ideal

(a) voltage controlled current source
(b) voltage controlled voltage source
(c) current controlled current source
(d) current controlled voltage source

**Exp. (b)**

$$V_o = A_d V_d$$

It is voltage controlled voltage source.

**Q.62** The circuit in the figure is a

(a) low-pass filter
(b) high-pass filter
(c) band-pass filter
(d) band-reject filter

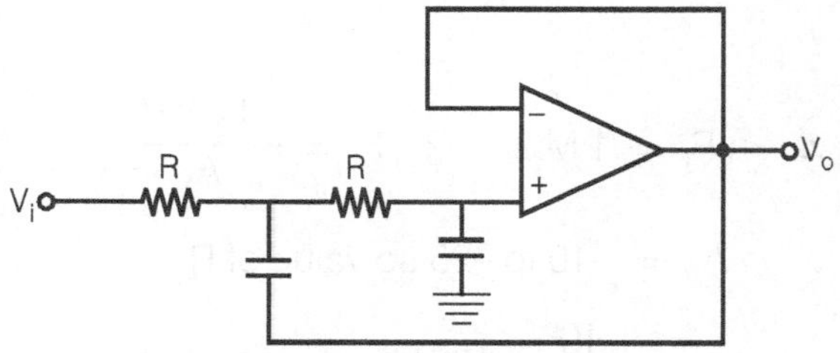


**Exp. (a)**

This is a low pass filter, because

At $\omega = \infty$ $\quad \dfrac{V_o}{V_{in}} = 0$

$\omega = 0$ $\quad \dfrac{V_o}{V_{in}} = 1$

**Q.63** The value of C required for sinusoidal oscillations of frequency 1 kHz in the circuit of the figure is

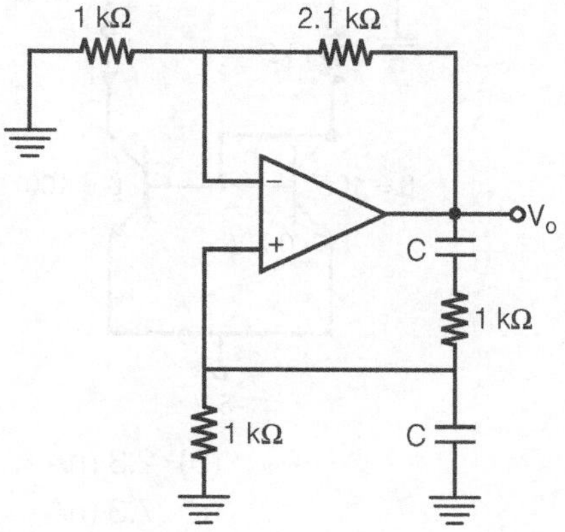


(a) $\frac{1}{2\pi}\mu F$

(b) $2\pi\mu F$

(c) $\frac{1}{2\pi\sqrt{6}}\mu F$

(d) $2\pi\sqrt{6}\mu F$

**Exp. (a)**

This is wien bridge oscillator. The frequency of oscillation is

$$2\pi f = \frac{1}{RC}$$

$$C = \frac{1}{2\pi Rf} = \frac{1}{2\pi}\mu F$$

**Q.64** In the op-amp circuit given in the figure, the load current $i_L$ is

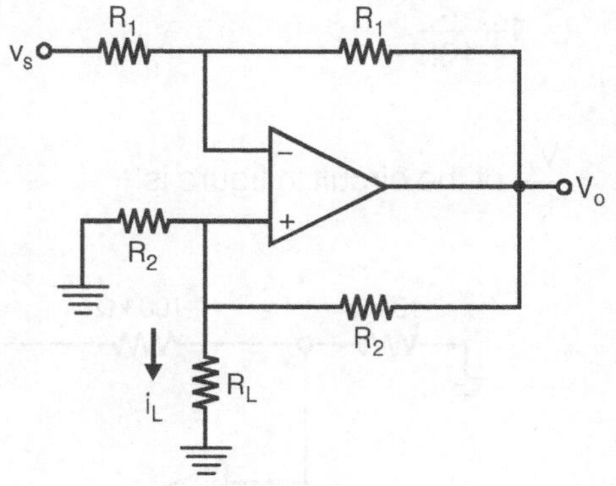


(a) $-\frac{V_S}{R_2}$

(b) $\frac{V_S}{R_2}$

(c) $-\frac{V_S}{R_L}$

(d) $\frac{V_S}{R_1}$

**Exp. (a)**

According to ideal op-amp $V_- = V_+$

Applying KCL at inverting terminal

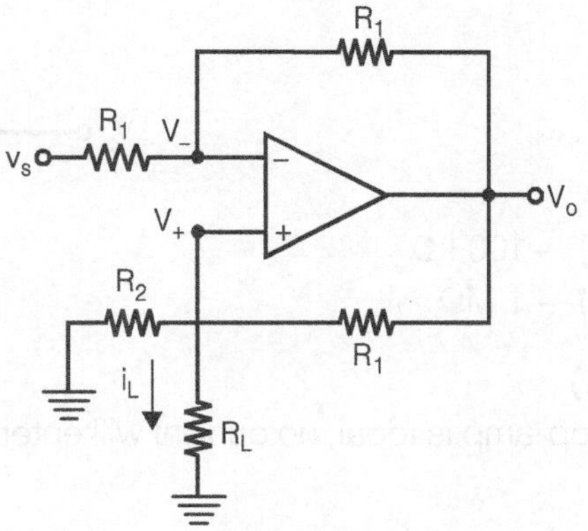


$$\frac{V_- - V_S}{R_1} + \frac{V_- - V_o}{R_1} = 0$$

$$2V_- - V_o = V_S$$

Applying KCL at non inverting

$$\frac{V_+}{R_2} + I_L + \frac{V_+ - V_o}{R_2} = 0$$

$$2V_+ - V_o + I_L R_2 = 0$$

Since $V_- = V_+$

$$V_S + I_L R_2 = 0$$

$$I_L = -\frac{V_S}{R_2}$$

**Q.65** Two perfectly matched silicon transistors are connected as shown in figure. The value of the current I is

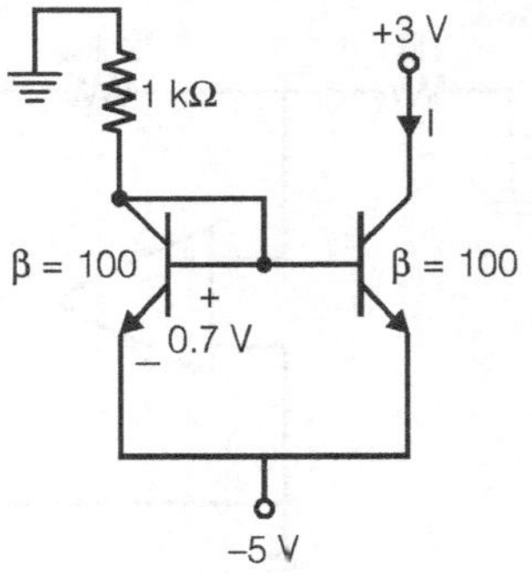


(a) 0 mA
(b) 2.3 mA
(c) 4.3 mA
(d) 7.3 mA

**Exp.** (c)

$$I_{C2} = I = \frac{I_R}{1+\frac{2}{\beta}}$$

$$I_R = \frac{-5+0.7}{1\times 10^3} = -4.3 \text{ mA}$$

$$I = \frac{4.3}{1+\frac{2}{100}} \cong 4.3 \text{ mA}$$

**Q.66** The input resistance $R_{in} = \frac{V_x}{I_x}$ of the circuit in figure is

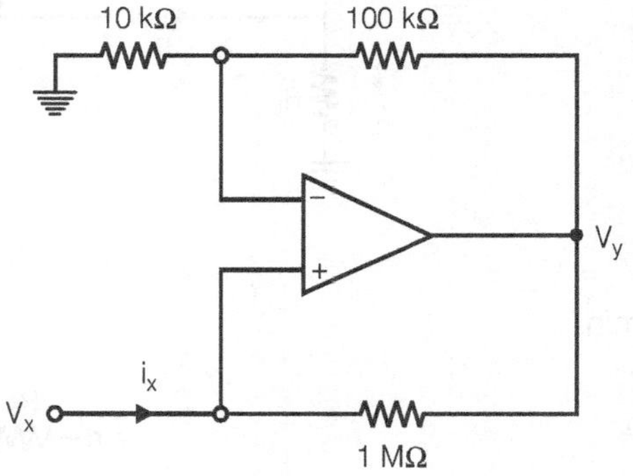


(a) +100 kΩ
(b) –100 kΩ
(c) +1 MΩ
(d) –1 MΩ

**Exp.** (b)

If op-amp is ideal, no current will enter in op-amp.

So current $i_x$ is

$$i_x = \frac{V_x - V_y}{1\times 10^6}$$

$$V_+ = V_- = V_x$$

$$\frac{V_x - V_y}{100\times10^3} + \frac{V_x - 0}{10\times10^3} = 0$$

$$V_x - V_y + 10\,V_x = 0$$

$$11\,V_x = V_y$$

$$i_x = \frac{V_x - 11V_x}{1\times10^6}$$

$$i_x = -\frac{10\,V_x}{10^6}$$

$$R_{in} = \frac{V_x}{i_x} = -\frac{10^6}{10} = -100\,k\Omega$$

**Q.67** In the Schmitt trigger circuit shown in figure, if $V_{CE(sat)} = 0.1$ V, the output logic low level ($V_{OL}$) is

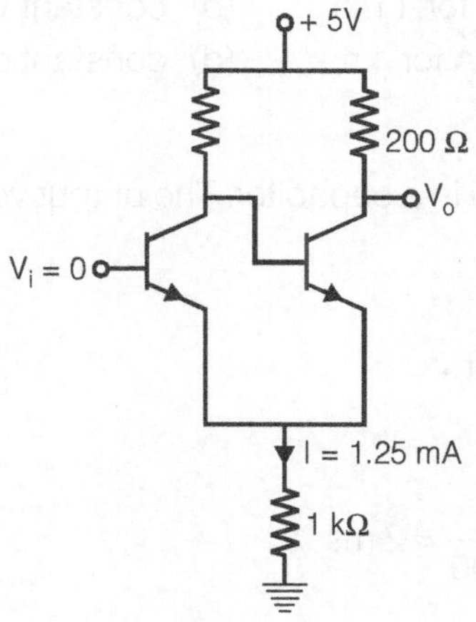


(a) 1.25 V (b) 1.35 V
(c) 2.50 V (d) 5.00 V

**Exp. (b)**

In the given circuit $V_i = 0$ V

So, transistor $Q_1$ is in cut off region and $Q_2$ is in saturation

$$5 - I_C R_C - V_{CE\,sat} - 1.25 = 0$$

$$5 - I_C R_C - 0.1 - 1.25 = 0$$

$$5 - I_C R_C = 1.35$$

$$V_o = 1.35 \qquad (V_o = 5 - I_C R_C)$$

**Q.68** If the differential voltage gain and the common mode voltage gain of a differential amplifier are 48 dB and 2 dB respectively, then its common mode rejection ratio is

(a) 23 dB (b) 25 dB
(c) 46 dB (d) 50 dB

**Exp. (c)**

$$\text{CMRR} = \frac{A_d}{A_c}$$

$$20\log \text{CMRR} = 20\log A_d - 20\log A_c = 48 - 2 = 46\text{ dB}$$

**Q.69** If the op-amp in the figure is ideal, the output voltage $V_{out}$ will be equal to

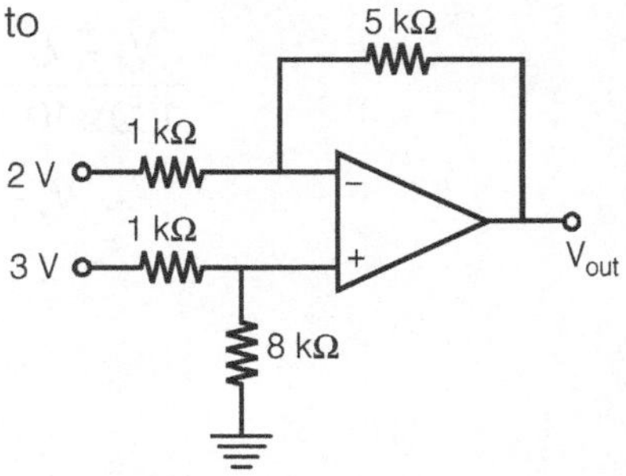


(a) 1 V

(b) 6 V

(c) 14 V

(d) 17 V

**Exp. (b)**

$$V_o = -\frac{5\,k\Omega}{1k\Omega}\times 2\,V + \left[1+\frac{5\,k\Omega}{1k\Omega}\right]\times 3\,V\times\frac{8\,k\Omega}{8\,k\Omega+1k\Omega} = 6\,V$$

**Q.70** An ideal sawtooth voltage waveform of frequency 500 Hz and amplitude 3 V is generated by charging a capacitor of 2 μF in every cycle. The charging requires

(a) constant voltage source of 3 V for 1 ms
(b) constant voltage source of 3 V for 2 ms
(c) constant current source of 1 mA for 1 ms
(d) constant current source of 3 mA for 2 ms

**Exp. (d)**

If a constant current is made to flow in a capacitor. The output voltage integration of input current and that is sawtooth waveform.

$$V_c = \frac{1}{C}\int_0^t i\,dt$$

$$T = \frac{1}{f} = \frac{1}{500} = 2\text{ ms}$$

$$3 = \frac{1}{2\times 10^{-6}}\int_0^{2\times 10^{-3}} i\,dt$$

$$i(2\times 10^{-3} - 0) = 6\times 10^{-6}$$

$$i = 3\text{ mA}$$

Thus charging requires 3 mA current source for 2 msec.

**Q.71** A 741 type op-amp has a gain bandwidth product of 1 MHz. A non inverting amplifier using this op-amp and having a voltage gain of 20 dB will exhibit a –3 dB bandwidth of

(a) 50 kHz
(b) 100 kHz
(c) $\frac{1000}{17}$ kHz
(d) $\frac{1000}{7.07}$ kHz

**Exp. (b)**

Let x be the gain and it is 20 dB.

Therefore $20\log x = 20$

$$x = 10$$

Gain Bandwidth is $10^6$ Hz, thus

$$\text{Bandwidth} = \frac{10^6}{\text{gain}} = \frac{10^6}{10} = 10^5\text{ Hz} = 100\text{ kHz}$$

**Q.72** Assuming the operational amplifier to be ideal, the gain $V_{out}/V_{in}$ for the circuit shown in figure is

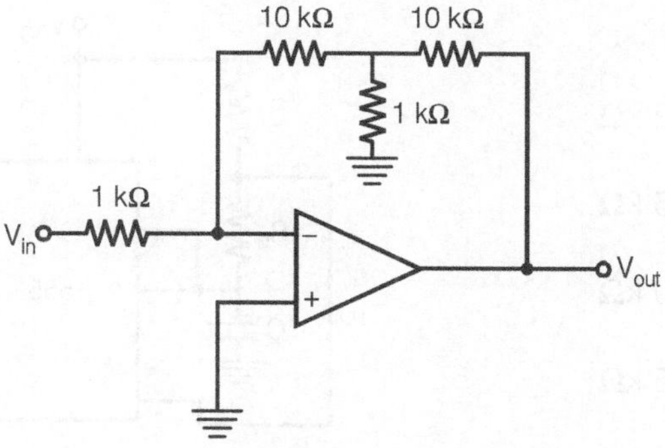


(a) −1 (b) −20
(c) −100 (d) −120

**Exp.** (d)

Let the voltages at input terminals of op-amp are $V_-$ and $V_+$ respectively.

So $$V_+ = V_- = 0$$

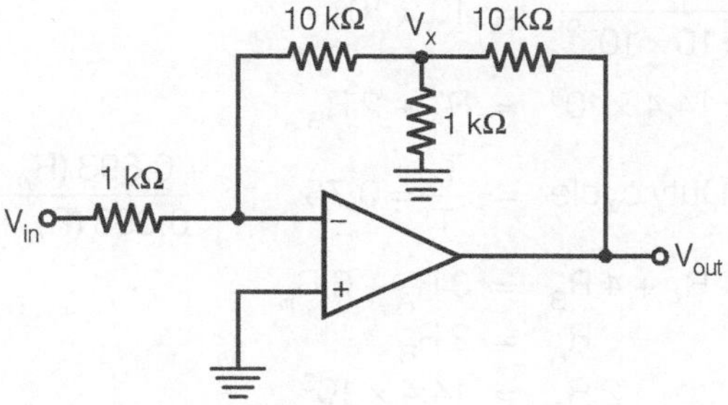


At negative terminal

$$\frac{0-V_{in}}{1}+\frac{0-V_x}{10} = 0$$

At node x $$\frac{V_x-0}{10}+\frac{V_x-V_o}{10}+\frac{V_x-0}{1} = 0$$

$$V_x + V_x - V_o + 10\,V_x = 0$$

$$12\,V_x = V_o$$

$$V_x = \frac{V_o}{12}$$

$$\frac{V_{in}}{1}+\frac{V_x}{10} = 0$$

$$V_{in} = -\frac{V_{out}}{120}$$

$$\frac{V_{out}}{V_{in}} = -120$$

**Q.73** The circuit of figure shows a 555 Timer IC connected as an astable multivibrator. The value of the capacitor C is 10 nF. The values of the resistors $R_A$ and $R_B$ for a frequency of 10 kHz and a duty cycle of 0.75 for the output voltage waveform are

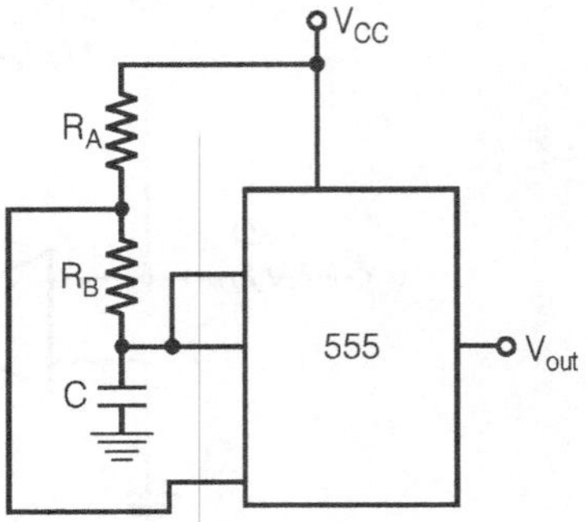


(a) $R_A = 3.62\ k\Omega$, $R_B = 3.62\ k\Omega$

(b) $R_A = 3.62\ k\Omega$, $R_B = 7.25\ k\Omega$

(c) $R_A = 7.21\ k\Omega$, $R_B = 3.60\ k\Omega$

(d) $R_A = 7.25\ k\Omega$, $R_B = 7.25\ k\Omega$

**Exp. (c)**

In the circuit, the capacitor charges through resistor $(R_A + R_B)$ and discharging through $R_B$.

$$T_C = 0.693\,(R_A + R_B)C$$

$$T_D = 0.693\,R_B\,C$$

$$f = \frac{1}{T} = \frac{1}{T_C + T_D} = \frac{1}{0.693\,(R_A + 2R_B)C}$$

$$\frac{1}{0.693\,(R_A + 2R_B) \times 10 \times 10^{-9}} = 10 \times 10^3$$

$$14.4 \times 10^3 = R_A + 2R_B$$

$$\text{Duty cycle} = \frac{T_C}{T} = 0.75\ ; \quad \frac{0.693\,(R_A + R_B)C}{0.693\,(R_A + 2R_B)C} = \frac{3}{4}$$

$$4R_A + 4R_B = 3R_A + 6R_B$$

$$R_A = 2R_B$$

$$2R_A = 14.4 \times 10^3$$

$$R_A = 7.21\ k\Omega$$

$$R_B = 3.60\ k\Omega$$

**Q.74** An amplifier using an opamp with a slew-rate SR=1 V/μsec has a gain of 40dB. If this amplifier has to faithfully amplify sinusoidal signals from dc to 20 kHz without introducing any slew-rate induced distortion, then the input signal level must not exceed.

(a) 795 mV
(b) 395 mV
(c) 79.5 mV
(d) 39.5 mV

**Exp. (c)**

Let the gain of op-amp be $A_v$ then

$$20 \log A_V = 40\ dB$$

$$A_V = 100$$

$$V_i = V_m \sin \omega t$$

$$V_o = A_V V_i = A_V V_m \sin \omega t$$

$$\frac{dV_o}{dt} = A_V V_m \omega = A_V V_m 2\pi f$$

$$\left(\frac{dV_o}{dt}\right)_{max} = A_V V_m \omega = A_V V_m 2\pi f$$

$$V_m = \frac{SR}{A_V \times 2\pi f} = \frac{1}{10^{-6} \times 100 \times 2\pi \times 20 \times 10^3} = 79.5\ mV$$

**Q.75** The ideal op-amp has the following characteristics

(a) $R_i = \infty, A = \infty, R_0 = 0$  (b) $R_i = 0, A = \infty, R_0 = 0$

(c) $R_i = \infty, A = \infty, R_0 = \infty$  (d) $R_i = 0, A = \infty, R_0 = \infty$

**Exp. (a)**

Ideal op-amp $R_i \to \infty$

$R_o \to 0$

$A \to \infty$

**Q.76** The inverting OP-AMP shown in the figure has an open-loop gain of 100. The closed-loop gain $V_0/V_s$ is

(a) –8

(b) –9

(c) –10

(d) –11

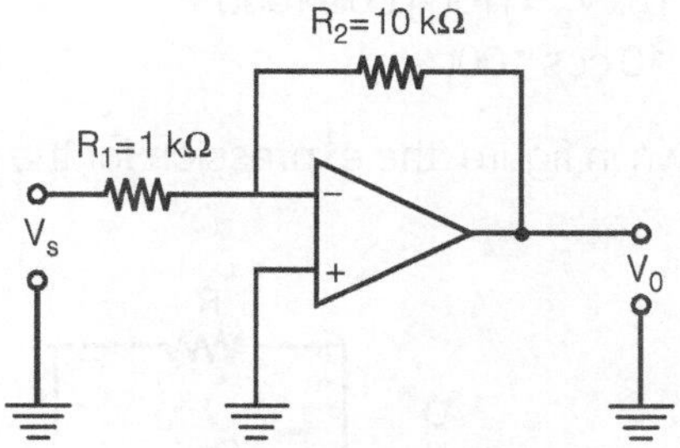


**Exp. (d)**

Let $V_-$ be the voltage of inverting terminal, since non inverting terminal is at ground.

The output voltage is

$$V_o = A_{OL} V_-$$

Applying KCL at inverting terminal

$$\frac{V_- - V_s}{R_1} + \frac{V_- - V_o}{R_2} = 0$$

$$\frac{V_o}{V_s} = A_{CL} = \frac{-R_2}{R_1 - \dfrac{R_2 + R_1}{A_{OL}}}$$

$$A_{CL} = \frac{-10\text{ k}\Omega}{1\text{k}\Omega - \dfrac{10\text{ k}\Omega + 1\text{k}\Omega}{100\text{ k}\Omega}} = \frac{-1000}{89} \approx -11$$

**Q.77** In the figure assume the OP-AMPs to be ideal. The output $V_o$ of the circuit is:

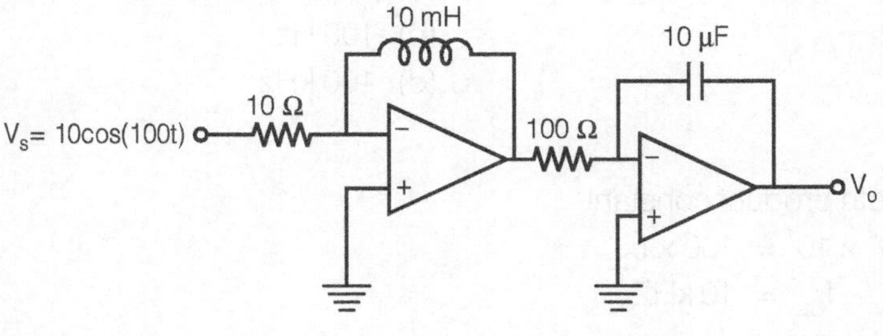


(a) 10cos (100t)  (b) $10\int_0^t \cos(100\tau)d\tau$

(c) $10^{-4}\int_0^t \cos(100\tau)d\tau$  (d) $10^{-4}\frac{d}{dt}\cos(100t)$

**Exp. (a)**

$$\frac{V_1}{V_s} = \frac{-\omega jL}{R} = -\frac{100\times10^{-3}}{10} = -\frac{9}{10}$$

$$V_1 = -\frac{jV_s}{10} = -j\cos 100t$$

Applying KCL at inverting terminal of second op-amp.

$$\frac{V_o}{V_i} = \frac{-\frac{1}{j\omega C}}{100} = \frac{1}{j\times100\times10\times10^{-6}\times100} = j10$$

$$V_o = j10\,V_2 = j10(-j\cos 100t)$$

$$V_o = 10\cos 100t$$

**Q.78** For the oscillator circuit shown in figure, the expression for the time period of oscillation can be given by (where $\tau = RC$)

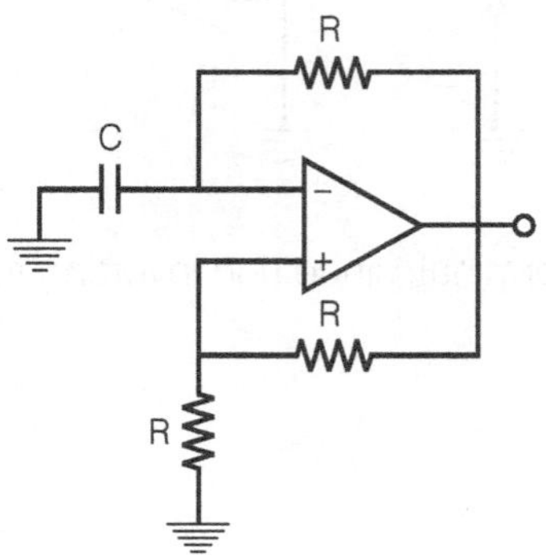


(a) $\tau\, ln\, 3$
(b) $2\tau\, ln\, 3$
(c) $\tau\, ln\, 2$
(d) $2\tau\, ln\, 2$

**Exp. (b)**

$$\tau = 2R_f C\, ln\left(\frac{2R_1+R_2}{R_2}\right)$$

$$\tau = 2\tau\, ln\, 3 \qquad \left(\therefore T = 2\tau\, ln\left(\frac{1+\beta}{1-\beta}\right)\right)$$

**Q.79** An op-amp has an open loop gain of $10^5$ and an and an open loop upper cut off frequency of 10 Hz. If this op-amp is connected as an amplifier with a closed loop pain of 100, then the new upper cut of frequency is

(a) 10 Hz
(b) 100 Hz
(c) 10 kHz
(d) 100 kHz

**Exp. (c)**

Since gain bandwidth product constant

$$10^5\times10 = 100\times f_{CL}$$

$$f_{CL} = 10\text{ kHZ}$$

**Q.80** An op-amp, having a slew rate of 62.8 V/μsec is connected in a voltage follower configuration. If the maximum amplitude of the input sinusoidal is 10 V, then the minimum frequency at which the slew rate limited distortion would set in at the output is

(a) 1.0 MHz
(b) 6.28 MHz
(c) 10 MHz
(d) 62.8 MHz

**Exp.** (a)

$$\text{Slew rate} = \frac{dV_o}{dt}$$

For voltage follower $V_o = V_{in}$

$$S.R = \frac{dV_{in}}{dt}$$

$$V_{in} = 10 \sin \omega t$$

$$\text{Slew rate} = \frac{d}{dt}(10 \sin \omega t)$$

$$\text{Slew rate} = 10\,\omega \cos \omega t$$

$$(\text{Slew rate/max}) = 10\,\omega = 62.8 \text{ volt/}\mu\text{ sec}$$

$$10 \times 2\,\pi F = 62.8 \times 10^6$$

$$f = \frac{62.8 \times 10^6}{62.8}$$

$$f = 1 \text{ MHz}$$

**Q.81** The block diagrams of two of half wave rectifiers are shown in the figure with the transfer characteristics of the rectifiers. It is desired to make full wave rectifier using above two half-wave rectifiers. The resultant circuit will be

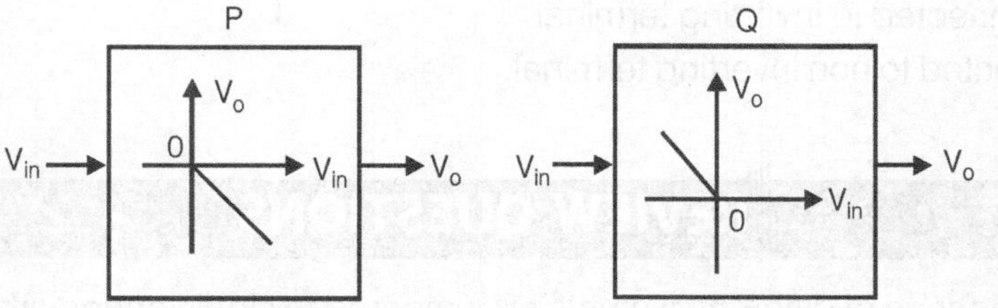


(a)
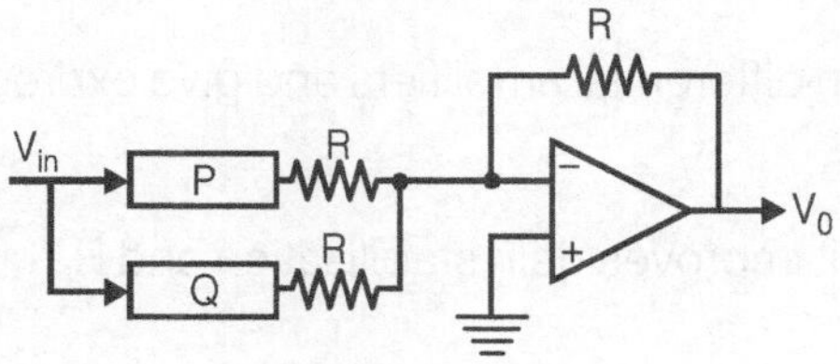


(b)
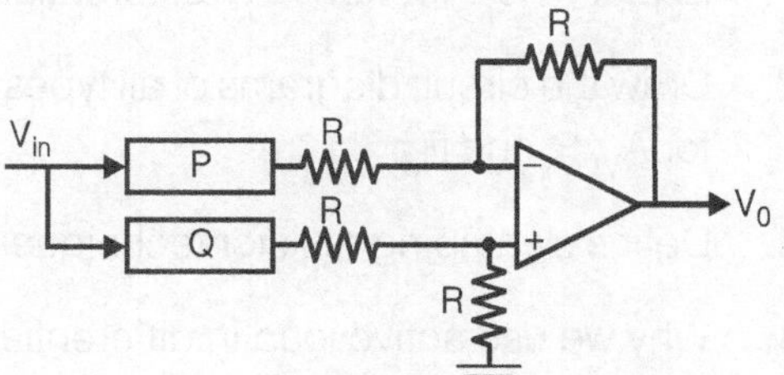


(c)
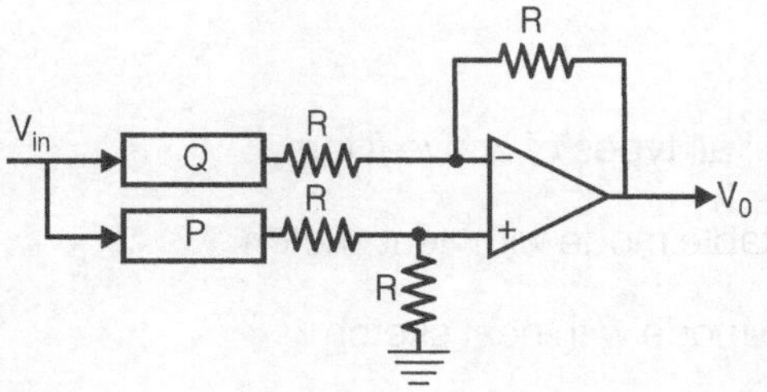


(d)
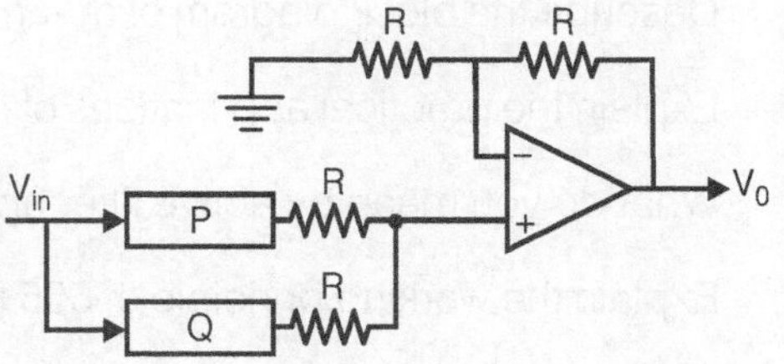


**Exp.** (b)

Let input $V_{in}$ is a sine wave shown

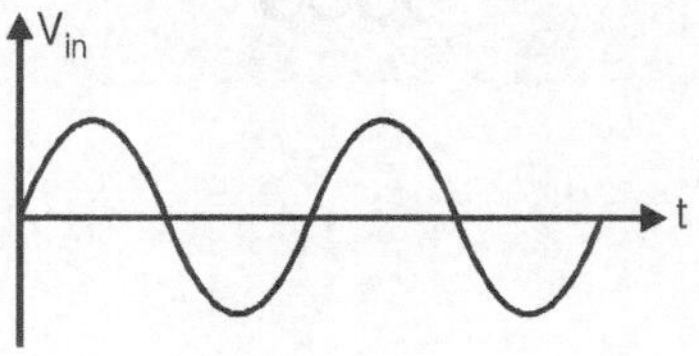

According to given transfer characteristics of rectifiers output of rectifier P is

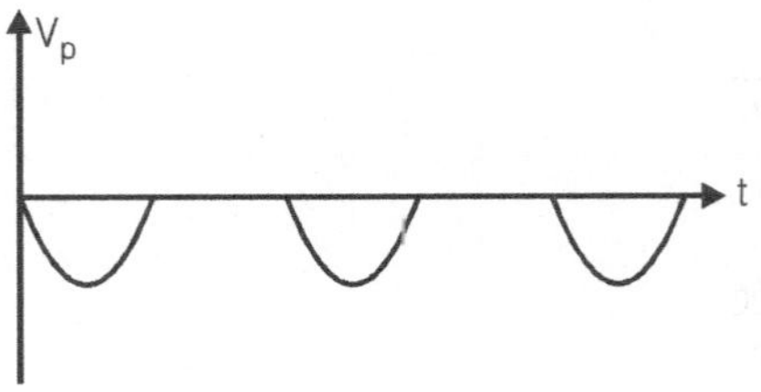


Similarly output of rectifier Q is

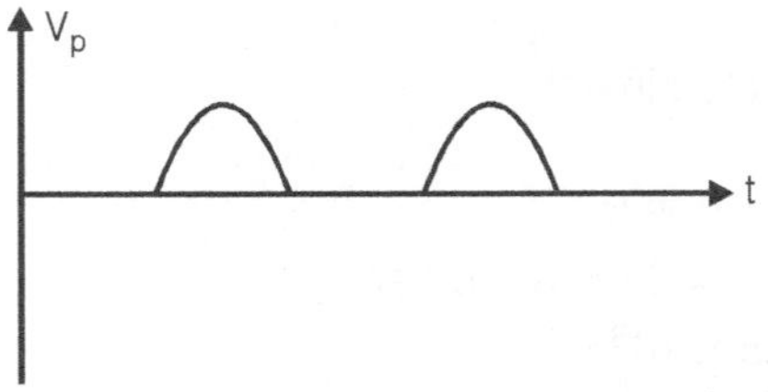


Output of full wave rectifier is

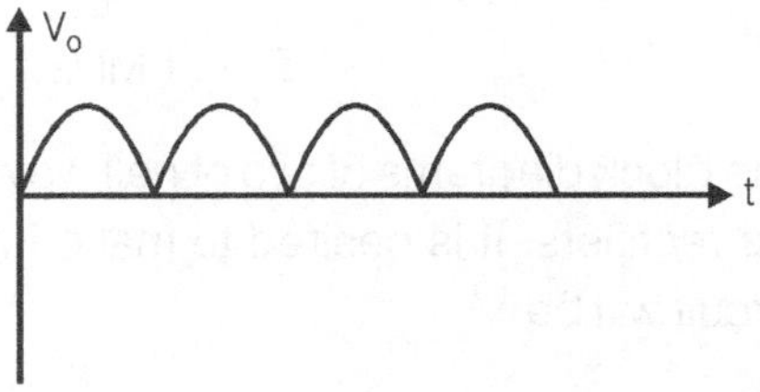


To get output $V_o$

$$V_o = k(-V_p + V_Q)$$

↓ gain of op-amp

So p should be connected to inverting terminal.
Q should be connected to non inverting terminal.

## REVIEW QUESTIONS

1. Explain the importance of differential amplifier compared to single ended amplifiers?
2. Draw the circuit diagrams of all types of configuration in differential amplifiers and give expressions for $A_V$, $R_i$ and $R_o$.
3. Define swamping resistor technique and explain how it improves gain stabilisation and $R_i$.
4. Why we use active load in differential amplifiers.
5 Describe the block diagram of op-amp and explain important characteristics of op-amp.
6. Explain the practical applications of op-amp.
7. What do you mean by active filter and explain about all types of active filters.
8. Explain the working principle of 555 timer in monostable mode with neat sketch.
9. Explain the working principle of 555 timer in astable mode with neat sketch.

OOOO

# CHAPTER 10

# FET BIASING AND SMALL SIGNAL ANALYSIS

## About This Chapter

- Study of dc load lines for FET and identify the Q point.
- Analyse various types of FET bias circuits by using the device maximum and minimum transfer characteristics to determine voltage and current levels.
- Design of fixed bias, self bias and voltae-divider bias circuits.
- Study of coupling and bypass capacitors in FET circuits, and draw ac equivalent circuit for FET.
- Study of ac lod line for a basic FET circuit.
- Sketch ac equivalent circuits for various FET common-source, common-drain and common-gate circuits.
- Analyse various CS, CD and CG FET circuits to determine input resistance, output resistance and voltage gain.
- Compare the performance of CS, CD and CG circuits.

# 10 FET BIASING AND SMALL SIGNAL ANALYSIS

## FET BIASING

## 10.1 Introduction

- The input controlling variable for a BJT transistor is a current level, whereas for the FET a voltage is the controlling variable.
- Field-effect transistors used in amplifier circuits must be biased into an on-state, so widely differing maximum and minimum transfer characteristics for a FET can result in a wide range of drain current levels.
- For reasonable upper and lower drain current limits in a FET circuit, fixed bias, self bias and voltage divider networks must be employed.
- Since the graphical approach is in general the most popular, the analysis will have a graphical orientation rather than use of direct mathematical techniques.
- The general relationships that can be applied to the dc analysis of all FET amplifiers are:
  (i) $I_G \approx 0$
  (ii) $I_D = I_S$
  (iii) For JFET and depletion type MOSFET we use $I_D = I_{DSS}\left[1-\frac{V_{GS}}{V_P}\right]^2$
  (iv) For enhancement-type MOSFETS, we use
  $$I_D = k(V_{GS} - V_T)^2$$

## 10.2 DC-Load Line and Bias Point

Consider the FET circuit and characteristics shown in Figure (10.1) and (10.2) respectively.

- The dc load line for a FET circuit is drawn upon the output characteristics of the device in exactly the same way as for a BJT circuit.
- Refer to the n-channel FET circuit in Figure (10.1). The source terminal is grounded, the gate is biased via resistor $R_G$ to a negative voltage ($-V_G$) and the drain terminal is supplied from $+V_{DD}$ via resistor $R_D$.

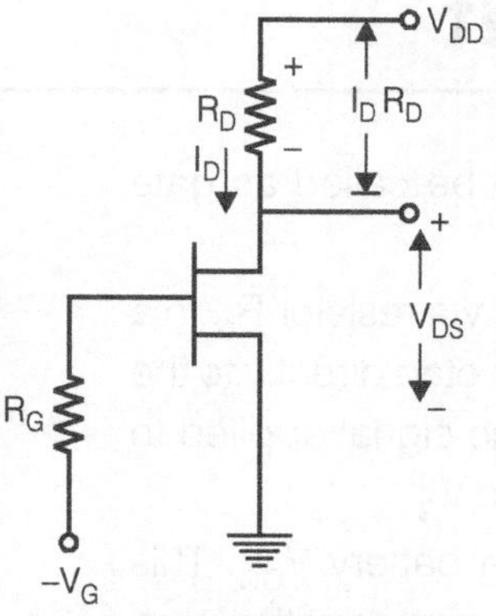


Figure 10.1 FET circuit

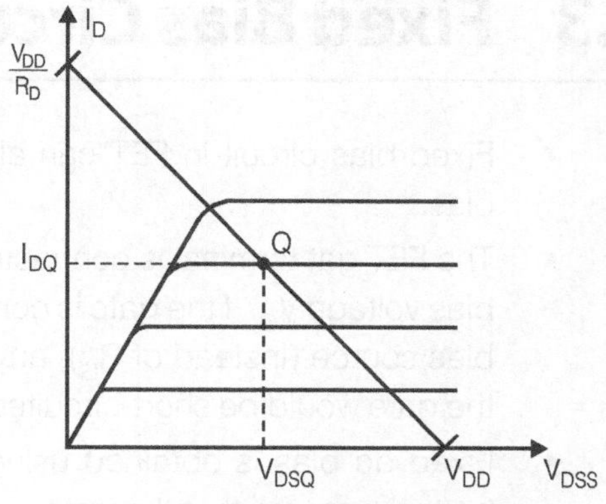


Figure 10.2 DC load line

- The DC load line for this circuit is a graph of drain current ($I_D$) versus drain-source voltage ($V_{DS}$) for a given value of drain resistance and a given supply voltage.
  The drain source voltage is given by
  $$V_{DS} = V_{DD} - I_D R_D$$
  Where $V_{DD}$ is the supply voltage and $I_D R_D$ is the voltage drop across $R_D$.
  Consider the following:
  $$\text{When} \quad I_D = 0, \quad V_{DS} = V_{DD}$$
  $$\text{When} \quad V_{DS} = 0, \quad I_D = \frac{V_{DD}}{R_D}$$
  By joining these two points, dc load line is obtained.

## OPERATING POINT Q

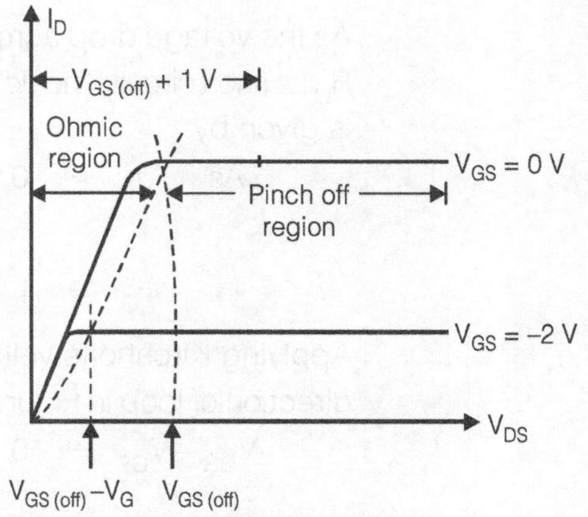


Figure 10.3 $V_{DS}$ versus $I_D$ graph

- Just like BJT circuit, the dc bias point, or quiescent point identifies the device current and terminal voltages when there is no ac input signal. When signal is applied to the gate, $I_D$ varies according to the instantaneous amplitude of the signal, producing a variation in $V_{DS}$.
- For a FET amplifier circuit, $V_{DS}$ must remain in the pinch-off region of the characteristics. That means it must not be allowed to fall below the gate-source cut-off voltage $V_{GS\,(off)}$, to avoid going into the channel ohmic region.
- Therefore, in a FET bias circuit, the drain-source voltage should always be a minimum of ($V_{GS\,(off)}$ + 1 V) as shown in above figure.
- When external bias voltage ($V_G$) is included, the pinch off voltage for that bias level on the device characteristics is ($V_{GS\,(off)} - V_G$).
  Therefore the minimum drain-source voltage may be
  $$\boxed{V_{DS(min)} = V_{GS(off)} - V_G + 1\,V}$$

# 10.3 Fixed Bias Circuit

- Fixed bias circuit in FET can also be called as gate bias.
- The FET gat terminal is connected via resistor $R_G$ to a bias voltage $V_G$. If the gate is connected directly to the bias source (instead of $R_G$), any ac signal applied to the gate would be short-circuited to $V_G$.
- Fixed dc bias is obtained using a battery $V_{GG}$. This battery ensures that the gate is always negative w.r.t. source and no current flows through resistor $R_G$ and gate terminal i.e. $I_G = 0$.
- The dc voltage drop across $R_G$ is equal to $I_G R_G$ i.e. 0 V.

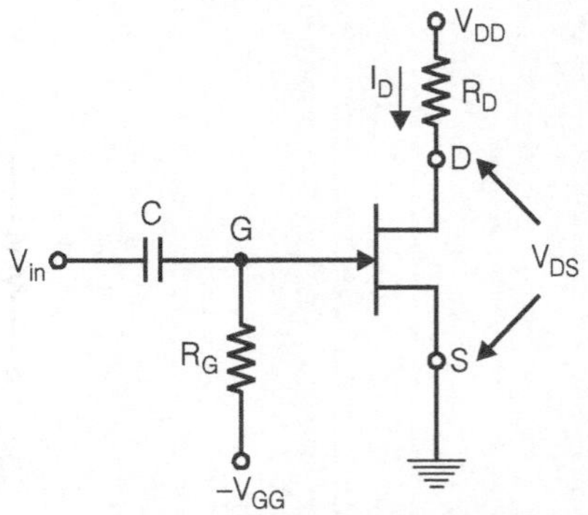

*Figure 10.4 Fixed bias circuit*

The gate-source voltage $V_{GS}$ is then

$$V_{GS} = -V_{GG}$$

The drain current causes a voltage drop across the drain resistor $R_D$ and is given as

$$V_{R_D} = I_D R_D$$

and output voltage

$$V_o = V_{DD} - I_D R_D$$

Since $V_{GG}$ is fixed value of dc supply and the magnitude of gate-to source voltage $V_{GS}$ is also fixed, hence this circuit is named as fixed bias circuit.

## Analysis

As the voltage drop across $R_G$ is 0 V. Assume $R_G$ as short circuit. The dc model of Figure (10.4) is given by

As $\quad I_G \approx 0$

$$V_{R_G} = I_G R_G = (0\text{ A}) R_G = 0\text{ V}$$

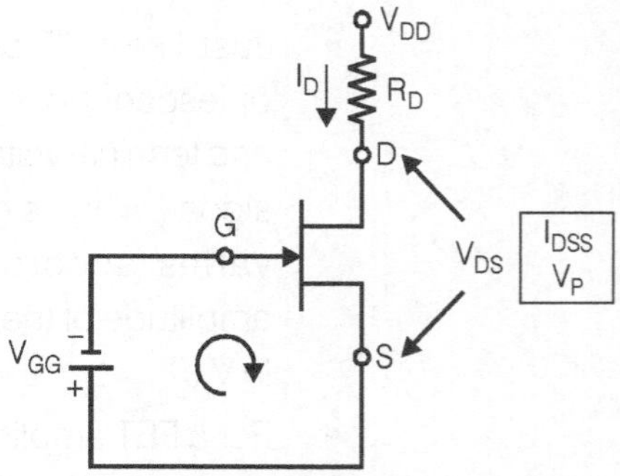

*Figure 10.5 DC analysis of fixed bias*

Applying Kirchhoffs voltage law in clockwise direction of loop in Figure (10.5) results in

$$-V_{GG} - V_{GS} = 0$$

$$\boxed{V_{GS} = -V_{GG}}$$

The resulting drain current $I_D$ is now calculated by Shockley's equation

$$I_D = I_{DSS}\left[1 - \frac{V_{GS}}{V_p}\right]^2$$

The drain-to-source voltage of the output section can be determined by applying Kirchhoffs voltage law.

$$+V_{DS} + I_D R_D - V_{DD} = 0$$

$$\boxed{V_{DS} = V_{DD} - I_D R_D}$$

$$\boxed{V_s = 0} \quad \text{as } R_s = 0$$

$$V_{DS} = V_D - V_S$$
$$V_D = V_{DS} + V_S = V_{DS} + 0\text{ V}$$
$$\boxed{V_D = V_{DS}}$$
$$V_{GS} = V_G - V_S$$
$$V_G = V_{GS} + V_S = V_{GS} + 0\text{ V}$$
$$\boxed{V_G = V_{GS}}$$

Since the configuration requires two dc supplies, its use is limited.

## TRANSFER CHARACTERISTICS

In Figure (10.6), the fixed level of $V_{GS}$ has been superimposed as a vertical line at $V_{GS} = -V_{GG}$ .At any point on the vertical line, the level of $V_{GS}$ is –VGG. The level of $I_D$ must simply be determined on this vertical line. The point where the two curves intersect is the common solution to the configuration-commonly referred as quiescent or operating point.

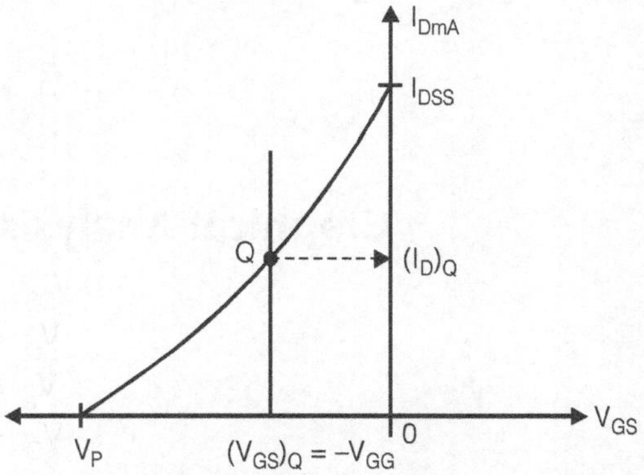


*Figure 10.6 Q point extraction for the fixed bias*

# 10.4 Self Bias

- The self bias circuit eliminates the requirement of two dc supplies.
- In this circuit, a resistor $R_s$ known as bias resistor. The dc component of drain current $I_D$ flowing through $R_s$ make a voltage drop across resistor $R_s$.

$$\therefore \quad V_s = I_D R_S$$

- The voltage drop across $R_s$ reduces the gate-to-source voltage required for FET operation. The resistor $R_s$, the feedback resistor prevents any variation in FET drain current.

$$V_{GS} = -I_D R_S$$

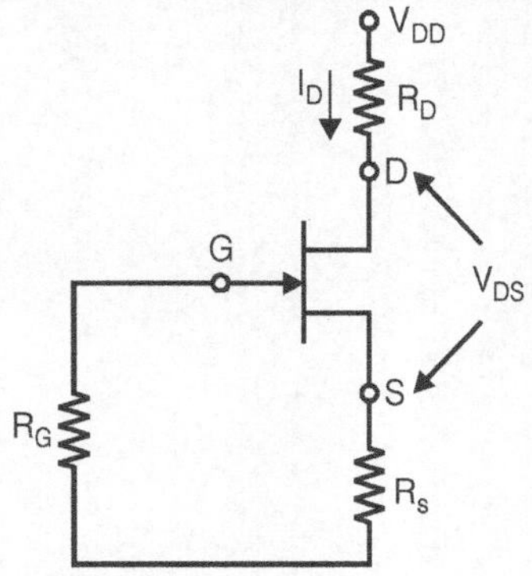


*Figure 10.7 Self bias circuit*

Since no gate current flows through the reverse bias gate-source, the gate current $I_G = 0$ therefore,

$$V_G = I_G R_G = 0\text{ V}$$

Self biasing of a FET stabilises its quiescent operating point against any change in its parameters like transconductance. Let the given FET be replaced by another FET having the double conductance then drain current will also try to be double but since any increase in voltage drop across $R_s$, therefore, gate source voltage, $V_{GS}$ becomes more negative and thus increase in drain current is reduced.

## Analysis

As the voltage drop across $R_G$ is 0 V. Assume $R_G$ as short circuit. The DC model of Figure (10.8) is given by

From the above figure,

$$V_{GS} = -I_D R_S$$

Assume $R_s = 1\,k\Omega$

For $V_{GS} = -1\,V$, $R_S = 1\,k\Omega$ } $I_D = 1\,mA$

For $V_{GS} = -2\,V$, $R_S = 1\,k\Omega$ } $I_D = 2\,mA$

For $V_{GS} = -3\,V$, $R_S = 1\,k\Omega$ } $I_D = 3\,mA$

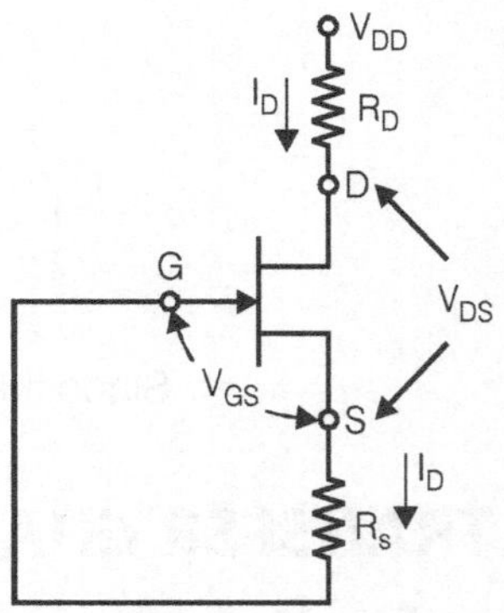


*Figure 10.8 DC model for self bias*

## Graphical Analysis

$$V_{DS} = V_{DD} - I_D (R_D + R_S)$$
$$V_D = V_{DD} - I_D R_D$$
$$V_s = I_D R_s$$
$$V_G = V_{GS} + I_D R_S$$

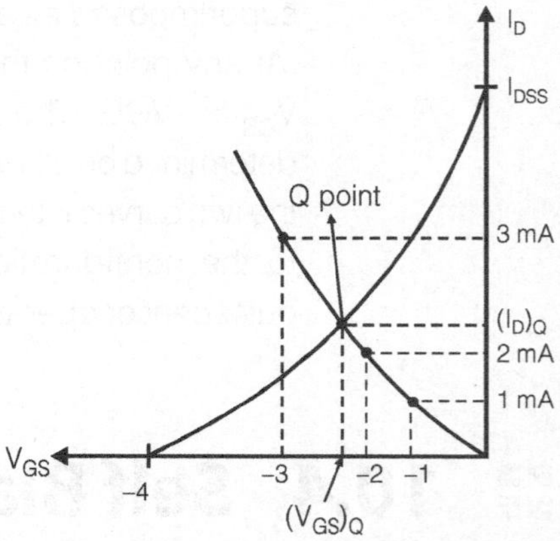


# 10.5 Voltage Divider Bias

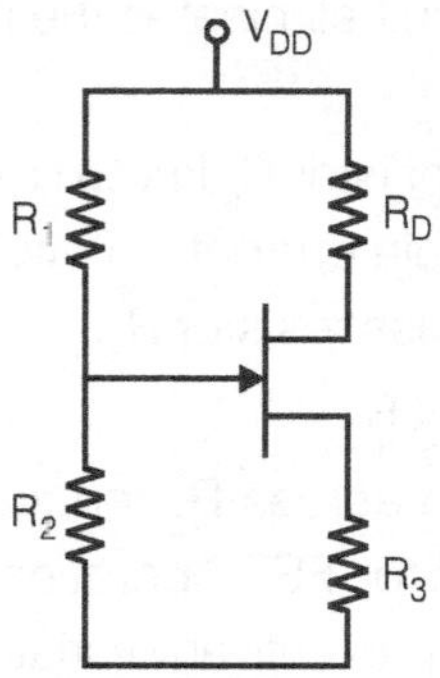


*Figure 10.9 Voltage divider bias for n-channel JFET*

For self bias circuit, it was seen that increasing the resistance of $R_s$ brings $I_{D\,(max)}$ and $I_{D\,(min)}$ closer together, but that increased $R_s$ values result in lower $I_D$ levels.

- Voltage divider bias allows $R_s$ to be increased without making $I_D$ very small. Figure (10.9) shows that a voltage-divider bias circuit combines the use of a source resistor ($R_s$) with a gate bias voltage ($V_G$).
- The gate bias voltage is derived from the supply by means of the voltage divider ($R_1$ and $R_2$) and the source voltage ($V_s$) depends upon $R_s$ and $I_D$.

- The gate terminal and the source terminal are both positive with respect to ground, and the gate-source voltage ($V_{GS}$) is the difference between $V_G$ and $V_s$.
- Because $V_s$ is larger than $V_G$, the FET source terminal is at a higher level than the gate. So the gate is negative with respect to the source and the gate-channel junctions are reverse biased.
- As in the case of a self biased circuit, the voltage drop across $R_s$ increases if $I_D$ increases. This produces an increase in $-V_{GS}$, which drives $I_D$ back toward its original level. A decrease in $I_D$ causes a reduction in $-V_{GS}$ which tends to increase $I_D$.

## Analysis

From the above figure

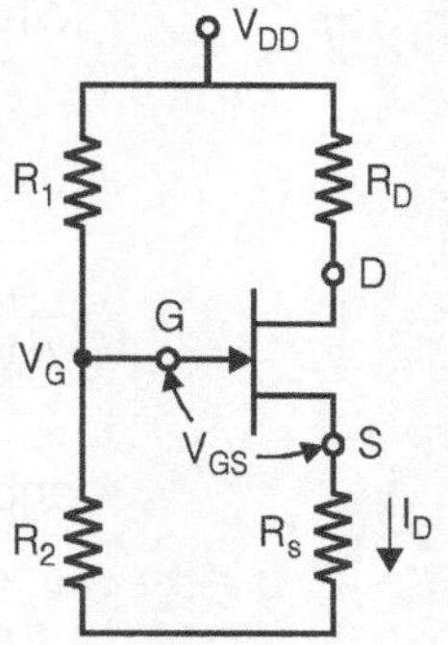


Gate voltage $V_G = \dfrac{V_{DD} \times R_2}{R_1 + R_2}$

$$V_G = V_{GS} + I_D R_S$$

$$\boxed{V_{GS} = V_G - I_D R_s}$$

From the above equation

If $V_{GS} = 0\text{ V}$    If $I_D = 0$

$I_D = \dfrac{V_G}{R_S}$    $V_{GS} = V_G$

## Graphical Analysis

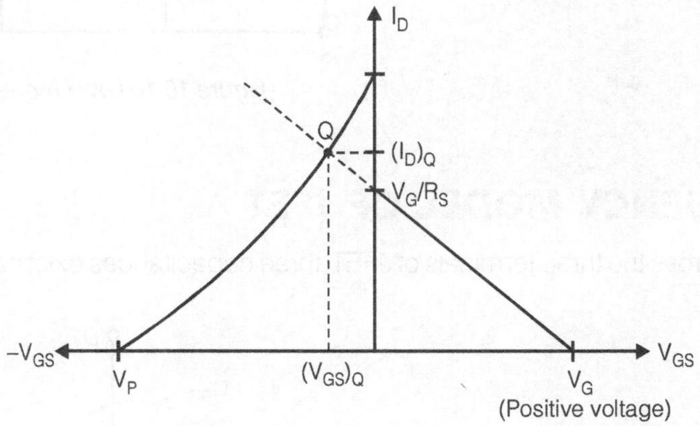


$$V_{DS} = V_{DD} - (R_D + R_S)$$
$$V_D = V_{DD} - I_D R_D$$
$$V_S = I_D R_S$$
$$V_G = V_{GS} + I_D R_S$$

# SMALL SIGNAL ANALYSIS OF FET

## 10.6 Introduction

### LOW FREQUENCY MODEL OF JFET

We have $I_D = f(V_{GS}, V_{DS})$

Where f denotes function

$$I_D = \frac{\Delta I_D}{\Delta V_{GS}} V_{GS} + \frac{\Delta I_D}{\Delta V_{DS}} V_{DS}$$

$\frac{\Delta I_D}{\Delta V_{GS}}$ can be replaced by $g_m$

and $\frac{\Delta I_D}{\Delta V_{DS}}$ can be replaced by $\frac{1}{r_d}$

$$\therefore\ I_D = g_m V_{GS} + \frac{V_{DS}}{r_d}$$

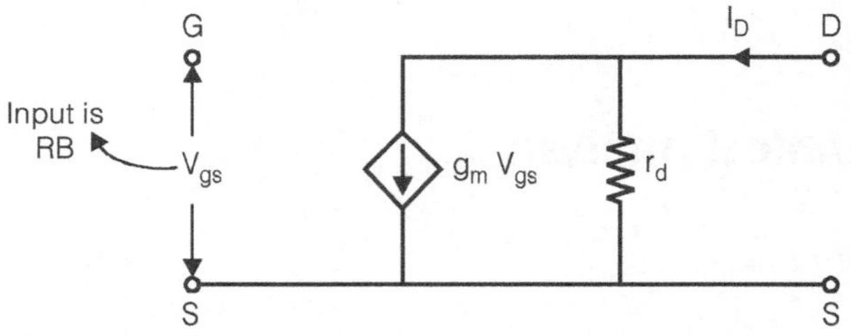


*Figure 10.10 Low Frequency model*

### HIGH FREQUENCY MODEL OF JFET

Across the three terminals of JFET, three capacitances exist called as interelectrode capacitances.

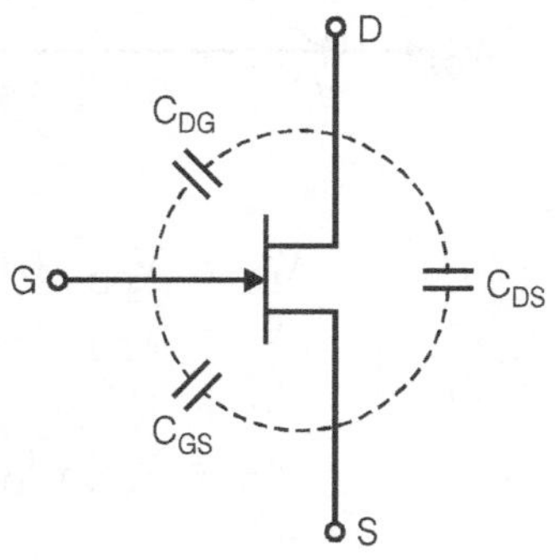


$C_{GS}$, $C_{DG}$, $C_{DS}$ are called interelectrode capacitances in JFET.

In high frequency analysis, these capacitors are included in the model. Therefore the model is represented as

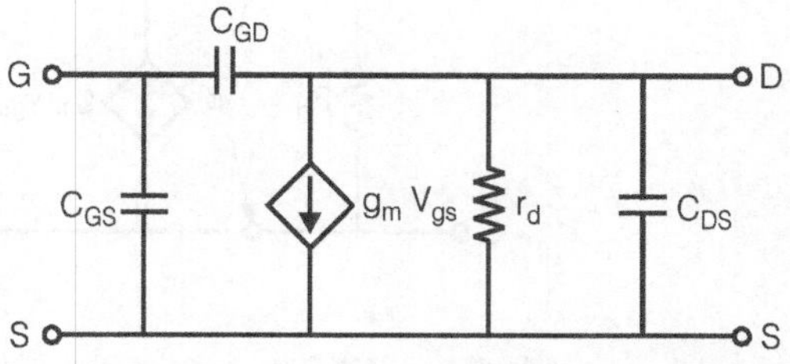


*Figure 10.11 High Frequency Model of JFET*

## Conclusion

For amplifier analysis, the JFET should be concentrated in the mid band range. In mid band range, the inter electrode capacitances are open, therefore high frequency model will be converted as low frequency model for amplifier analysis.

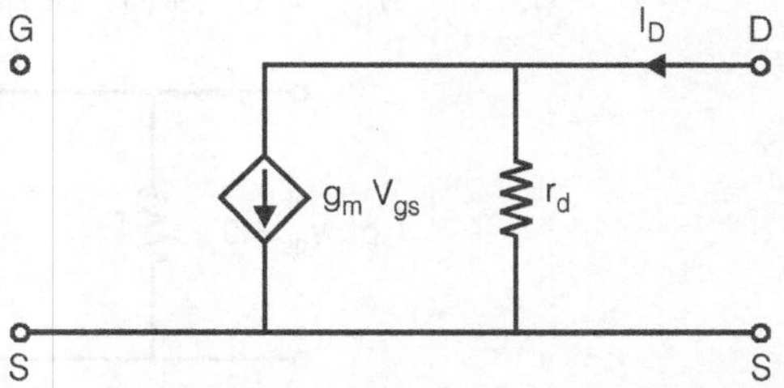


*Figure 10.12 JFET Amplifier model where $C_{GS}$, $C_{GD}$, $C_{DS}$ are open*

# 10.7 Common Source Bypass JFET Amplifier

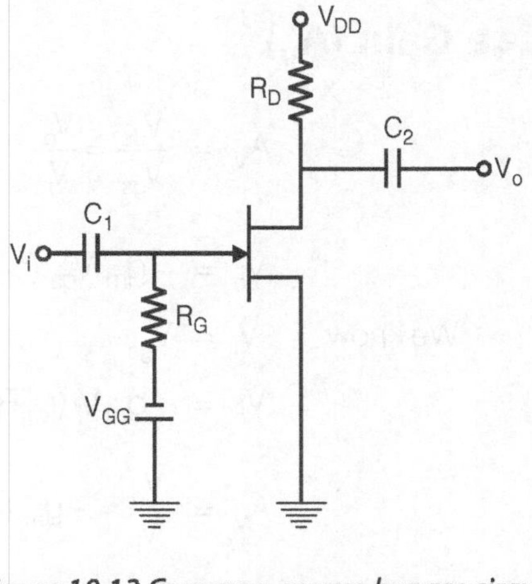


*Figure 10.13 Common source bypass circuit*

The ac equivalent model for the Figure (10.13) is

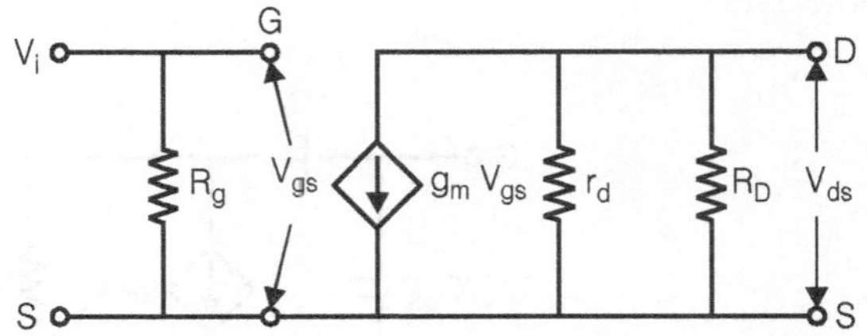


## (i) Input Impedance ($Z_i$)

$$Z_i = R_g$$

## (ii) Output Impedance ($Z_o$)

The output impedance $Z_o$ is the impedance measured looking from the output side with input voltage $V_i = 0$.

i.e. $V_{gs} = 0$ and hence $g_m V_{gs} = 0$

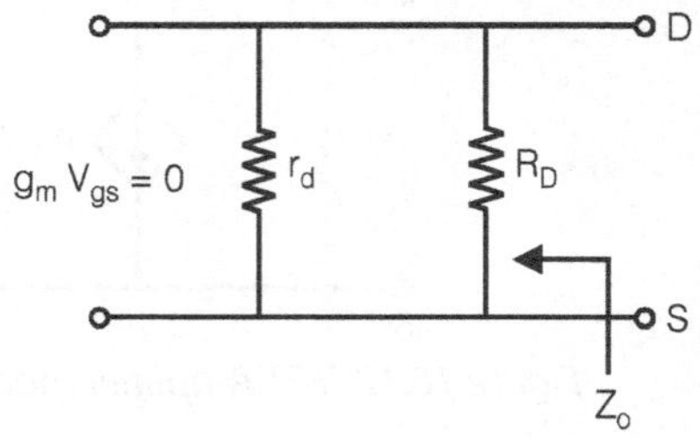


$$Z_o = R_D \| r_d$$

If $r_d >> R_D$

$$Z_o \approx R_D$$

## (iii) Voltage Gain ($A_V$)

$$A_V = \frac{V_{ds}}{V_{gs}} = \frac{V_o}{V_i}$$

$$V_o = -g_m V_{gs} (r_d \| R_D)$$

We know $V_i = V_{gs}$

$$V_o = -g_m V_i (r_d \| R_D)$$

$$A_V = \frac{V_o}{V_i} = -g_m (r_d \| R_D)$$

If $r_d >> R_D$

$$A_V \cong -g_m R_D$$

# 10.8 Common Source Unbypass JFET Amplifier

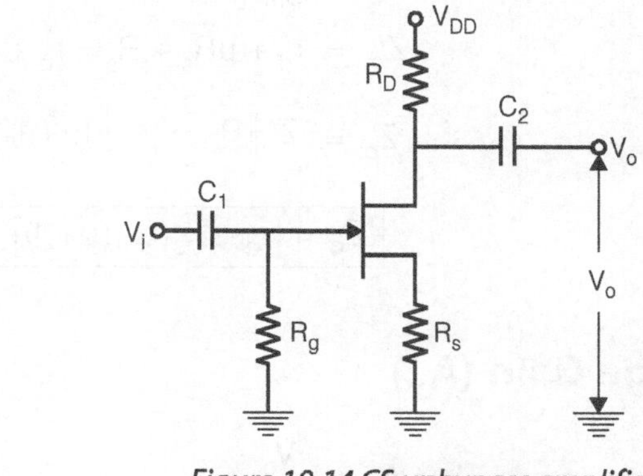


*Figure 10.14 CS unbypass amplifier*

The ac equivalent model for Figure (10.14) is given by

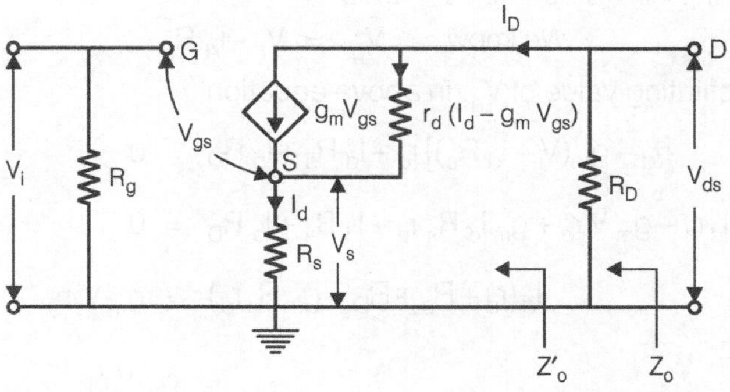


## (i) Input Impedance ($Z_i$)

$$Z_i = R_g$$

## (ii) Output Impedance ($Z_o$)

$$Z_o = Z'_o \| R_D$$

Where $$Z'_o = \left.\frac{V_o}{I_d}\right|_{V_i=0}$$

$$V_o = (I_d - g_m\ V_{gs})\ r_d + I_d\ R_s$$

$$V_{gs} = V_i - I_d\ R_s$$

If $$V_i = 0$$

$$V_{gs} = -I_d\ R_s$$

Substitute $V_{gs}$ value in $V_o$ equation

$$V_o = \{I_d - g_m(-I_d\ R_s)\}\ r_d + I_d\ R_s$$

$$= I_d\ (r_d + g_m\ R_s\ r_d + R_s)$$

$$Z'_o = \frac{V_o}{I_d} = r_d + g_m R_s r_d + R_s$$

$$\mu = g_m r_d \quad \text{(Amplification factor)}$$

$$Z'_o = r_d + \mu R_s + R_s = r_d + R_s(\mu + 1)$$

$$Z_o = Z'_o \| R_D = \{r_d + (\mu + 1)R_s\} \| R_D$$

$$\boxed{Z_o = \{r_d + R_s(g_m r_d + 1)\} \| R_D}$$

## (iii) Voltage Gain ($A_V$)

$$A_V = \frac{V_o}{V_i}$$

$$V_o = -I_d R_D$$

$$(I_d - g_m V_{gs}) r_d + I_d R_s + I_d R_D = 0$$

We know $V_{gs} = V_i - I_d R_s$

Substituting value of $V_{gs}$ in above equation

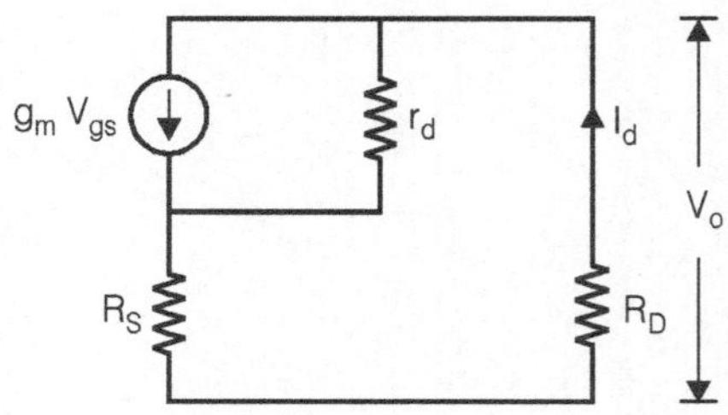


$$\{I_d - g_m(V_i - I_d R_s)\} r_d + I_d R_s + I_d R_D = 0$$

$$I_d r_d - g_m V_i r_d + g_m I_d R_s r_d + I_d R_s + I_d R_D = 0$$

$$I_d(r_d + R_s + R_D + g_m R_s r_d) = g_m V_i r_d$$

$$I_d = \frac{g_m V_i r_d}{r_d + R_s + R_D + g_m R_s r_d}$$

Substitute $I_d$ value in the equation

$$V_o = -I_d R_D$$

$$V_o = \frac{-g_m V_i r_d R_D}{r_d - R_s + R_D + g_m R_s r_d}$$

$$A_V = \frac{V_o}{V_i} = \frac{-g_m r_d R_D}{r_d + R_s + R_D + g_m R_s r_d}$$

$$A_V = \frac{V_o}{V_i} = \frac{-g_m R_D}{1 + g_m R_s + \dfrac{R_s + R_D}{r_d}}$$

If $r_d >> R_S + R_D$

$$\boxed{A_V = \frac{V_o}{V_i} = \frac{-g_m R_D}{1 + g_m R_s}}$$

# 10.9 Common Drain JFET Amplifier

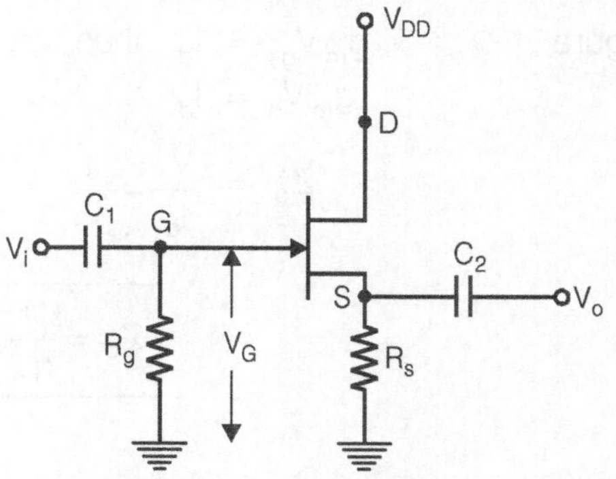


*Figure 10.15 Common Drain amplifier*

The ac equivalent model for Figure (10.15) is shown below:

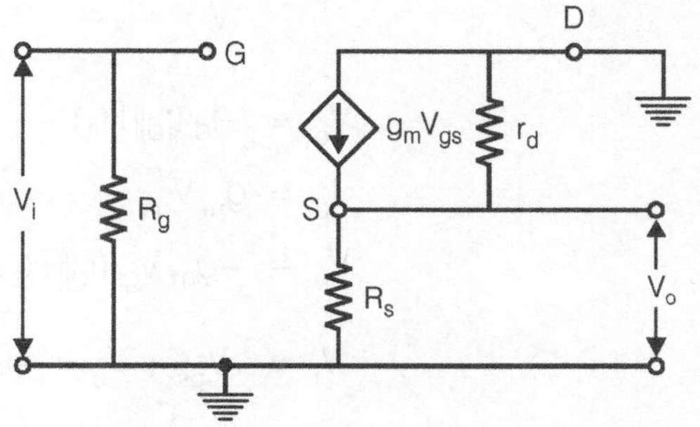


The re modified ac model is given by

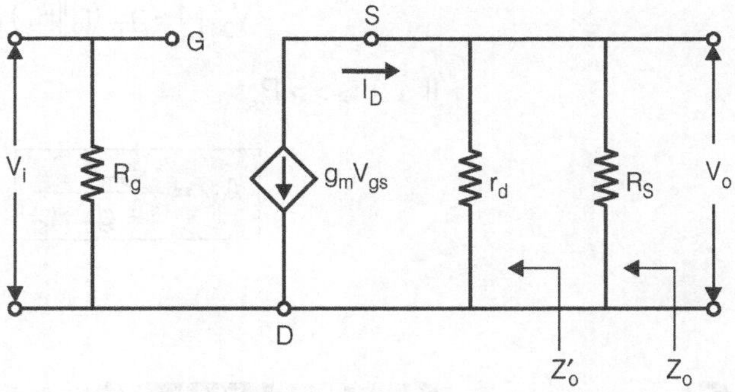


## (i) Input Impedance ($Z_i$)

$$\boxed{Z_i = R_g}$$

## (ii) Output Impedance ($Z_o$)

$$Z_o = Z'_o \| R_s$$

Where $$Z'_o = \left.\frac{V_o}{I_d}\right|_{V_i=0}$$

$$V_i + V_{gs} - V_o = 0$$

When $V_i = 0$

$$V_{gs} = V_o$$

From the figure $g_m V_{gs} = I_d$ then

$$g_m V_o = I_d$$

$$Z'_o = \frac{V_o}{I_d} = \frac{1}{g_m}$$

$$\boxed{Z_o = \frac{1}{g_m} \| R_s}$$

### (iii) Voltage Gain ($A_V$)

$$A_V = \frac{V_o}{V_i}$$

$$V_o = -I_d\,(r_d \| R_s)$$

$$I_d = g_m V_{gs}$$

$$V_o = -g_m V_{gs}(r_d \| R_s)$$

$$V_i = -V_{gs} + V_o = -V_{gs} + \left[-g_m V_{gs}(r_d \| R_s)\right]$$

$$A_V = \frac{-g_m V_{gs}(r_d \| R_s)}{-V_{gs}\left[1+g_m (r_d \| R_s)\right]} = \frac{g_m(r_d \| R_s)}{1+g_m(r_d \| R_s)}$$

If $r_d >> R_s$

$$\boxed{A_V = \frac{g_m R_s}{1+g_m R_s}}$$

## 10.10 Common Gate JFET Amplifier

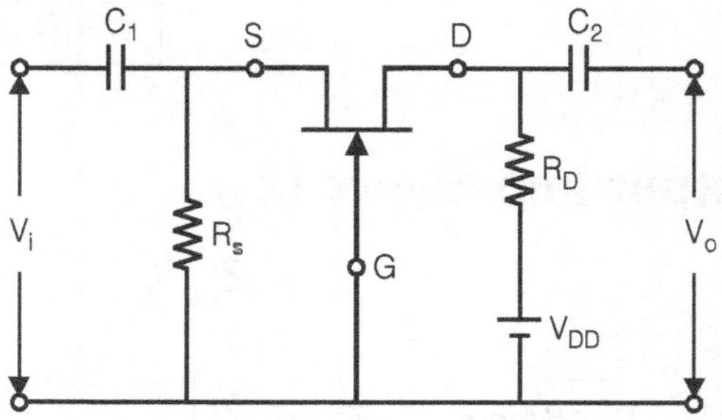


*Figure 10.16 CG JFET amplifier*

The ac equivalent model for the Figure (10.16) is shown below:

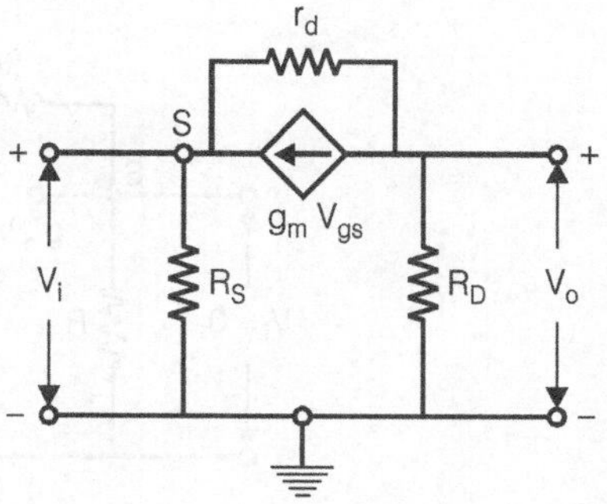


The remodified ac model is given by

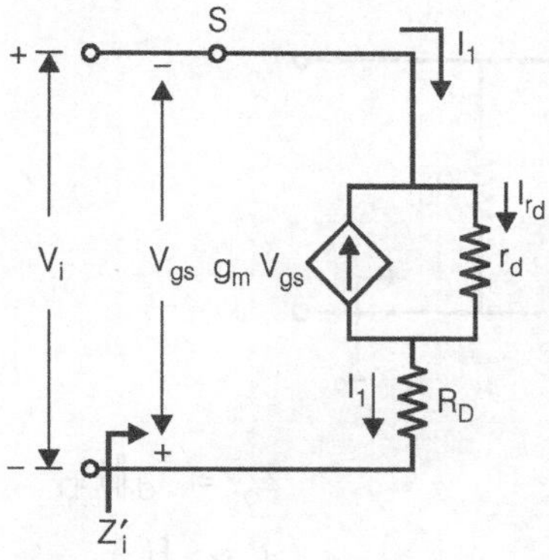


## (i) Input Impedance ($Z_i$)

$$Z_i = R_s \| Z_i'$$

$$I_1 = I_{r_d} - g_m V_{gs}$$

$$I_{r_d} = \frac{V_{r_d}}{r_d}$$

$$V_{r_d} = -V_{gs} - I_1 R_D$$

$$I_1 = \frac{-V_{gs} - I_1 R_D}{r_d} - g_m V_{gs} = -\frac{V_{gs}}{r_d} - \frac{I_1 R_D}{r_d} - g_m V_{gs} = -\left[\frac{1}{r_d} + g_m\right] V_{gs} - I_1 \frac{R_D}{r_d}$$

$$I_1\left[1 + \frac{R_D}{r_d}\right] = -\left[\frac{1}{r_d} + g_m\right] V_{gs}$$

$$Z_i' = \frac{V_i}{I_1} = -\frac{V_{gs}}{I_1}$$

$$Z_i' = \frac{\left[1 + \frac{R_D}{r_d}\right]}{\left[g_m + \frac{1}{r_d}\right]} = \frac{r_d + R_D}{1 + g_m r_d} \quad ; \quad Z_i = Z_i' \| R_s = R_s \| \frac{[r_d + R_D]}{1 + g_m r_d}$$

For $r_d >> R_D$

$$\boxed{Z_i = R_s \| [1/g_m]}$$

## (ii) Output Impedance ($Z_o$)

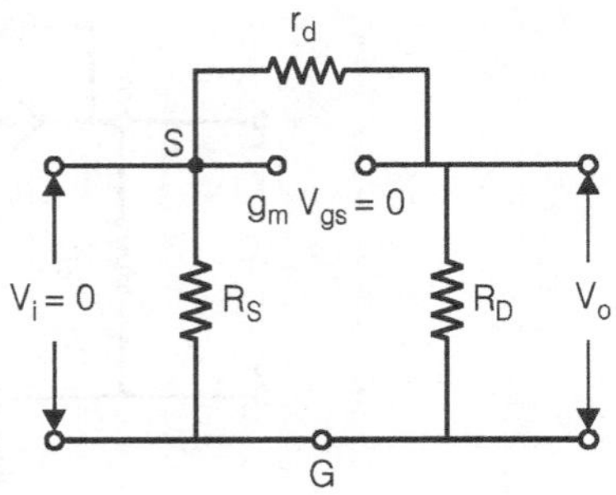


At output side the equivalent circuit is

$\Rightarrow$

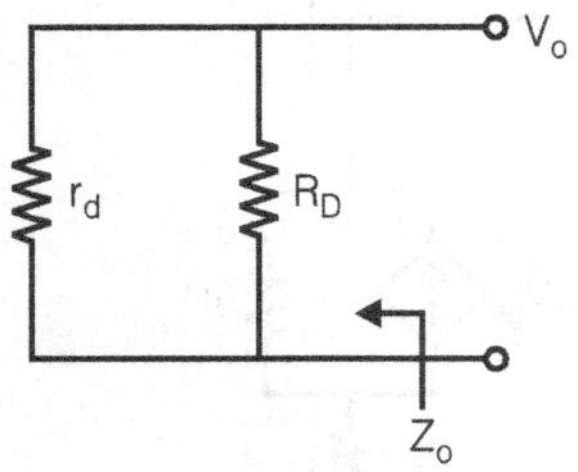


$$Z_o = r_d \| R_D$$

$$r_d >> R_D$$

$$Z_o \approx R_D$$

## (iii) Voltage Gain ($A_V$)

$$A_V = \frac{V_o}{V_i}$$

From figure

$$V_i = -V_{gs}$$

$$V_o = I_d R_D$$

From the figure

$$I_d + g_m V_{gs} + I_{r_d} = 0$$

$$I_d = -g_m V_{gs} - I_{r_d}$$

$$I_{r_d} = \frac{V_o - V_i}{r_d}$$

$$= \frac{V_o + V_{gs}}{r_d}$$

$$I_d = -g_m V_{gs} - \left[\frac{V_o + V_{gs}}{r_d}\right]$$

$$= -g_m V_{gs} - \frac{V_o}{r_d} - \frac{V_{gs}}{r_d}$$

$$\therefore \quad V_o = -g_m V_{gs} R_D - V_o \frac{R_D}{r_d} - V_{gs} \frac{R_D}{r_d}$$

$$\left[1+\frac{R_D}{r_d}\right] V_o = -\left[g_m R_D + \frac{R_D}{r_d}\right] V_{gs}$$

$$A_V = \frac{V_o}{V_i} = \frac{V_o}{-V_{gs}}$$

$$A_V = \frac{\left[g_m R_D + \frac{R_D}{r_d}\right]}{\left[1+\frac{R_D}{r_d}\right]}$$

$$A_V = \frac{g_m r_d R_D + R_D}{r_d + R_D}$$

$$r_d >> R_D$$

$$A_V = \frac{g_m r_d R_D}{r_d}$$

$$\boxed{A_v \approx g_m R_D}$$

## 10.11 Comparison Between CS Bypass, CS Unbypass, CD, CG Amplifiers

### CD, CG, Amplifiers

| Characteristic | $Z_i$ | $Z_o$ | $A_V$ |
|---|---|---|---|
| CS bypass | $R_g$ | $r_d \| R_D$ | $-g_m (r_d \| R_D)$ |
| CS unbypass | $R_g$ | $\{r_d + (1+\mu)R_s\} \| R_D$ | $-\frac{g_m R_D}{1+g_m R_s}$ |
| CD | $R_g$ | $R_s \| \frac{1}{g_m}$ | $\frac{g_m R_s}{1+g_m R_s}$ |
| CG | $R_s \| \frac{1}{g_m}$ | $r_d \| R_D$ | $g_m (r_d \| R_D)$ |

## SUMMARY

- Field effect transistor is a unipolar and voltage control of device.
- The terminals drain, gate and source of a FET are identical to collector, base and emitter of a BJT.
- Since the input junction is reverse biased in JFET, the current drawn is very small and so $Z_i$ is very high.
- The FET can be used as amplifier in CS or CD configurations CS is similar to CE and CD is similar to CC configurations in a BJT.
- There are various biasing circuits in FET like fixed bias, self bias and voltage divider bias.
- When FET is operated below pinch off voltage it acts as a voltage variable resistor.

## GATE AND IES QUESTIONS

1. The valve of $V_{GS}$ and $V_G$ for the circuit shown in figure is

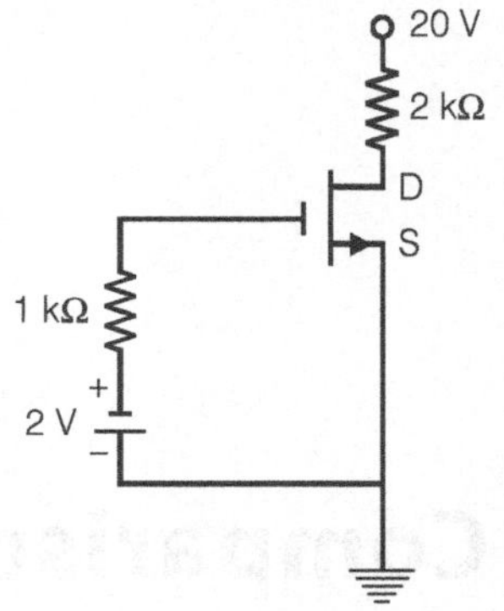


(a) –2 V, –2 V

(b) 2 V, –2 V

(c) 2 V, –2 V

(d) –2 V, 2 V

**Exp. (a)**

It is a fixed bias circuit

$$(V_{GS})_Q = -V_{GG} = -2\text{ V}$$
$$V_G = V_{GS} + V_s \qquad \text{(Where } V_s = 0\text{)}$$
$$V_{GS} = V_G = -2\text{ V}$$

2. Consider an N channel MOSFET with parameters $k_n = 0.25\text{ mA/V}^2$, $V_{Th} = 1_r$, $\gamma = 0$, $C_{gd} = 0.04$ pF and $C_{gs} = 0.2$ pF. The transistor is biased at $V_{GN} = 3$ V. The unity gain band width of an FET is

(a) 500 MHz
(b) 350 MHz
(c) 332 MHz
(d) 663 MHz

**Exp. (d)**

The transconductance is

$$g_m = 2k_n(V_{GN} - V_{TN}) = 1\text{ mA/V}$$

Unity gain bandwidth is

$$f_T = \frac{g_m}{2\pi(C_{gs} + C_{gd})} = 663\text{ MHz}$$

3. Given for an FET, $g_m = 95$ mA/V, total capacitance = 500 pF. For a voltage gain of –30. The bandwidth will be

(a) 19 MHz
(b) 6.3 MHz
(c) 100 kHz
(d) 3 MHz

Exp. (b)

$$\frac{g_m}{C} = A_V \times BW$$

$$\frac{95\times10^{-3}}{500\times10^{-12}} = 30 \times BW$$

$$BW = 6.3 \text{ MHz}$$

Common Data Q.4 and Q.5

Given parameters

$C_G = 0.01\ \mu F, C_c = 0.5\ \mu F$

$C_S = 2\ \mu F, R_{si} = 10\ k\Omega,$

$R_G = 1\ m\Omega, R_o = 4.7\ k\Omega$

$R_S = 1\ k\Omega, R_L = 2.2\ k\Omega,$

$I_{DSS} = 8\ mA, V_{GS} = -2\ V$

$V_P = -4\ V, r_d = \infty, A_v = -3$

$C_{gd} = 2\ pF, C_{gs} = 4\ pF, C_{ds} = 0$

$C_{\omega i} = 5\ pF, C_{\omega o} = 6\ pF$

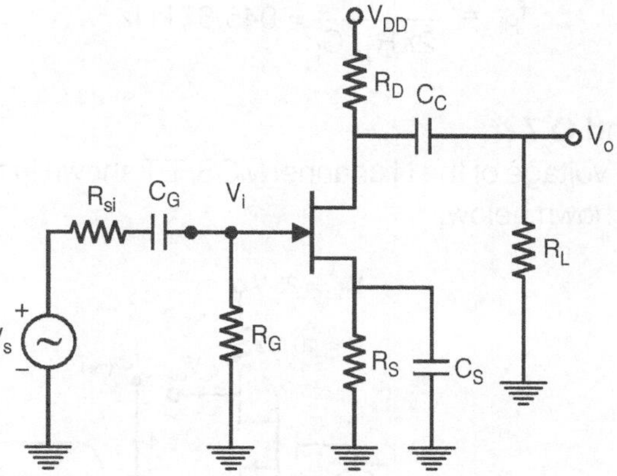


4. Find the lower cut-off frequency due to $C_s$ only
   (a) 183.7 Hz
   (b) 233.5 Hz
   (c) 238.7 Hz
   (d) 188.5 Hz

Exp. (c)

$$f_{LS} = \frac{1}{2\pi R_{eq} C_S}$$

$$R_{eq} = R_S \| 1/g_m$$

$$g_m = g_{mo}\left(1 - \frac{V_{GS}}{V_P}\right)$$

$$g_{mo} = 2\frac{I_{DSS}}{V_P} = \frac{2\,(8\text{ mA})}{4} = 4\text{ mS}$$

$$R_{eq} = R_S \| 1/g_m = 1\text{k}\Omega \,\Big\|\, \frac{1}{2\text{ mS}} = 333.33\ \Omega$$

$$f_{LS} = \frac{1}{2\pi(333.33)2\times10^{-6}} = 238.73\text{ Hz}$$

5. The higher cut-off frequency due to input capacitors only
   (a) 949.5 Hz
   (b) 945.6 Hz
   (c) 983.3 Hz
   (d) 987.5 Hz

Exp. (b)

$$f_{Hi} = \frac{1}{2\pi R_{Th} C_i}$$

$$R_{Th} = R_{si} \| R_G$$

$$C_i = C_{wi} + C_{gs} + C_{mi}$$
$$C_{mi} = \text{Miller capacitance at input} = (1 - A_V)\, C_{gd}$$
$$A_V = -3$$
$$C_i = C_{wi} + C_{be} + (1 - A_V)\, C_{bc}$$
$$C_i = 5\text{ pF} + 4\text{ pF} + (1 + 3)\, 2\text{ pF} = 17\text{ pF}$$
$$R_{Th} = 10\,\text{k}\Omega \,\|\, 1\,\text{M}\Omega = 9.9\,\text{k}\Omega$$
$$f_{Hi} = \frac{1}{2\pi R_{Th} C_i} = 945.67\text{ kHz}$$

**Common Data Q.6 and Q.7**

Assume that threshold voltage of the N channel MOSFET shown in figure is +0.75 V. The output characteristic of the MOSFET are also shown below:

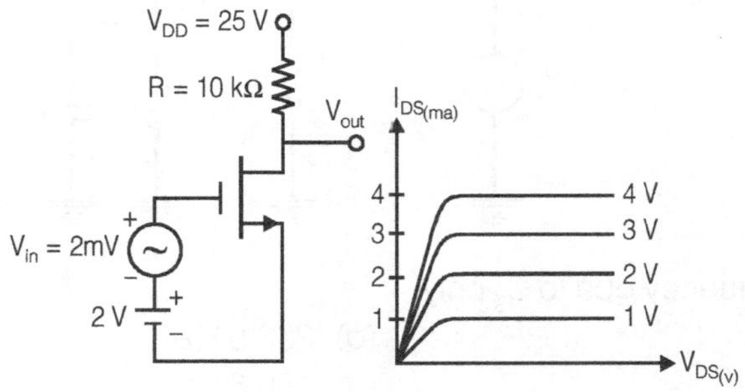


6. The transcondutance of the MOSFET is
   (a) 0.75 mS (b) 1 mS
   (c) 2 mS (d) 10 mS

**Exp. (b)**

$$g_m = \frac{\Delta I_{DS}}{\Delta V_{GS}} = \frac{(4-3)\text{ mA}}{(4-3)\text{ V}} = 1\text{mS}$$

7. The voltage of the amplifier is
   (a) +5 (b) –7.5
   (c) +10 (d) –10

**Exp. (d)**

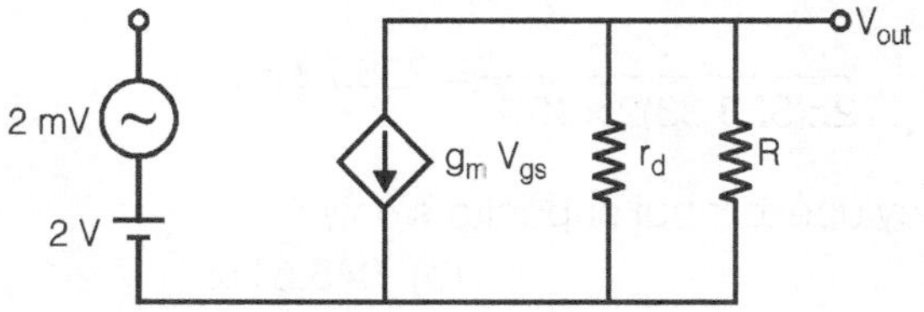


$$r_d >> R$$
$$V_{out} = -g_m V_{gs} R = -1 \times 10^{-3} \times 2 \times 10^{-3} \times 10 \times 10^{-3} = -20\text{ mV}$$
$$\text{Voltage gain} = \frac{V_{out}}{V_{in}} = \frac{-20\text{ mV}}{2\text{ mV}} = -10$$

**8.** An n channel JFET has $I_{DSS}$ = 1 mA and $V_p$ = 5 V. It maximum transconductance is

(a) 0.4 mS (b) 10 mS
(c) 15 mS (d) 5 mS

**Exp. (a)**

$$I_{DSS} = 1 \text{ mA}, V_p = 5\text{ V}$$

Maximum transconductance

$$g_{mo} = \frac{2I_{DSS}}{V_p} = \frac{2 \times 1\text{mA}}{5} = 0.4 \text{ mS}$$

**9.** For the circuit shown in the following figure, transistors $M_1$ and $M_2$ are identical NMOS transistors. Assume that $M_2$ is in saturation and output is unloaded.

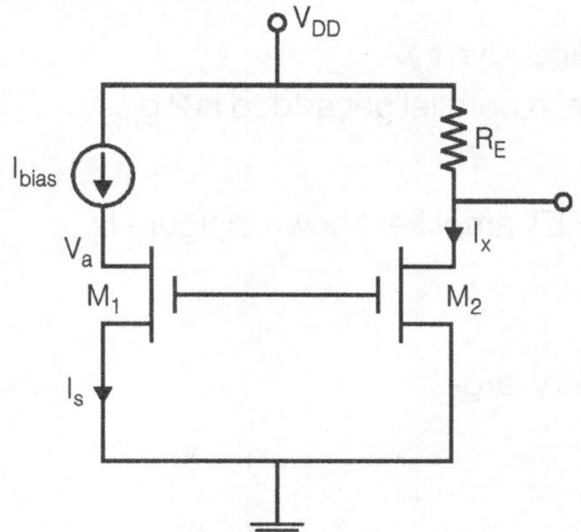


(a) $I_x = I_{bias} + I_s$

(b) $I_x = I_{bias}$

(c) $I_x = I_{bias} - I_s$

(d) $I_x = I_{bias} - \left(V_{DD} - \frac{V_{out}}{R_E}\right)$

**Exp. (b)**

It is current mirror.

**10.** The drain current of a MOSFET in saturation is given by $I_D = k(V_{GS} - V_T)^2$ where k is a constant. The magnitude of the tranconductance $g_m$ is

(a) $\frac{k(V_{GS} - V_T)^2}{V_{DS}}$ (b) $2k(V_{GS} - V_T)$

(c) $\frac{I_D}{V_{GS} - V_{DS}}$ (d) $\frac{k(V_{GS} - V_T)^2}{V_{GS}}$

**Exp. (b)**

$$g_m = \left.\frac{\partial I_d}{\partial V_{GS}}\right|_{V_{DS}} = \frac{\partial k(V_{GS} - V_T)^2}{\partial V_{GS}} = 2k(V_{GS} - V_T)$$

**11.** In the MOSFET amplifier of figure, the signal output $V_1$ and $V_2$ obey the relationship

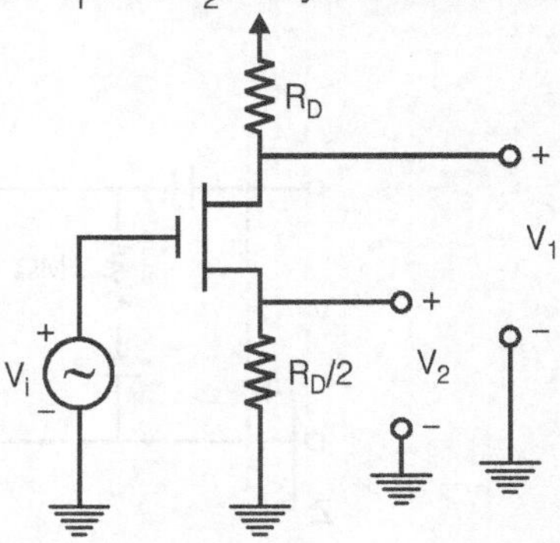


(a) $V_1 = \frac{V_2}{2}$

(b) $V_1 = -\frac{V_2}{2}$

(c) $V_1 = 2V_2$

(d) $V_1 = -2V_2$

**Exp. (d)**

$$\frac{V_1}{V_2} = \frac{-R_D}{R_D/2} = -2$$

$$V_1 = -2V_2$$

**12.** Two identical FETS, each characterized by the parameters $g_m$ and $r_d$ are connected in parallel. The composite FET is then characterized by the parameters

(a) $\frac{g_m}{2}$ and $2r_d$
(b) $\frac{g_m}{2}$ and $2\frac{r_d}{2}$
(c) $2g_m$ and $\frac{r_d}{2}$
(d) $2g_m$ and $2r_d$

**Exp. (c)**

Two identical resistance in parallel give $r_d/2$.
Two identical transconductance in parallel get added is $2g_m$.

**13.** The voltage gain $A_V = \frac{V_o}{V_i}$ of JFET amplifier shown in figure is

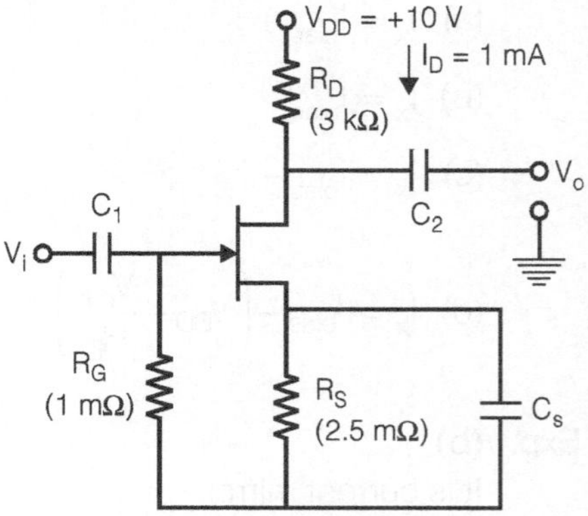


$I_{DSS}$ = 10 mA, $V_p$ = –5 V
(Assume $C_1$, $C_2$ and $C_S$ to be very large)

(a) +18
(b) +6
(c) –18
(d) –6

**Exp. (d)**

$$g_m = \frac{2I_{DSS}}{V_p}\left(1-\frac{V_{GS}}{V_p}\right)$$

$$V_G = 0,\quad V_S = I_D R_S = 1\text{ mA} \times 2.5\text{ k} = 2.5\text{ V}$$

$$V_{GS} = V_G - V_S = 0 - 2.5\text{ V} = -2.5\text{ V}$$

$$g_m = \frac{2\times 10\text{ mA}}{5}\left(1-\frac{(-2.5)}{(-5)}\right) = 2\text{ mS}$$

$$A_V = -g_m R_D = -2\text{ mS} \times 3\text{ k} = -6$$

**Common Data Q.14 and Q.15**

Given $r_d$ = 20 kΩ, $I_{DSS}$ = 10 mA, $V_p$ = –8 V

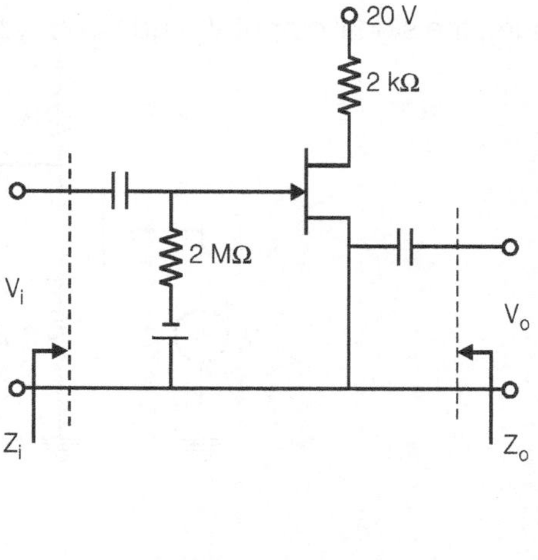

14. $Z_i$ and $Z_o$ of the circuit are respectively
(a) 2 mΩ and 2 kΩ
(b) infinity and 2 MΩ
(c) 2 MΩ and 20/11 kΩ
(d) infinity and 20/11 kΩ

Exp. (c)

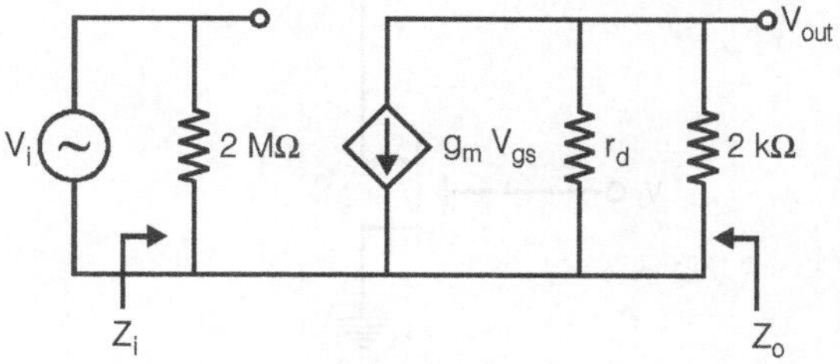


$$Z_i = 2\ \text{M}\Omega$$

$$Z_o = r_d \parallel R_D = 20\ \text{k}\Omega \parallel 2\ \text{k}\Omega = \frac{20}{11}\ \text{k}\Omega$$

15. $I_D$ and $V_{DS}$ under DC conditions are respectively
(a) 5.625 mA and 8.75 V
(b) 1.875 mA and 5 V
(c) 4.5 mA and 11 V
(d) 6.25 mA and 7.5 V

Exp. (a)

DC condition

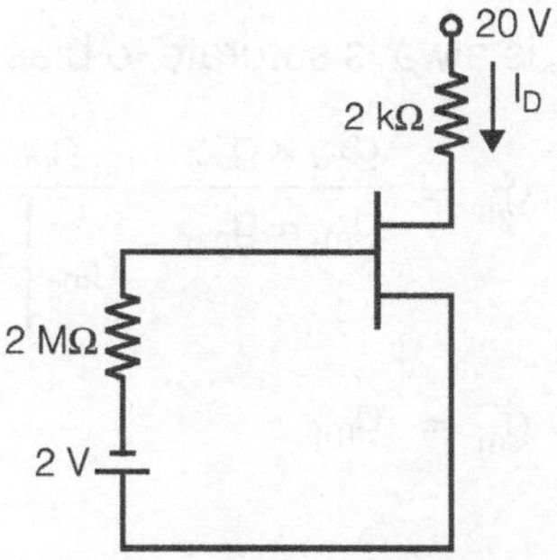


$$V_{GS} = -2\ \text{V}$$

$$I_D = I_{DSS}\left(1-\frac{V_{GS}}{V_p}\right)^2 = 10\left[1-\left(\frac{-2}{-8}\right)\right]^2 = 5.625\ \text{mA}$$

$$V_{DS} = V_{DD} - I_D R_D = 20 - 5.625\ \text{mA} \times 2\ \text{k}\Omega$$
$$= 8.75\ \text{V}$$

16. Tranconductance in milli siemens (ms) and voltage gain of the amplifier are respectively
(a) 1.875 ms and 3.41
(b) 1.875 ms and –3.41
(c) 3.3 ms and –6
(d) 3.3 ms and 6

Exp. (b)

$$g_m = \frac{2\sqrt{I_D I_{DSS}}}{V_p} = \frac{2}{8}\sqrt{5.625\ \text{mA} \times 10\ \text{mA}} = 1.875\ \text{mS}$$

$$A_V = -g_m (r_d \parallel R_D) = -1.875\ \text{mS} \times \frac{20}{11}\ \text{k}\Omega = -3.41$$

17. Two identical NMOS transistors $M_1$ and $M_2$ are connected as shown below. $V_{bias}$ is chosen so that both transistors are in saturation. The equivalent $g_m$ of the pair is defined to be $\frac{\partial I_{out}}{\partial V_i}$ at constant $V_{out}$. The equivalent $\rho_m$ of the pair is

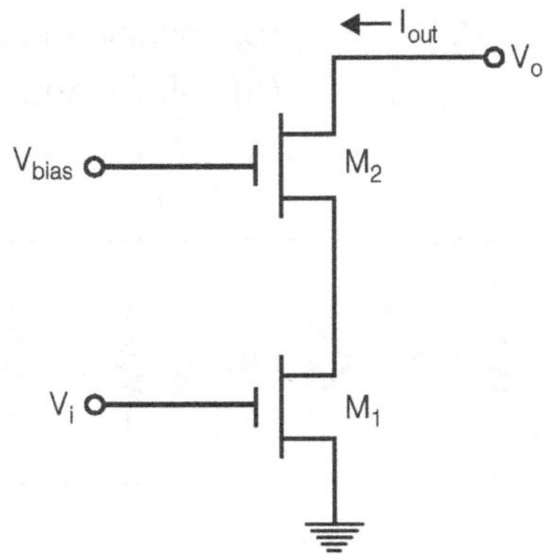


(a) sum of individual $g'_{ms}$ of transistors
(b) product of individual $g'_{ms}$
(c) nearly equal to $g_m$ of M1
(d) nearly equal to $g_m/g_o$ of $M_2$

Exp. (c)

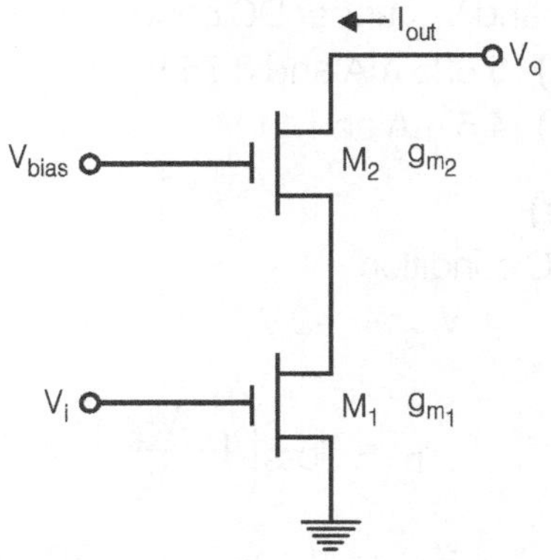


$$\frac{1}{g_m} = \frac{1}{g_{m_1}} + \frac{1}{g_{m_2}}$$

$M_2$ is always saturate to bias but $g_m$, change according to $V_i$.

$$g_m = \frac{g_{m_1} \times g_{m_2}}{g_{m_1} + g_{m_2}} = \frac{g_{m_1} \times g_{m_2}}{g_{m_2}\left[1+\frac{g_{m_1}}{g_{m_2}}\right]} \qquad g_{m_2} >> g_{m_1}$$

$$g_m = g_{m_1}$$

## REVIEW QUESTIONS

1. Derive an expression for $Z_i$, $Z_o$ and $A_V$ of a common source FET amplifier with and with out source resistance included in the circuit
2. Derive an expression for $Z_i$, $Z_o$ and $A_V$ of common drain FET amplifier?
3. Derive an expression for $Z_i$, $Z_o$ and $A_V$ of a common gate FET amplifier?
4. Explain fixed bias arrangement in FET and calculate the Q-point with graphical analysis.
5 Explain self bias arrangement in FET and calculate the Q point?
6. Give the advantage of voltage divider bias compared to fixed and self bias.

OOOO

# CHAPTER 11

# MULTIVIBRATORS AND TIME BASE GENERATORS

## About This Chapter

In this chapter, we try to concentrate on:

- Bistable multivibrator
- Schmitt trigger
- Monostable multivibrator
- Astable multivibrator
- UJT Sweep generator
- Bootstrap sweep generator
- Millers sweep circuit

# 11 MULTIVIBRATORS AND TIME BASE GENERATORS

## MULTIVIBRATORS

## 11.1 Introduction

- A circuit which can oscillate at a number of frequencies is called a multivibrator. Multi means many; vibrator means oscillator.
- Multivibrators are basically regenerative circuits comprising of two cross-coupled active devices like bipolar junction transistors.
- Basically there are three types of multivibrators:
  - Bistable multivibrator
  - Mono stable multivibrator
  - Astable multivibrator.
- Each of these multivibrators has two states. As the names indicate, a bistable circuit has got two stable states.
- A monostable circuit has got only one stable and other state being quasi stable. The astable circuit has got no stable state, both the states being quasi stable.
- Stable state of a multivibrator is the state in which the device can stay permanently. Only when a proper external triggering signal is applied, it will change its state.
- Quasi stable state means temporarily stable state. After a pre-determined time, the device will automatically come out of the quasi stable state.
- A bistable multivibrator is the basic memory element. It is used to perform many digital operations such as counting and storing of binary data.
- Monostable multivibrator finds applications in pulse circuits.
- Astable circuit is used as an oscillator to generate square waves.

# 11.2 Bistable Multivibrator

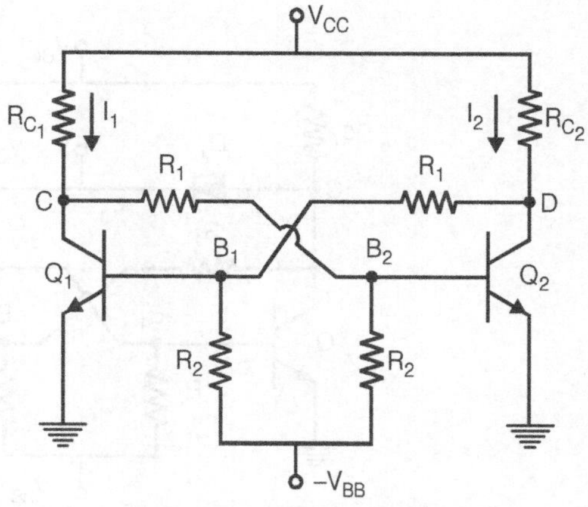


*Figure 11.1 Bistable multivibrator*

## Operation

- Consider $Q_1$, $Q_2$ are identical npn transistors. $R_1$, $R_2$, $R_{C_1}$, $R_{C_2}$ are resistors where $R_{C_1} = R_{C_2}$. $V_{CC}$ and $V_{BB}$ are supply voltage and fixed bias voltage.
- From the Figure (11.1), it is clear that the output of each transistor is coupled to the input of the other transistor.
- Since the transistor are identical, their quiescent currents would be the same, unless the loop gain is greater than unity.
- Suppose, the loop gain is greater than unity and let current $I_1$ is greater than $I_2$.
- If $I_1$ increases slightly, the voltage drop across the collector resistance $R_{C_1}$ increases. Since $V_{CC}$ is fixed, the voltage of collector point C decreases.
- It will reduces the base current of $Q_2$. In turn, decreases the collector current of $Q_2$ ($I_2$).
- If $I_2$ decreases, the voltage drop $I_2 R_{C_2}$ decreases. Hence the voltage at point D increases.
- Due to increase of $V_D$, the base current of $Q_1$ increases. This again increases the collector current ($I_1$) of $Q_1$.
- That means, $I_1$ further increases. $I_1 R_{C_1}$ drop further increases, $V_C$ further decreases. The base current of $Q_2$ further decreases, with the result that $I_2$ further decreases.
- Finally, $I_2$ would become practically zero, cutting off the transistor $Q_2$. At the same time transistor $Q_1$ would conduct heavily and driven in to saturation. Hence $Q_2$ OFF and $Q_1$ ON.
- In a similar way, if $I_2 > I_1$ then $Q_2$ ON and $Q_1$ OFF.
- Thus when $Q_1$ is ON, $Q_2$ is OFF (or) when $Q_1$ is OFF, $Q_2$ is ON. It is observed that both transistors are not ON or OFF simultaneously.
- Therefore there are two stable states, in which the transistors can remain indefinitely. Hence the name bistable multivibrator.

## Collector-Catching Diodes

- We have seen that the collector voltage $V_C$ of the OFF transistor decreases progressively, as the load across the transistor increases.

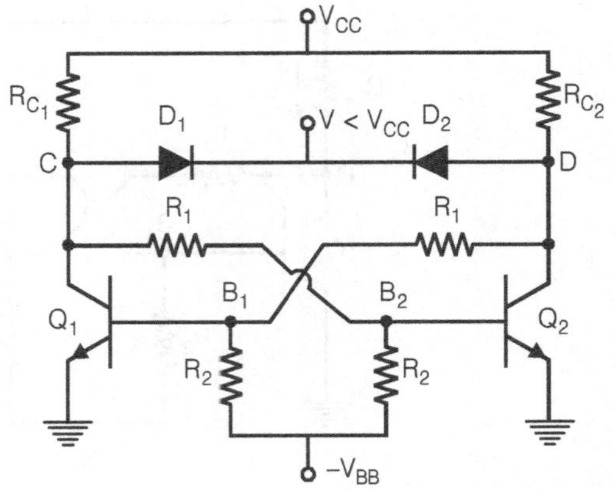


*Figure 11.2 Collector catching diodes in bistable multivibrator*

- That means, the collector voltage is dependent upon the load resistor. When the binary is used to drive other circuits, it become necessary to ensure that the collector voltage is independent of $R_L$.
- This is achieved by using collector catching diodes as shown in Figure (11.2).

## Operation

- The diodes $D_1$ and $D_2$ are used to clamp the collector voltage to a voltage $V < V_{CC}$. Let $Q_1$ be OFF and $Q_2$ be ON.
- Since V is positive which is less than $V_{CC}$, diode $D_2$ becomes reverse biased, as the potential of D is equal to $V_{CE}$ i.e. zero.
- As $Q_1$ is OFF, the potential of C tends to rise to almost $V_{CC}$. Since $V_{CC} > V$, diode $D_1$ becomes forward biased and conducts readily.
- As the reference voltage V is fixed, the collector voltage $V_C$ of $Q_1$ is clamped to V volts. Thus resistor $R_L$ is shunted across the collector of $Q_1$. The collector voltage $V_C$ remains steady at V. Therefore collector voltage is thus made independent of $R_L$.

## Commutating Capacitors

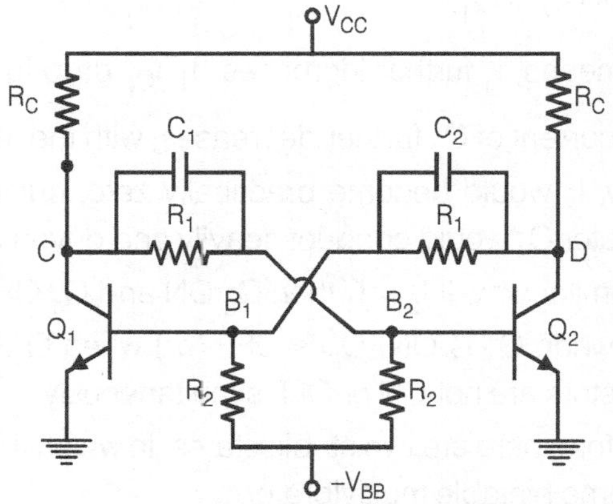


*Figure 11.3 Commutating capacitors*

- The bistable multivibrator has got two stable states and that it can remain in either of its two stable states indefinitely. It can change state only when a triggering signal such as a pulse from some external source is applied.
- When a triggering signal is applied. Conduction has to transfer from one device to another. The transition time is defined as the interval during which conduction transfer from one transistor to another.
- The reason for this transition time is, even though the input signal at the base of the transistor may be transferred to the collector with zero rise time, the signal at the collector of the transistor can not be transferred to the base of the other transistor instantaneously.
- The reason is the input capacitance $C_i$ present at the base of the transistor makes the $R_1$-$R_2$ attenuator act as an uncompensated attenuator, so it will have a finite rise time $t_r = (R_1 \| R_2)C_i$.
- The transition time may be reduced by compensating this attenuator by introducing a small capacitor in parallel with the coupling resistors $R_1$ and $R_1$ of the binary shown in Figure (11.3).

**Commutating capacitors are introduced to increase the speed of operation of the device, they are called speed-up capacitors or transpose capacitors.**

## 11.3 Schmitt Trigger

- Figure (11.4) shows the circuit diagram of an emitter-coupled bistable multivibrator using n-p-n transistors.
- It differs from the basic binary circuit in that the resistive coupling between the output of $Q_2$ and the input of $Q_1$ of the basic circuit is missing, although the collector of $Q_1$ and the base of $Q_2$ are coupled in the usual manner.
- The emitters of $Q_1$ and $Q_2$ are joined, and they are grounded through a common resistor $R_e$. The base of $Q_1$ is connected to a voltage source $V_i$.

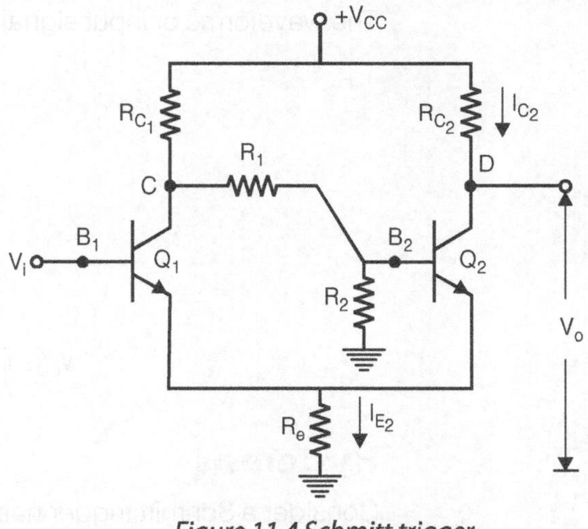

*Figure 11.4 Schmitt trigger*

### Operation

- Let us assume a sinusoidal voltage $V_i$ is applied to transistor $Q_1$.

$$\text{Where} \quad V_i = V_m \sin \omega t$$

- When $V_i = 0$, the transistor $Q_1$ will be cut-off (OFF) and $Q_2$ will get base drive from the supply voltage $V_{CC}$ and hence it conducts, but it would be conducting in the active region
- Due to the flow of current $I_{E_2}$ through $R_e$, there is a voltage drop $I_{E_2} R_e$.
- Assuming $I_{E_2} = I_{C_2}$, we have $V_E = I_{C_2} R_e$. It is clear from the circuit, that for $Q_1$ to conduct, $V_i$ should rise to the voltage $(V_E + V_\gamma)$ where $V_\gamma$ is the cut in voltage of $Q_1$.

- As $V_i < V_E + V_\gamma$, $Q_1$ can not conduct and $Q_1$ starts conducting when $V_i = V_E + V_\gamma$ i.e. $V_i = I_{C_2} R_e + V_\gamma$.
- As $Q_1$ conducts, the potential at collector terminal C progressively decreases (due to increasing $I_{C_1} R_{C_1}$ drop), and because of feedback. The potential of point $B_2$ progressively diminishes, with the result that $Q_2$ becomes OFF.
- This value of the input voltage $V_i$ which makes $Q_1$ conduct is termed as upper trigger point or upper trip point (UTP), denoted by $V_A$

$$\therefore \quad V_A = V_{E_2} + V_\gamma$$

or $$\boxed{V_A\,(UTP) = I_{C_2} R_e + V_\gamma}$$

- As $V_i$ decreases more and more, $V_{B_1}$ decreases, the potential at point C progressively increases. Which makes the potential at $B_2$ increases.
- As $V_{B_2}$ becomes equal to $V_{BE\,(act)} + V_E$, $Q_2$ starts conducting and $Q_1$ goes OFF.
- This value of the input voltage which makes $Q_2$ conduct again is termed as lower trigger point (or lower trip point). It is denoted as $V_B$

$$\therefore \quad V_B = V_{BE\,(act)} + V_E$$

or $$\boxed{V_B(LTP) = V_{BE(act)} + I_{C_1} R_e}$$

The waveforms of input signal $V_i$ and output voltage $V_o$ are as shown:

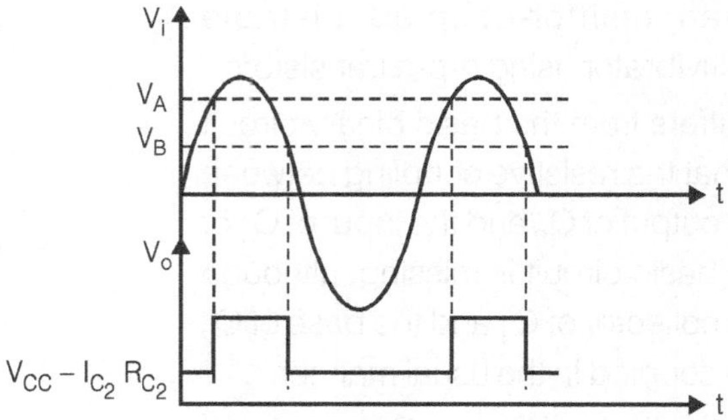


## Hysteresis

Consider a Schmitt trigger has the following specifications.

$V_{CC} = 12\text{ V}$, UTP = 6 V, LTP = 3 V and

$I_C = 2\text{ mA}$ and $R_{C_1} = R_{C_2} = 5\text{k}\Omega$

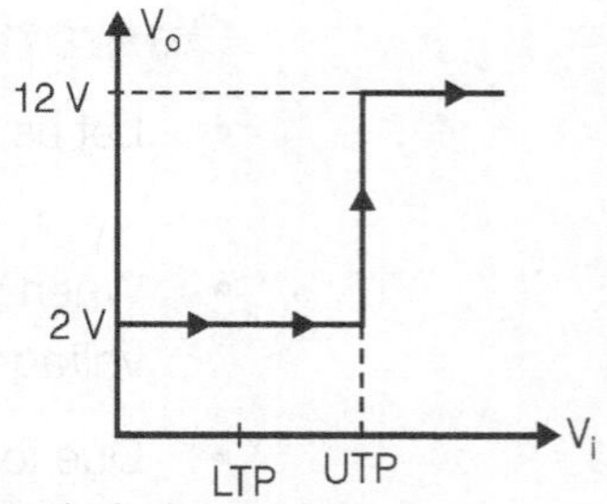


(i) When $Q_2$ conducts and $Q_1$ is OFF, we have

Output $V_o = V_{CC} - I_{C_2} R_{C_2} = 12 - 2 \times 5 = 2\text{ V}$

When $Q_2$ is OFF, $I_{C_2} = 0$ and hence potential at D is $V_{CC}$

$$V_o = V_{CC} = 12\text{ V}$$

When input voltage is increasing from zero, the transfer characteristic is as shown: Between $V_i = 0$ V and $V_i$ = UTP, the output remains constant at 2 V. During this period, $Q_2$ is ON. For $V_i$ = UTP, $Q_1$ begins to conduct. As long as $Q_1$ is ON, $Q_2$ is OFF.

(ii) As $V_i$ increases beyond UTP, reaches its positive peak and falls to LTP, output remains at 12 V.
When $V_i$ = LTP, $Q_2$ again begins to conduct and $Q_1$ goes OFF. At $V_i$ = LTP, output falls instantly to 2 V and remains at that level.
The transfer characteristic when $V_i$ decreases from its positive peak is

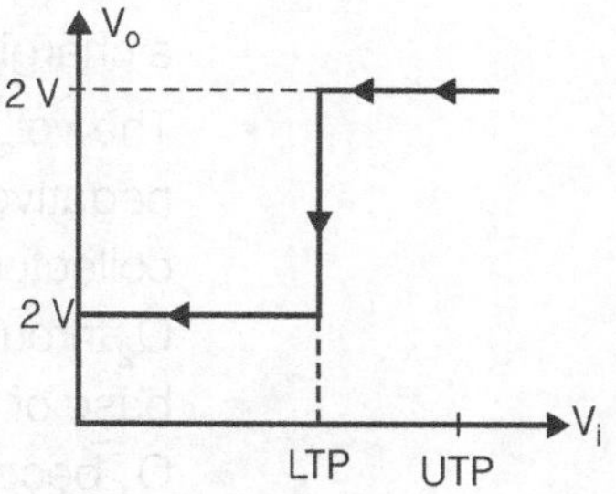


The total transfer characteristic (graph of $V_o$ versus $V_i$) can be plotted.

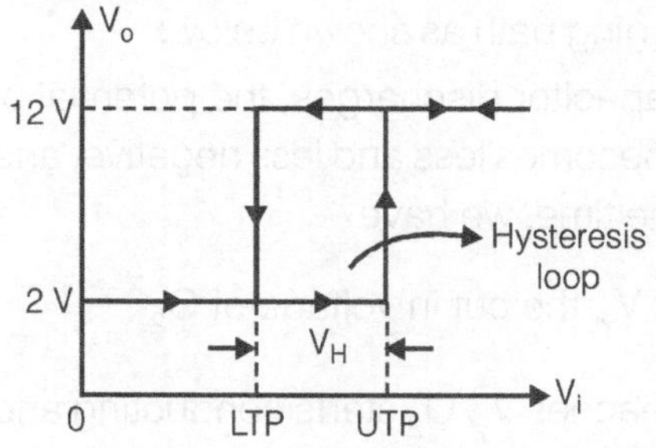


$V_H$ = Hysteresis voltage = UTP – LTP.

# 11.4 Mono Stable Multivibrator

- Monostable multivibrator has got only one permanent stable state, the other state being quasi stable.
- When an external triggering pulse is applied, it switches from the stable to the quasi-stable state. It remains in the quasi stable state for a short duration, and automatically switches back to its original stable state, without any triggering pulse.
- The monostable multivibrator is also referred as 'one shot' or univibrator. Since only one triggering signal is required to revert to the original stable state.
- Monostable multi is called a 'pulse generator' because when the output of monostable multi remains in quasi stable state, is a pulse of duration T, whose value depends on circuit components.

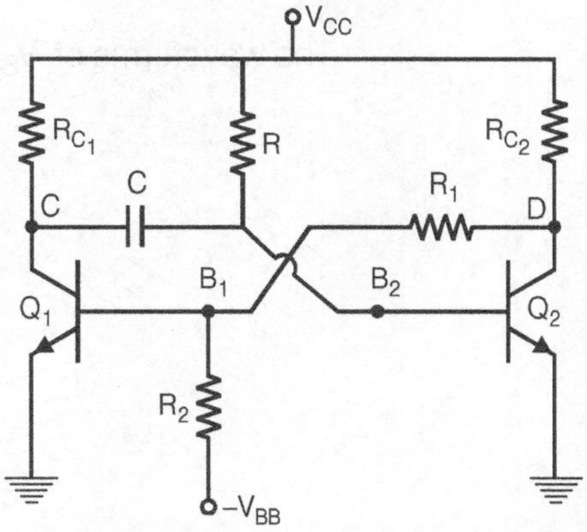


*Figure 11.5 Mono stable multivibrator*

## Operation

- Assume $Q_1$ and $Q_2$ are identical npn transistors, where $Q_1$ is OFF and $Q_2$ is ON. $R_1$ and $R_2$ are connected to OFF transistor and the capacitor C is connected to ON transistor.
- Under normal conditions, the supply voltage $V_{CC}$ provides enough base drive to the transistor $Q_2$ through resistance R, with the result that $Q_2$ goes in to saturation.

- With $Q_2$ ON and $Q_1$ OFF, the capacitor finds a charging path as shown below:

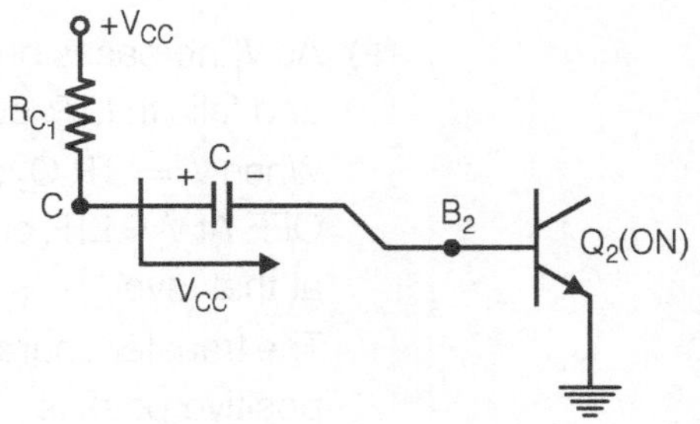


- The voltage across the capacitor is $V_{CC}$. If a negative triggering pulse is applied to the collector of $Q_1$. It is transmitted to the base of $Q_2$ through the capacitor and hence makes the base of $Q_2$ negative. Then $Q_2$ goes OFF and $Q_1$ becomes ON. This is quasi stable state.
- When $Q_1$ ON and $Q_2$ OFF, the capacitor C finds a discharging path as shown below:

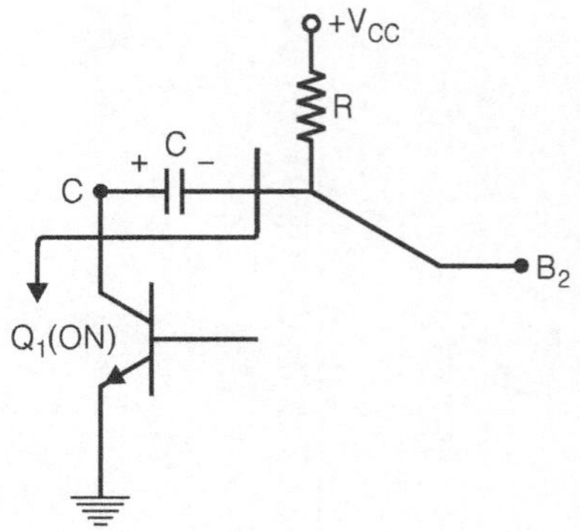


- As the capacitor discharges, the potential of point $B_2$ becomes less and less negative, and after some time, we have

  $V_{B_2} = V_\gamma$, the cut in voltage of $Q_2$

- As $V_{B_2}$ reaches $V_\gamma$, $Q_2$ starts conducting and gets saturated. When $Q_2$ becomes ON, $Q_1$ becomes OFF. Thus original stable state of the multivibrator is restored.
- The interval during which the quasi stable state of the multi i.e. $Q_2$ remains OFF is dependent upon the rate at which the capacitor C discharge. This duration of the quasi-stable state is termed as delay time or pulse width or gate time. It is denoted as T.

The waveforms of $V_{B_2}$, $V_{C_1}$, $V_{C_2}$ and $V_{B_1}$ is shown below:

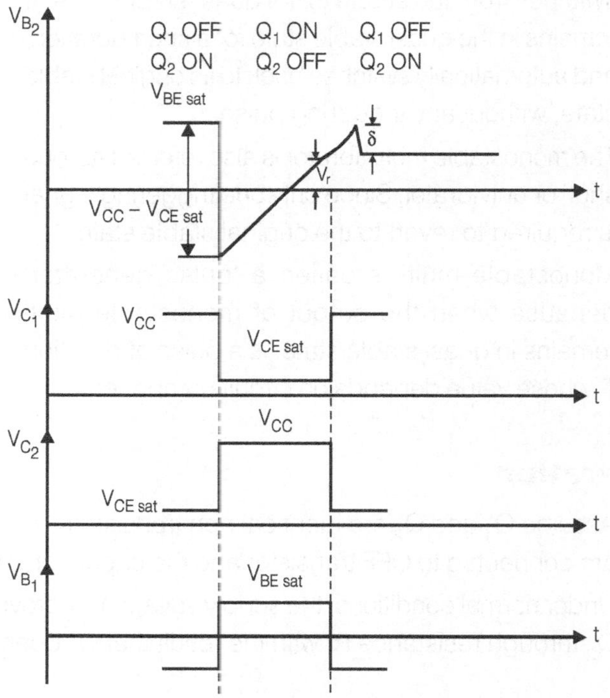

## Expression for The Gatewidth (T)

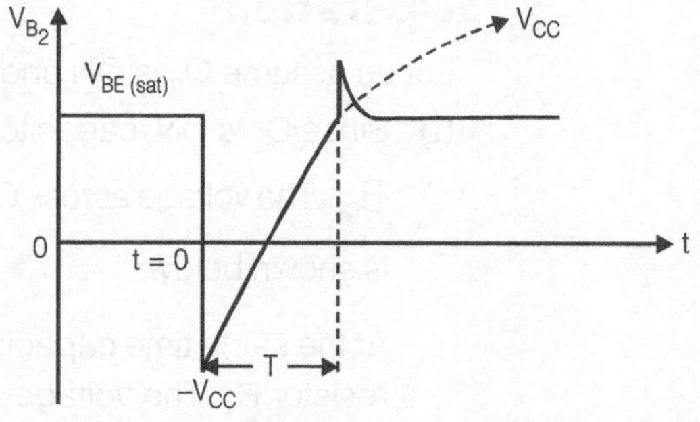

*Figure 11.6 Voltage variation at the base of $Q_2$ during the quasi stable state*

At t = 0, initial value of $V_{B_2}$ is $-V_{CC}$.

As capacitor discharges, the voltage $V_{B_2}$ rises exponentially and should attain the value of $+V_{CC}$. But at t = T, $Q_2$ becomes ON and $V_B$ takes the value $V_\gamma$, the cut in voltage of $Q_2$.
The exponentially increasing voltage $V_B$ is given as

$$V_{B_2} = V_{final} - (V_{final} - V_{initial})e^{-t/RC}$$

Where
$$V_{final} = +V_{CC}$$
$$V_{initial} = -V_{CC}$$
$$V_{B_2} = V_{CC} - [V_{CC} - (-V_{CC})]e^{-t/RC}$$

at
$$t = T$$
$$V_{B_2} = V_\gamma = 0$$
$$0 = V_{CC} - 2\,V_{CC}\,e^{-T/RC}$$
$$2\,e^{-T/RC} = 1$$
$$e^{T/RC} = 2$$
$$\frac{T}{RC} = ln2 = 0.69$$

$$\boxed{\text{Gatewidth } T = 0.69\,RC}$$

# 11.5 Astable Multivibrator

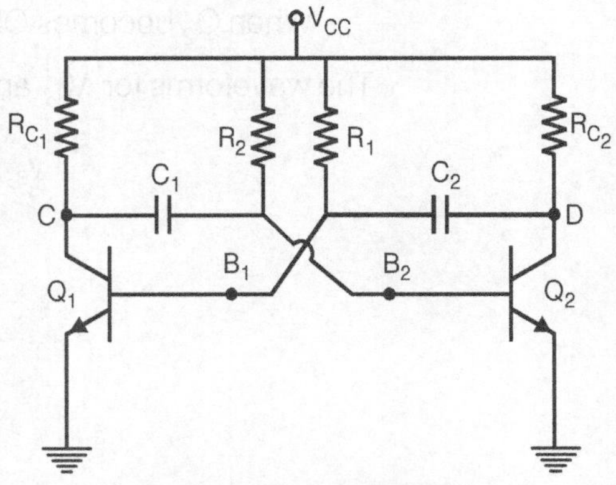

*Figure 11.7 Astable multivibrator*

- The astable multivibrator can also be called as free-running oscillator, widely employed for the generation of square-waves or pulses.
- It is a two-stage RC coupled amplifier with the output of first stage coupled to the input of the second stage and the output of second stage coupled to the input of the first stage.
- An astable multi has two quasi stable states, and it keeps on switching between these two states, by itself. No external triggering signal is needed.
- The two amplifier stages of an astable multivibrator are regeneratively cross-coupled by capacitors.

## Operation

Let us assume $Q_2$ is ON and $Q_1$ is OFF.

(i) Since $Q_2$ is ON, capacitor $C_1$ charges through resistor $R_{C_1}$. The voltage across $C_1$ is $V_{CC}$. The equivalent circuit is shown below:

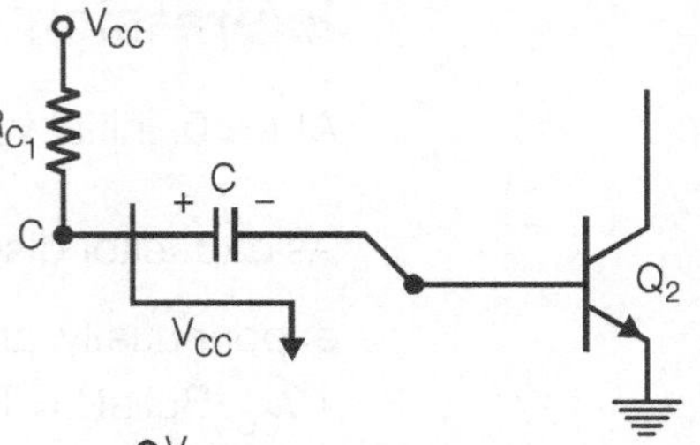


At the same time capacitor $C_2$ discharges through resistor $R_1$. The voltage across $C_2$ when it is about to start discharging is $V_{CC}$ with polarity as shown:

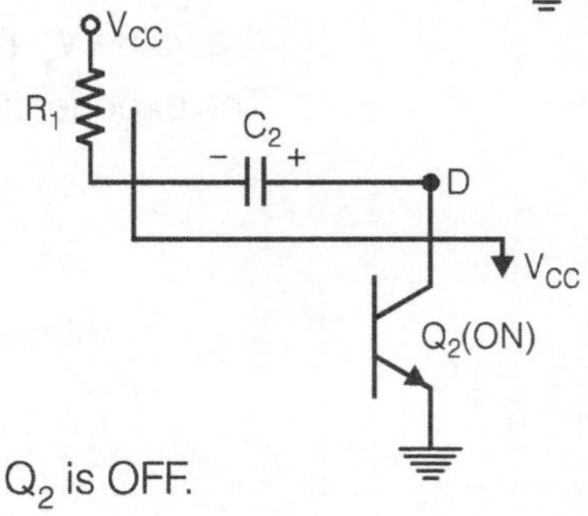


As capacitor $C_2$ discharges more and more, the potential of point $B_1$ becomes more and more positive and $V_{B_1}$ becomes equal to $V_\gamma$, the cut in voltage of $Q_1$.

If $V_{B_1} > V_\gamma$, transistor $Q_1$ starts conducting. When $Q_1$ is ON $Q_2$ is OFF.

Similarly, suppose $Q_1$ ON and $Q_2$ OFF. The capacitor $C_2$ charges through resistor $R_{C_2}$ and capacitor $C_1$ discharges through resistor $R_2$. The charging and discharging paths are as shown below:

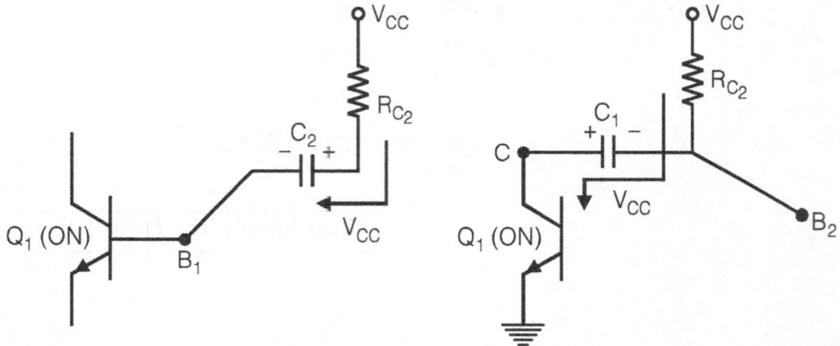


As capacitor $C_1$ discharges more and more, the potential at $B_2$ becomes more and more positive. If $V_{B_2} > V_\gamma$, transistor $Q_2$ starts conducting.

When $Q_2$ becomes ON, $Q_1$ becomes OFF.

The waveforms for $V_{B_2}$ and $V_C$ are shown below:

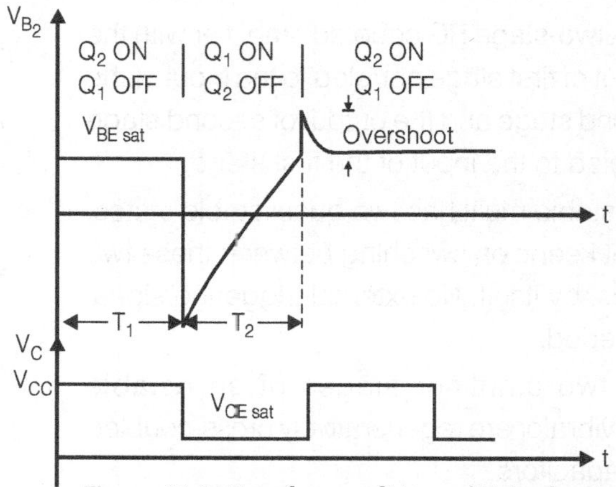


*Figure 11.8 Waveforms of $V_{B_2}$ and $V_C$ in the circuit*

## Expression for T

Consider the waveform of $V_{B_2}$ shown below:

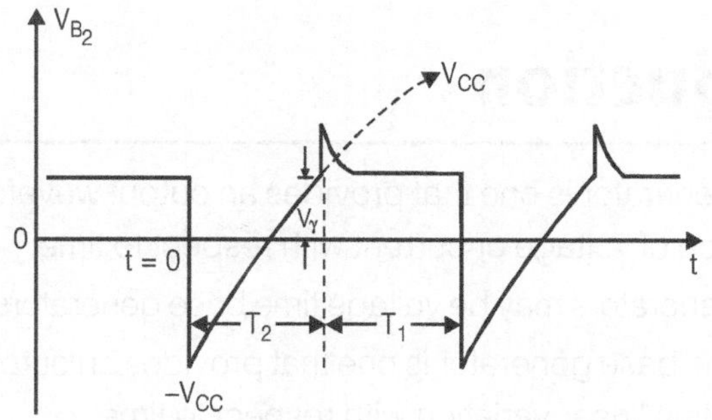


In the waveform $T_1$ = charging time

$T_2$ = discharging time

The time period $T = T_1 + T_2$

The expression for exponentially rising voltage is

$$V_{B_2} = V_f - (V_f - V_{initial})e^{-t/R_2C_1}$$

We have $V_{initial} = -V_{CC}$

$V_{final} = V_{CC}$

$$\therefore \quad V_{B_2} = V_{CC} - (V_{CC} - (-V_{CC}))e^{-t/R_2C_1} = V_{CC} - 2V_{CC}\, e^{-t/R_2C_1}$$

At $t = T_2$ we have $V_{B_2} = V_\gamma = 0$

$$V_{CC} - 2V_{CC}\, e^{-T_2/R_2C_1} = 0$$

$$1 - 2e^{-T_2/R_2C_1} = 0$$

$$e^{T_2/R_2C_1} = 2$$

$$\frac{T_2}{R_2C_1} = ln 2 = 0.69$$

or $T_2 = 0.69\, R_2\, C_1$

Similarly $T_1 = 0.69\, R_1\, C_2$

Period of oscillations

$$T = T_1 + T_2 = 0.69\, R_1\, C_2 + 0.69\, R_2\, C_1$$

$$= 0.69\,(R_1\, C_2 + R_2\, C_1)$$

If $R_1 = R_2 = R \;;\; C_1 = C_2 = C$

Then $\boxed{T = 1.38\, RC}$

Frequency of oscillation

$$\boxed{f = \frac{1}{T}}$$

# TIME BASE GENERATORS

## 11.6 Introduction

- Time base generator is one that provides an output waveform, a portion of which exhibits a linear variation of voltage or current with respect to time.
- Time base generators may be voltage time base generators or current time base generators.
- A voltage time base generator is one that provides an output voltage waveform, a portion of which exhibits a linear variation with respect to time.
- A current time-base generator is one that provides an output current waveform, a portion of which exhibits a linear variation with respect to time.
- Time base generators find use in CRO's, radar, television etc.
- In CRO, for deflecting the electron beam horizontally across the screen, the applied voltage makes the electron beam sweep across the screen, it is termed as sweep voltage.
- Such type of circuit generating the sweep voltage is termed as sweep generator.
- There are two types of sweep generators:
  - Voltage sweep generators
  - Current sweep generators.

## 11.7 Sweep Voltage

Figure 11.9 shows the typical waveform of a time-base voltage.

- The voltage starting from some initial value increases linearly with time to a maximum value after which it returns again to its initial value.
- The time taken by the wave to reach the maximum value, starting from the initial value, is termed as sweep time ($T_s$).
- The time during which it returns to the initial value is termed as return time or restoration time or fly-back time ($T_r$).
- A sweep voltage may be a sawtooth or triangular wave also.

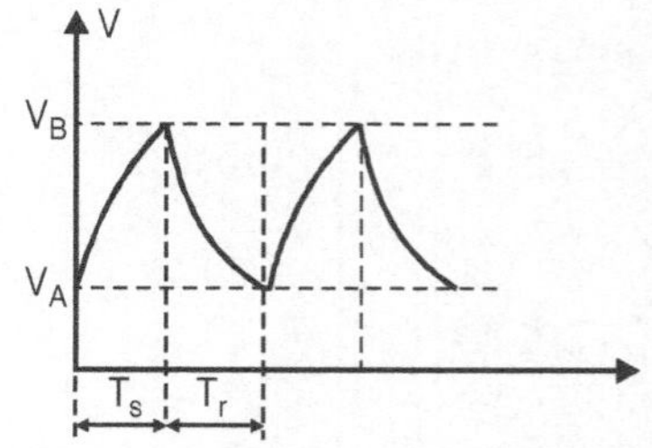

*Figure 11.9 Time base voltage*

The deviation of a signal from linearity can be defined in terms of three errors:

(i) sweep error

(ii) displacement error

(iii) transmission error.

### Sweep Error

Sweep error can also be called as slope error.

Slope error is defined as the difference between initial slope and final slope expressed as

$$e_s = \frac{\text{Initial slope} - \text{Final slope}}{\text{Initial slope}} = \frac{\left.\frac{dV_o}{dt}\right|_{t=0} - \left.\frac{dV_o}{dt}\right|_{t=T_s}}{\left.\frac{dV_o}{dt}\right|_{t=0}}$$

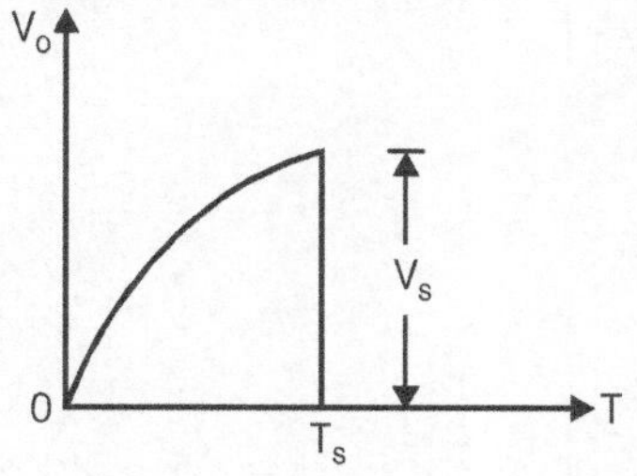


### Displacement Error

The displacement error $e_d$ is defined as "maximum difference between the actual sweep voltage and the linear sweep which passe through the beginning and end points of the actual sweep divide by amplitude of the sweep at the end of sweep time".

$$e_d = \frac{(V_{s_1} - V_{s_2})_{max}}{V_{s_1}}$$

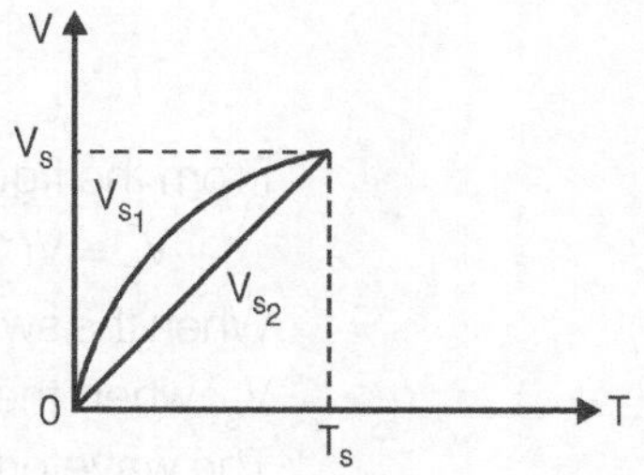


### Transmission Error

When a ramp signal is transmitted through a high pass circuit, the output falls away from the input. This deviation is expressed as transmission error $e_r$.

$$e_t = \frac{V_{s_2} - V_{s_1}}{V_{s_2}}$$

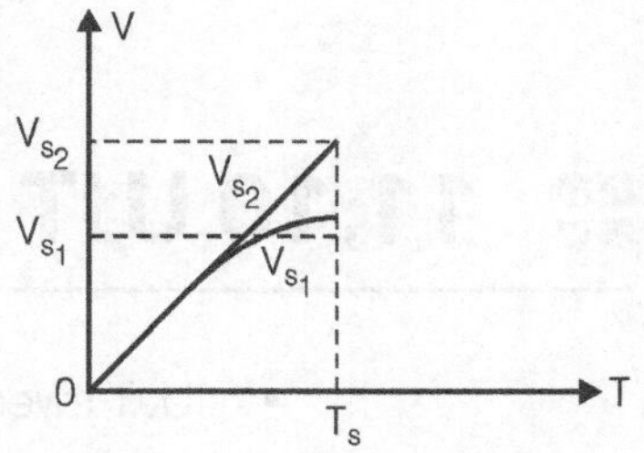


It is defined as the difference between the input and output divided by the input at the end of the sweep.

## 11.8 Methods of Generating a Time Base Waveform

### Exponential Charging

In this method, a capacitor is charged through a resistor to a voltage which is small compared to the supply voltage. This can be achieved by

- exponential sweep circuit
- UTJ relaxation oscillator.

### Miller Circuit

In this method, an integrator is used to convert an input step voltage into a ramp waveform.

### Bootstrap Circuit

In this method, a constant current is passed through a capacitor and the voltage across the capacitor is a ramp.

# 11.9 Exponential Sweep Circuit

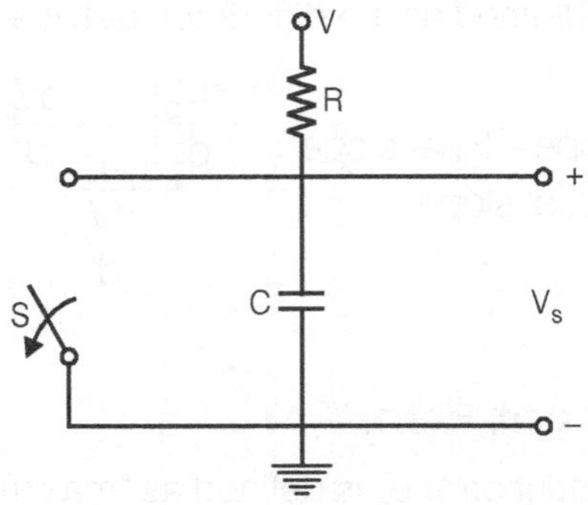


*Figure 11.10 Exponential sweep circuit*

From the figure, the output across the capacitor C is given as

$V_s = V(1 - e^{-t/RC})$

When the switch is open, the capacitor charges to a voltage $V_s$, when the switch is closed, the capacitor discharges. The waveform is shown below:

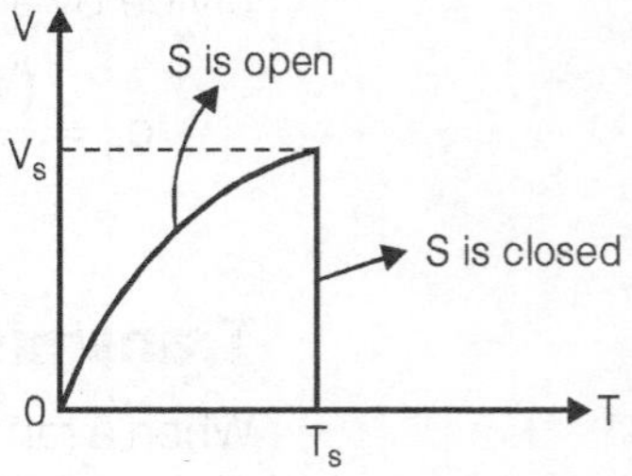


# 11.10 UJT Sweep Generator

- UJT sweep generator is also known as relaxation oscillator, which generates non sinusoidal oscillations.
- Figure (11.11) shows the UJT sweep circuit consists of a UJT, a capacitor and a resistor.
- The capacitor C charges through the resistance R from the supply voltage V as long as $V_E < V_P$ (UJT is OFF).
- As the capacitor voltage $V_s$ rises the value $V_p$, UJT readily conducts. When UJT becomes ON, capacitor discharges and its voltage falls.

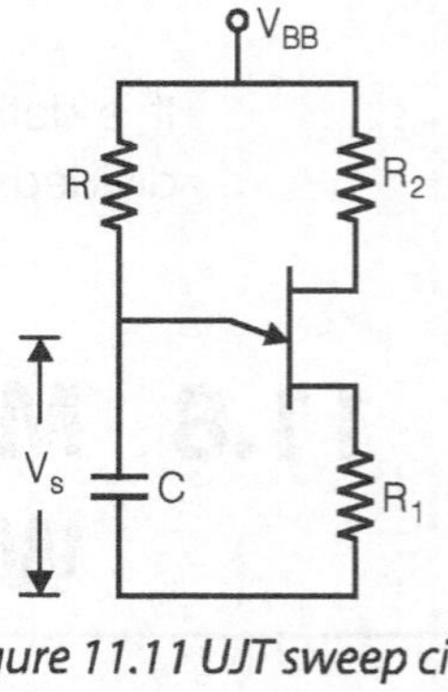


*Figure 11.11 UJT sweep circuit*

- When the voltage falls to the valley point $V_V$, the UJT becomes OFF and the capacitor charges again to $V_p$.

  $T_s$ = Sweep time

  $T_r$ = Return time

  $V_p$ = Peak voltage

  $V_V$ = Valley voltage

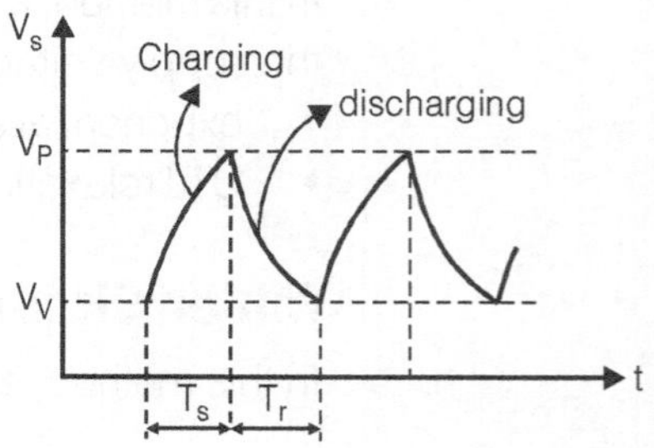


- A sawtooth waveform of voltage across C is generated because of charging and discharging of capacitor C.

## Expression for Frequency (F)

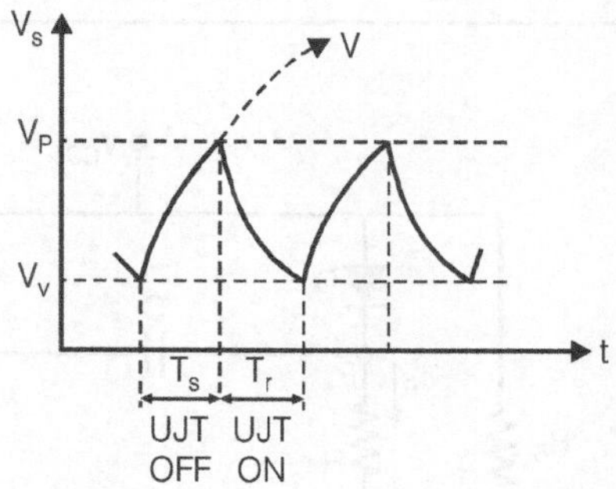


The capacitor charges and increases exponentially given by

$$V_s = V_{final} - (V_{final} - V_{initial})\, e^{-t/RC}$$

Where $V_{initial} = V_v$

$V_{final} = V$

At $t = T_s$, we have $V_s = V_p$

$$V_p = V - (V - V_V)e^{-T_s/RC}$$

$$(V - V_v)e^{-T_s/RC} = V - V_p$$

$$e^{-T_s/RC} = \frac{V - V_p}{V - V_v}$$

$$e^{T_s/RC} = \frac{V - V_v}{V - V_p}$$

$$\frac{T_s}{RC} = ln\left[\frac{V - V_v}{V - V_p}\right]$$

$$T_s = RC\, ln\left(\frac{V - V_v}{V - V_p}\right)$$

As $V_V << V$

$$T_s = RC\, ln\frac{V}{V - V_p}$$

Frequency of oscillation

$$f = \frac{1}{T_s} = \frac{1}{RC\,ln\left(\frac{V}{V - V_p}\right)} = \frac{1}{RC\,ln\left(\frac{1}{1 - V_p/V}\right)}$$

Where $\frac{V_p}{V} = \eta$ = intrinsic stand-off ratio

$$\therefore f = \frac{1}{RC\,ln\left(\frac{1}{1 - \eta}\right)}$$

# 11.11 Bootstrap Sweep Generator

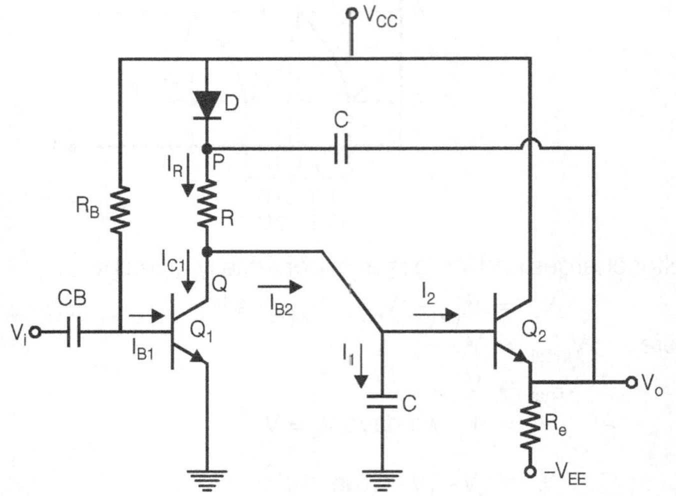


*Figure 11.12 Bootstrap sweep circuit*

- The input to transistor $Q_1$ is the gating waveform from a pulse generator.
- Before the application of gating waveform at t = 0, $Q_1$ is in saturation because it gets enough base drive from $V_{CC}$ through $R_B$.
- The voltage across the capacitor which is also the voltage at the collector of $Q_1$ and the base of $Q_2$ is $V_{CE\,(sat)}$.
- Since $Q_2$ is acting as an emitter follower and it is conducting. The voltage at emitter of $Q_2$ which is the output voltage $V_o$ is less than base voltage of $Q_2$.

$$V_o = V_{CE\,(sat)} - V_{BE\,2}$$

- This output voltage is a small negat ve voltage i.e.

For npn Si,
$$V_{CE\,(sat)} = 0.2\text{ V}$$
$$V_{BE\,2} = 0.7\text{ V}$$
$$\therefore \quad V_o = 0.2 - 0.7\text{ V} = -0.5\text{ V}$$

- The emitter of $Q_2$ is coupled to the collector of $Q_1$ through the capacitor $C_1$.
- The point 'P' becomes negative w.r.t $V_{CC}$. Diode D readily conducts where $V_p = V_{CC}$.
- When the input $V_i$ goes negative, $Q_1$ becomes OFF. The potential of 'Q' rises. This voltage increase at Q is transmitted to 'P' through $Q_2$ and capacitor $C_1$.

∴ The potential at 'P' rises by the some amount. This is the principle of bootstrap. Thus

$$V_p = V_{CC} + V_Q$$

Let $I_R$ denote the current through $R_C$

We have
$$I_R = \frac{V_P - V_Q}{R_C} = \frac{V_{CC}}{R_C}$$

As $V_{CC}$, $R_C$ are fixed in magnitude the ratio $\left(\frac{V_{CC}}{R_C}\right)$ is constant. Therefore $I_R$ is a constant current.

From the circuit, $I_R = I_{C_1} + I_{B_2}$

As the transistor $Q_1$ is cut-off, therefore $I_R = I_{B_2}$.

Again $I_{B_2}$ is divided as $I_1$ and $I_2$. $I_1$ current flows through capacitor C and $I_2$ current flows through $Q_2$ transistor.

Here $I_2 = 0$, because $Q_2$ is an emitter follower, its input impedance is very high and hence

$$I_{B_2} = I_1$$

As this current flows through the capacitor C, a ramp voltage develops.

Thus input and output waveforms are shown below:

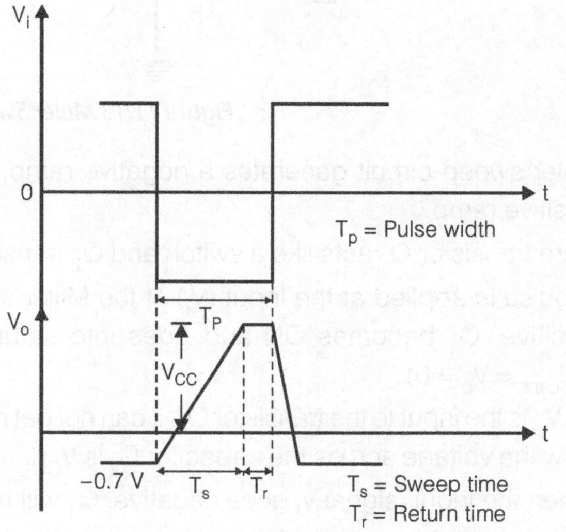


## Basic Principle

Consider a low pass RC circuit

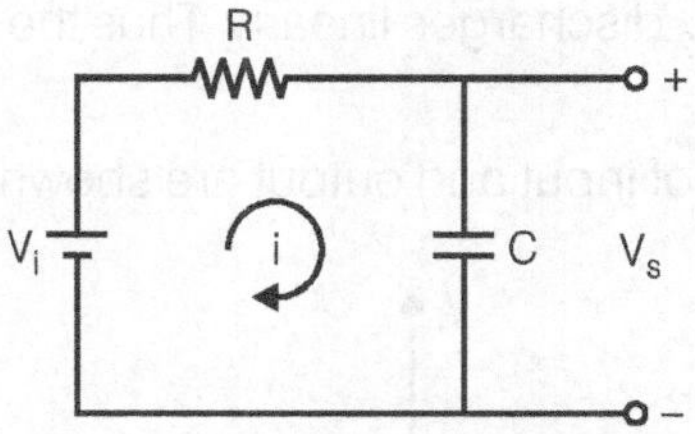


We have from the circuit

$$V_s = \frac{1}{C}\int i\,dt$$

If i (charging current) is maintained as constant I then

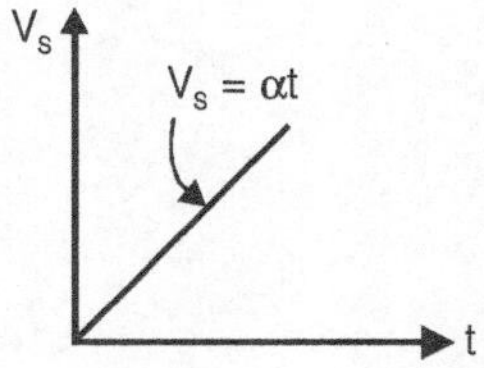


$$V_s = \frac{I}{C}\int dt = \left(\frac{I}{C}\right)t$$

Since I and C are fixed, the ratio I/C is constant

Let $\qquad \frac{I}{C} = \propto$

$\therefore \quad V_s = \alpha t \quad$ which represents ramp voltage

# 11.12 Miller Sweep Circuit

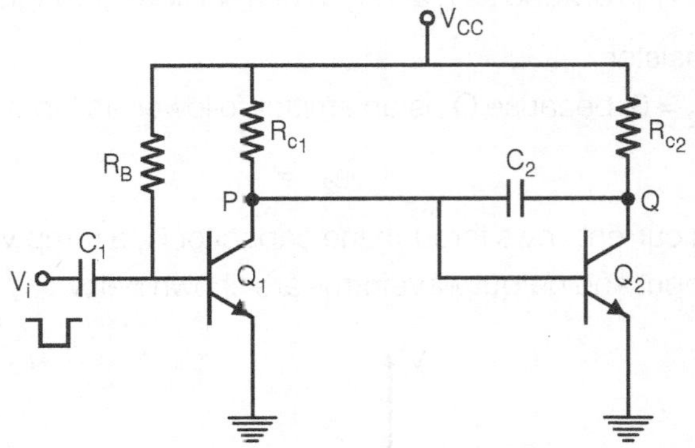


*Figure 11.13 Miller Sweep Circuit*

- Miller sweep circuit generates a negative ramp, where as Bootstrap circuit generates a positive ramp.
- Here transistor $Q_1$ acts like a switch and $Q_2$ transistor acts as a high gain amplifier.
- A pulse is applied at the input ($V_i$) of the Miller sweep circuit. When the input signal $V_i$ is positive. $Q_1$ becomes ON and goes into saturation. The potential of point P is zero ($V_{CE\ sat} = V_p = 0$).
- As $V_p$ is the input to the transistor $Q_2$, it can not get necessary base drive. Therefore $V_Q = V_{CC}$. Now the voltage across the capacitor $C_2$ is $V_{CC}$.
- When the input signal $V_i$ goes negative, $Q_1$ will be cut-off (OFF). The potential at P rise to $V_{CC}$, but due to capacitor $C_2$ present in the circuit, $V_p$ can not rise to $V_{CC}$ instantaneously.
- $V_p$ rises gradually, and as a result the voltage across capacitor falls gradually.
- As $Q_1$ OFF, $Q_2$ becomes ON, because $V_p$ is increasing to $V_{CC}$.
- As $V_p$ increases gradually, $V_Q$ decreases gradually. The voltage across the capacitor falls.
- Therefore $C_2$ discharges linearly. Thus the voltage $V_s$ across capacitor $C_2$ is a negative going ramp.

The waveforms of input and output are shown below:

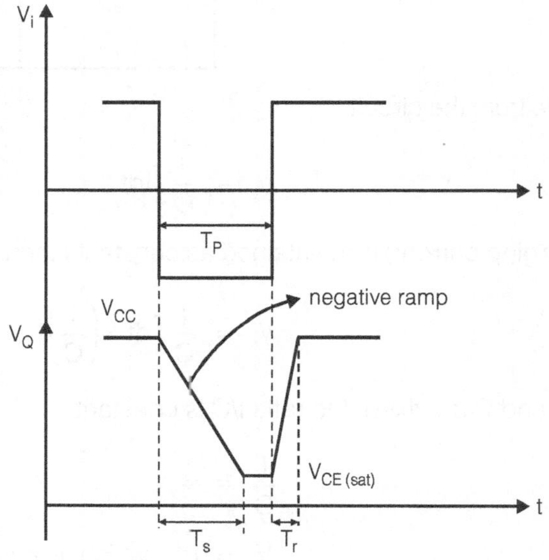

# SUMMARY

- A bistable circuit has got two stable states. It is the basic memory element. It is used to perform many digital operations such as storing and counting of binary data.
- A monostable circuit has got only one stable state and other state being quasi stable. It finds applications in pulse circuits.
- A stable circuit has got no stable states, both the states being quasi stable. It can be used as an oscillator to generate square waves.
- The diodes used in a bistable multivibrator to maintain a constant output swing are called collector-catching diodes.
- The transition time is defined as the interval during which conduction transfers from one transistor to another.
- The transition time may be reduced by shunting the coupling resistors with capacitors are called commutating capacitors.
- Commutating capacitors also called speed up capacitors or transpose capacitors are used to increase the speed of operation.
- The application of an external signal to induce a transition from one state to the other is called triggering.
- The Schmitt trigger is called an emitter-coupled binary.
- Schmitt trigger exhibits hysteresis when loop gain is greater than one.
- Schmitt trigger is used as an amplitude comparator and as a squaring circuit.
- A time base generator is to provide an output waveform a portion of which exhibits a linear variation of voltage or current with time.
- Time base generators are used in CRO's, TV and radar systems.
- The time during which the output increases linearly is called sweep time and the time required by the sweep voltage to return to the initial value is called restoration time or return time.
- The deviation errors from linearity is given by
  - slope or sweep error $e_s$
  - displacement error $e_d$
  - transmission error $e_t$.
- In exponential charging a capacitor is charged through a resistor from the supply voltage to generate sweep voltage.
- In bootstrap circuit, current is maintained constant across a fixed resistor in series with a capacitor. In a bootstrap time base generator, a non inverting amplifier with unity gain is required.
- Bootstrap time base generator produces a positive-going sweep.
- In Miller circuit, an integrator is used to convert a step into ramp. In Miller time base generator, an inverting amplifier with a gain of infinity is required.
- Miller time-base generator produces a negative-going sweep.

## GATE AND IES QUESTIONS

**Q.1** Consider the following bistable multivibrator circuit.

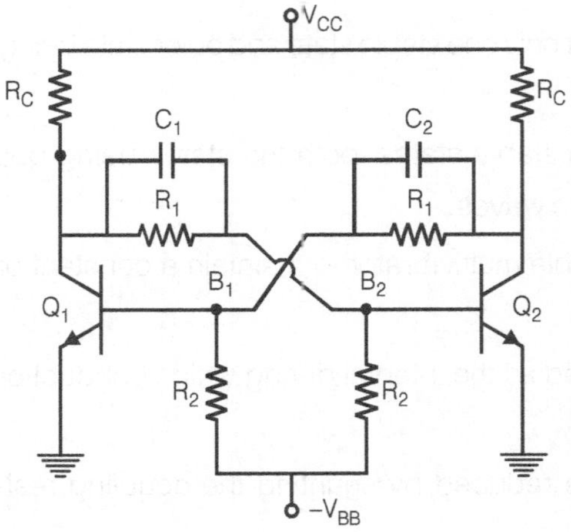


Which of the following statement is correct?
In the above circuit, condensers $C_1$ and $C_2$ are used.

(a) For passing the noise.
(b) To compensate the collector-emitter capacitance.
(c) To speed up the switching action.
(d) To provide negative feedback path.

**Exp.** (c)
Commutating capacitors are introduced to increase the speed of operation of the device.

**Q.2** Match **List-I** (Circuit) with **List-II** (Application) and select the correct answer using the codes given below:

**List-I**
A. Monostable multivibrator
B. Bistable multivibrator
C. Clamping circuit
D. Schmitt trigger

**List-II**
1. Comparator
2. dc level translator
3. Delay
4. Voltage controlled oscillator
5. Counter

**Codes:**

| | A | B | C | D |
|---|---|---|---|---|
| (a) | 4 | 3 | 1 | 2 |
| (b) | 3 | 5 | 2 | 1 |
| (c) | 4 | 5 | 2 | 1 |
| (d) | 5 | 4 | 1 | 2 |

**Exp.** (b)

Monostable multivibrator — delay
Bistable multivibrator — Counter
Clamping circuit — dc level translator
Schmitt trigger — Comparator

**Q.3** **Assertion (A):** A monostable multivibrator can be used to alter the pulse width of a repetitive pulse train.
**Reason (R):** Monostable multivibrator has a single stable state.
(a) Both A and R are true and R is the correction explanation of A
(b) Both A and R are true but R is NOT the correct explanation of A
(c) A is true but R is false
(d) A is false but R is true

**Exp.** (b)
Monostable is a pulse stretcher.
Monostable has one stable state.

**Q.4** **Assertion (A):** Miller sweep circuit producing sawtooth waveform is a relaxation oscillator.
**Reason (R):** The active device alternately supplies power to the load and relaxes when it is cut-off.
(a) Both A and R are true and R is the correction explanation of A
(b) Both A and R are true but R is NOT the correct explanation of A
(c) A is true but R is false
(d) A is false but R is true

**Exp.** (d)
Miller sweep circuit produces negative ramp.
It relaxes when the device is cut-off.

**Q.5** **Assertion (A):** While using UJT for the generation of sawtooth voltage, it has to be biased so that it can work within the range of V-I characteristic.
**Reason (R):** With in a range of its V-I characteristics, the UJT has negative resistance and can be used as an oscillator.
(a) Both A and R are true and R is the correction explanation of A
(b) Both A and R are true but R is NOT the correct explanation of A
(c) A is true but R is false
(d) A is false but R is true

**Exp.** (a)
To generate sawtooth voltage, it has to be biased so that it can work with in the range of V-I characteristics, where UJT has negative resistance and can be used as an oscillator.

**Q.6** A relaxation oscillator is one which
(a) has two stable states
(b) oscillates continuously
(c) relaxes indefinitely
(d) produces non-sinusoidal output

**Exp.** (d)
A relaxation oscillator produces non sinusoidal output.

**Q.7** Consider the following statements:
In order to generate square waveform a sinusoidal input signal one can use:
**1.** Schmitt trigger circuit.
**2.** Clippers and amplifiers
**3.** Monostable multivibrators
Which of the above statements is/are correct?
(a) 1 alone
(b) 1 and 2
(c) 2 and 3
(d) 1, 2 and 3

**Exp.** (a)
Schmitt trigger circuit can only produce square wave.

**Q.8** Match **List-I** (Circuit name) with **List-II** (Characteristics) and select the correct answer using the codes given below:

| List-I | List-II |
|---|---|
| A. Schmitt trigger | 1. It needs a pulse transformer. |
| B. Monostable multivibrator | 2. It is used to generate gating pulse whose width can be controlled |
| C. Astable multivibrator | 3. It is a bistable circuit |
| D. Blocking oscillator | 4. It has no stable state. |

**Codes:**

| | A | B | C | D |
|---|---|---|---|---|
| (a) | 3 | 2 | 4 | 1 |
| (b) | 2 | 3 | 1 | 4 |
| (c) | 3 | 2 | 1 | 4 |
| (d) | 2 | 3 | 4 | 1 |

**Exp.** (a)

Schmitt trigger — It is a bistable circuit

Monostable multivibrator — It is used to generate gating pulse whose width can be controlled

Astable multivibrator — It has no stable state

Blocking oscillator — It needs a pulse transformer.

**Q.9** Consider the following circuit

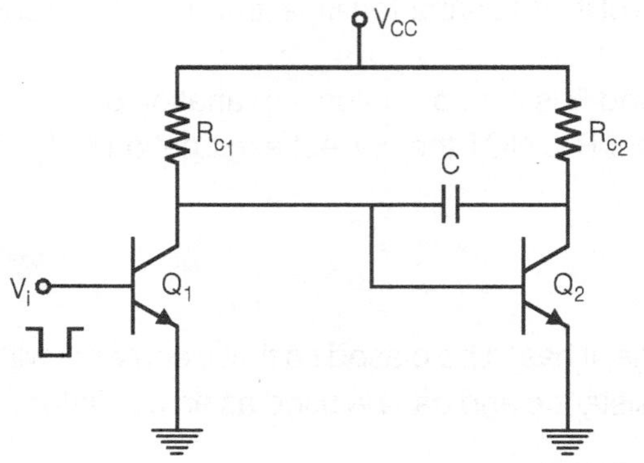


What is the circuit shown above?

(a) Miller sweep
(b) Bootstrap sweep
(c) Schmitt trigger
(d) Triangular wave generator

**Exp.** (a)

It is a Miller sweep circuit which generates negative ramp.

**Q.10** A Miller time base generator produces a ______ going sweep where as a bootstrap time base generator produces a ______ going sweep

(a) negative, positive
(b) positive, negative
(c) negative, negative
(d) positive, positive

**Exp.** (a)

Miller sweep generates negative going ramp. Where as Bootstrap time base generator produces a positive going ramp.

## REVIEW QUESTIONS

1. Draw the circuit diagram of a bistable multivibrator and explain the operation of it? Give the importance of collector catching diodes and commutating capacitors.
2. Draw the circuit diagram of Schmitt trigger, and explain the working of a Schmitt trigger? Define the terms UTP and LTP and explain about hysteresis loop.
3. With the help of a neat circuit diagram and waveforms, explain the working of a monostable multivibrator.
4. Derive an expression for gate width of a monostable multivibrator.
5. With the help of a neat circuit diagram and waveforms, explain the working of an astable multivibrator.
6. Derive an expression for frequency of oscillation of an astable multivibrator.
7. Why time base generators called sweep circuits? Give the applications of time base generators?
8. Define voltage time base generator and current time base generator.
9. Define the terms slope error, displacement error and transmission error.
10. With the help of a neat circuit diagram and waveforms, explain the working of a UJT relaxation oscillator.
11. With the help of a neat circuit diagram explain the working of a transistor Miller time base generator.
12. With the help of a neat circuit diagram explain the working of a transistor Bootstrap time base generator.

OOOO